C. SELECTED RULES AND SUGGESTIONS FOR SI USAGE

1. Care must be taken to use the correct case for symbols, units, and prefixes (e.g., m for meter or milli, M for mega).

2. For numbers having five or more digits, the digits should be placed in groups of three separated by a space, counting both to the left and to the right of the decimal point (e.g., 61 354.982 03). The space is not required for four-digit numbers. Spaces are used instead of commas to avoid confusion—many countries use the comma as the decimal marker.

3. In compound units formed by multiplication, use the product dot (e.g., $N \cdot m$).

4. Division may be indicated by a solidus (m/s), or a negative exponent with a product dot ($m \cdot s^{-1}$).

5. Avoid the use of prefixes in the denominator (e.g., km/s is preferred over m/ms). The exception to this rule is the prefix k in the base unit kg (kilogram).

Strength of Materials

THIRD EDITION

Strength of Materials

Ferdinand L. Singer

EMERITUS, NEW YORK UNIVERSITY

Andrew Pytel

THE PENNSYLVANIA STATE UNIVERSITY

HARPER & ROW, PUBLISHERS, New York
Cambridge, Philadelphia, San Francisco,
London, Mexico City, São Paulo, Sydney

1817

Sponsoring editor: Charlie Dresser
Project Editor: Celine Keating
Production Manager: Marion A. Palen
Compositor: Science Typographers Incorporated
Printer and Binder: The Maple Press Company
Art Studio: J & R Services
Cover Design: Helen Iranyi

STRENGTH OF MATERIALS

Library of Congress Cataloging in Publication Data
Singer, Ferdinand Leon, Date-
   Strength of materials.
   Includes index.
   1. Strength of materials.   I.   Pytel, Andrew,
joint author.   II.   Title.
TA405.S45   1980   620.1'1   79-18222
ISBN 0-06-046229-9

To Evelyn, Joan, Karen, Lucy,
Leslie, Lori, and John

Contents

Preface

Successful machine or structural design is practically impossible without a thorough mastery of engineering mechanics and strength of materials. In the modern engineering curricula, the importance of these subjects is emphasized by a comprehensive study of their fundamental and advanced concepts. This revised SI* edition attempts to explain lucidly and rigorously the theory and application of these concepts.

During the 1960s, engineering education in this country became increasingly mathematically oriented. In some instances, courses in strength of materials were even *replaced* by courses such as continuum mechanics and elasticity theory. The current philosophy of engineering education recognizes the importance of design in the engineering curriculum. Therefore, applied courses such as strength of materials again emerge as important components in the education of engineers in virtually all areas of speciality.

The unique feature of the third edition of this text, as compared with the previous editions, is that it is written completely in SI units. Although we are currently going through a transition period when the practicing engineer may need to be familiar with more than one system of units, the worldwide movement to SI units will probably make this transition period short-lived. The authors agree with the philosophy that

*SI is the official abbreviation for the international system of units, Le Système International d'Unités.

fundamental texts such as this one should be written completely in SI units. Once the *fundamentals* have been thoroughly mastered in one system of units, extensions to other systems should not be difficult.

This edition retains the general plan and features of the earlier editions. The major emphasis is still on elastic analysis, although an extensive coverage of inelastic analysis is included. After much consideration, inelastic analysis was again incorporated into a coordinated continuous treatment of residual stress and limit analysis. As in the second edition, a modernized exposition of the double-integration method is given which greatly broadens and simplifies its application to the deflection of beams. Also, energy methods, such as Castigliano's theorem and virtual work, have been retained so that, with the double-integration and area-moment methods, there is an exceptionally complete discussion of deflection methods. There is also a short but effective presentation of moment distribution embodying a modernized sign treatment which should simplify its application. Each of these topics is fairly independent of the others; thus one or more may be omitted if time is limited.

Other features of this text include an expanded treatment of shear flow; an extended analysis of the states of stress and strain with emphasis on the application of Mohr's circle to strain gage and rosette analysis; and relegation of riveted and welded connections to the latter part of the book, where this subject does not interrupt the continuity of basic principles. In addition, revisions in design codes have necessitated updating the discussion of column theory.

These topics, as well as the others in the text, are presented in a manner that should relieve the instructor of the burden of detailed explanation. Principles are developed by a consistent plan which first relates stresses to deformations, then applies the equations of static equilibrium, and finally satisfies the boundary conditions.

Primarily, the point of view of students has been retained and their special problems kept in mind. We have made every effort to make a fundamental principle perfectly understood, but in clear, concise language. The physical significance of fundamental concepts, and the assumptions and limitations made in developing them, are carefully discussed so that memorization is reduced to a minimum. The summaries appended to most chapters are intended to give the student a concise statement of key elements which should be useful in review and postcollege work. Rules of sign have also been simplified by assigning the positive sense to all quantities to which the adjectives *up*, *above*, or similar terms may be applied; for negative signs, the converse is true.

Numerous illustrative problems show in detail how principles are applied. The explanations are complete—nothing is taken for granted. Throughout, the equation or principle to be applied is stated in brackets on the left side of the equation. In the solution, values are substituted in the respective order in which the symbols appear in the equation. This procedure enables the reader to follow readily the various steps of the solution without continually referring to the body of the text.

The almost 1000 problems contained in this edition are either SI versions of problems from the previous edition or completely new problems. In either case, they have been carefully chosen so as to illustrate the fundamental concepts without overburdening the student with tedious numerical computation. The importance of free-body diagrams in the solution of problems in strength of materials continues to be emphasized. The problems have been arranged approximately in the order of their difficulty, and answers to two-thirds of them have been given; the others may on occasion be used for quizzes.

The numbering plan used enables the reader to locate quickly any cross reference. With this plan, all articles, figures, equations, tables, and problems are preceded by the numeral of the chapter in which they appear and are numbered consecutively throughout each chapter. Figures for assigned problems are given the number of the problem to which they refer in order to simplify correlation of a problem figure with corresponding problem data.

The authors wish to acknowledge their indebtedness to their colleagues all over the nation for their many valuable suggestions for this edition. To identify them individually would make too lengthy a list (with possibly an inadvertent omission), but individually each has received our thanks. However, a special debt is due to Dr. Jean Landa, whose assistance in the preparation of this edition is greatly appreciated. Although great care was taken to eliminate errors, it is inevitable that some will still be found. The authors appreciate being informed about these and welcome any comments that readers may care to offer.

FERDINAND L. SINGER
ANDREW PYTEL

List of Symbols and Abbreviations

A	area
A'	partial area of beam section
$\bar{a}, \bar{b}$	coordinates of centroid of moment diagram caused by simply supported loads
b	breadth, width
c	distance from neutral axis to extreme fiber
D, d	diameter
E	modulus of elasticity in tension or compression
e	eccentricity, natural base of logarithms
f	frequency
f_c	unit compressive stress in concrete
f_s	unit tensile stress in reinforcing steel
G	modulus of rigidity (i.e., modulus of elasticity in shear)
g	gravitational acceleration ($9.81 \ m/s^2$)
h	height, depth of beam
I	moment of inertia of area
I_{NA}	moment of inertia with respect to neutral axis

$\bar{I}$	centroidal moment of inertia
J	polar moment of inertia
$\bar{J}$	centroidal polar moment of inertia
K	stress concentration factor
k	spring constant, radius of gyration
L	length
L_e	effective length for columns
M	bending moment
m	mass
N	normal force, factor of safety
n	ratio of moduli of elasticity
P	force, concentrated load, hoop tension
$\mathcal{P}$	power
P_{cr}	critical load for columns
P_{uv}, P_{xy}	products of inertia
p	pressure per unit area
Q	statical or first moment of area
q	shear flow
R	reaction, resultant force, radius
r	radius, radius of gyration
S	section modulus (I/c)
σ	unit stress, normal stress
σ_b	unit bearing stress
σ_c	unit compressive stress
σ_{cr}	critical unit stress in column formula
σ_f	unit flexural stress
σ_r	unit radial stress
σ_w	allowable stress
σ_t	unit tensile stress, unit tangential stress
$\sigma_x, \sigma_y, \sigma_z$	unit normal stress in x, y, and z directions, respectively
σ_{yp}	stress at yield point
T	torque, temperature
t	thickness, tangential deviation
τ	unit shearing stress
τ_{xy}	unit shearing stress in x–y plane
u, v, w	rectangular coordinates
V	vertical shearing force
v	velocity
W	total weight or load
w	weight or load per unit of length
X, Y, Z	orthogonal components of a force
x, y, z	rectangular coordinates
$\bar{x}, \bar{y}, \bar{z}$	coordinates of centroid or center of gravity

y	deflection of beam
α	temperature coefficient of linear expansion
$\alpha, \beta, \gamma \cdots$	angles
γ	unit shearing strain
δ	total elongation or contraction; deflection of beam; maximum deflection of column
δ_{st}	static deflection
ϵ	unit tensile or compressive strain
$\epsilon_x, \epsilon_y, \epsilon_z$	unit tensile or compressive strain in the x, y, and z direction, respectively
θ	total angle of twist, slope angle for elastic curve
ρ	radius of curvature, variable radius, mass density
ν	Poisson's ratio
ω	angular velocity
CG	center of gravity
deg	degrees
DF	distribution factor
FS	factor of safety
FEM	fixed end moment
ID	inner diameter
NA	neutral axis
OD	outer diameter
PL	proportional limit
YP	yield point

Simple
Stress

1-1 INTRODUCTION

Strength of Materials extends the study of forces that was begun in *Engineering Mechanics*, but there is a sharp distinction between the two subjects. Fundamentally, the field of mechanics covers the relations between forces acting on rigid bodies; in *statics*, the bodies are in equilibrium, whereas in *dynamics*, they are accelerated but can be put in equilibrium by applying correctly placed inertia forces.

In contrast to mechanics, strength of materials deals with the relations between externally applied loads and their internal effects on bodies. Moreover, the bodies are no longer assumed to be ideally rigid; the deformations, however small, are of major interest. The properties of the material of which a structure or machine is made affect both its choice and the dimensions that will satisfy the requirements of strength and rigidity.

The difference between mechanics and strength of materials can be further emphasized by the following example. It is a simple problem in statics to determine the force required at the end of a crowbar to pry up a given load (Fig. 1–1). A moment summation about the fulcrum determines *P*. This statics solution assumes the crowbar to be both rigid enough and strong enough to permit the desired action. In strength of

1

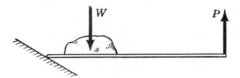

Figure 1–1.   Crowbar must neither break nor bend excessively.

materials, however, the solution must extend further. We must investigate the bar itself to be sure that it will neither break nor be so flexible that it bends without lifting the load.

Throughout this book we shall study the principles that govern these two fundamental concepts of *strength* and *rigidity*. In this first chapter we start with simple axial loadings; later we consider twisting loads and bending loads; and finally we discuss simultaneous combinations of these three basic types of loadings.

1–2   ANALYSIS OF INTERNAL FORCES

Consider a body of any shape acted upon by the forces shown in Fig. 1–2. In engineering mechanics, we would start by determining the resultant of the applied forces to determine whether or not the body remains at rest. If the resultant is zero, we have static equilibrium—a condition generally prevailing in structures. If the resultant is not zero, we may apply inertia forces to bring about dynamic equilibrium. Such cases are discussed later under dynamic loading. For the present we shall consider only cases involving static equilibrium.

In strength of materials, we make an additional investigation of the internal distribution of the forces. This is done by passing an exploratory section a–a through the body and exposing the internal

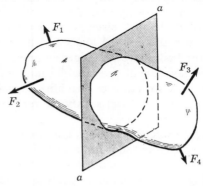

Figure 1–2.   Exploratory section a–a through loaded member.

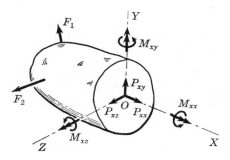

Figure 1–3. Components of internal effects on exploratory section a–a.

forces acting on the exploratory section that are necessary to maintain equilibrium of a free-body diagram of either segment. In general, the internal forces reduce to a force and a couple which, for convenience, are resolved into components normal and tangent to the section, as shown in Fig. 1–3.

The origin of the reference axes is always at the centroid which is the key reference point of the section. Although we are not yet ready to show why this is so, we shall prove it as we progress; in particular, we shall prove it for normal forces in the next article. If the X axis is normal to the section, the section is known as the X surface or, more briefly, the X face. The orientation of the Y and Z axes in the plane of the section is usually chosen to coincide with the principal axes of inertia.

The notation used in Fig. 1–3 identifies both the exploratory section and the direction of the force or moment component. The first subscript denotes the face on which the component acts; the second subscript indicates the direction of the particular component. Thus P_{xy} is the force on the X face acting in the Y direction.

Each component reflects a different effect of the applied loads on the member and is given a special name, as follows:

P_{xx} *Axial force.* This component measures the pulling (or pushing) action over the section. A pull represents a tensile force which tends to elongate the member, whereas a push is a compressive force which tends to shorten it. It is often denoted by P.

P_{xy}, P_{xz} *Shear force.* These are components of the total resistance to sliding the portion to one side of the exploratory section past the other. The resultant shear force is usually designated by V, and its components by V_y and V_z to identify their directions.

M_{xx} *Torque*. This component measures the resistance to twisting the member and is commonly given the symbol T.

M_{xy}, M_{xz} *Bending moments*. These components measure the resistance to bending the member about the Y or Z axes and are often denoted merely by M_y or M_z.

From the preceding discussion, it is evident that the internal effect of a given loading depends upon the selection and orientation of the exploratory section. In particular, if the loading acts in one plane, say the XY plane as is frequently the case, the six components in Fig. 1–3 reduce to only three, viz., the axial force P_{xx} (or P), the shear force P_{xy} (or V), and the bending moment M_{xz} (or M). Then, as shown in Fig. 1–4a, these components are equivalent to the single resultant force R. A little reflection will show that if the exploratory section had been oriented differently, like $b–b$ in Fig. 1–4b where it is perpendicular to R, the shearing effect on the section would reduce to zero and the tensile effect would be at a maximum.

The purpose of studying strength of materials is to ensure that the structures used will be safe against the maximum internal effects that may be produced by any combination of loading. We shall learn as our study proceeds that it is not always possible or convenient to select an exploratory section that is perpendicular to the resultant load; instead, we may have to start by analyzing the effects acting on a section like $a–a$ in Figs. 1–2 and 1–4a, and then learn how these effects combine to

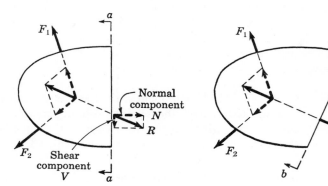

(a) Normal and shear
    components on arbitrary
    section $a – a$.

(b) When exploratory section $b – b$ is
    perpendicular to resultant R
    of applied loads, only normal
    forces are produced.

Figure 1–4.

produce maximum internal effects like those on section b–b in Fig. 1–4b. This procedure we shall study later in Chapter 9, which deals with combined stresses. For the present, we restrict our study to conditions of loading in which the section of maximum internal effect is evident by inspection.

1-3 SIMPLE STRESS

One of the basic problems of the engineer is to select the proper material and correctly use and proportion it so as to enable a structure or machine to do most efficiently what it is designed to do. For this purpose, it is essential to determine the strength, stiffness, and other properties of materials. A tabulation of the average properties of common metals is given in Appendix B, Table B–1, on page 634.

Let us consider two bars of equal length but different materials, suspended from a common support as shown in Fig. 1–5. If we knew nothing about the bars except that they could support the indicated maximum loads [500 N (Newtons) for bar 1 and 5000 N for bar 2], we could not tell which material is stronger. Of course, bar 2 supports a greater load, but we cannot compare strengths without having a common basis of comparison. In this instance, the cross-sectional areas are needed. So let us further specify that bar 1 has a cross-sectional area of 10 mm² and bar 2 has an area of 1000 mm². Now it is simple to compare their strengths by reducing the data to load capacity per unit area. Here we note that the unit strength of bar 1 is

$$\sigma_1 = \frac{500 \text{ N}}{10 \text{ mm}^2} = \frac{500 \text{ N}}{10 \times 10^{-6} \text{ m}^2} = 50 \times 10^6 \text{ N/m}^2$$

and bar 2 has a unit strength

$$\sigma_2 = \frac{5000 \text{ N}}{1000 \text{ mm}^2} = \frac{5000 \text{ N}}{1000 \times 10^{-6} \text{ m}^2} = 5 \times 10^6 \text{ N/m}^2$$

Thus the material of bar 1 is ten times as strong as the material of bar 2.

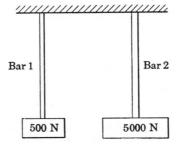

Figure 1-5. Bars supporting maximum loads.

The unit strength of a material is usually defined as the stress* in the material. Stress is expressed symbolically as

$$\sigma = \frac{P}{A} \tag{1-1}$$

where σ (Greek lowercase letter *sigma*) is the stress or force per unit area, P is the applied load, and A is the cross-sectional area. Observe that maximum stress in tension or compression occurs over a section normal to the load, as indicated in Fig. 1–4b. Shearing stress is discussed in the next section.

From Eq. (1–1) it can be seen that the units for stress are the units of force divided by the units of area. In SI (which is the official abbreviation for the international system of units, *Le Système International d'Unités*), force is measured in newtons (N) and area is measured in square meters (m^2). Thus the units for stress are newtons per square meter (N/m^2). Frequently, one newton per square meter is referred to as one pascal (Pa). Since the prefix M (read as "mega") refers to multiples of 10^6 in SI, in the above example, the stress in bar 1 may be expressed as 50 MN/m^2 (or 50 MPa) and that in bar 2 as 5 MN/m^2 (or 5 MPa).

Even as simple an expression as Eq. (1–1) requires careful discussion. Dividing load by area does not give the stress at all points in the cross-sectional area; it merely determines the *average* stress. A more exact definition of stress is obtained by dividing the differential load dP by the differential area over which it acts:

$$\sigma = \frac{dP}{dA} \tag{1-1a}$$

Next let us see under what conditions $\sigma = P/A$ will accurately define the stress at all points of the cross section. The condition under which the stress is constant or uniform is known as *simple stress*. We shall show now that a uniform stress distribution can exist only if the resultant of the applied loads passes through the centroid of the cross section.[†]

Suppose that a cutting plane isolates the lower half of one of the bars in Fig. 1–5. Then, as shown in Fig. 1–6, the resisting forces over the cut section must balance the applied load P. A typical resisting force is dP. Applying the conditions of equilibrium, we obtain

$$[\Sigma Z = 0] \qquad P = \int dP = \int \sigma \, dA$$

$$[\Sigma M_y = 0] \qquad Pb = \int x \, dP = \int x(\sigma \, dA)$$

---

*Some engineers use the terms *stress* or *total stress* as synonymous with load or force, and *unit stress* or *stress intensity* when referring to the intensity of load per unit area. In this book, *stress* will always denote *force per unit area*.

[†] There are certain exceptions to this rule; they are caused by stress concentration (see p. 512) and by abrupt changes in cross section, and at points in the vicinity of the applied loads (see p. 7).

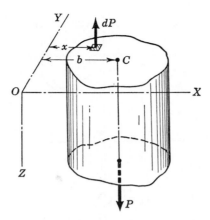

Figure 1–6. For uniform stress, P must pass through the centroid C.

If we specify that the stress distribution is to be constant over the cut section, σ may be written outside the integrals in the above equations to obtain

$$P = \sigma \int dA = \sigma A$$

and therefore,

$$Pb = (\sigma A)b = \sigma \int x \, dA$$

Then, canceling the common factor σ, we obtain

$$b = \frac{\int x \, dA}{A} = \bar{x}$$

from which the coordinate b of the point C is recognized as being the x coordinate of the centroid of the section. By taking a moment summation about the X axis, we could similarly show that $\bar{y}$ defines the y coordinate of C. We conclude that a uniform stress distribution is obtained only when the resultant of the applied loads passes through the centroid of that surface.

It does not follow, however, that positioning the load through the centroid of the section *always* results in a uniform stress distribution. For example, in Fig. 1–7 is shown the profile of a flat bar of constant thickness. The load P is applied at the center line of the bar. At sections $b–b$ and $f–f$, the stress distribution is uniform and illustrates the principle discussed earlier; but at the other indicated sections the stresses are not uniform.

At section $e–e$, the stress distribution is not uniform because the line of action of P obviously does not pass through the centroid of the section. Nor are the stresses uniformly distributed all across sections $c–c$ and $d–d$ because, although the action line of P does pass through the

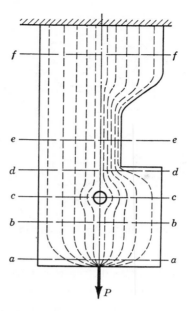

Figure 1-7. Exceptions to uniform stress distribution occur at sections *a–a*, *c–c*, *d–d*, and *e–e*.

centroids of these sections, here there are abrupt changes in section. At such sections, the stresses are usually highly localized and can be determined only by the mathematical theory of elasticity or some experimental method such as photoelasticity. Also, the stress is not uniform across section *a–a* because here the section is too close to the point where the load is applied. Unless a section is located at a distance from the end of the rod at least equal to the minimum width of the rod, we will not obtain a uniform stress distribution.*

In order to visualize why sections *c–c*, *d–d*, and *a–a* do not have uniform stress, imagine that the applied force *P* produces stress lines which radiate out from the load and distribute themselves throughout the body as indicated by the dashed lines in the figure. Although this concept is not actually correct, it does indicate the existence of stress concentration wherever the shape of the body interferes with the "free flow" of the stress lines. The bunching of these lines about the hole in section *c–c*, and around the sharp corner of section *d–d*, which indicates stress concentration, contrasts with the relatively smooth flow of stress around the radius between sections *e–e* and *f–f*.

*See S. Timoshenko and J. N. Goodier, *Theory of Elasticity*, 2nd ed., McGraw-Hill, New York, 1951, p. 33.

ILLUSTRATIVE PROBLEMS

101.   An aluminum tube is rigidly fastened between a bronze rod and a steel rod as shown in Fig. 1–8a. Axial loads are applied at the positions indicated. Determine the stress in each material.

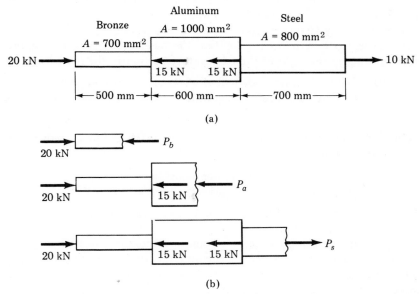

(a)

(b)

Figure 1–8.

Solution: To calculate the stress in each section, we must first determine the axial load in each section. The appropriate free-body diagrams are shown in Fig. 1-8b, from which we determine the axial load in each section to be $P_b = 20$ kN (compression), $P_a = 5$ kN (compression), and $P_s = 10$ kN (tension). The stresses in each section are

$$\left[\sigma = \frac{P}{A} \right] \qquad \sigma_b = \frac{20 \text{ kN}}{700 \text{ mm}^2} = \frac{20 \times 10^3 \text{ N}}{700 \times 10^{-6} \text{ m}^2}$$

$$= 28.6 \times 10^6 \text{ N/m}^2 = 28.6 \text{ MPa} \qquad Ans.$$

$$\sigma_a = \frac{5 \text{ kN}}{1000 \text{ mm}^2} = \frac{5 \times 10^3 \text{ N}}{1000 \times 10^{-6} \text{ m}^2}$$

$$= 5 \times 10^6 \text{ N/m}^2 = 5 \text{ MPa} \qquad Ans.$$

$$\sigma_s = \frac{10 \text{ kN}}{800 \text{ mm}^2} = \frac{10 \times 10^3 \text{ N}}{800 \times 10^{-6} \text{ m}^2}$$

$$= 12.5 \times 10^6 \text{ N/m}^2 = 12.5 \text{ MPa} \qquad Ans.$$

The stresses in the bronze and aluminum sections are compressive, whereas the stress in the steel section is tensile.

Note that neither the lengths of the sections nor the materials from which the sections are made affects the calculation of the stresses.

As you can see from this example, the first step in calculating the stress in a member is to determine the internal force carried by the member. This determination is accomplished by the analysis of *correctly drawn* free-body diagrams. Note that in this example, it would have been easier to determine the load in the steel section by taking the section lying to the *right* of the exploratory section in the steel.

102. For the truss shown in Fig. 1–9a, determine the stress in members AC and BD. The cross-sectional area of each member is 900 mm².

Solution: The three assumptions used in the elementary analysis of trusses are as follows:

1. Weights of the members are neglected.
2. All connections are smooth pins.
3. All external loads are applied directly to the pin joints.

Using the above three assumptions, the members of the truss may be analyzed as *two-force members*—the internal force system carried by any member reduces to simply a single force (tension or compression) acting along the line of the member.

The free-body diagram of the entire truss is shown in Fig. 1–9a. An equilibrium analysis of this free-body diagram results in the following values for the external reactions: $A_y = 40$ kN, $H_y = 60$ kN, and $H_x = 0$.

To determine the force in member AC, we pass an imaginary cutting plane which isolates joint A (section ①, Fig. 1–9a). The free-body diagram of joint A is shown in Fig. 1–9b. Here, AB and AC represent the forces in members AB and AC, respectively. Note that both members have been assumed to be in tension. Analyzing the free-body diagram in Fig. 1–9b,

$$[\Sigma Y = 0] \qquad (\uparrow +) \quad A_y + \tfrac{3}{5}AB = 0$$
$$AB = -\tfrac{5}{3}A_y = -\tfrac{5}{3}(40) = -66.7 \text{ kN}$$
$$[\Sigma X = 0] \qquad (\xrightarrow{+}) \quad AC + \tfrac{4}{5}AB = 0$$
$$AC = -\tfrac{4}{5}AB = -\tfrac{4}{5}(-66.7) = 53.4 \text{ kN}$$

The minus sign indicates that the 66.7 kN force in member AB is

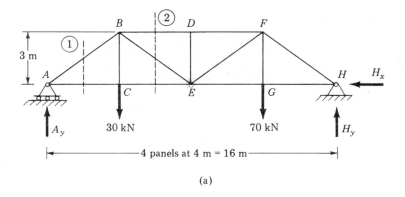

-4 panels at 4 m = 16 m-

(a)

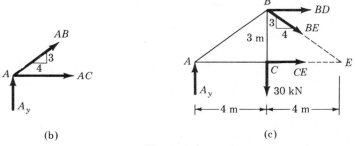

(b)   (c)

Figure 1-9.

compressive. The force in member AC is 53.4 kN, tension.

To determine the force in member BD, we pass an imaginary cutting plane which exposes the force in member BD (section ②, Fig. 1-9a). The free-body diagram of the portion of the truss to the left of section ② is shown in Fig. 1-9c. (The portion of the truss to the right of section ② could also have been used.) The forces in members BD, BE, and CE are assumed to be tensile. To calculate the force BD, we eliminate the forces BE and CE by taking a moment summation about their point of intersection, E, and write

$$[\Sigma M_E = 0] \qquad \circlearrowleft \quad -A_y(8) + 30(4) - BD(3) = 0$$
$$3BD = -8A_y + 120 = -8(40) + 120$$
$$= -200$$
$$BD = -66.7 \text{ kN}$$

Therefore, the force in member BD is 66.7 kN, compression.

The stresses in members AC and BD, are

$$\left[\sigma = \frac{P}{A}\right] \quad \sigma_{AC} = \frac{53.4 \text{ kN}}{900 \text{ mm}^2} = \frac{53.4 \times 10^3 \text{ N}}{900 \times 10^{-6} \text{ m}^2}$$
$$= 59.3 \times 10^6 \text{ N/m}^2 = 59.3 \text{ MPa} \qquad \textit{Ans.}$$
$$\text{(tension)}$$

$$\sigma_{BD} = \frac{66.7 \text{ kN}}{900 \text{ mm}^2} = \frac{66.7 \times 10^3 \text{ N}}{900 \times 10^{-6} \text{ m}^2}$$
$$= 74.1 \times 10^6 \text{ N/m}^2 = 74.1 \text{ MPa} \qquad \textit{Ans.}$$
$$\text{(compression)}$$

In truss analysis, the method of analyzing a single joint, as shown in Fig. 1–9b, is referred to as the *method of joints*. The analysis of a section of the truss composed of two or more joints, as shown in Fig. 1–9c, is called the *method of sections*. It must be reemphasized that the force internal to a member of a truss lies along the line of the member only because sufficient assumptions are made which reduce all members to two-force members. As discussed in Art. 1–2, the internal forces for an arbitrarily loaded member are considerably more complicated than simply an axial force.

PROBLEMS

103. Determine the largest weight W which can be supported by the two wires shown in Fig. P–103. The stresses in wires AB and AC are not to exceed 100 MPa and 150 MPa, respectively. The cross-sectional areas of the two wires are 400 mm^2 for wire AB and 200 mm^2 for wire AC.

Ans. $W = 33.5$ kN

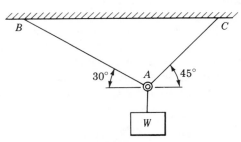

Figure P–103.

104. For the truss shown in Fig. P–104, calculate the stresses in members DF, CE, and BD. The cross-sectional area of each member is 1200 mm^2. Indicate tension (T) or compression (C).

Ans. $DF = 188$ MPa (C); $CE = 113$ MPa (T);
$BD = 80.1$ MPa (C)

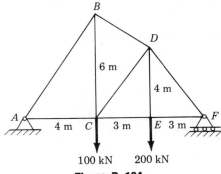

Figure P-104.

105.   For the truss shown in Fig. P–105, determine the cross-sectional areas of bars *BE*, *BF*, and *CF* so that the stresses will not exceed 100 MN/m² in tension or 80 MN/m² in compression. A reduced stress in compression is specified to avoid the danger of buckling.

   Ans. $A_{BE} = 625$ mm²; $A_{BF} = 427$ mm²; $A_{CF} = 656$ mm²

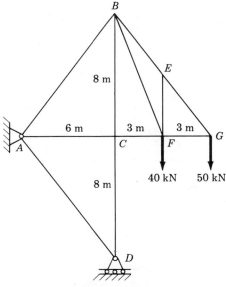

Figure P-105.

106.   The bars of the pin-connected frame in Fig. P–106 are each 30 mm by 60 mm in section. Determine the maximum load P that can be applied without exceeding the allowable stresses specified in Problem 105.

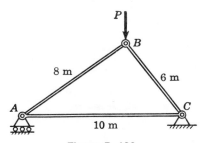

Figure P–106.

107.   A cast-iron column supports an axial compressive load of 250 kN. Determine the inside diameter of the column if its outside diameter is 200 mm and the limiting compressive stress is 50 MPa.

108.   Determine the outside diameter of a hollow steel tube that will carry a tensile load of 500 kN at a stress of 140 MN/m². Assume the wall thickness to be one-tenth of the outside diameter.

Ans. $D = 107$ mm

109.   Part of the landing gear for a light plane is shown in Fig. P–109. Determine the compressive stress in the strut AB caused by a landing reaction $R = 20$ kN. Strut AB is inclined at 53.1° with BC. Neglect weights of the members.   *Ans.* $\sigma = 65.7$ MN/m²

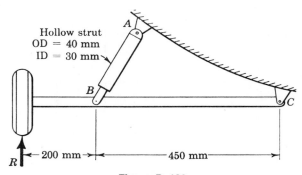

Figure P–109.

110.   A steel tube is rigidly attached between an aluminum rod and a bronze rod as shown in Fig. P–110. Axial loads are applied at the positions indicated. Find the maximum value of P that will not exceed a stress in aluminum of 80 MPa, in steel of 150 MPa, or in bronze of 100 MPa.

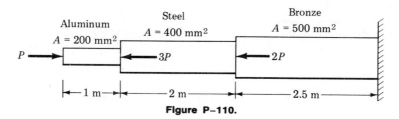

Figure P–110.

111. A homogeneous 150-kg bar AB carries a 2-kN force as shown in Fig. P–111. The bar is supported by a pin at B and a 10-mm-diameter cable CD. Determine the stress in the cable.

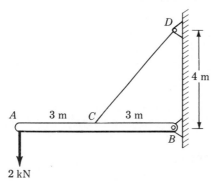

Figure P–111.

112. Determine the weight of the heaviest cylinder which can be placed in the position shown in Fig. P–112 without exceeding a stress of 50 MN/m² in the cable BC. Neglect the weight of bar AB. The cross-sectional area of cable BC is 100 mm².

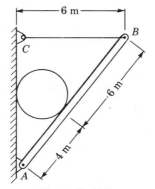

Figure P–112.

113. A 1000-kg homogeneous bar AB is suspended from two cables AC and BD, each with cross-sectional area 400 mm^2, as shown in Fig. P–113. Determine the magnitude P and location x of the largest additional force which can be applied to the bar. The stresses in the cables AC and BD are limited to 100 MPa and 50 MPa, respectively.

Ans. $P = 50.2$ kN; $x = 0.602$ m

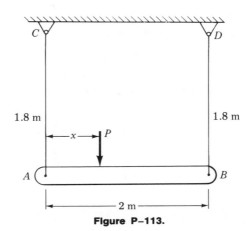

Figure P–113.

1–4   SHEARING STRESS

Shearing stress differs from both tensile and compressive stress in that it is caused by forces acting along or parallel to the area resisting the forces, whereas tensile and compressive stresses are caused by forces perpendicular to the areas on which they act. For this reason, tensile and compressive stresses are frequently called *normal stresses*, whereas a shearing stress may be called a *tangential stress*.

A shearing stress is produced whenever the applied loads cause one section of a body to tend to slide past its adjacent section. Several examples are shown in Fig. 1–10. In (a) the rivet resists shear across its cross-sectional area, whereas in the clevis at (b) the bolt resists shear across two cross-sectional areas; case (a) may be called *single shear* and case (b) *double shear*. In (c) a circular slug is about to be punched out of a plate; the resisting area is similar to the milled edge of a coin. In each case, the shear occurs over an area parallel to the applied load. This may be called *direct shear* in contrast to the *induced shear* that may occur over sections inclined with the resultant load, as was illustrated in Fig. 1–4a.

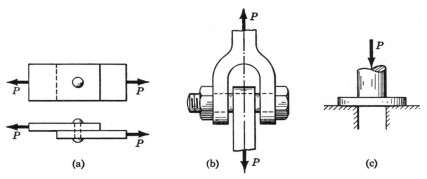

Figure 1-10. Examples of shear.

The discussion concerning uniform normal stresses in the preceding article might lead us to conclude also that a uniform shearing stress will exist when the resultant shearing force V passes through the centroid of the cross section being sheared. If this were true, the shearing stress τ (Greek lowercase letter *tau*) could be found from

$$\tau = \frac{V}{A} \tag{1-2}$$

Actually, the shearing stress across a section is practically never uniformly distributed (e.g., see Art. 5-7), so Eq. (1-2) must be interpreted as giving merely the *average* shearing stress. This does not limit the usefulness of Eq. (1-2) provided we use an average shearing stress that takes into account the actual nonuniform distribution. Moreover, the shearing stress distribution does approach uniformity when both the distance between the applied shearing loads and the depth of the shearing area are small. These are the conditions that prevail in Fig. 1-10 and in the following problems.

PROBLEMS

114. As in Fig. 1-10c, a hole is to be punched out of a plate having an ultimate shearing stress of 300 MPa. (a) If the compressive stress in the punch is limited to 400 MPa, determine the maximum thickness of plate from which a hole 100 mm in diameter can be punched. (b) If the plate is 10 mm thick, compute the smallest diameter hole which can be punched. *Ans.* (a) $t = 33.3$ mm; (b) $d = 30.0$ mm

115.   The end chord of a timber truss is framed into the bottom chord as shown in Fig. P–115. Neglecting friction, (a) compute dimension *b* if the allowable shearing stress is 900 kPa; and (b) determine dimension *c* so that the bearing stress does not exceed 7 MPa.

Ans.   (a) $b = 321$ mm; (b) $c = 41.2$ mm

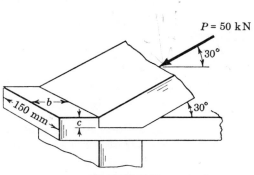

Figure P–115.

116.   In the landing gear described in Problem 109, the bolts at *A* and *B* are in single shear and the one at *C* is in double shear. Compute the required diameter of these bolts if the allowable shearing stress is 50 MN/m^2.

117.   A 750-mm pulley, loaded as shown in Fig. P–117, is keyed to a shaft of 50-mm diameter. Determine the width *b* of the 75-mm-long key if the allowable shearing stress is 70 MPa.   *Ans.* $b = 11.4$ mm

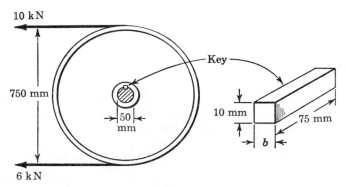

Figure P–117.

118. The bell crank shown in Fig. P–118 is in equilibrium. (a) Determine the required diameter of the connecting rod AB if its axial stress is limited to 100 MN/m². (b) Determine the shearing stress in the pin at D if its diameter is 20 mm.

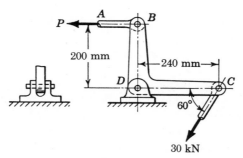

200 mm

240 mm

30 kN

60°

Figure P–118.

119. The mass of the homogeneous bar AB shown in Fig. P–119 is 2000 kg. The bar is supported by a pin at B and a smooth vertical surface at A. Determine the diameter of the smallest pin which can be used at B if its shear stress is limited to 60 MPa. The detail of the pin support at B is identical to that of the pin support at D shown in Fig. P–118.   *Ans.* $d = 14.9$ mm

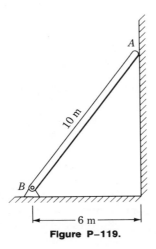

10 m

6 m

Figure P–119.

120. Two blocks of wood, 50 mm wide and 20 mm thick, are glued together as shown in Fig. P–120. (a) Using the free-body diagram concept illustrated in Fig. 1–4a, determine the shear load and from it the shearing stress on the glued joint if $P = 6000$ N. (b) Generalize the procedure of part (a) to show that the shearing stress on a plane inclined at an angle θ to a transverse section of area A is $\tau = P \sin 2\theta/2A$.

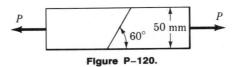

Figure P–120.

121. A rectangular piece of wood, 50 mm by 100 mm in cross section, is used as a compression block as shown in Fig. P–121. Determine the maximum axial force P which can be safely applied to the block if the compressive stress in the wood is limited to 20 MN/m² and the shearing stress parallel to the grain is limited to 5 MN/m². The grain makes an angle of 20° with the horizontal, as shown. (*Hint*: Use the results of Problem 120.)   *Ans.* $P = 77.8$ kN

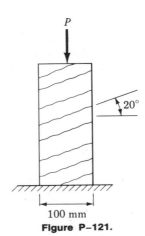

Figure P–121.

1–5   BEARING STRESS

Bearing stress differs from compressive stress in that the latter is the internal stress caused by a compressive force whereas the former is a contact pressure between separate bodies. Some examples of bearing stress are the soil pressure beneath piers and the forces on bearing plates. We now consider the contact pressures between an axle and its bearing, or between a rivet or bolt and the contact surface of the plate against which it pushes.

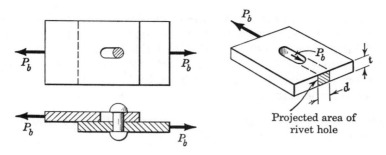

Projected area of
rivet hole

Figure 1–11.   Exaggerated bearing deformation of upper plate.

$$P_b = A_b\sigma_b = (td)\sigma_b$$

In Fig. 1–11, the result of an excessive bearing stress is to cause yielding of the plate or of the rivet, or both. The intensity with which the rivet bears against the rivet hole is not constant, but actually varies from zero at the edges of the hole to a maximum directly in back of the rivet. The difficulty inherent in a variable stress distribution is avoided by the common practice of assuming the bearing stress σ_b to be uniformly distributed over a reduced area which is the projected area of the rivet hole. Then the bearing load is expressed by

$$P_b = A_b\sigma_b = (td)\sigma_b \tag{1–3}$$

This result is analogous to that for a cylinder subjected to a uniform internal pressure (see the next article, especially Fig. 1–14). There, as we shall see, the net force is equal to the uniform pressure multiplied by the projected area.

ILLUSTRATIVE PROBLEM

122.   Figure 1–12 shows a W460 × 97 beam riveted to a W610 × 125 girder by two 100 × 90 × 10-mm angles with 19-mm-diameter

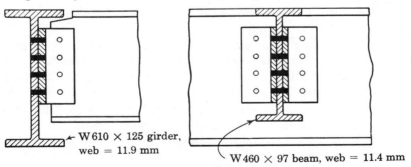

W 610 × 125 girder, web = 11.9 mm

W 460 × 97 beam, web = 11.4 mm

Figure 1–12.   Strength of beam and girder connection.

rivets. (Refer to Appendix B for the properties of structural sections.) For the shop-driven rivets that attach the angles to the beam, assume $\tau = 80$ MPa and $\sigma_b = 170$ MPa. For the field-driven rivets (riveted on the job), assume $\tau = 70$ MPa and $\sigma_b = 140$ MPa. The web of the girder is 11.9 mm thick, and the web of the beam is 11.4 mm thick. Determine the allowable end reaction.

Solution: At the girder, the shearing resistance is that of 8 field-driven rivets in single shear; hence we have

$$[P = A\tau] \qquad P = 8\left(\frac{\pi}{4}\right)(19 \times 10^{-3})^2(70 \times 10^6) = 159 \text{ kN}$$

The bearing resistance at the girder depends on the minimum thickness of the connection, which in this case is the 10-mm thickness of the clip angle. We obtain for 8 field-driven rivets in bearing:

$$[P = A\sigma_b] \qquad P = 8(19 \times 10^{-3})(10 \times 10^{-3})(140 \times 10^6)$$
$$= 213 \text{ kN}$$

At the beam, there are 4 shop-driven rivets in double shear, giving a total of 8 single-shear areas. With an allowable shearing stress of 80 MPa, this makes the shear resistance greater here than at the girder.

The bearing resistance at the beam depends on the web thickness of the beam. Since this is smaller than the combined thickness of the two clip angles, for the 4 rivets in bearing, we obtain

$$[P = A\sigma_b] \qquad P = 4(19 \times 10^{-3})(11.4 \times 10^{-3})(170 \times 10^6)$$
$$= 147 \text{ kN}$$

The safe beam reaction is the smallest of the above values, that is, 147 kN; it is limited by the bearing of the shop-driven rivets against the W460 × 97 beam.

PROBLEMS

123. In Fig. 1–11, assume that a 20-mm-diameter rivet joins the plates which are each 100 mm wide. (a) If the allowable stresses are 140 MN/m^2 for bearing in the plate material and 80 MN/m^2 for shearing of the rivet, determine the minimum thickness of each plate. (b) Under the conditions specified in part (a), what is the largest average tensile stress in the plates. *Ans.* (a) 8.98 mm; (b) 35.0 MN/m^2

124. The lap joint shown in Fig. P–124 is fastened by three 20-mm-diameter rivets. Assuming that $P = 50$ kN, determine (a) the shearing stress in each rivet, (b) the bearing stress in each plate, and (c) the maximum average tensile stress in each plate. Assume that the applied load P is distributed equally among the three rivets.

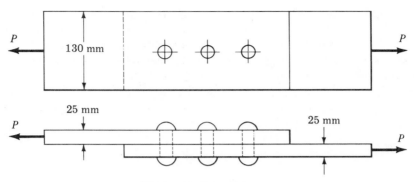

Figures P–124 and P–125.

125.   For the lap joint in Problem 124, determine the maximum safe load P which may be applied if the shearing stress in the rivets is limited to 60 MPa, the bearing stress in the plates to 110 MPa, and the average tensile stress in the plate to 140 MPa.   *Ans.*   56.5 kN

126.   In the clevis shown in Fig. 1–10b, on page 17, determine the minimum bolt diameter and the minimum thickness of each yoke that will support a load $P = 55$ kN without exceeding a shearing stress of 70 MPa and a bearing stress of 140 MPa.

127.   A 22.2-mm-diameter bolt having a diameter at the root of the threads of 18.6 mm is used to fasten two timbers as shown in Fig. P–127. The nut is tightened to cause a tensile load in the bolt of 34 kN. Determine (a) the shearing stress in the head of the bolt, (b) the shearing stress in the threads, and (c) the outside diameter of the washers if their inside diameter is 28 mm and the bearing stress is limited to 6 MPa.

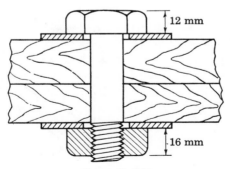

Figure P–127.

128. Figure P–128 shows a roof truss and the detail of the riveted connection at joint B. Using allowable stresses of $\tau = 70$ MPa and $\sigma_b = 140$ MPa, how many 19-mm-diameter rivets are required to fasten member BC to the gusset plate? Member BE? What is the largest average tensile or compressive stress in BC and BE?

Ans.   For BC, 7 rivets; for BE, 5 rivets

129. Repeat Problem 128 if the rivet diameter is 22 mm and all other data remain unchanged.

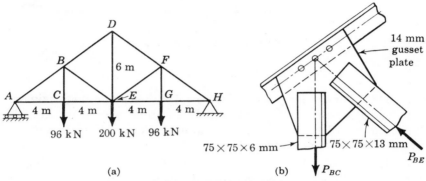

Figures P–128 and P–129.

1–6   THIN-WALLED CYLINDERS

A cylindrical tank carrying a gas or fluid under a pressure of p N/m² is subjected to tensile forces which resist the bursting forces developed across longitudinal and transverse sections. Consider first a typical longitudinal section $A–A$ through the pressure-loaded cylinder in Fig. 1–13a. A free-body diagram of the half-cylinder isolated by the cutting plane $A–A$ is shown in Fig. 1–13b.

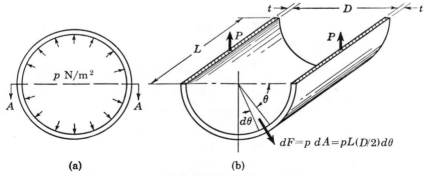

Figure 1–13.   Analytical determination of bursting force F.

The elementary force acting normal to an element of the cylinder located at an angle θ from the horizontal diameter is

$$dF = p\ dA = pL\frac{D}{2}\ d\theta$$

A similar force (not shown) acts on the symmetrically placed element on the other side of the vertical center line. Since the horizontal components of such pairs of forces cancel out, the bursting force F is the summation of the vertical components of these elementary forces:

$$F = \int_0^\pi \left(pL\frac{D}{2}\ d\theta\right)\sin\theta = pL\frac{D}{2}\left[-\cos\theta\right]_0^\pi$$

which reduces to

$$F = pDL$$

It is apparent that the total bursting force F, acting normal to the cutting plane $A\!-\!A$, is resisted by the equal forces P acting on each cut surface of the cylinder wall. Applying a vertical summation of forces, we obtain

$$[\Sigma V = 0]\qquad F = pDL = 2P \tag{1–4}$$

A simpler method of determining the bursting force F is indicated in Fig. 1–14. Here the lower half of the cylinder is occupied by a fluid. Since a fluid transmits pressure equally in all directions, the pressure distribution on the cylinder is the same as that in Fig. 1–13. From the accompanying free-body diagram, it is apparent that the bursting force F, acting over the flat surface of the fluid, equals the pressure intensity p multiplied by the area DL over which it acts, or $F = pDL$ as before.

The stress in the longitudinal section that resists the bursting force F is obtained by dividing it by the area of the two cut surfaces. This gives

$$\left[\sigma = \frac{F}{A}\right]\qquad \sigma_t = \frac{pDL}{2tL} = \frac{pD}{2t} \tag{1–5}$$

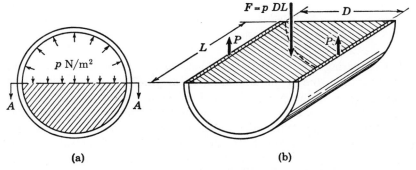

Figure 1–14.   Direct evaluation of bursting force F.

This stress is usually called the *tangential stress* because it acts tangent to the surface of the cylinder; other common names are circumferential stress, hoop stress, and girth stress. The stress computed by Eq. (1–5) is the *average* stress; for cylinders having a wall thickness equal to 1/10 or less of the inner radius, it is practically equal to the maximum stress at the inside surface. (See Art. 13–11 for the stress distribution in thick-walled cylinders.)

If we consider next a free-body diagram of a transverse section (Fig. 1–15), we see that the bursting force acting over the end of the cylinder is resisted by the resultant P of the tearing forces acting over the transverse section. The area of a transverse section is the wall thickness multiplied by the mean circumference, or $\pi(D + t)t$; if t is small compared to D, it is closely approximated by πDt. Thus we obtain

$$[P = F] \qquad \pi Dt\sigma_l = \frac{\pi D^2}{4}p$$

or

$$\sigma_l = \frac{pD}{4t} \tag{1–6}$$

where σ_l denotes what is called the longitudinal stress because it acts parallel to the longitudinal axis of the cylinder.

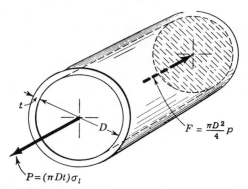

Figure 1–15. Bursting force on a transverse section.

Comparing Eqs. (1–5) and (1–6) shows that the longitudinal stress is one-half the value of the tangential stress. In effect, this is equivalent to stating that, if the pressure in a cylinder is raised to the bursting point, failure will occur along a longitudinal section or longitudinal seam of the cylinder. When a cylindrical tank is composed of two sheets riveted together, as in Fig. 1–16, the strength of the longitudinal joint should be twice the strength of the girth joint. In other words if, as is often the case, the longitudinal joint is not twice as strong as the girth

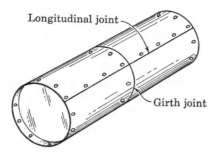

Figure 1–16.

joint, the permissible internal pressure will depend on the strength of the longitudinal joint.

   Equations (1–5) and (1–6) have been developed primarily to determine the relation stated in the above paragraph, *not as equations to be memorized*. It is generally best to compute the stresses by determining the resisting load P from a free-body diagram and then computing the stress by using $\sigma = P/A$. For this purpose, Fig. 1–14 is replaced by the equivalent skeleton diagram in Fig. 1–17, which also establishes the relation $2P = pDL$.

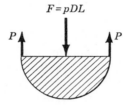

Figure 1–17.

   When the ends of the cylinder are not squared off as in Fig. 1–15, but are rounded or dished as in Fig. 1–18, the bursting force on a transverse section may still be computed as the product of the internal pressure multiplied by the projected area of the transverse section. Thus, using the concept discussed in connection with Fig. 1–14, we may imagine the volume between the transverse section A–A and the

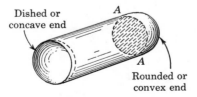

Figure 1–18.

rounded end in Fig. 1–18 to be full of a fluid. The resultant longitudinal force will equal the product of the pressure intensity multiplied by the shaded area of the transverse section.

As another application of the concept of a fluid to transmit pressure, consider a pump chamber cast in several parts, with projecting flanges that are bolted together as shown in Fig. 1–19. The bursting force to be resisted by the bolts in section $A–A$ is proportional to the cross-sectional area at $A–A$ and is expressed by $F_1 = p(\pi D_1^2/4)$; similarly, the bursting force resisted by the bolts in section $B–B$ is $F_2 = p(\pi D_2^2/4)$.

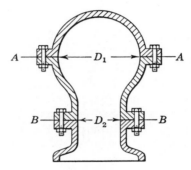

Figure 1–19.

The principles just discussed for determining the tangential stress in thin-walled cylinders may also be applied to computing the contact pressure exerted by hoops shrunk upon cylinders or the tensile stress developed in a thin rotating ring. In the latter case, for example, the bursting force is generated by the centrifugal force developed in one-half of the ring. Its value may be obtained (Fig. 1–20) by assuming the mass of the half-ring concentrated at its center of gravity, whence we have

$$F = m\bar{r}\omega^2 \qquad\qquad (a)$$

in which ω is the angular velocity in radians per second and m is the

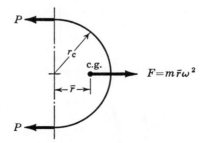

Figure 1–20.   Free-body diagram of one-half of rotating ring.

mass of one-half of the ring. For a thin ring, m is given by

$$m = \rho V = \rho \pi A r_c$$

where ρ (Greek lowercase letter *rho*) is the mass per unit volume of the ring, A is the cross-sectional area of the ring, and r_c is the radius of the mean circumference. From mechanics, the value of $\bar{r}$ for a semicircular ring is $\bar{r} = 2r_c/\pi$. Substituting these values reduces Eq. (a) to

$$F = (\rho A \pi r_c)\left(\frac{2r_c}{\pi}\right)\omega^2 = 2\rho A v^2 \tag{b}$$

where $v = r_c \omega$ is the peripheral velocity of the ring.

From equilibrium of the free-body diagram in Fig. 1–20 we have

$$2P = F$$

Hence the stress is

$$\sigma = \frac{P}{A} = \frac{\rho A v^2}{A} = \rho v^2 \tag{c}$$

Thus the stress varies directly with the mass density and the square of the peripheral velocity. In applying Eq. (c), care must be taken to use consistent units. With ρ in kilograms per cubic meter and v in meters per second, σ will be determined in newtons per square meter.

ILLUSTRATIVE PROBLEM

130. A large pipe, called a penstock in hydraulic work, is 1.5 m in diameter. Here it is composed of wooden staves bound together by steel hoops, each 300 mm² in cross-sectional area, and is used to conduct water from a reservoir to a powerhouse. If the maximum tensile stress permitted in the hoops is 130 MPa, what is the maximum spacing between hoops under a head of water of 30 m? (The mass density of water is 1000 kg/m³.)

Solution: The pressure corresponding to a head of water of 30 m is given by

$$[p = \rho g h] \qquad p = (1000 \text{ kg/m}^3)(9.81 \text{ m/s}^2)(30 \text{ m})$$
$$= 294 \times 10^3 \text{ N/m}^2 = 294 \text{ kPa}$$

If the maximum spacing between hoops is denoted by L, then, as shown in Fig. 1–21, each hoop must resist the bursting force on the length L. Since the tensile force in a hoop is given by $P = A\sigma$, we obtain from the free-body diagram

$$[pDL = 2P]$$
$$(294 \times 10^3)(1.5)L = 2(300 \times 10^{-6})(130 \times 10^6)$$

whence

$$L = 0.177 \text{ m} = 177 \text{ mm} \qquad \textit{Ans.}$$

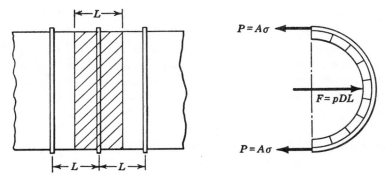

Figure 1-21. Spacing of hoops in a penstock.

PROBLEMS

131. Show that the stress in a thin-walled spherical shell of diameter D and wall thickness t subjected to internal pressure p is given by $\sigma = pD/4t$.

132. A cylindrical pressure vessel is fabricated from steel plates which have a thickness of 20 mm. The diameter of the pressure vessel is 500 mm and its length is 3 m. Determine the maximum internal pressure which can be applied if the stress in the steel is limited to 140 MPa. If the internal pressure were increased until the vessel burst, sketch the type of fracture which would occur.   *Ans.*   11.2 MPa

133. Find the limiting peripheral velocity of a rotating steel ring if the allowable stress is 140 MN/m² and the mass density of steel is 7850 kg/m³. At what angular velocity will the stress reach 200 MN/m² if the mean radius is 250 mm?   *Ans.*   134 m/s; 640 rad/s

134. A water tank is 8 m in diameter and 12 m high. If the tank is to be completely filled, determine the minimum thickness of the tank plating if the stress is limited to 40 MPa.   *Ans.*   11.8 mm

135. The strength per meter of the longitudinal joint in Fig. 1–16 is 480 kN, whereas for the girth joint it is 200 kN. Determine the maximum diameter of the cylindrical tank if the internal pressure is 1.5 MN/m².

136. A pipe carrying steam at 3.5 MPa has an outside diameter of 450 mm and a wall thickness of 10 mm. A gasket is inserted between the flange at one end of the pipe and a flat plate used to cap the end. How many 40-mm diameter bolts must be used to hold the cap on if the allowable stress in the bolts is 80 MPa, of which 55 MPa is the initial stress? What circumferential stress is developed in the pipe? Why is it necessary to tighten the bolts initially, and what will happen if the steam pressure should cause the stress in the bolts to be twice the value of the initial stress?   *Ans.*   17 bolts; 75.3 MPa

137.   A spiral-riveted penstock 1.5 m in diameter is made of steel plate 10 mm thick. The pitch of the spiral or helix is 3 m. The spiral seam is a single-riveted lap joint consisting of 20-mm-diameter rivets. Using $\tau = 70$ MPa and $\sigma_b = 140$ MPa, determine the spacing of the rivets along the seam for a water pressure of 1.25 MPa. Neglect end thrust. What is the circumferential stress?

Ans.   43.7 mm; 93.8 MPa

138.   Repeat Problem 137, using a 2-m-diameter penstock fastened with 30-mm-diameter rivets, with all other data remaining unchanged.

139.   The tank shown in Fig. P–139 is fabricated from 10-mm steel plate. Determine the maximum longitudinal and circumferential stresses caused by an internal pressure of 1.2 MPa.

Ans.   17.9 MPa; 60 MPa

140.   The tank shown in Fig. P–140 is fabricated from steel plate. Determine the minimum thickness of plate which may be used if the stress is limited to 40 MN/m² and the internal pressure is 1.5 MN/m².

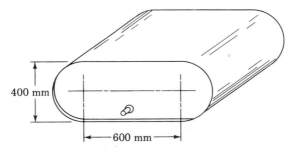

400 mm

600 mm

Figures P–139 and P–140.

SUMMARY

Axial loads result in a uniform stress distribution that may be determined from

$$\sigma = \frac{P}{A} \qquad (1\text{–}1)$$

Shearing stresses and bearing stresses are also computed by dividing the load by the resisting area, but the results represent average values. In particular, the bearing area of a rivet against a plate is given by the projected area of the rivet hole.

The stresses in thin-walled cylinders subjected to internal pressure are most readily obtained by applying the conditions of equilibrium to a free-body diagram of either a longitudinal or a transverse section, depending on whether circumferential or longitudinal stress is involved. The resisting forces thereby exposed are assumed to be uniformly distributed over the corresponding cut surfaces.

2

Simple Strain

2-1 INTRODUCTION

Our main concern in the preceding chapter was the strength of a material, i.e., the relations between load, area, and stress. We now consider the other major field of strength of materials—the changes in shape, i.e., deformations, that accompany a loading. Although we limit ourselves here to axially loaded bars, the principles and methods developed apply equally well to the more complex cases of twisting or bending. In particular, we shall learn how to apply the geometric relations between elastic deformations which, in combination with the conditions of equilibrium and the relations between loads and deformations, will enable us to solve statically indeterminate problems.

2-2 STRESS-STRAIN DIAGRAM

The strength of a material is not the only criterion that must be considered in designing structures. The stiffness of a material is frequently of equal importance. To a lesser degree, such properties as hardness, toughness, and ductility determine the selection of a material. These properties are determined by making tests on the materials and comparing the results with established standards. Although a complete description of these tests is the province of materials testing and hence will not be given here, one of the tests (the tension test of steel) and its results will be considered because it helps to develop several important basic concepts.

If a specimen of structural steel is gripped between the jaws of a testing machine and the load and the extension in a specified length are observed simultaneously, we can plot these observations on a graph on which the ordinates represent *load* and the abscissae represent *extension.*

Figure 2–1 represents such a graph. Notice that we did not plot load against extension; rather, unit load or stress was plotted against unit elongation, technically known as *strain.* Only by reducing observed values to a unit basis can the properties of one specimen be compared with those of other specimens. The diagram in Fig. 2–1 is called a *stress-strain diagram,* the name being taken from the coordinates.

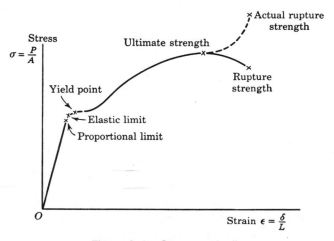

Figure 2–1. Stress–strain diagram.

Strain

To obtain the unit deformation or strain, ϵ, it would seem obvious to divide the elongation δ by the length L in which it was measured, thereby obtaining

$$\epsilon = \frac{\delta}{L} \tag{2-1}$$

The strain so computed, however, measures only the average value of strain. The correct expression for strain at any position is

$$\epsilon = \frac{d\delta}{dL} \tag{2-1a}$$

where $d\delta$ is the differential elongation of the differential length dL. Thus, Eq. (2–1a) determines the average strain in a length so small that the strain must be constant over that length. However, under certain conditions the strain may be assumed constant and its value computed from Eq. (2–1). These conditions are as follows:

1. The specimen must be of constant cross section.
2. The material must be homogeneous.
3. The load must be axial, that is, produce uniform stress.

Finally, note that since strain represents a change in length divided by the original length, strain is a dimensionless quantity. However, it is common to use units of meters per meter (m/m) when referring to strain. In engineering work, strains of the order of 1.0×10^{-3} m/m are frequently encountered.

Proportional limit

From the origin O to a point called the *proportional limit*, Fig. 2–1 shows the stress-strain diagram to be a straight line. From this we deduce the well-known relation, first postulated by Robert Hooke* in 1678, that stress is proportional to strain. Notice carefully that this proportionality does not extend throughout the diagram; it ends at the proportional limit. Beyond this point, the stress is no longer proportional to the strain. The proportional limit is important because all subsequent theory involving the behavior of elastic bodies is based upon a stress-strain proportionality.† This assumption places an upper limit on the usable stress a material may carry. This is also our first indication that the proportional limit, and not the ultimate strength, is the maximum stress to which a material may be subjected. We shall return to this observation later when we discuss working stress and the factor of safety.

Other concepts developed from the stress-strain curve are the following: (1) The *elastic limit,* that is, the stress beyond which the material will not return to its original shape when unloaded but will retain a permanent deformation called *permanent set*. (2) *Yield point*, at which there is an appreciable elongation or yielding of the material without any corresponding increase of load; indeed, the load may actually decrease while the yielding occurs. However, the phenomenon of yielding is peculiar to structural steel; other grades of steels and steel

*Robert Hooke's famous law *Ut tensio sic vis*, i.e., "As strain, so force," related total strain to total force and did not recognize a limit to this proportionality.
†The stress-strain diagram of many materials is actually a curve on which there is no definite proportional limit. In such cases, the stress-strain proportionality is assumed to exist up to a stress at which the strain increases at a rate 50% greater than shown by the initial tangent to the stress-strain diagram.

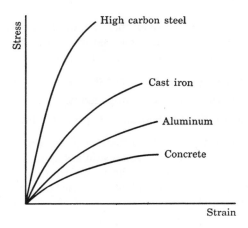

Figure 2-2.   Comparative stress–strain diagrams for different materials.

alloys or other materials do not possess it, as is indicated by the typical stress-strain curves of these materials shown in Fig. 2–2. These curves, incidentally, are typical for a first loading of materials that contain appreciable residual stresses produced by manufacturing or aging processes. After repeated loading, these residual stresses are removed and the stress-strain curves become practically straight, as can be demonstrated in the testing laboratory. (3) *Yield strength*, closely associated with yield point. For materials which do not have a well-defined yield point, yield strength is determined by the offset method. This consists of drawing a line parallel to the initial tangent of the stress-strain curve, this line being started at an arbitrary offset strain, usually of 0.2% or 0.002 m/m. As shown in Fig. 2–3, the intersection of this line with the stress-strain curve is called the yield strength. (4) Ultimate stress, or *ultimate strength* as it is more commonly called, which is the highest ordinate on the stress-strain curve. (5) *Rupture strength*, or the stress at failure. For structural steel it is somewhat lower than ultimate strength because the rupture strength is computed by dividing the rupture load by the original cross-sectional area which, although convenient, is incorrect. The error is caused by a phenomenon known as *necking*. As failure occurs, the material stretches very rapidly and simultaneously narrows down, as shown in Fig. 2–4, so that the rupture load is actually distributed over a smaller area.* If the rupture area is measured after failure occurs, and divided into the rupture load, the result is a truer value of the actual failure stress. Although this is considerably higher than the ultimate strength, the ultimate strength is commonly taken as the maximum stress of the material.

*For reasons that are explained in Art. 13–4, the actual failure is caused by shear, resulting in the cuplike rupture shown.

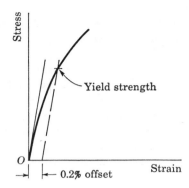

Figure 2–3.   Yield strength determined by offset method.

Figure 2–4.   Necking, or narrowing, of steel specimen at failure.

Working stress and factor of safety

The working stress is defined as the actual stress the material has when under load. It is almost synonymous with *allowable stress,* which is the maximum safe stress a material may carry. In actual design, the allowable stress σ_w should be limited to values not exceeding the proportional limit so as not to invalidate the stress-strain relation of Hooke's law on which all subsequent theory is based. However, since the proportional limit is difficult to determine accurately, it is customary to base the allowable stress on either the yield point or the ultimate strength, divided by a suitable number N, called the factor of safety:

$$\sigma_w = \frac{\sigma_{yp}}{N_{yp}} \quad \text{or} \quad \sigma_w = \frac{\sigma_{ult}}{N_{ult}} \tag{2–2}$$

The yield point is selected as the basis for determining σ_w in structural steel because it is the stress at which a prohibitively large permanent set may occur. For other materials, the allowable stress is usually based on the ultimate strength.

Many factors must be considered in selecting the allowable stress. This selection should not be made by the novice, as will become apparent in a moment; usually the allowable stress is set by a group of experienced engineers and embodied in various building codes and specifications. A short discussion of the factors governing the selection of an allowable stress starts with the observation that in many materials the proportional limit is about one-half the ultimate strength. To avoid accidental overloading, an allowable stress of one-half the proportional limit is usually specified for dead loads gradually applied. The term dead loads generally refers to the weight of the structure or to loads which, once applied, are not removed. An allowable stress set in this way corresponds to a factor of safety of 4 and is recommended for materials that are known to be quite uniform and homogeneous. For other materials, like wood, in which unpredictable nonuniformities (such as knotholes) may occur, larger factors of safety are desirable. The dynamic effect of suddenly applied loads also requires higher factors of safety. As a rule, factors of safety are not directly specified; rather, allowable stresses are set for different materials under different conditions of use, and these stresses are used by the designer.

2-3 HOOKE'S LAW: AXIAL DEFORMATION

Let us return now to a consideration of the straight-line portion of the stress-strain diagram in Fig. 2–1. The slope of that line is the ratio of stress to strain. It is called the *modulus of elasticity* and is denoted by E:

$$\text{slope of stress-strain curve} = E = \frac{\sigma}{\epsilon}$$

which is usually written in the form

$$\sigma = E\epsilon \quad (\text{Hooke's Law}) \tag{2-3}$$

In this form it is known as Hooke's law. Originally Hooke's law specified merely that stress was proportional to strain, but Thomas Young in 1807 introduced a constant of proportionality that came to be known as Young's modulus. Eventually this name was superseded by the phrase *modulus of elasticity*.

From Hooke's law, Eq. (2–3), it can be seen that the units for modulus of elasticity E are identical to the units for stress σ—recall that strain ϵ is a dimensionless quantity. As an illustration, the modulus of elasticity for steel is approximately 200×10^9 N/m^2 (200×10^9 Pa). Using the SI prefix G (read as "giga") to denote multiples of 10^9, this can be expressed as 200 GN/m^2 (200 GPa).

A convenient variation of Hooke's law is obtained by replacing σ by its equivalent P/A and replacing ϵ by δ/L, so that Eq. (2–3) becomes

$$\frac{P}{A} = E\frac{\delta}{L}$$

or

$$\delta = \frac{PL}{AE} = \frac{\sigma L}{E} \quad \left(\text{Important Equation}\right) \tag{2-4}$$

Equation (2–4) expresses the relation among the total deformation δ, the applied load P, the length L, the cross-sectional area A, and the modulus of elasticity E. The unit of deformation δ has the same unit as length L, since the units of σ and E, being equivalent, cancel out of the equation. Note that Eq. (2–4) is subject to all the restrictions previously discussed in connection with the equations it combines. For convenience, let us restate these restrictions:

1. The load must be axial.

2. The bar must have a constant cross section and be homogeneous.

3. The stress must not exceed the proportional limit.

Shearing deformation

Shearing forces cause a shearing deformation, just as axial forces cause elongations, but with an important difference. An element subject to tension undergoes an increase in length; an element subject to shear does not change the length of its sides, but undergoes a change in shape from a rectangle to a parallelogram, as shown in Fig. 2–5.

The action may be visualized for the present as equivalent to the infinitesimal sliding of infinitely thin layers past each other, thereby resulting in the total shearing deformation δ_s in the length L. The actual action is more complex than that pictured and will be discussed more fully in Art. 9–9.

The average shearing strain is found by dividing δ_s by L. In Fig. 2–5, this defines $\tan \gamma = \delta_s / L$. However, since the angle γ is usually very small, $\tan \gamma \approx \gamma$ and we obtain

$$\gamma = \frac{\delta_s}{L} \qquad (2-5)$$

More precisely, the shearing strain is defined as the angular change between two perpendicular faces of a differential element.

The relation between shearing stress and shearing strain, assuming Hooke's law to apply to shear, is

$$\tau = G\gamma \qquad (2-6)$$

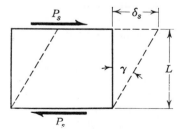

Figure 2–5. Shear deformation.

in which G represents the modulus of elasticity in shear, more commonly called the *modulus of rigidity*. The relation between the shearing deformation and applied shearing forces is then expressed by

$$\delta_s = \frac{VL}{A_s G} \tag{2–7}$$

in which V is the shearing force acting over the shearing area A_s. Note the similarity of this result with Eq. (2–4).

ILLUSTRATIVE PROBLEMS

201. Compute the total elongation caused by an axial load of 100 kN applied to a flat bar 20 mm thick, tapering from a width of 120 mm to 40 mm in a length of 10 m as shown in Fig. 2–6. Assume $E = 200 \times 10^9 \ \text{N/m}^2$.

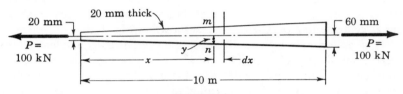

Figure 2–6.

Solution: Since the cross-sectional area is not constant, Eq. (2–4) does not apply directly. However, it may be used to find the elongation in a differential length for which the cross-sectional area is constant. Then the total elongation is the sum of these infinitesimal elongations.

At section $m–n$, the half width y (mm) at a distance x (m) from the left end is found from geometry to be

$$\frac{y - 20}{x} = \frac{60 - 20}{10} \quad \text{or} \quad y = (4x + 20) \ \text{mm}$$

and the area at that section is

$$A = 20(2y) = (160x + 800) \ \text{mm}^2$$

At section $m–n$, in a differential length dx, the elongation may be found from Eq. (2–4):

$$\left[\delta = \frac{PL}{AE} \right] \quad d\delta = \frac{(100 \times 10^3) \ dx}{(160x + 800)(10^{-6})(200 \times 10^9)}$$

$$= \frac{0.500 \ dx}{160x + 800}$$

from which the total elongation is

$$\delta = 0.500 \int_0^{10} \frac{dx}{160x + 800} = \frac{0.500}{160} \left[\ln(160x + 800) \right]_0^{10}$$

$$= (3.13 \times 10^{-3}) \ln \frac{2400}{800} = 3.44 \times 10^{-3} \text{ m} = 3.44 \text{ mm} \qquad Ans.$$

202.   Two steel bars AB and BC support a load $P = 30$ kN as shown in Fig. 2-7a. Area of AB is 300 mm^2; area of BC is 500 mm^2. If $E = 200$ GPa, compute the horizontal and vertical components of the movement of B.

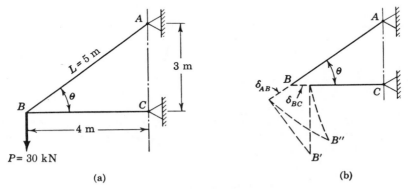

(a)   (b)

Figure 2-7.

Solution: We begin by computing the deformations produced in each bar by P. From statics, we obtain $P_{AB} = 50$ kN (tension) and $P_{BC} = 40$ kN (compression). The corresponding deformations are

$$\left[\delta = \frac{PL}{AE} \right] \qquad \delta_{AB} = \frac{(50 \times 10^3)(5000)}{(300 \times 10^{-6})(200 \times 10^9)}$$

$$= 4.17 \text{ mm lengthening}$$

$$\delta_{BC} = \frac{(40 \times 10^3)(4000)}{(500 \times 10^{-6})(200 \times 10^9)}$$

$$= 1.60 \text{ mm shortening}$$

To analyze the effect of these deformations on the movement of B, imagine first that bars AB and BC are disconnected at B so that they undergo the deformations pictured (greatly exaggerated) in Fig. 2-7b. To refasten the bars, rotate them about A and C to meet at B''. However, the arcs generated in these rotations are so small that they may be effectively replaced by straight lines drawn perpendicular to AB and BC, respectively; these lines, intersecting at B', determine the

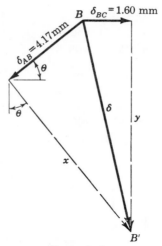

Figure 2–8.

effective final position of B. The deformations δ_{AB} and δ_{BC} are drawn to a larger scale in Fig. 2–8, in which the total movement of B is the vector BB' or δ directed as shown.

From Fig. 2–8 it is evident that the horizontal component of δ is

$$\delta_h = \delta_{BC} = 1.60 \text{ mm rightward} \qquad Ans.$$

However, δ_h is also equal to the algebraic sum of the horizontal components of δ_{AB} and the unknown length x, so that

$$\delta_h = x \sin \theta - \delta_{AB} \cos \theta$$

from which

$$1.60 = x\left(\tfrac{3}{5}\right) - 4.17\left(\tfrac{4}{5}\right), \qquad x = 8.23 \text{ mm}$$

This value of x is used to determine y, i.e., δ_v, which is the sum of the vertical components of δ_{AB} and x:

$$\delta_v = \delta_{AB} \sin \theta + x \cos \theta$$
$$= 4.17\left(\tfrac{3}{5}\right) + 8.23\left(\tfrac{4}{5}\right) = 9.09 \text{ mm down} \qquad Ans.$$

If we return to Fig. 2–7b, we may now compute the magnitude of the angle through which bars AB and BC rotate. We obtain

$$\left[\theta = \frac{s}{r} \right] \quad \alpha_{AB} = \frac{x}{L_{AB}} = \frac{8.23}{5000} = 1.65 \times 10^{-3} \text{ rad} = 0.0945°$$

and

$$\alpha_{BC} = \frac{y}{L_{BC}} = \frac{9.09}{4000} = 2.27 \times 10^{-3} \text{ rad} = 0.130°$$

These rotations are so small that it is justifiable to assume that the directions of δ_{AB} and δ_{BC} coincide with the original directions of bars AB and BC.

PROBLEMS

203. During a stress–strain test, the unit deformation at a stress of 35 MN/m^2 was observed to be 167×10^{-6} m/m and at a stress of 140 MN/m^2 it was 667×10^{-6} m/m. If the proportional limit was 200 MN/m^2, what is the modulus of elasticity? What is the strain corresponding to a stress of 80 MN/m^2? Would these results be valid if the proportional limit were 150 MN/m^2. Explain.

   Ans. $E = 210 \times 10^9$ N/m^2; $\epsilon = 381 \times 10^{-6}$ m/m

204. A uniform bar of length L, cross-sectional area A, and unit mass ρ is suspended vertically from one end. Show that its total elongation is $\delta = \rho g L^2 / 2E$. If the total mass of the bar is M, show also that $\delta = MgL/2AE$.

205. A steel rod having a cross-sectional area of 300 mm^2 and a length of 150 m is suspended vertically from one end. It supports a tensile load of 20 kN at the lower end. If the unit mass of steel is 7850 kg/m^3 and $E = 200 \times 10^3$ MN/m^2, find the total elongation of the rod. (*Hint:* Use the results of Problem 204.)   *Ans.* $\delta = 54.3$ mm

206. A steel wire 10 m long, hanging vertically supports a tensile load of 2000 N. Neglecting the weight of the wire, determine the required diameter if the stress is not to exceed 140 MPa and the total elongation is not to exceed 5 mm. Assume $E = 200$ GPa.

207. A steel tire, 10 mm thick, 80 mm wide, and of 1500.0 mm inside diameter, is heated and shrunk onto a steel wheel 1500.5 mm in diameter. If the coefficient of static friction is 0.30, what torque is required to twist the tire relative to the wheel? Neglect the deformation of the wheel. Use $E = 200$ GPa.   *Ans.* $T = 75.0$ kN · m

208. An aluminum bar having a cross-sectional area of 160 mm^2 carries the axial loads at the positions shown in Fig. P–208. If $E = 70$ GPa, compute the total deformation of the bar. Assume that the bar is suitably braced to prevent buckling.

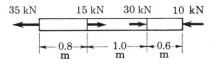

Figures P–208 and P–209.

209.   Solve Problem 208 if the magnitudes of the loads at the ends are interchanged, i.e., if the load at the left end is 10 kN and that at the right end is 35 kN.   *Ans.* $\delta = 1.61$ mm (contraction)

210.   An aluminum tube is fastened between a steel rod and a bronze rod as shown in Fig. P–210. Axial loads are applied at the positions indicated. Find the value of P that will not exceed a maximum overall deformation of 2 mm or a stress in the steel of 140 MN/m², in the aluminum of 80 MN/m², or in the bronze of 120 MN/m². Assume that the assembly is suitably braced to prevent buckling and that $E_s = 200 \times 10^3$ MN/m², $E_a = 70 \times 10^3$ MN/m², and $E_b = 83 \times 10^3$ MN/m².

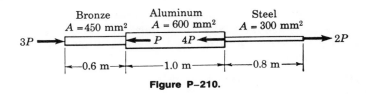

Figure P–210.

211.   The rigid bars shown in Fig. P–211 are separated by a roller at C and pinned at A and D. A steel rod at B helps support the load of 50 kN. Compute the vertical displacement of the roller at C.

Ans.   2.82 mm

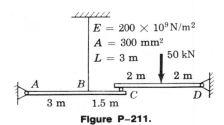

Figure P–211.

212.   A uniform concrete slab of mass M is to be attached, as shown in Fig. P–212, to two rods whose lower ends are initially at the same level. Determine the ratio of the areas of the rods so that the slab will remain level after it is attached to the rods.   *Ans.* $A_a/A_s = 8.57$

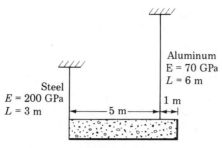

Figure P-212.

213.   The rigid bar AB, attached to two vertical rods as shown in Fig. P-213, is horizontal before the load P is applied. If the load $P = 50$ kN, determine its vertical movement.

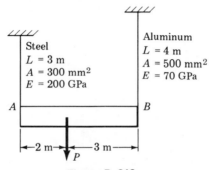

Figure P-213.

214.   The rigid bars AB and CD shown in Fig. P-214 are supported by pins at A and C and the two rods. Determine the maximum force P which can be applied as shown if its vertical movement is limited to 5 mm. Neglect the weights of all members.

Ans. $P = 76.3$ kN

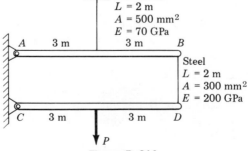

Figure P-214.

215. A round bar of length L tapers uniformly from a diameter D at one end to a smaller diameter d at the other. Determine the elongation caused by an axial tensile load P.

Ans. $\delta = 4PL/\pi EDd$

216. A uniform slender rod of length L and cross-sectional area A is rotating in a horizontal plane about a vertical axis through one end. If the unit mass of the rod is ρ, and it is rotating at a constant angular velocity of ω rad/s, show that the total elongation of the rod is $\rho\omega^2 L^3/3E$.

217. As shown in Fig. P–217, two aluminum rods AB and BC, hinged to rigid supports, are pinned together at B to carry a vertical load $P = 20$ kN. If each rod has a cross-sectional area of 400 mm^2 and $E = 70 \times 10^3$ MN/m^2, compute the elongation of each rod and the horizontal and vertical displacements of point B. Assume $\alpha = 30°$ and $\theta = 30°$. *Ans.* $\delta_h = 0.412$ mm; $\delta_v = 3.57$ mm

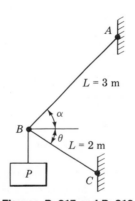

Figures P–217 and P–218.

218. Solve Problem 217 if rod AB is of steel, with $E = 200 \times 10^3$ MN/m^2. Assume $\alpha = 45°$ and $\theta = 30°$; all other data remain unchanged.

219. A round bar of length L, tapering uniformly from a diameter D at one end to a smaller diameter d at the other, is suspended vertically from the large end. If ρ is the unit mass, find the elongation caused by its own weight. Use this result to determine the elongation of a cone suspended from its base.

Ans. $\delta = \dfrac{\rho g L^2 (D + d)}{6E(D - d)} - \dfrac{\rho g L^2 d^2}{3ED(D - d)}$

2–4 POISSON'S RATIO: BIAXIAL AND TRIAXIAL DEFORMATIONS

Another type of elastic deformation is the change in transverse dimensions accompanying axial tension or compression. Experiments show that if a bar is lengthened by axial tension, there is a reduction in the transverse dimensions. Siméon D. Poisson showed in 1811 that the ratio of the *unit* deformations or strains in these directions is constant for stresses within the proportional limit. Accordingly this ratio is named after him; it is denoted by ν and defined by

$$\nu = -\frac{\epsilon_y}{\epsilon_x} = -\frac{\epsilon_z}{\epsilon_x} \qquad (2\text{–}8)$$

where ϵ_x is the strain due only to stress in the X direction, and ϵ_y and ϵ_z are the strains induced in the perpendicular directions. The minus sign indicates a decrease in transverse dimensions when ϵ_x is positive, as in the case of tensile elongation.

Poisson's ratio permits us to extend Hooke's law of uniaxial stress to the case of biaxial stress. Thus, if an element is subjected simultaneously to tensile stresses in the X and Y directions, the strain in the X direction due to the tensile stress σ_x is σ_x/E. Simultaneously the tensile stress σ_y will produce lateral contraction in the X direction of the amount $\nu\sigma_y/E$, so the resultant unit deformation or strain in the X direction will be

$$\epsilon_x = \frac{\sigma_x}{E} - \nu\frac{\sigma_y}{E} \qquad (2\text{–}9)$$

Similarly the total strain in the Y direction is

$$\epsilon_y = \frac{\sigma_y}{E} - \nu\frac{\sigma_x}{E} \qquad (2\text{–}10)$$

If desired, Eqs. (2–9) and (2–10) can be solved to express the stresses in terms of the strains as follows:

$$\sigma_x = \frac{(\epsilon_x + \nu\epsilon_y)E}{1 - \nu^2}; \quad \sigma_y = \frac{(\epsilon_y + \nu\epsilon_x)E}{1 - \nu^2} \qquad (2\text{–}11)$$

A further extension of the above discussion results in the following expressions for strains caused by simultaneous action of triaxial tensile stresses:

$$\left.\begin{array}{l} \epsilon_x = \dfrac{1}{E}\left[\sigma_x - \nu(\sigma_y + \sigma_z)\right] \\[2mm] \epsilon_y = \dfrac{1}{E}\left[\sigma_y - \nu(\sigma_z + \sigma_x)\right] \\[2mm] \epsilon_z = \dfrac{1}{E}\left[\sigma_z - \nu(\sigma_x + \sigma_y)\right] \end{array}\right\} \qquad (2\text{–}12)$$

All the above equations are valid for compressive effects also; it is only necessary to assign positive signs to elongations and tensile stresses, and, conversely, negative signs to contractions and compressive stresses.

An important relation* between the constants E, G, and ν for a given material is expressed by

$$G = \frac{E}{2(1 + \nu)} \qquad (2-13)$$

which is useful for computing values of ν when E and G have been determined. Common values of Poisson's ratio are 0.25 to 0.30 for steel, approximately 0.33 for most other metals, and 0.20 for concrete.

ILLUSTRATIVE PROBLEMS

220. A specimen of any given material is subjected to a uniform triaxial stress. Determine the theoretical maximum value of Poisson's ratio.

Solution: If we add the three relations in Eq. (2–12), we obtain

$$\epsilon_x + \epsilon_y + \epsilon_z = \frac{1 - 2\nu}{E}(\sigma_x + \sigma_y + \sigma_z) \qquad (a)$$

For a uniform triaxial stress, we have $\epsilon_x = \epsilon_y = \epsilon_z = \epsilon$, and $\sigma_x = \sigma_y = \sigma_z = \sigma$. Hence Eq. (a) reduces to

$$\epsilon = (1 - 2\nu)\left(\frac{\sigma}{E}\right)$$

Since both ϵ and σ must be of the same sign, it follows that $(1 - 2\nu)$ must be positive; that is,

$$1 - 2\nu \geqq 0$$

from which

$$\nu \leqq \tfrac{1}{2} \qquad Ans.$$

221. A solid aluminum shaft of 80-mm diameter fits concentrically in a hollow steel tube. Compute the minimum internal diameter of the steel tube so that no contact pressure exists when the aluminum shaft carries an axial compressive load of 400 kN. Assume $\nu = \tfrac{1}{3}$ and $E_a = 70 \times 10^9 \text{ N/m}^2$.

Solution: The axial compressive stress in the aluminum is

$$\left[\sigma = \frac{P}{A} \right] \qquad \sigma_x = -\frac{400 \times 10^3}{\frac{\pi}{4}(0.080)^2} = -79.6 \text{ MN/m}^2$$

*This relation is proved in Art. 9–11.

For uniaxial stress, the transverse strain is

$$\left[\epsilon_y = -\nu\epsilon_x = -\nu\frac{\sigma_x}{E} \right] \qquad \epsilon_y = -\frac{1}{3}\left(\frac{-79.6 \times 10^6}{70 \times 10^9} \right)$$

$$= 379 \times 10^{-6}\,\text{m/m}$$

Therefore, the required diametrical clearance is

$$\left[\delta = \epsilon L \right] \qquad \delta_y = (379 \times 10^{-6})(80) = 0.0303\,\text{mm}$$

The required internal diameter of the tube is found by adding this clearance to the original diameter of the aluminum shaft, thus giving

$$D = 80 + 0.0303 = 80.0303\,\text{mm} \qquad Ans.$$

PROBLEMS

222. A solid cylinder of diameter d carries an axial load P. Show that its change in diameter is $4P\nu/\pi Ed$.

223. A rectangular aluminum block is 100 mm long in the X direction, 75 mm wide in the Y direction, and 50 mm thick in the Z direction. It is subjected to a triaxial loading consisting of a uniformly distributed tensile force of 200 kN in the X direction and uniformly distributed compressive forces of 160 kN in the Y direction and 220 kN in the Z direction. If $\nu = \frac{1}{3}$ and $E = 70$ GPa, determine a single distributed loading in the X direction that would produce the same Z deformation as the original loading. *Ans.* 410 kN tension

224. A welded steel cylindrical drum made of 10-mm plate has an internal diameter of 1.20 m. By how much will the diameter be changed by an internal pressure of 1.5 MPa? Assume that Poisson's ratio is 0.30 and $E = 200$ GPa.

225. A 50-mm-diameter steel tube with a wall thickness of 2 mm just fits in a rigid hole. Find the tangential stress if an axial compressive load of 10 kN is applied. Assume $\nu = 0.30$ and $E = 200 \times 10^9\,\text{N/m}^2$. Neglect the possibility of buckling.

226. A 150-mm-long bronze tube, closed at its ends, is 80 mm in diameter and has a wall thickness of 3 mm. It fits without clearance in an 80-mm hole in a rigid block. The tube is then subjected to an internal pressure of 4.00 MN/m^2. Assuming $\nu = \frac{1}{3}$ and $E = 83 \times 10^3\,\text{MN/m}^2$, determine the tangential stress in the tube.
Ans. 8.89 MN/m^2

227. A 200-mm-long aluminum tube, closed at its ends, is 100 mm in diameter with a wall thickness of 2 mm. If the tube just fits between two rigid walls at zero internal pressure, determine the longitudinal and tangential stresses at an internal pressure of 4.00 MN/m^2. Assume $\nu = \frac{1}{3}$ and $E = 70 \times 10^9\,\text{N/m}^2$.

2–5 STATICALLY INDETERMINATE MEMBERS

There are certain combinations of axially loaded members in which the equations of static equilibrium are not sufficient for a solution. This condition exists in structures where the reactive forces or the internal resisting forces over a cross section exceed the number of independent equations of equilibrium. Such cases are called *statically indeterminate* and require the use of additional relations which depend upon the elastic deformations in the members. The cases are so varied that they can best be described by sample problems illustrating the following general principles:

1. To a free-body diagram of the structure, or a part of it, apply the equations of static equilibrium.

2. If there are more unknowns than independent equations of equilibrium, obtain additional equations from the geometric relations between the elastic deformations produced by the loads. To define these relations clearly, you will find it helpful to draw a sketch that exaggerates the magnitudes of the elastic deformations.

ILLUSTRATIVE PROBLEMS

228. The short concrete post in Fig. 2–9 is reinforced axially with six symmetrically placed steel bars, each 600 mm² in area. If the applied load P is 1000 kN, compute the stress developed in each material. Use

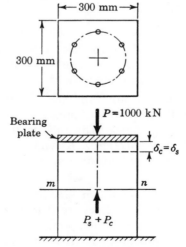

Figure 2-9. Reinforced concrete post.

the following moduli of elasticity: for steel, $E_s = 200 \times 10^9 \text{ N/m}^2$; for concrete, $E_c = 14 \times 10^9 \text{ N/m}^2$.

Solution: As shown in Fig. 2–9, the applied load and the resisting forces on any transverse section $m-n$ form a collinear system. From equilibrium, we have

$$[\Sigma F = 0] \qquad P_s + P_c = 1 \times 10^6 \text{ N} \tag{a}$$

Since no other equations of static equilibrium are available to indicate in what proportion the load is distributed to each material, we consider the elastic deformation of the structure. By symmetry, it is evident that the bearing plate causes the steel and concrete to deform equally. Hence, applying $\delta = \sigma L / E$ to these equal deformations, we obtain

$$[\delta_s = \delta_c] \qquad \left(\frac{\sigma L}{E}\right)_s = \left(\frac{\sigma L}{E}\right)_c$$

from which, by canceling out the equal lengths of steel and concrete and substituting the moduli of elasticity, we have the following relation between the stresses:

$$\sigma_s = \frac{E_s}{E_c} \sigma_c = \frac{200}{14} \sigma_c = 14.3 \sigma_c \tag{b}$$

Equation (b) may be called the governing relation for stress. Note that it depends only upon the fact that both materials deform equally. This relation is independent of the loads or areas and remains valid provided the proportional limit of neither material is exceeded.

We may now use $P = \sigma A$ to rewrite Eq. (a):

$$\sigma_s\left[6(600 \times 10^{-6})\right] + \sigma_c\left\{\left[(300 \times 300) - 6(600)\right] \times 10^{-6}\right\} = 1 \times 10^6$$

$$\sigma_s(3.6 \times 10^{-3}) + \sigma_c(86.4 \times 10^{-3}) = 1 \times 10^6$$

whence, replacing σ_s by $14.3\sigma_c$ from Eq. (b), we obtain

$$14.3\sigma_c(3.6 \times 10^{-3}) + \sigma_c(86.4 \times 10^{-3}) = 1 \times 10^6$$

$$\sigma_c = 7.25 \times 10^6 \text{ N/m}^2 = 7.25 \text{ MN/m}^2 \qquad Ans.$$

and from Eq. (b)

$$\sigma_s = 14.3\sigma_c = 104 \text{ MN/m}^2 \qquad Ans.$$

229.   In the preceding problem, assume the allowable stresses to be $\sigma_s = 120 \text{ MN/m}^2$ and $\sigma_c = 6 \text{ MN/m}^2$. Compute the maximum safe axial load P which may be applied.

Solution: The unwary student may substitute the allowable stresses only in the equation of static equilibrium. This would be incorrect because it would not consider the equal deformations of the materials. From Eq. (b) in the preceding problem, we saw that equal deformations produce the following governing relation between the stresses:

$\sigma_s = 14.3\sigma_c$

From this relation we note that if the concrete were stressed to its limit of 6 MN/m², the corresponding stress in the steel would be

$\sigma_s = (14.3)6 = 85.8 \text{ MN/m}^2$

Therefore, the steel could not be stressed to its limit of 120 MN/m² without overstressing the concrete. The actual working stresses are thereby determined to be $\sigma_c = 6$ MN/m² and $\sigma_s = 85.8$ MN/m². These values are substituted in the equation of static equilibrium [Eq. (*a*) in the preceding problem] to obtain

$$P = P_s + P_c = \sigma_s A_s + \sigma_c A_c$$
$$= (85.8 \times 10^6)(3.6 \times 10^{-3}) + (6 \times 10^6)(86.4 \times 10^{-3})$$
$$= 827 \text{ kN} \quad Ans.$$

230. A copper rod is inserted into a hollow aluminum cylinder. The copper rod projects 0.130 mm as shown in Fig. 2–10. What maximum load P may be applied to the bearing plate? Use the data in the accompanying table.

	COPPER	ALUMINUM
Area (mm²)	1200	1800
E (GPa)	120	70
Allowable stress (MPa)	140	70

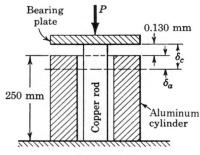

Figure 2–10.

Solution: To find the governing relation between the stresses, we consider the elastic deformations, which are shown exaggerated in Fig. 2–10. We obtain

$$\left[\delta_c = \delta_a + (0.130 \times 10^{-3}) \text{ m} \right]$$
$$\left(\frac{\sigma L}{E} \right)_c = \left(\frac{\sigma L}{E} \right)_a + (0.130 \times 10^{-3})$$
$$\frac{\sigma_c(0.25)}{120 \times 10^9} = \frac{\sigma_a(0.25)}{70 \times 10^9} + (0.130 \times 10^{-3})$$

from which

$$\sigma_c = 1.71\sigma_a + (62.4 \times 10^6) \qquad (a)$$

From this governing relation for the stresses, we note that using $\sigma_a = 70$ MPa will overstress the copper to 182 MPa. Therefore copper governs, and the corresponding stress in the aluminum is determined from Eq. (*a*) to be

$$(140 \times 10^6) = 1.71\sigma_a + (62.4 \times 10^6) \qquad \sigma_a = 45.4 \text{ MPa}$$

The total safe load is given by

$$P = P_c + P_a = \sigma_c A_c + \sigma_a A_a$$

whence, substituting the working stresses just determined, we obtain

$$P = (140 \times 10^6)(1200 \times 10^{-6}) + (45.4 \times 10^6)(1800 \times 10^{-6})$$
$$= 250 \text{ kN} \qquad Ans.$$

231. A horizontal bar of negligible mass, hinged at A in Fig. 2–11a and assumed rigid, is supported by a bronze rod 2 m long

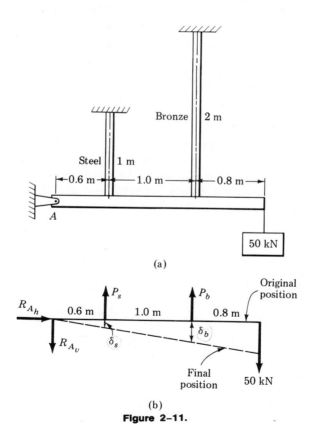

(a)

(b)

Figure 2–11.

and a steel rod 1 m long. Using the data in the accompanying table, compute the stress in each rod.

	STEEL	BRONZE
Area (mm²)	600	300
E (GN/m²)	200	83
Proportional limit (MN/m²)	240	140

Solution: The free-body diagram of the bar in Fig. 2–11b shows it to be statically indeterminate to the first degree; that is, there is one more unknown force than can be found from the equations of static equilibrium. A moment summation about A gives one relation between the forces in the rods:

$$[\Sigma M_A = 0] \qquad 0.6P_s + 1.6P_b = 2.4(50 \times 10^3) \qquad (a)$$

Another relation between these forces is obtained from the elastic deformations of the rods. From the similar triangles formed in Fig. 2–11b (recall that the horizontal bar is assumed rigid), we obtain

$$\frac{\delta_s}{0.6} = \frac{\delta_b}{1.6} \quad \text{or} \quad \frac{1}{0.6}\left(\frac{PL}{AE}\right)_s = \frac{1}{1.6}\left(\frac{PL}{AE}\right)_b,$$

$$\frac{1}{0.6}\frac{P_s(1)}{(600)(200)} = \frac{1}{1.6}\frac{P_b(2)}{(300)(83)}$$

whence

$$P_s = 3.61P_b \qquad (b)$$

Note that the areas need not be expressed in square meters, since the conversion factor of 10^{-6} appears on both sides of the equation and thus will cancel out. Similarly the factor 10^9 in the moduli of elasticity is not shown in the foregoing equation since it also cancels.

Solving Eq. (a) and (b), we obtain

$$P_s = 115 \text{ kN}$$
$$P_b = 31.9 \text{ kN}$$

Computing the stresses,

$$\sigma_s = \frac{P_s}{A_s} = \frac{115 \times 10^3}{600 \times 10^{-6}} = 192 \times 10^6 \text{ N/m}^2$$
$$= 192 \text{ MN/m}^2 \qquad Ans.$$
$$\sigma_b = \frac{P_b}{A_b} = \frac{31.9 \times 10^3}{300 \times 10^{-6}} = 106 \times 10^6 \text{ N/m}^2$$
$$= 106 \text{ MN/m}^2 \qquad Ans.$$

Since both stresses are less than the proportional limits, the answers may be accepted. If the steel stress, for example, had exceeded the proportional limit, the results would not have been valid and a redesign would be required. Perhaps the simplest redesign would be to increase the length of the steel rod, thus making it less rigid. It may be noted here that generally the most rigid parts of an indeterminate structure carry the most load. This is a fundamental principle in the theory of indeterminate structures and is known as the *principle of ridigities.**

PROBLEMS

232.   A steel bar 50 mm in diameter and 2 m long is surrounded by a shell of cast iron 5 mm thick. Compute the load that will compress the combined bar a total of 1 mm in the length of 2 m. For steel, $E = 200 \times 10^9$ N/m^2, and for cast iron, $E = 100 \times 10^9$ N/m^2.

Ans.   P = 240 kN

233.   A reinforced concrete column 250 mm in diameter is designed to carry an axial compressive load of 400 kN. Using allowable stresses of $\sigma_c = 6$ MPa and $\sigma_s = 120$ MPa, determine the required area of reinforcing steel. Assume that $E_c = 14$ GPa and $E_s = 200$ GPa.

Ans. $A_s = 1320$ mm^2

234.   A timber block 250 mm square is reinforced on each side by a steel plate 250 mm wide and t mm thick. Determine the thickness t so that the assembly will support an axial load of 1200 kN without exceeding a maximum timber stress of 8 MN/m^2 or a maximum steel stress of 140 MN/m^2. For timber, $E = 10 \times 10^3$ MN/m^2; for steel, $E = 200 \times 10^3$ MN/m^2.

235.   A rigid block of mass M is supported by three symmetrically spaced rods as shown in Fig. P–235. Each copper rod has an area of 900 mm^2; $E = 120$ GPa; and the allowable stress is 70 MPa. The steel rod has an area of 1200 mm^2; $E = 200$ GPa; and the allowable stress is 140 MPa. Determine the largest mass M which can be supported.   *Ans. $M = 22.3 \times 10^3$ kg*

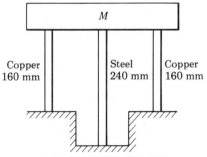

Figures P–235 and P–236.

*See J. I. Parcel and G. A. Maney, *Statically Indeterminate Stresses*, 2nd ed., Wiley, New York, 1936, p. 109.

236. In Problem 235, how should the length of the steel rod be changed so that each material will be stressed to its allowable limit?

237. The lower ends of the three bars in Fig. P–237 are at the same level before the rigid homogeneous 18 Mg block is attached. Each steel bar has an area of 600 mm^2 and $E = 200$ GN/m^2. For the bronze bar, the area is 900 mm^2 and $E = 83$ GN/m^2. Find the stress developed in each bar.   *Ans.* $\sigma_s = 124$ MN/m^2; $\sigma_b = 32.0$ MN/m^2

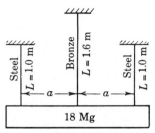

Figure P–237.

238. The rigid platform in Fig. P–238 has negligible mass and rests on two aluminum bars, each 250.00 mm long. The center bar is steel and is 249.90 mm long. Find the stress in the steel bar after the center load $P = 400$ kN is applied. Each aluminum bar has an area of 1200 mm^2 and $E = 70$ GPa. The steel bar has an area of 2400 mm^2 and $E = 200$ GPa.

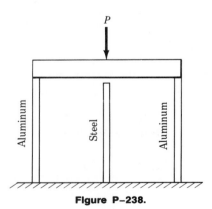

Figure P–238.

239. Three steel eye-bars, each 100 mm by 25 mm in section, are to be assembled by driving 20-mm-diameter drift pins through holes drilled in the ends of the bars. The center-line spacing between the holes is 10 m in the outer two bars but is 1.25 mm shorter in the middle bar.

Find the shearing stress developed in the drift pins. Neglect local deformation at the holes and use $E_s = 200$ GPa.

Ans. $\tau = 66.2$ MPa

240.   As shown in Fig. P–240, three steel wires, each 30 mm² in area, are used to lift a mass M. Their unstretched lengths are 19.994 m, 19.997 m, and 20.000 m. (a) If $M = 600$ kg, what stress exists in the longest wire? (b) If $M = 200$ kg, determine the stress in the shortest wire. Use $E = 200$ GN/m².

Figure P–240.

241.   The assembly in Fig. P–241 consists of a rigid bar AB (having negligible mass) pinned at O and attached to the aluminum rod and the steel rod. In the position shown, the bar AB is horizontal and there is a gap $\Delta = 4$ mm between the lower end of the aluminum rod and its pin support at D. Find the stress in the steel rod when the lower end of the aluminum rod is pinned to the support at D.

Ans. $\sigma_s = 174$ MPa

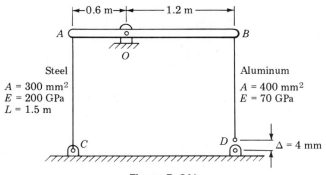

Figure P–241.

242. A homogeneous rod of constant cross section is attached to unyielding supports. It carries an axial load P applied as shown in Fig. P–242. Prove that the reactions are given by $R_1 = Pb/L$ and $R_2 = Pa/L$. (Note that these reactions are equivalent to those of a simply supported beam carrying a concentrated load.)

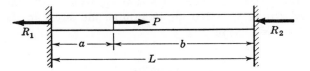

Figure P–242.

243. A homogeneous bar with a cross-sectional area of 500 mm² is attached to rigid supports. It carries the axial loads $P_1 = 25$ kN and $P_2 = 50$ kN, applied as shown in Fig. P–243. Determine the stress in the segment BC. (*Hint*: Use the results of Problem 242, and compute the reactions caused by P_1 and P_2 acting separately. Then use the principle of superposition to compute the reactions when both loads are applied.)

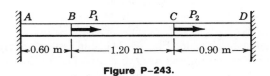

Figure P–243.

244. The bar shown in Fig. P–244 is firmly attached to unyielding supports. Find the stress caused in each material by applying an axial load $P = 200$ kN.

Ans. $\sigma_a = 62.8$ MN/m²; $\sigma_s = 120$ MN/m²

245. Refer to Problem 244. What maximum load P can be applied without exceeding an allowable stress of 70 MPa for aluminum or 120 MPa for steel? Can a larger load P be carried if the length of the aluminum rod is changed, the length of the steel portion being kept the same? If so, determine this length.

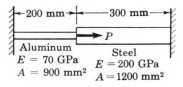

Figures P–244 and P–245.

246. A rod is composed of three segments shown in Fig. P–246 and carries the axial loads P_1 = 120 kN and P_2 = 50 kN. Determine the stress in each material if the walls are rigid.

Ans. σ_s = 122 MN/m²

247. Solve Problem 246 if the left wall yields 0.60 mm.

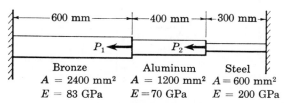

Figures P–246 and P–247.

248. A steel tube 2.5 mm thick just fits over an aluminum tube 2.5 mm thick. If the contact diameter is 100 mm, determine the contact pressure and tangential stresses when the outward radial pressure on the aluminum tube is p = 4 MN/m². Here, E_s = 200 × 10⁹ N/m², and E_a = 70 × 10⁹ N/m².

Ans. p_c = 2.96 MN/m²; σ_s = 59.2 MN/m²; σ_a = 20.8 MN/m²

249. In Problem 248, assume that there is a radial clearance of 0.01 mm between the tubes before the internal pressure of 4 MN/m² is applied to the aluminum tube. Solve for the contact pressure and tangential stresses.

250. In the assembly of the bronze tube and steel bolt shown in Fig. P–250, the pitch of the bolt thread is 0.80 mm and the cross-sectional area of the bronze tube is 900 mm² and of the steel bolt is 450 mm². The nut is turned until there is a compressive stress of 30 MN/m² in the bronze tube. Find the stress in the bronze tube if the nut is then given one additional turn. How many turns of the nut will reduce this stress to zero? Take E as in Problem 246.

Ans. σ_b = 75.4 MN/m²; 1.66 turns

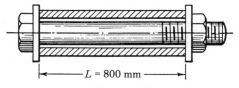

Figure P–250.

251.   As shown in Fig. P–251, a rigid beam with negligible mass is pinned at O and supported by two rods, identical except for length. Determine the load in each rod if $P = 30$ kN.

Ans. $P_A = 9.10$ kN; $P_B = 11.94$ kN

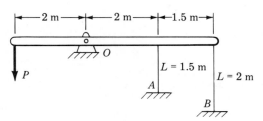

Figure P–251.

252.   As shown in Fig. P–252, a rigid beam with negligible mass is pinned at one end and supported by two rods. The beam was initially horizontal before the load P was applied. Find the vertical movement of P if $P = 120$ kN.   *Ans.*   2.92 mm

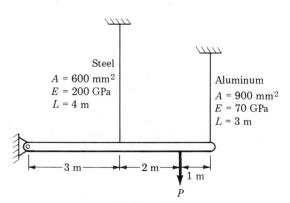

Figure P–252.

253.   A rigid bar of negligible mass, pinned at one end, is supported by a steel rod and a bronze rod as shown in Fig. P–253. What maximum load P can be applied without exceeding a stress in the steel of 120 MN/m² or in the bronze of 70 MN/m²?

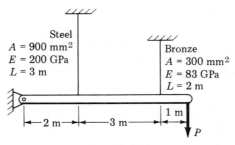

Figure P–253.

254. Shown in Fig. P–254 is a section through a balcony. The total uniform load of 600 kN is supported by three rods of the same area and material. Compute the load in each rod. Assume the floor to be rigid, but note that it does not necessarily remain horizontal.

Ans. $P_B = 183$ kN

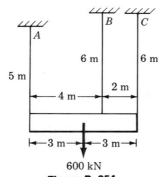

Figure P–254.

255. Three rods, each with an area of 300 mm², jointly support the load of 10 kN, as shown in Fig. P–255. Assuming there was no slack or stress in the rods before the load was applied, find the stress in each rod. Here, $E_s = 200 \times 10^9$ N/m² and $E_b = 83 \times 10^9$ N/m².

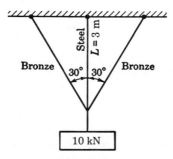

Figure P–255.

256. Three bars, AB, AC, and AD, are pinned together to support a load $P = 20$ kN as shown in Fig. P–256. Horizontal movement is prevented at joint A by the short horizontal strut AE. Determine the stress in each bar and the force in the strut AE. For the steel bar, $A = 200$ mm^2 and $E = 200$ GPa. For each aluminum bar, $A = 400$ mm^2 and $E = 70$ GPa. *Ans.* $P_{AE} = 180$ N

257. Refer to the data in Problem 256, and determine the maximum value of P that will not exceed an aluminum stress of 40 MPa or a steel stress of 120 MPa.

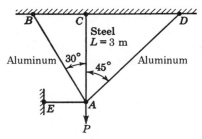

Figures P–256 and P–257.

2–6 THERMAL STRESSES

It is well known that changes in temperature cause bodies to expand or contract, the amount of the linear deformation, δ_T, being expressed by the relation

$$\delta_T = \alpha L(\Delta T) \tag{2–14}$$

in which α is the coefficient of linear expansion, usually expressed in units of meters per meter per degree of temperature change, L is the length, and ΔT is the temperature change. Substituting these units in Eq. (2–14) shows the dimensional unit of δ_T to be the same as that of the length L.

If a temperature deformation is permitted to occur freely, as by the use of expansion joints, no load or stress will be induced in the structure. But in some cases it may not be feasible to permit these temperature deformations; the result is that internal forces are created which resist them. The stresses caused by these internal forces are known as *thermal stresses*.

A general procedure for computing the loads and stresses caused when temperature deformation is prevented is outlined in these steps:

1. Imagine the structure relieved of all applied loads and constraints so that temperature deformations can occur freely. Represent these deformations on a sketch, and exaggerate their effect.

2. Now imagine sufficient loads applied to the structure to restore it to the specified conditions of restraint. Represent these loads and corresponding load deformations on the sketch for step 1.

3. The geometric relations between the temperature and load deformations on the sketch give equations which, together with the equations of static equilibrium, may be solved for all unknown quantities.

The following examples illustrate these steps applied in several different types of problems.

ILLUSTRATIVE PROBLEMS

258. A steel rod 2.5 m long is secured between two walls. If the load on the rod is zero at 20°C, compute the stress when the temperature drops to −20°C. The cross-sectional area of the rod is 1200 mm², $\alpha = 11.7~\mu\text{m}/(\text{m} \cdot {}^\circ\text{C})$, and $E = 200$ GN/m². Solve, assuming (a) that the walls are rigid and (b) that the walls spring together a total distance of 0.500 mm as the temperature drops.

Figure 2–12. Rigid walls.

Solution:

Part a. Imagine the rod is disconnected from the right wall. Temperature deformations can then freely occur. A temperature drop causes the contraction represented by δ_T in Fig. 2–12. To reattach the rod to the wall will evidently require a pull P to produce the load deformation δ_P. From the sketch of deformations, we see that $\delta_T = \delta_P$, or, in equivalent terms,

$$\alpha(\Delta T)L = \frac{PL}{AE} = \frac{\sigma L}{E}$$

whence

$$\sigma = E\alpha(\Delta T) = (200 \times 10^9)(11.7 \times 10^{-6})(40) = 93.6 \times 10^6 \text{ N/m}^2$$
$$= 93.6 \text{ MN/m}^2 \quad Ans.$$

Note that L cancels out of the above equation, indicating that the stress is independent of the length of the rod.

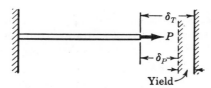

Figure 2–13. Nonrigid walls.

Part b.   When the walls spring together, Fig. 2–13 shows that the free temperature contraction is equal to the sum of the load deformation and the yield of the walls. Hence

$$\delta_T = \delta_P + \text{yield}$$

whence, replacing the deformations by equivalent terms, we obtain

$$\alpha L(\Delta T) = \frac{\sigma L}{E} + \text{yield}$$

or

$$(11.7 \times 10^{-6})(2.5)(40) = \frac{\sigma(2.5)}{200 \times 10^9} + (0.5 \times 10^{-3})$$

from which we obtain

$$\sigma = 53.6 \text{ MN/m}^2 \qquad Ans.$$

Notice that the yield of the walls reduces the stress considerably, and also that the length of the rod does not cancel out as in Part *a*.

259.   A rigid block having a mass of 5 Mg is supported by three rods symmetrically placed, as shown in Fig. 2–14. Determine the stress in each rod after a temperature rise of 40°C. The lower ends of the rods are assumed to have been at the same level before the block was

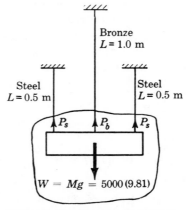

Figure 2–14. Free-body diagram.

attached and the temperature changed. Note that symmetry dictates that the block will remain horizontal. Use the data in the accompanying table.

	EACH STEEL ROD	BRONZE ROD
Area (mm^2)	500	900
E (N/m^2)	200×10^9	83×10^9
$\alpha\,[\,\mu m/\,(m\cdot{}^\circ C)]$	11.7	18.9

Solution: With the block detached, the original lower position of the rods is as shown in Fig. 2–15. With the rods free of any constraint, a temperature rise will cause the temperature deformations δ_{T_s} and δ_{T_b} in the steel and bronze, respectively. When the rods are attached to the rigid block after the temperature change has occurred, assume their final horizontal level to be as shown. To attach them to the block, it will be necessary to pull their expanded ends through the load deformations δ_{P_s} and δ_{P_b} by means of the loads P_s and P_b in the steel and bronze, respectively. The free-body diagram of the block in Fig. 2–14 represents the equal and opposite effects of the forces exerted by the rods upon the block.

   From the deformations shown in Fig. 2–15, we obtain the following geometric relation between the deformations:

$$\delta_{T_s} + \delta_{P_s} = \delta_{T_b} + \delta_{P_b}$$

or

$$(\alpha L\ \Delta T)_s + \left(\frac{PL}{AE}\right)_s = (\alpha L\ \Delta T)_b + \left(\frac{PL}{AE}\right)_b$$

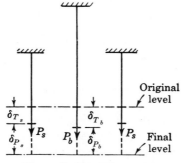

Figure 2–15.   Deformations.

whence, substituting the given data, we have

$$(11.7 \times 10^{-6})(0.5)(40) + \frac{P_s(0.5)}{(500 \times 10^{-6})(200 \times 10^9)}$$

$$= (18.9 \times 10^{-6})(1)(40) + \frac{P_b(1)}{(900 \times 10^{-6})(83 \times 10^9)}$$

Simplifying the above equation, we obtain

$$P_s - 2.68 P_b = 104 \times 10^3 \text{ N} \qquad (a)$$

Another relation between P_s and P_b is obtained from the free-body diagram in Fig. 2–14.

$$[\Sigma Y = 0] \qquad 2P_s + P_b = (5000)(9.81) = 49.05 \times 10^3 \text{ N} \qquad (b)$$

Solving Eqs. (a) and (b) yields

$$P_s = 37.0 \text{ kN}$$
$$P_b = -25.0 \text{ kN}$$

The negative sign for P_b means that the load P_b acts oppositely to that assumed; that is, the bronze rod is actually in compression and suitable provision must be made to prevent buckling.

The stresses are

$$\left[\sigma = \frac{P}{A} \right] \qquad \sigma_s = \frac{37.0 \times 10^3}{500 \times 10^{-6}}$$

$$= 74.0 \text{ MN/m}^2 \quad \text{(tension)} \qquad Ans.$$

$$\sigma_b = \frac{25.0 \times 10^3}{900 \times 10^{-6}}$$

$$= 27.8 \text{ MN/m}^2 \quad \text{(compression)} \qquad Ans.$$

260.   Using the data in Problem 259, determine the temperature rise necessary to cause all the applied load to be supported by the steel rods.

Solution: Instead of trying to use the results obtained in the solution of Problem 259, we apply the three steps outlined in Art. 2–6. Imagine the rods disconnected from the block and hanging freely, as in Fig. 2–16. A temperature rise causes the temperature deformations δ_{T_s} and δ_{T_b}.

Since the bronze rod is to carry no load, the final level of the steel rods must coincide with the unstressed expanded length of the bronze. If the rods are to be at the same final level, the steel rods must go through a load deformation δ_P caused by the pulls P_s, each of which must be equal to one-half of the weight, or $\frac{1}{2}(5000)(9.81) = 24.53$ kN.

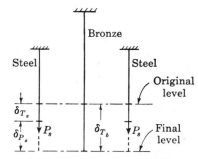

Figure 2–16.   Bronze rod supporting no load.

From Fig. 2–16, the geometric relation between the deformations is

$$\delta_{T_b} = \delta_{T_s} + \delta_{P_s}$$

which can be written as

$$(\alpha L \, \Delta T)_b = (\alpha L \, \Delta T)_s + \left(\frac{PL}{AE} \right)_s$$

whence

$$(18.9 \times 10^{-6})(1)(\Delta T) = (11.7 \times 10^{-6})(0.5)(\Delta T)$$
$$+ \frac{(24.53 \times 10^3)(0.5)}{(500 \times 10^{-6})(200 \times 10^9)}$$
$$\Delta T = 9.4°C \qquad Ans.$$

It is evident that a greater temperature rise will cause the bronze to push against the rigid block, thereby causing compression of the bronze. This confirms the result obtained in Problem 259, where the temperature rise was given to be 40°C.

PROBLEMS

261. A steel rod with a cross-sectional area of 150 mm² is stretched between two fixed points. The tensile load at 20°C is 5000 N. What will be the stress at −20°C? At what temperature will the stress be zero? Assume $\alpha = 11.7 \; \mu m/(m \cdot °C)$ and $E = 200 \times 10^9 \; N/m^2$.

Ans. $\sigma = 127 \; MN/m^2$; $T = 34.2°C$

262. A steel rod is stretched between two rigid walls and carries a tensile load of 5000 N at 20°C. If the allowable stress is not to exceed 130 MN/m² at −20°C, what is the minimum diameter of the rod? Assume $\alpha = 11.7 \; \mu m/(m \cdot °C)$ and $E = 200 \; GPa$.

263. Steel railroad rails 10 m long are laid with a clearance of 3 mm at a temperature of 15°C. At what temperature will the rails just touch? What stress would be induced in the rails at that temperature if there were no initial clearance? Assume $\alpha = 11.7$ μm/(m·°C) and $E = 200$ GPa.

264. At a temperature of 90°C, a steel tire 10 mm thick and 75 mm wide that is to be shrunk onto a locomotive driving wheel 1.8 m in diameter just fits over the wheel, which is at a temperature of 20°C. Determine the contact pressure between the tire and wheel after the assembly cools to 20°C. Neglect the deformation of the wheel caused by the pressure of the tire. Assume $\alpha = 11.7$ μm/(m·°C) and $E = 200 \times 10^9$ N/m².

265. At 130°C, a bronze hoop 20 mm thick whose inside diameter is 600 mm just fits snugly over a steel hoop 15 mm thick. Both hoops are 100 mm wide. Compute the contact pressure between the hoops when the temperature drops to 20°C. Neglect the possibility that the inner ring may buckle. For steel, $E = 200$ GPa and $\alpha = 11.7$ μm/(m·°C). For bronze, $E = 83$ GPa and $\alpha = 19$ μm/(m·°C).

Ans. $p = 2.86$ MN/m²

266. At 20°C, a rigid slab having a mass of 55 Mg is placed upon two bronze rods and one steel rod as shown in Fig. P–266. At what temperature will the stress in the steel rod be zero? For the steel rod, $A = 6000$ mm², $E = 200 \times 10^9$ N/m², and $\alpha = 11.7$ μm/(m·°C). For each bronze rod, $A = 6000$ mm², $E = 83 \times 10^9$ N/m², and $\alpha = 19.0$ μm/(m·°C).   *Ans. $T = 129$°C*

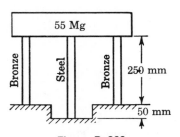

Figure P–266.

267. At 20°C, there is a gap $\Delta = 0.2$ mm between the lower end of the bronze bar and the rigid slab supported by two steel bars, as shown in Fig. P–267. Neglecting the mass of the slab, determine the stress in each rod when the temperature of the assembly is increased to 100°C. For the bronze rod, $A = 600$ mm², $E = 83 \times 10^9$ N/m², and

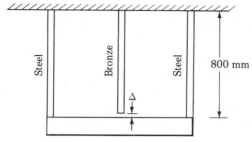

Figure P–267.

$\alpha = 18.9 \ \mu\mathrm{m}/(\mathrm{m} \cdot {}^\circ\mathrm{C})$. For each steel rod, $A = 400 \ \mathrm{mm}^2$, $E = 200 \times 10^9$ N/m^2, and $\alpha = 11.7 \ \mu\mathrm{m}/(\mathrm{m} \cdot {}^\circ\mathrm{C})$.

268. An aluminum cylinder and a bronze cylinder are centered and secured between two rigid slabs by tightening two steel bolts, as shown in Fig. P–268. At 10°C no axial load exists in the assembly. Find the stress in each material at 90°C. For the aluminum cylinder, $A = 1200 \ \mathrm{mm}^2$, $E = 70 \times 10^9 \ \mathrm{N}/\mathrm{m}^2$, and $\alpha = 23 \ \mu\mathrm{m}/(\mathrm{m} \cdot {}^\circ\mathrm{C})$. For the bronze cylinder, $A = 1800 \ \mathrm{mm}^2$, $E = 83 \times 10^9 \ \mathrm{N}/\mathrm{m}^2$, and $\alpha = 19.0$ $\mu\mathrm{m}/(\mathrm{m} \cdot {}^\circ\mathrm{C})$. For each steel bolt, $A = 500 \ \mathrm{mm}^2$, $E = 200 \times 10^9 \ \mathrm{N}/\mathrm{m}^2$, and $\alpha = 11.7 \ \mu\mathrm{m}/(\mathrm{m} \cdot {}^\circ\mathrm{C})$.   *Ans.* $\sigma_s = 33.7 \ \mathrm{MN}/\mathrm{m}^2$

269. Resolve Problem 268 assuming there is a 0.05 mm gap between the right end of the bronze cylinder and the rigid slab at 10°C.

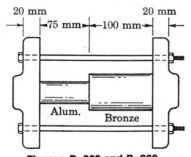

Figures P–268 and P–269.

270. A steel cylinder is enclosed in a bronze sleeve; both simultaneously support a vertical compressive load of 250 kN which is applied to the assembly through a horizontal bearing plate. The lengths of the cylinder and sleeve are equal. Compute (a) the temperature change that will cause a zero load in the steel, and (b) the temperature change that will cause a zero load in the bronze. For the steel cylinder,

$A = 7200$ mm^2, $E = 200$ GPa, and $\alpha = 11.7$ μm/(m·°C). For the bronze sleeve, $A = 12 \times 10^3$ mm^2, $E = 83$ GPa, and $\alpha = 19.0$ μm/(m·°C).

271. A bronze sleeve is slipped over a steel bolt and is held in place by a nut that is tightened "finger-tight." Compute the temperature change which will cause the stress in the bronze to be 20 MPa. For the steel bolt, $A = 450$ mm^2, $E = 200$ GPa, and $\alpha = 11.7$ μm/(m·°C). For the bronze sleeve, $A = 900$ mm^2, $E = 83$ GPa, and $\alpha = 19.0$ μm/(m·°C).

272. For the sleeve-bolt assembly described in Problem 271, assume the nut is tightened to produce an initial stress of 15×10^6 N/m^2 in the bronze sleeve. Find the stress in the bronze sleeve after a temperature rise of 70°C.   *Ans.* 38.2 MN/m^2

273. The composite bar shown in Fig. P–273 is firmly attached to unyielding supports. An axial load $P = 200$ kN is applied at 20°C. Find the stress in each material at 60°C. Assume $\alpha = 11.7$ μm/(m·°C) for steel and 23.0 μm/(m·°C) for aluminum.

Ans. $\sigma_a = 18.7$ MN/m^2; $\sigma_s = 181$ MN/m^2

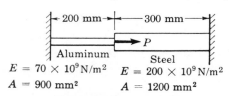

$E = 70 \times 10^9$ N/m^2 $E = 200 \times 10^9$ N/m^2
$A = 900$ mm^2 $A = 1200$ mm^2

Figures P–273 and P–274.

274. At what temperature will the aluminum and steel segments in Problem 273 have numerically equal stresses?

275. A rod is composed of the three segments shown in Fig. P–275. If the axial loads P_1 and P_2 are each zero, compute the stress induced in each material by a temperature drop of 30°C if (a) the walls are rigid and (b) the walls spring together by 0.300 mm. Assume $\alpha = 18.9$ μm/(m·°C) for bronze, 23.0 μm/(m·°C) for aluminum, and 11.7 μm/(m · °C) for steel.

Ans.   (a) $\sigma_s = 118$ MPa; (b) $\sigma_a = 40.0$ MPa

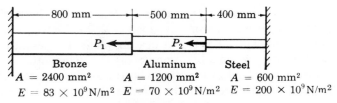

Bronze   Aluminum   Steel
$A = 2400$ mm^2 $A = 1200$ mm^2 $A = 600$ mm^2
$E = 83 \times 10^9$ N/m^2 $E = 70 \times 10^9$ N/m^2 $E = 200 \times 10^9$ N/m^2

Figures P–275 and P–276.

276. Solve Problem 275 if P_1 and P_2 each equal 50 kN and the walls yield 0.300 mm when the temperature drops 50°C.

277. The rigid bar AB is pinned at O and connected to two rods as shown in Fig. P–277. If the bar AB is horizontal at a given temperature, determine the ratio of the areas of the two rods so that the bar AB will be horizontal at any temperature. Neglect the mass of bar AB.   *Ans.* $A_s/A_a = 0.516$

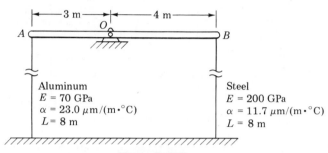

Figure P–277.

278. A rigid horizontal bar of negligible mass is connected to two rods as shown in Fig. P–278. If the system is initially stress-free, determine the temperature change that will cause a tensile stress of 60 MPa in the steel rod.

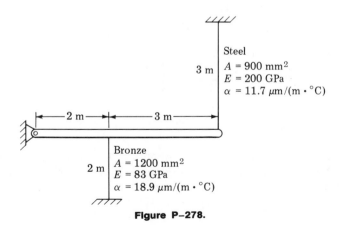

Figure P–278.

279. For the assembly shown in Fig. P–279, determine the stress in each of the two vertical rods if the temperature rises 40°C after the load $P = 50$ kN is applied. Neglect the deformation and mass of the horizontal bar AB.

> *Ans.* $\sigma_s = 134$ MPa (tension); $\sigma_a = 11.3$ MPa (compression)

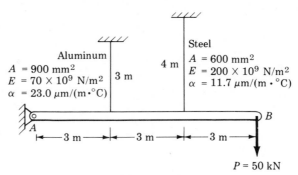

Aluminum
$A = 900$ mm²
$E = 70 \times 10^9$ N/m²
$\alpha = 23.0\ \mu m/(m \cdot °C)$

3 m

4 m

Steel
$A = 600$ mm²
$E = 200 \times 10^9$ N/m²
$\alpha = 11.7\ \mu m/(m \cdot °C)$

A

B

|← 3 m →|← 3 m →|← 3 m →|

$P = 50$ kN

Figure P–279.

280. The lower ends of the three steel rods shown in Fig. P–280 are at the same level before the force $P = 600$ kN is applied to the horizontal rigid slab. For each rod, $A = 2000$ mm², $\alpha = 11.7\ \mu m/(m \cdot °C)$, and $E = 200 \times 10^9$ N/m². Determine the relationship between the force in rod C and the change in temperature ΔT, measured in degrees Celsius. Neglect the mass of the rigid slab.

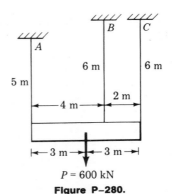

A

B C

5 m

6 m   6 m

4 m

2 m

|← 3 m →|← 3 m →|

$P = 600$ kN

Figure P–280.

281. Four steel bars jointly support a mass of 15 Mg as shown in Fig. P–281. Each bar has a cross-sectional area of 600 mm². Find the load carried by each bar after a temperature rise of 50°C. Assume $\alpha = 11.7\ \mu m/(m \cdot °C)$ and $E = 200 \times 10^9$ N/m².

> *Ans.* $P_A = P_D = 21.5$ kN; $P_B = P_C = 67.3$ kN

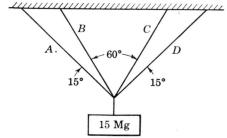

Figures P–281 and P–282.

282. Solve Problem 281 if bars A and D are steel and bars B and C are aluminum. For aluminum, $\alpha = 23.0 \ \mu m/(m \cdot °C)$ and $E = 70 \times 10^9 \ N/m^2$.

SUMMARY

Axial loads cause uniform stress distribution that is computed by

$$\sigma = \frac{P}{A} \tag{1-1}$$

and elongations determined from

$$\delta = \frac{PL}{AE} \tag{2-4}$$

Remember that Eq. (2–4) is valid only for homogeneous materials of constant cross section, axially loaded to stresses below the proportional limit.

Axial loads cause changes in lateral dimensions which are determined by Poisson's ratio. For cases in which such lateral deformations are restricted, Hooke's law for biaxial or triaxial stresses must be used.

Structures that are composed of two or more materials or are statically indeterminate are solved by applying the equations of static equilibrium in combination with additional equations obtained from the geometric relations between the elastic deformations.

Thermal stresses are computed by determining the relations between the thermal deformations

$$\delta_T = \alpha L(\Delta T) \tag{2-14}$$

and the elastic deformations which are used, in combination with the equations of static equilibrium, to solve the various problems that may be encountered.

3

Torsion

3-1 INTRODUCTION AND ASSUMPTIONS

In this chapter we shall consider the derivation and application of the twisting or torsion problem only in connection with circular shafts or closed thin-walled tubes. The twisting of noncircular shafts is so complex that we will only state the formulas that are used.

Torsion is our introduction to the problems of variable stress. Although the general theory of these problems is complex, its application consists of little more than substituting given values in the formulas soon to be derived, and is fairly simple.

The general technique used in all cases of nonuniform stress distribution is outlined in these steps:

1. From a study of the elastic deformations produced by a specified load, plus the application of Hooke's law, determine the relations between stresses that are compatible with the deformations. Such relations are known as the *equations of compatibility*.

2. By applying the conditions of equilibrium to a free-body diagram of a portion of the body, obtain additional relations between the stresses. These relations, resulting from a study of the equilibrium between externally applied loads and the internal resisting forces over an exploratory section, are called the *equations of equilibrium*.

3. Be sure that the solution of the equations in steps 1 and 2 is consistent with the loading conditions at the surface of the body. This is known as *satisfying the boundary conditions*.

In the theory of elasticity, it is shown that a solution satisfying these three steps is unique; that is, it is the only possible solution.

In deriving the torsion formulas, we make the following assumptions. These assumptions may be proved mathematically, and some may be demonstrated experimentally. The first two apply only to shafts of circular section.

   1.   Circular sections remain circular.

   2.   Plane sections remain plane and do not warp.

   3.   The projection upon a transverse section of straight radial lines in the section remains straight.

   4.   Shaft is loaded by twisting couples in planes that are perpendicular to the axis of the shaft.

   5.   Stresses do not exceed the proportional limit.

3–2   DERIVATION OF TORSION FORMULAS

Figure 3–1 shows two views of a solid circular shaft. If a torque T is applied at the ends of the shaft, a fiber AB on the outside surface, which is originally straight, will be twisted into a helix AC as the shaft is twisted through the angle θ. This helix is formed as follows:

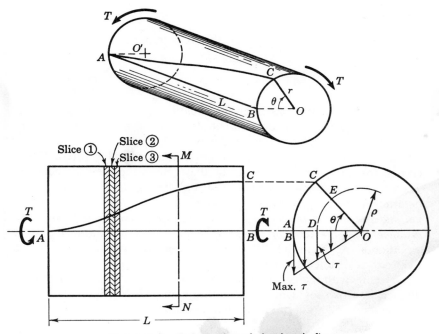

Figure 3–1.   Deformation of circular shaft.

Imagine the shaft to consist of innumerable thin slices, each of which is rigid and joined to adjacent slices by elastic fibers. Slice ② will rotate past slice ① until the elastic fibers joining them are deformed enough to create a resisting torque which balances the applied torque. When this happens, slices ① and ② will act as a rigid unit and transmit the torque to slice ③; this slice will rotate enough so that the elastic fibers joining it and slice ② develop a resisting torque equal to the applied torque. This type of deformation proceeds throughout the length L of the shaft. The helix AC is the line joining the original reference line AB on these slices as they become infinitely thin. This description of the twisting action is idealized, but the resulting helix is accurately described; actually, all such slices start rotating simultaneously relative to each other as soon as the torque is applied, the angle of rotation θ becoming larger as the applied torque is increased.

Consider now any internal fiber located a radial distance ρ from the axis of the shaft. From assumption 3 in Art. 3–1, the radius of such a fiber also rotates through the angle θ, causing a total shearing deformation δ_s equal to DE. The length of this deformation is the arc of a circle whose radius is ρ and which is subtended by the angle of θ radians; the length is given by

$$\delta_s = DE = \rho\theta \tag{a}$$

The unit deformation of this fiber is

$$\gamma = \frac{\delta_s}{L} = \frac{\rho\theta}{L} \tag{b}$$

The shearing stress at this typical fiber is determined from Hooke's law to be

$$\tau = G\gamma = \left(\frac{G\theta}{L}\right)\rho \tag{c}$$

Equation (c) may be called the equation of compatibility, since the stresses expressed by it are compatible with the elastic deformations. Note that each of the terms in the parentheses in this equation is a constant which does not depend upon the particular internal fiber chosen for analysis; the product of these terms represents a constant. Therefore we conclude that the shearing stress at any internal fiber is determined by the product of a constant and a variable radial distance; i.e., *the stress distribution along any radius varies linearly with the radial distance from the axis of the shaft.* Figure 3–1 illustrates the stress variation along the radius OB; the maximum stress occurs at the outside fiber and is denoted by max. τ.

In line with the general procedure outlined in Art. 3–1, the shaft is divided into two segments by a cutting plane $M–N$. Figure 3–2 shows the free-body diagram of the left-hand portion.

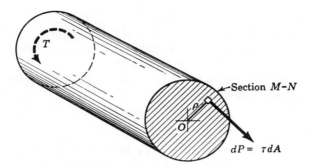

Figure 3–2. Free-body diagram of Fig. 3–1.

A differential area of section $M-N$ at a radial distance ρ from the axis of the shaft carries the differential resisting load $dP = \tau\, dA$. By considering an area infinitesimally small, we may assume the stress to be uniform over such an area. Since the function of this resisting load dP is to produce resistance to the applied torque T, the load must be directed perpendicular to the radius ρ in order to produce the maximum effect. It is true, but difficult to prove here, that in circular sections dP is directed perpendicular to ρ; nevertheless, we may take it as axiomatic that loads always distribute themselves as efficiently as possible. It is this variation of the principle of the conservation of energy that determines the direction of dP as perpendicular to ρ so that it produces maximum torsional resistance.

To satisfy the condition of static equilibrium, we apply $\Sigma M = 0$, or the fact that the applied torque T equals the resisting torque T_r. The resisting torque T_r is the sum of the resisting torques developed by all differential loads dP:

$$T = T_r = \int \rho\, dP = \int \rho(\tau\, dA)$$

Replacing τ by its value from Eq. (c) gives

$$T = \frac{G\theta}{L} \int \rho^2\, dA$$

or, since $\int \rho^2\, dA = J$, the polar moment of intertia of the cross section,

$$T = \frac{G\theta}{L} J$$

This is usually written,*

$$\theta = \frac{TL}{JG} \qquad\qquad (3\text{–}1)$$

---

*Note the similarity of Eq. (3–1) and the equation for linear deformation $\delta = PL/AE$. This similarity will make the two equations easier to remember.

In order for θ to be in the proper units of radians, T must be in
$N \cdot m$ and L in m; J of course is in m^4, and G is in N/m^2. If we wish to
express θ in degrees, we multiply the right-hand member of Eq. (3–1) by
the unit fraction, 180 deg/π rad = 57.3 deg/rad.

By replacing the product $G\theta/L$ in Eq. (c) by its equivalent value
T/J from Eq. (3–1), we obtain

$$\tau = \frac{T\rho}{J} \tag{3–2}$$

This is called the torsion formula. The formula that determines the
maximum shearing stress is a more common form of the torsion
formula. It is obtained by replacing ρ by the radius r of the shaft:

$$\textbf{Max. } \tau = \frac{Tr}{J} \tag{3–2a}$$

Note that since Hooke's law was used in deriving these equations,
the stresses must not exceed the shearing proportional limit[*]; also, these
formulas are applicable only to circular shafts, either solid or hollow.[†]

The values of polar moments of inertia for circular shafts are given
in Fig. 3–3. Using these values, we obtain the following modifications of
the torsion formula:

Solid shaft: $\textbf{Max. } \tau = \dfrac{2T}{\pi r^3} = \dfrac{16T}{\pi d^3}$ $\tag{3–2b}$

Hollow shaft: $\textbf{Max. } \tau = \dfrac{2TR}{\pi(R^4 - r^4)} = \dfrac{16TD}{\pi(D^4 - d^4)}$ $\tag{3–2c}$

In many practical applications, shafts are used to transmit power.
From dynamics, it is known that the power $\mathscr{P}$ transmitted by a constant
torque T rotating at a constant angular speed ω is given by

$$\mathscr{P} = T\omega$$

where ω is measured in radians per unit time. If the shaft is rotating
with a frequency of f revolutions per unit time, $\omega = 2\pi f$, and we have

$$\mathscr{P} = T2\pi f$$

Thus the torque can be expressed as

$$T = \frac{\mathscr{P}}{2\pi f} \tag{3–3}$$

---

[*]Equation (3–2a) is sometimes used to determine the shearing stress at
rupture. Although the proportional limit is exceeded, the fictitious shearing stress
so obtained is called the *torsional modulus of rupture*. It is used to compare the
ultimate strengths of specimens of various materials and diameters.

[†]A satisfactory formula for determining the maximum shearing stress in
rectangular shafts is

$$\tau = \frac{T}{ab^2}\left(3 + 1.8\frac{b}{a}\right)$$

where a is the long side and b the short side of the rectangular section.

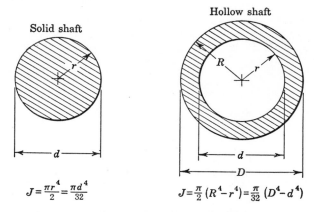

Figure 3-3.   Polar moments of inertia.

With $\mathcal{P}$ measured in watts $(1 \text{ W} = 1 \text{ N} \cdot \text{m/s})$ and f in revolutions per second (r/s), the above equation will determine the torque T in newton-meters. This value of T may be used in Eq. (3–2) to obtain the maximum shearing stress and in Eq. (3–1) to determine the angle of twist.

ILLUSTRATIVE PROBLEMS

301.   A solid shaft in a rolling mill transmits 20 kW at 2 r/s. Determine the diameter of the shaft if the shearing stress is not to exceed 40 MN/m² and the angle of twist is limited to 6° in a length of 3 m. Use $G = 83 \text{ GN/m}^2$.

Solution: This problem illustrates a design that must possess sufficient strength as well as rigidity. We start by applying Eq. (3–3) to determine the torque:

$$\left[T = \frac{\mathcal{P}}{2\pi f} \right] \qquad T = \frac{20 \times 10^3}{2\pi(2)} = 1590 \text{ N} \cdot \text{m}$$

To satisfy the condition of strength, we apply the torsion formula, Eq. (3–2b):

$$\left[\tau = \frac{16T}{\pi d^3} \right] \qquad 40 \times 10^6 = \frac{16(1590)}{\pi d^3}$$

from which

$$d^3 = 202 \times 10^{-6} \text{ m}^3 = 202 \times 10^3 \text{ mm}^3 \quad \text{and} \quad d = 58.7 \text{ mm}$$

We next apply the angle of twist relation, Eq. (3–1), to determine the diameter necessary to satisfy the requirement of rigidity. In degrees, this is

$$\theta = \frac{TL}{JG} \times 57.3 \quad \text{or} \quad J = \frac{TL}{\theta G} \times 57.3$$

whence

$$\frac{\pi d^4}{32} = \frac{1590(3)(57.3)}{(6)(83 \times 10^9)}$$

From this

$$d^4 = 5.59 \times 10^{-6} \text{ m}^4 = 5.59 \times 10^6 \text{ mm}^4 \quad \text{and} \quad d = 48.6 \text{ mm}$$

The larger diameter, $d = 58.7$ mm, will satisfy both strength and rigidity.

302. Two solid shafts of different materials are rigidly fastened together and attached to rigid supports as shown in Fig. 3–4. The aluminum segment is 75 mm in diameter, and $G_a = 28 \times 10^9 \text{ N/m}^2$. The steel segment has a diameter of 50 mm and $G_s = 83 \times 10^9 \text{ N/m}^2$. The torque, $T = 1000 \text{ N}\cdot\text{m}$, is applied at the junction of the two segments. Compute the maximum shearing stress developed in the assembly.

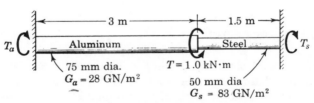

Figure 3–4. Statically indeterminate composite shaft.

Solution: This problem is statically indeterminate in that we do not know how the applied torque is apportioned to each segment. The procedure we follow is exactly the same as that discussed in Art. 2–5 for statically indeterminate axially loaded members. Applying the conditions of static equilibrium and of geometric compatibility, we obtain first

$$[\Sigma M = 0] \qquad T_s + T_a = T = 1000 \qquad (a)$$

Another relation between T_s and T_a is obtained from the condition that each segment has the same angular deformation, so that $\theta_s = \theta_a$. Applying Eq. (3–1) gives

$$\left[\left(\frac{TL}{JG}\right)_s = \left(\frac{TL}{JG}\right)_a\right]$$

$$\frac{T_s(1.5)}{\dfrac{\pi(0.050)^4}{32}(83 \times 10^9)} = \frac{T_a(3)}{\dfrac{\pi(0.075)^4}{32}(28 \times 10^9)}$$

from which

$$T_s = 1.17T_a \qquad\qquad (b)$$

Solving Eqs. (a) and (b), we obtain

$$T_a = 461 \text{ N·m} \quad \text{and} \quad T_s = 539 \text{ N·m}$$

Applying the torsion formula, we find the stresses to be

$$\left[\tau = \frac{16T}{\pi d^3}\right] \qquad \tau_a = \frac{16(461)}{\pi(0.075)^3} = 5.57 \times 10^6 \text{ N/m}^2$$

$$= 5.57 \text{ MN/m}^2$$

$$\tau_s = \frac{16(539)}{\pi(0.050)^3} = 22.0 \times 10^6 \text{ N/m}^2$$

$$= 22.0 \text{ MN/m}^2$$

303.   A steel shaft with a constant diameter of 50 mm is loaded as shown in Fig. 3–5 by torques applied to gears fastened to it. Using $G = 83 \times 10^3 \text{ MN/m}^2$, compute in degrees the relative angle of rotation between gears A and D.

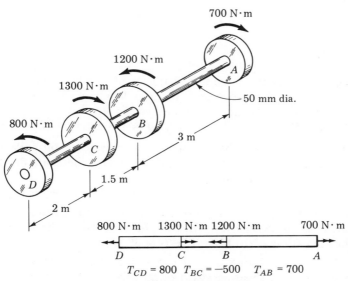

Figure 3–5.   Angular deformations.

Solution: The use of double-headed vectors to denote torques, as shown in the lower portion of Fig. 3–5, simplifies determining the torque transmitted by each segment of the shaft. All we need do is pass a section through each segment and apply the conditions of equilibrium to a free body between that section and one end, say, D. Then, relative to D, the torques transmitted by each segment are $T_{AB} = 700 \text{ N·m}$

clockwise, $T_{BC} = 500$ N·m counterclockwise, and $T_{CD} = 800$ N·m clockwise.

The total angular deformation is the algebraic sum of the deformation in each segment. Arbitrarily denoting clockwise deformation as positive, we apply Eq. (3–1), remembering to convert radians to degrees. Doing this, and recognizing the common value of JG, gives

$$\left[\theta_{A/D} = \sum \frac{TL}{JG}\right]$$

$$\theta_{A/D} = \frac{1}{JG}\sum TL \times 57.3$$

$$= \frac{57.3}{\frac{\pi(0.050)^4}{32}(83 \times 10^9)}\left[700(3) - 500(1.5) + 800(2)\right]$$

which gives

$$\theta_{A/D} = 3.32° \quad Ans.$$

The positive result indicates that the net rotation of A relative to D is clockwise.

PROBLEMS

304. What is the minimum diameter of a solid steel shaft that will not twist through more than 3° in a 6-m length when subjected to a torque of 14 kN·m? What maximum shearing stress is developed? Use $G = 83$ GN/m². *Ans.* $d = 118$ mm; $\tau = 43.4$ MN/m²

305. A solid steel shaft 5 m long is stressed to 60 MPa when twisted through 4°. Using $G = 83$ GPa, compute the shaft diameter. What power can be transmitted by the shaft at 20 r/s? *Ans.* $d = 104$ mm; $\mathscr{P} = 1.67$ MW

306. Determine the length of the shortest 2-mm-diameter bronze wire which can be twisted through two complete turns without exceeding a shearing stress of 70 MPa. Use $G = 35$ GPa.

307. A steel marine propeller is to transmit 4.5 MW at 3 r/s without exceeding a shearing stress of 50 MN/m² or twisting through more than 1° in a length of 25 diameters. Compute the proper diameter if $G = 83$ GN/m².

308. Show that a hollow circular shaft whose inner diameter is half the outer diameter has a torsional strength equal to $\frac{15}{16}$ of that of a solid shaft of the same outside diameter.

309. A steel shaft with a constant diameter of 60 mm is loaded by torques applied to gears attached to it as shown in Fig. P–309. Using $G = 83$ GN/m², determine the relative angle of twist of gear D relative to gear A.   *Ans.* $\theta_{D/A} = 2.17°$

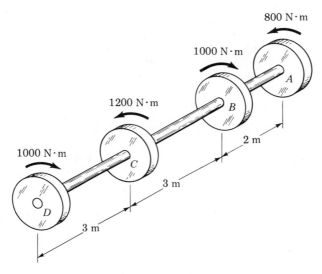

Figure P–309.

310. Determine the maximum torque that can be applied to a hollow circular steel shaft of 100-mm outside diameter and 70-mm inside diameter without exceeding a shearing stress of 60×10^6 N/m² or a twist of 0.5 deg/m. Use $G = 83 \times 10^9$ N/m².

311. A stepped steel shaft consists of a hollow shaft 2 m long, with an outside diameter of 100 mm and an inside diameter of 70 mm, rigidly attached to a solid shaft 1.5 m long, and 70 mm in diameter. Determine the maximum torque that can be applied without exceeding a shearing stress of 70 MN/m² or a twist of 2.5 deg in the 3.5-m length. Use $G = 83$ GN/m².   *Ans.* $T = 4.01$ kN·m

312. A flexible shaft consists of a 5-mm-diameter steel wire encased in a stationary tube that fits closely enough to impose a frictional torque of 2 N·m/m. Determine the maximum length of the shaft if the shearing stress is not to exceed 140 MPa. What will be the angular rotation of one end relative to the other end? Use $G = 83$ GPa.   *Ans.* $L = 1.72$ m; $\theta = 33.3°$

313.   The steel shaft shown in Fig. P–313 rotates at 3 r/s with 30 kW taken off at *A*, 15 kW removed at *B*, and 45 kW applied at *C*. Using $G = 83 \times 10^9 \text{ N/m}^2$, find the maximum shearing stress and the angle of rotation of gear *A* relative to gear *C*.

Ans.   Max. $\tau = 64.9 \text{ MN/m}^2$; $\theta = 8.23°$

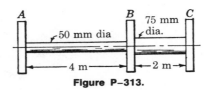

Figure P–313.

314.   A solid steel shaft is loaded as shown in Fig. P–314. Using $G = 83 \text{ GN/m}^2$, determine the required diameter of the shaft if the shearing stress is limited to 60 MN/m² and the angle of rotation at the free end is not to exceed 4 deg.

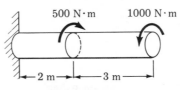

Figure P–314.

315.   A 5-m steel shaft rotating at 2 r/s has 70 kW applied at a gear that is 2 m from the left end where 20 kW are removed. At the right end, 30 kW are removed and another 20 kW leaves the shaft at 1.5 m from the right end. (a) Find the uniform shaft diameter so that the shearing stress will not exceed 60 MN/m². (b) If a uniform shaft diameter of 100 mm is specified, determine the angle by which one end of the shaft lags behind the other end. Use $G = 83 \text{ GN/m}^2$.

Ans. $d = 69.6 \text{ mm}$; $\theta = 0.448°$

316.   A round steel shaft 3 m long tapers uniformly from a 60-mm diameter at one end to a 30-mm diameter at the other end. Assuming that no significant discontinuity results from applying Eq. (3–1) over each infinitesimal length, compute the angular twist for the entire length when the shaft is transmitting a torque of 170 N · m. Use $G = 83 \times 10^3 \text{ MN/m}^2$.   *Ans.* $\theta = 1.29°$

317.   A hollow bronze shaft of 75 mm outer diameter and 50 mm inner diameter is slipped over a solid steel shaft 50 mm in diameter and of the same length as the hollow shaft. The two shafts are then fastened rigidly together at their ends. Determine the maximum shearing stress developed in each material by end torques of 3 kN·m. For bronze, $G = 35$ GN/m²; for steel, $G = 83$ GN/m².

Ans. $\tau_b = 28.5$ MN/m²; $\tau_s = 45.1$ MN/m²

318.   A solid compound shaft is made of three different materials and is subjected to two applied torques as shown in Fig. P–318. (a) Determine the maximum shearing stress developed in each material. (b) Find the angle of rotation of the free end of the shaft. Use $G_a = 28$ GN/m², $G_s = 83$ GN/m², and $G_b = 35$ GN/m².

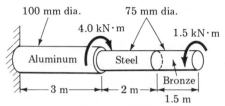

100 mm dia.   75 mm dia.

4.0 kN·m   1.5 kN·m

Aluminum   Steel

Bronze

3 m   2 m   1.5 m

Figure P–318.

319.   The compound shaft shown in Fig. P–319 is attached to rigid supports. For the bronze segment AB, the diameter is 75 mm, $\tau \leqslant 60$ MN/m², and $G = 35$ GN/m². For the steel segment BC, the diameter is 50 mm, $\tau \leqslant 80$ MN/m², and $G = 83$ GN/m². If $a = 2$ m and $b = 1.5$ m, compute the maximum torque T that can be applied.

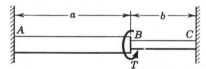

a   *b*

A   *B*   *C*

T

Figures P–319 and P–320.

320.   In Problem 319, determine the ratio of lengths b/a so that each material will be stressed to its permissible limit. What torque T is required?   *Ans.* $b/a = 1.19$; $T = 6.93$ kN·m

321.   A compound shaft consisting of an aluminum segment and a steel segment is acted upon by two torques as shown in Fig. P–321. Determine the maximum permissible value of T subject to the following conditions: $\tau_s \leqslant 100$ MPa, $\tau_a \leqslant 70$ MPa, and the angle of rotation of the free end limited to 12°. Use $G_s = 83$ GPa and $G_a = 28$ GPa.

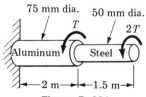

75 mm dia.   50 mm dia.

T $2T$

Aluminum   Steel

2 m   1.5 m

Figure P–321.

322. A torque T is applied, as shown in Fig. P–322, to a solid shaft with built-in ends. Prove that the resisting torques at the walls are $T_1 = Tb/L$ and $T_2 = Ta/L$. How would these values be changed if the shaft were hollow?

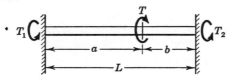

Figure P–322.

323. A shaft 100 mm in diameter and 3 m long, with built-in ends, is subjected to a clockwise torque of 4 kN·m applied 1 m from the left end, and to another clockwise torque of 16 kN·m applied 2 m from the left end. Compute the maximum shearing stress developed in each segment of the shaft. (*Hint*: Use the results of Problem 322 and apply superposition.)

324. A shaft composed of segments AC, CD, and DB is fastened to rigid supports and loaded as shown in Fig. P–324. For steel, $G = 83$ GN/m²; for aluminum, $G = 28$ GN/m²; and for bronze, $G = 35$ GN/m². Determine the maximum shearing stress developed in each segment.   *Ans.* $T_B = 472$ N·m; $\tau_{Al} = 9.3$ MN/m²

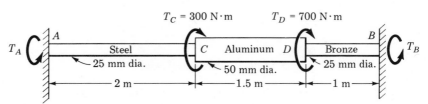

Figure P–324.

325. The two steel shafts shown in Fig. P–325, each with one end built into a rigid support, have flanges rigidly attached to their free ends. The shafts are to be bolted together at their flanges. However,

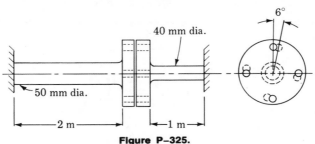

Figure P–325.

initially there is a 6° mismatch in the location of the bolt holes, as shown in the figure. Determine the maximum shearing stress in each shaft after the shafts are bolted together. Use $G = 83$ GN/m^2 and neglect deformations of the bolts and flanges.

3-3 FLANGED BOLT COUPLINGS

A commonly used connection between two shafts is a flanged bolt coupling. It consists of flanges rigidly attached to the ends of the shafts and bolted together, as in Fig. 3–6. The torque is transmitted by the shearing force P created in the bolts.

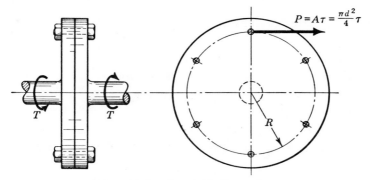

Figure 3–6. Flanged bolt coupling.

Assuming that the stress is uniformly distributed, the load in any bolt is given by the simple stress equation $P = A\tau$ and equals $(\pi d^2/4)\tau$. It acts through the center of the bolt and tangent to the bolt circle. The torque resistance of one bolt is PR, where R is the radius of the bolt circle. Therefore, for any number of bolts, n, the torque capacity of the coupling is expressed by

$$T = PRn = \frac{\pi d^2}{4}\tau Rn \tag{3-4}$$

Occasionally a coupling has two concentric rows of bolts, as in Fig. 3–7. Letting the subscript 1 refer to bolts on the outer circle and subscript 2 refer to bolts on the inner circle, the torque capacity of the coupling is

$$T = P_1 R_1 n_1 + P_2 R_2 n_2 \tag{3-5}$$

The relation between P_1 and P_2 can be determined from the fact that the comparatively rigid flanges cause shear deformations in the bolts which are proportional to their radial distances from the shaft axis.

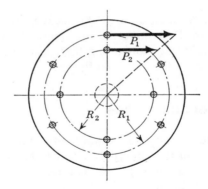

Figure 3–7. Coupling with two concentric bolt circles.

Thus, the shearing strains are related by

$$\frac{\gamma_1}{R_1} = \frac{\gamma_2}{R_2} \tag{a}$$

Using Hooke's law for shear, $G = \tau/\gamma$, we have

$$\frac{\tau_1}{G_1 R_1} = \frac{\tau_2}{G_2 R_2} \quad \text{or} \quad \frac{P_1/A_1}{G_1 R_1} = \frac{P_2/A_2}{G_2 R_2} \tag{b}$$

If the bolts on the two circles have the *same area*, $A_1 = A_2$, and if the bolts are made of the *same material*, $G_1 = G_2$, the relation between P_1 and P_2 reduces to

$$\frac{P_1}{R_1} = \frac{P_2}{R_2} \tag{3–6}$$

This is the case shown in Fig. 3–7. Using the relation between P_1 and P_2, Eq. (3–5) will determine the torque capacity of the coupling.

A similar procedure may be used for three or more concentric bolt circles. As we shall see in Chapter 12, this situation occurs in eccentrically loaded riveted connections.

PROBLEMS

326. A flanged bolt coupling consists of eight steel 20-mm-diameter bolts spaced evenly around a bolt circle 300 mm in diameter. Determine the torque capacity of the coupling if the allowable shearing stress in the bolts is 40 MN/m^2.   *Ans.* $T = 15.1$ kN $\cdot$ m

327. A flanged bolt coupling is used to connect a solid shaft 90 mm in diameter to a hollow shaft 100 mm in outside diameter and 90 mm in inside diameter. If the allowable shearing stress in the shafts and the bolts is 60 MN/m^2, how many 10-mm-diameter steel bolts must be used on a 200-mm-diameter bolt circle so that the coupling will be as strong as the weaker shaft?

328.   A flanged bolt coupling consists of six 10-mm-diameter steel bolts on a bolt circle 300 mm in diameter, and four 10-mm-diameter steel bolts on a concentric bolt circle 200 mm in diameter, as shown in Fig. 3–7. What torque can be applied without exceeding a shearing stress of 60 MPa in the bolts?   *Ans.* $T = 5.50\ \text{kN} \cdot \text{m}$

329.   Determine the number of 10-mm-diameter steel bolts that must be used on the 300-mm bolt circle of the coupling described in Problem 328 to increase the torque capacity to $8\ \text{kN} \cdot \text{m}$.   *Ans.*   10 bolts

330.   Solve Problem 328 if the diameter of the bolts used on the 200-mm bolt circle is changed to 20 mm.

331.   In a rivet group subjected to a twisting couple T, show that the torsion formula $\tau = T\rho/J$ can be used to find the shearing stress τ at the center of any rivet. Let $J = \Sigma A \rho^2$, where A is the area of a rivet at the radial distance ρ from the centroid of the rivet group.

332.   A plate is fastened to a fixed member by four 20-mm diameter rivets arranged as shown in Fig. P–332. Compute the maximum and minimum shearing stress developed. (*Hint:* Use the results of Problem 331.)

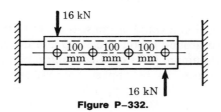

Figure P–332.

333.   Six 20-mm-diameter rivets fasten the plate in Fig. P–333 to the fixed member. Using the results of Problem 331, determine the average shearing stress caused in each rivet by the 40-kN loads. What

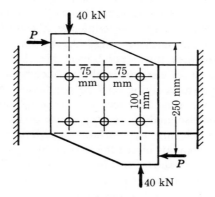

Figure P–333.

additional loads P can be applied before the average shearing stress in any rivet exceeds 60 MN/m²?

Ans.   Max. $\tau = 45.9$ MN/m²; $P = 55.4$ kN

334. The plate shown in Fig. P–334 is fastened to the fixed member by three 10-mm-diameter rivets. Compute the value of the loads P so that the average shearing stress in any rivet does not exceed 70 MPa. (*Hint*: Use the results of Problem 331.)

Ans. $P = 7.12$ kN

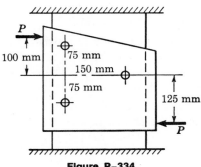

Figure P–334.

335. A flanged bolt coupling consists of six 10-mm-diameter steel bolts evenly spaced around a bolt circle 300 mm in diameter, and four 20-mm-diameter aluminum bolts on a concentric bolt circle 200 mm in diameter. What torque can be applied without exceeding a shearing stress of 60 MN/m² in the steel or 40 MN/m² in the aluminum? Use $G_s = 83$ GN/m² and $G_a = 28$ GN/m².   *Ans.* $T = 5.94$ kN · m

3–4 LONGITUDINAL SHEARING STRESS

So far in our discussion of torsional stress, we have considered only the shearing stress on transverse sections. However, a longitudinal shearing stress is also induced which is perpendicular and numerically equal to the transverse torsional shearing stress. As we shall see again in Art. 5–7, this fact illustrates the general principle that a shearing stress acting on one face of an element is always accompanied by a numerically equal shearing stress acting on a perpendicular face.

To demonstrate the existence of a longitudinal shearing stress, consider the element isolated by two transverse planes, two longitudinal planes through the axis, and two surfaces at different radii as shown in Fig. 3–8a. Taking moments about the axis gh of the enlarged free-body diagram of this element shown in Fig. 3–8b, we see that equilibrium is possible only if a longitudinal shearing stress τ' acts in addition to the torsional stress τ. Multiplying these stresses by the areas of the faces

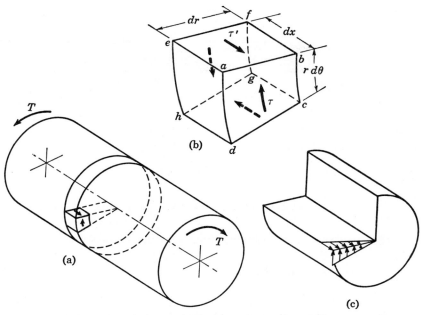

Figure 3–8. Equivalence of longitudinal and torsional shearing stress.

over which they act, we obtain

$$\left[\Sum M_{gh} = 0\right] \qquad (\tau\ dr\ r\ d\theta)\ dx - (\tau'\ dr\ dx)r\ d\theta = 0$$

Canceling out the common product $r\ d\theta\ dr\ dx$, we see that

$$\tau' = \tau$$

A pictorial view illustrating this equivalence of longitudinal and torsional shearing stresses is shown in Fig. 3–8c, where a portion of the shaft has been removed to clarify the concept.

3–5   TORSION OF THIN-WALLED TUBES; SHEAR FLOW

Although the torsion of noncircular shafts requires advanced methods, a simple approximate solution is possible for the special case of thin-walled tubes. In Fig. 3–9a we consider a tube of arbitrary shape with a variable wall thickness t which is relatively small compared with the dimensions of the cross section. Figure 3–9b shows, enlarged, a free body of a typical element of length ΔL cut from this tube. The torsional stress τ_1 across the thickness t_1 induces a numerically equal longitudinal stress, as was proved in the preceding article. Similarly, across the thickness t_2, a different torsional shearing stress τ_2 is accompanied by a numerically equal longitudinal stress.

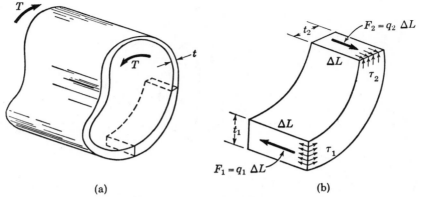

Figure 3-9. Shear flow in a thin-walled tube.

The resultants of these longitudinal shearing stresses are

$$F_1 = q_1 \Delta L \quad \text{and} \quad F_2 = q_2 \Delta L \tag{a}$$

where the symbol q represents $\int_{-t/2}^{t/2} \tau \, dt$. The term q is called the *shear flow* and is a convenient concept in cases where the precise distribution of shearing stress across a thickness is unknown or unimportant. Considering longitudinal equilibrium of the element, we have

$$q_1 \Delta L = q_2 \Delta L \quad \text{or} \quad q_1 = q_2 \tag{b}$$

The equivalence of the shear flow at two arbitrary positions means that the shear flow must be constant around the cross section of the tube. In fact, the name shear flow is based on the mathematical analogy between shear flow and the obviously constant flow of an incompressible fluid around a closed channel whose boundaries are the inner and outer walls of the tube.

To relate shear flow to the applied torque T, consider Fig. 3–10. Over any infinitesimal length dL, the tangential force is $q \, dL$, and its contribution toward resisting the torque is measured by its moment $r(q \, dL)$ about any convenient center O. Since the moment of the torsional couple T is independent of a moment center, on equating T to the summation of such contributions, we have

$$T = \int rq \, dL \tag{c}$$

Instead of carrying out this integration, we note that $r \, dL$ is twice the area of the shaded triangle whose base is dL and whose altitude is r. Consequently, since q is constant, the value of the integral is q times twice the area A enclosed by the center line of the tube wall, or

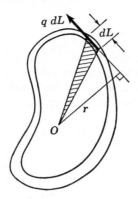

Figure 3–10.

$$T = 2Aq \qquad\qquad\qquad\qquad (3\text{–}7)$$

Finally, the average shearing stress across any thickness t is given by

$$\tau = \frac{q}{t} = \frac{T}{2At} \qquad\qquad\qquad\qquad (3\text{–}8)$$

ILLUSTRATIVE PROBLEM

336. A tube has the semicircular shape shown in Fig. 3–11. If stress concentration at the corners is neglected, what torque will cause a shearing stress of 40 MN/m²?

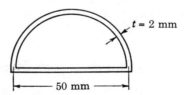

Figure 3–11.

Solution: In applying Eq. (3–8), observe that A is the area enclosed by the centerline of the tube. Thus we obtain

$$[T = 2At\tau] \qquad T = 2\left(\frac{\pi r^2}{2}\right)t\tau$$
$$= 2\left[\frac{\pi}{2}(0.025)^2\right](0.002)(40 \times 10^6)$$
$$= 157 \text{ N} \cdot \text{m} \qquad Ans.$$

PROBLEMS

337. A torque of $600\ \mathrm{N\cdot m}$ is applied to the rectangular section shown in Fig. P–337. Determine the wall thickness t so as not to exceed a shear stress of 60 MPa. What is the shear stress in the short sides? Neglect stress concentration at the corners.

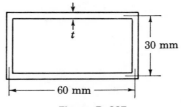

Figure P–337.

338. A tube 3 mm thick has the elliptical shape shown in Fig. P–338. What torque will cause a shearing stress of $60\ \mathrm{MN/m^2}$?

Ans. $T = 3.18\ \mathrm{kN\cdot m}$

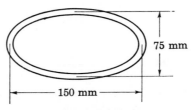

Figure· P–338.

339. A tube 3 mm thick has the shape shown in Fig. P–339. Find the shearing stress caused by a torque of $700\ \mathrm{N\cdot m}$ if dimension $a = 75$ mm.

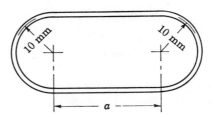

Figures P–339 and P–340.

340. Find dimension a in Problem 339 if a torque of $600\ \mathrm{N\cdot m}$ causes a shearing stress of $70\ \mathrm{MN/m^2}$.   *Ans.* $a = 55.7\ \mathrm{mm}$

341. Derive the torsion formula $\tau = T\rho/J$ for a solid circular section by assuming the section is composed of a series of concentric thin circular tubes. Assume that the shearing stress at any point is proportional to its radial distance.

3–6   HELICAL SPRINGS

The close-coiled helical spring in Fig. 3–12 is elongated by an axial load P. The spring is composed of a wire or round rod of diameter d wound into a helix of mean radius R. The helix angle is small, so that any coil of the spring may be considered as lying approximately in a plane perpendicular to the axis of the spring.

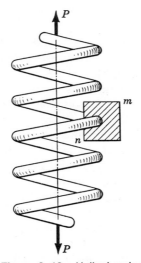

Figure 3–12. Helical spring.

To determine the stresses produced by P, we follow the general procedure of passing an exploratory cutting plane m–n through any typical section as shown and then determining the resisting forces required for equilibrium. We then analyze the stress distribution that creates these resisting forces.

Figure 3–13a shows the free-body diagram of the upper half of the spring. To balance the applied axial load P, the exposed shaded cross section of the spring must provide the resistance P_r equal to P. The free body is now in equilibrium as far as a vertical and horizontal summation of forces is concerned. To complete equilibrium, however, a moment summation must also equal zero. It is evident that P and P_r, being equal, opposite, and parallel, create a couple of magnitude PR which

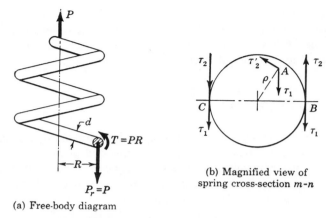

(a) Free-body diagram

(b) Magnified view of
spring cross-section m-n

Figure 3-13.   Analysis of helical spring.

can be balanced only by an opposite couple. This resisting couple is
created by a torsional shearing stress distributed over the cross section
of the spring; it is represented by $T = PR$.

The magnified view of the cross section in Fig. 3-13b shows the
stress distribution that created the resisting forces. Two types of shearing stress are produced: (1) direct shearing stresses like τ_1, uniformly
distributed over the spring section and creating the resisting load P_r that
passes through the centroid of the section; and (2) variable torsional
shearing stresses like τ_2 caused by the twisting couple $T = PR$. The
torsional stresses τ_2 vary in magnitude with their radial distance from
the centroid and are directed perpendicular to the radius, as at A. The
resultant shearing stress is the vector sum of the direct and torsional
shearing stresses. At B, the stresses are oppositely directed, and the
resultant stress is the difference between τ_2 and τ_1. At the inside fiber C,
however, the two stresses are collinear and in the same sense; their sum
produces the maximum stress in the section. The maximum stress
always occurs at the inside element of the spring wire. Is there any
position on the diameter BC at which the shearing stress is zero? If so,
how can you locate it?

To summarize the foregoing discussion, the maximum shearing
stress occurs at the inside element and is given by the sum of the direct
shearing stress $\tau_1 = P/A$ and the maximum value of the torsional
shearing stress $\tau_2 = Tr/J$, or

$$\tau = \tau_1 + \tau_2 = \frac{4P}{\pi d^2} + \frac{16(PR)}{\pi d^3}$$

This may be written

$$\tau = \frac{16PR}{\pi d^3}\left(1 + \frac{d}{4R}\right) \tag{3-9}$$

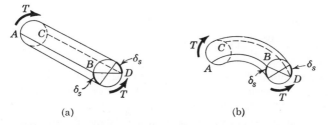

Figure 3-14.   Torsion of straight and of curved segments.

Examination of Eq. (3–9) shows that the ratio $d/4R$ is small for a spring composed of a wire of relatively small diameter wound on a spring with a large radius; this indicates that in such cases the maximum stress is caused primarily by torsion of the spring wire. On the other hand, heavy coil springs, such as those used on railroad cars, are made of wire with a relatively large diameter d in comparison with R, the mean radius of the spring; in these springs the effect of direct shearing stress is 14% or more of the total stress and cannot be disregarded.

It should be noted that the above discussion contains an error because the torsion formula derived for use with straight bars was applied to a curved bar. This error is of significance in heavy springs and is explained in Fig. 3–14. In the straight bar in Fig. 3–14a, torsion produces the same shearing deformation δ_s on fibers AB and CD. The shearing strain, $\gamma = \delta_s/L$, is the same at B and D, since the elements AB and CD have the same original length. A different situation, however, exists in the curved bar in Fig. 3–14b. Although fibers AB and CD undergo the same shearing deformation, the shearing strain at B on the inside element is greater than at D on the outside element because of the shorter initial length of AB. Therefore, since stress is proportional to strain, the shearing stress on the inner fibers of a curved bar is greater than on the outer fibers. This fact is not taken into account in Eq. (3–9). Of course, the importance of this error depends upon how greatly elements AB and CD differ in original length. Evidently this difference depends on how sharply curved the spring wire is, i.e., upon the ratio of d to R. A. M. Wahl has developed the following formula that takes account of the initial curvature of the spring wire*:

$$\text{Max. } \tau = \frac{16PR}{\pi d^3}\left(\frac{4m-1}{4m-4} + \frac{0.615}{m}\right) \qquad (3\text{–}10)$$

where $m = 2R/d = D/d$, the ratio of the mean diameter of the spring to the diameter of the spring wire. In light springs, where the ratio m is large, the first term in the parentheses approaches unity. Compare with

*See A. M. Wahl, Stresses in heavy closely coiled helical springs. *Trans. A.S.M.E.* **51**, paper No. APM-51-17.

Eq. (3–9), which may be rewritten in the following form:

$$\textbf{Max. } \tau = \frac{16PR}{\pi d^3}\left(1 + \frac{0.5}{m}\right) \tag{3–9a}$$

For heavy springs which are sharply curved and in which m is not so large, Eq. (3–10) emphasizes and corrects the error in Eq. (3–9).

Factors 0.5 and 0.615 differ in Eqs. (3–9a) and (3–10) largely because the direct shearing stress is not actually distributed uniformly over the cross section. We shall see later (Art. 5–7), when discussing horizontal shearing stress in beams, that for a circular cross section the maximum shearing stress produced is approximately $\frac{4}{3}$ times the average shearing stress and varies from 1.23 at the outside edges to 1.38 at the center. The factor 0.615 in Eq. (3–10) results from multiplying 0.5 by 1.23.

Note that springs are made of special steels and bronzes in which the allowable shearing stresses range from 200 to 800 MPa.

Spring deflection

Practically all the spring elongation, measured along its axis, is caused by torsional deformation of the spring wire. If we temporarily assume all the spring in Fig. 3–15 to be rigid except the small length dL, the end A will rotate to D through the small angle $d\theta$. Because $d\theta$ is

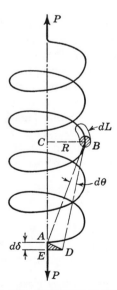

Figure 3–15. Deflection of helical spring.

small, the arc $AD = AB \cdot d\theta$ may be considered as a straight line perpendicular to AB, whence, from the similarity of triangles ADE and BAC, we obtain

$$\frac{AE}{AD} = \frac{BC}{AB}$$

or

$$\frac{d\delta}{AB \cdot d\theta} = \frac{R}{AB}$$

whence

$$d\delta = R \, d\theta \tag{a}$$

Applying Eq. (3–1), we may replace $d\theta$ by its equivalent value in terms of the torque and length:

$$d\delta = R \frac{(PR) \, dL}{JG} \tag{b}$$

which is integrated to give the total elongation contributed by all elements of the spring:

$$\delta = \frac{PR^2 L}{JG} \tag{c}$$

Replacing L by $2\pi Rn$, which is the length of n coils of radius R, and J by $\pi d^4/32$, we obtain

$$\delta = \frac{64 PR^3 n}{Gd^4} \tag{3–11}$$

This expression for spring deflection neglects the deformations caused by direct shear:

$$\delta' = \frac{PL}{A_s G} = \frac{P(2\pi Rn)}{\dfrac{\pi d^2}{4} G} = \frac{8PRn}{Gd^2} \tag{3–12}$$

This latter deformation, however, is generally negligible compared to the value of δ given by Eq. (3–11) and consequently is usually ignored. Equation (3–11) is also used to compute the deflection in compression springs provided the coils are not spaced so closely that they touch when the load is applied.

ILLUSTRATIVE PROBLEM

342. A load P is supported by two steel springs arranged in series as shown in Fig. 3–16. The upper spring has 20 turns of 20-mm-diameter wire on a mean diameter of 150 mm. The lower spring consists of 15 turns of 10-mm-diameter wire on a mean diameter of 130 mm.

Figure 3–16.

Determine the maximum shearing stress in each spring if the total deflection is 80 mm and $G = 83 \text{ GN}/\text{m}^2$.

Solution: The total deflection is the sum of the deflection in each spring. By applying Eq. (3–11), we find the load P to be

$$\left[\delta = \sum \frac{64PR^3n}{Gd^4} \right]$$

$$0.080 = \frac{64P}{83 \times 10^9} \left[\frac{(0.075)^3(20)}{(0.020)^4} + \frac{(0.065)^3(15)}{(0.010)^4} \right]$$

$$P = 233 \text{ N}$$

Knowing P, we can now find the stresses. For the upper spring, $m = 2R/d = 2(0.075)/0.020 = 7.5$; $4m = 30.0$. Applying Wahl's formula, Eq. (3–10), we obtain

$$\text{Max. } \tau = \frac{16PR}{\pi d^3} \left(\frac{4m - 1}{4m - 4} + \frac{0.615}{m} \right)$$

$$\text{Max. } \tau = \frac{16(223)(0.075)}{\pi(0.020)^3} \left(\frac{30 - 1}{30 - 4} + \frac{0.615}{7.5} \right)$$

$$= 12.7 \text{ MN}/\text{m}^2 \qquad Ans.$$

Similarly for the lower spring where $m = 2(0.065)/0.010 = 13$ and $4m = 52$, we find

$$\text{Max. } \tau = \frac{16(223)(0.065)}{\pi(0.010)^3} \left(\frac{52 - 1}{52 - 4} + \frac{0.615}{13} \right)$$

$$= 81.9 \text{ MN}/\text{m}^2 \qquad Ans.$$

If we had used Eq. (3–9) to compute these maximum shearing stresses, the results would have been $11.4 \text{ MN}/\text{m}^2$ in the upper spring and $76.7 \text{ MN}/\text{m}^2$ in the lower spring. Thus, the approximate formula gives results that are 10.2% and 6.35% lower than the more precise Wahl formula.

PROBLEMS

343. Determine the maximum shearing stress and elongation in a helical steel spring composed of 20 turns of 20-mm-diameter wire on a mean radius of 80 mm when the spring is supporting a load of 2 kN. Use Eq. (3–10) and $G = 83$ GN/m^2.

Ans. Max. $\tau = 121$ MN/m^2; $\delta = 98.7$ mm

344. What is the maximum elongation of the spring in Problem 343 if the spring is made of phosphor bronze for which $G = 42$ GN/m^2 and is stressed to 140 MN/m^2? Use Eq. (3–10).

345. A helical spring is made by wrapping steel wire 20 mm in diameter around a forming cylinder 150 mm in diameter. Compute the number of turns required to permit an elongation of 100 mm without exceeding a shearing stress of 140 MPa. Use Eq. (3–9) and $G = 83$ GPa.

Ans. $n = 17.9$ turns

346. Compute the maximum shearing stress developed in a phosphor bronze spring having a mean diameter of 200 mm and consisting of 24 turns of 20-mm-diameter wire when the spring is stretched 100 mm. Use Eq. (3–10) and $G = 42$ GN/m^2.

347. A clutch is activated by six helical springs symmetrically spaced. Each spring consists of 12 turns of steel wire 10 mm in diameter and has a mean diameter of 50 mm. Determine the load exerted against the clutch plate by a contraction of 40 mm in the springs. What is the maximum shearing stress in the springs? Use Eq. (3–9) and $G = 83$ GN/m^2.

348. Two steel springs arranged in series as shown in Fig. P–348 support a load P. The upper spring has 12 turns of 25-mm-diameter wire on a mean radius of 100 mm. The lower spring consists of 10 turns of 20-mm-diameter wire on a mean radius of 75 mm. If the maximum shearing stress in either spring must not exceed 200 MN/m^2, compute

Figure P–348.

the maximum value of P and the total elongation of the assembly. Use Eq. (3–10) and $G = 83 \text{ GN/m}^2$. Compute the equivalent spring constant by dividing the load by the total elongation.

349. A load P is supported by two concentric steel springs arranged as shown in Fig. P–349. The inner spring consists of 30 turns of 20-mm-diameter wire on a mean diameter of 150 mm; the outer spring has 20 turns of 30-mm wire on a mean diameter of 200 mm. Compute the maximum load that will not exceed a shearing stress of 140 MPa in either spring. Use Eq. (3–9) and $G = 83 \text{ GPa}$.

Ans. $P = 9.05 \text{ kN}$

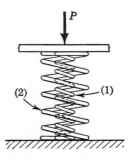

Figures P–349 and P–350.

350. If the inner spring in Problem 349 is made of phosphor bronze, with $G = 42 \text{ GN/m}^2$, compute the maximum shearing stress in each spring resulting from a load $P = 5 \text{ kN}$. Use Eq. (3–10).

351. A rigid plate of negligible mass rests on the central spring in Fig. P–351 which is 20 mm higher than the symmetrically located outer springs. Each of the outer springs consists of 18 turns of 10-mm wire on a mean diameter of 100 mm. The central spring has 24 turns of 20-mm wire on a mean diameter of 150 mm. If a load $P = 5 \text{ kN}$ is now applied to the plate, determine the maximum shearing stress in each spring. Use Eq. (3–9) and $G = 83 \text{ GN/m}^2$.

Ans.   Central spring: Max. $\tau = 170 \text{ MN/m}^2$

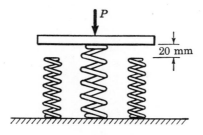

Figures P–351 and P–352.

352. Solve Problem 351 if the outer springs are made of phosphor bronze for which $G = 42 \text{ GN/m}^2$. Can you predict qualitatively the effect of this change upon the stresses?

353. A rigid bar, hinged at one end, is supported by two identical springs as shown in Fig. P–353. Each spring consists of 20 turns of 10-mm wire having a mean diameter of 150 mm. Compute the maximum shearing stress in the springs, using Eq. (3–9). Neglect the mass of the rigid bar.   *Ans.*   Max. $\tau = 46.5 \text{ MN/m}^2$

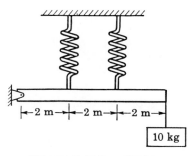

Figures P–353 and P–354.

354. If each spring in Problem 353 consists of 16 turns of 10-mm wire on a mean diameter of 160 mm, determine the largest mass that may be supported at the free end without exceeding a shearing stress of 140 MN/m^2 in either spring. Use Eq. (3–9).

355. As shown in Fig. P–355, a homogeneous 50-kg rigid block is suspended by three springs whose lower ends were originally at the same level. Each steel spring has 24 turns of 10-mm-diameter wire on a mean diameter of 100 mm, and $G = 83 \text{ GN/m}^2$. The bronze spring has 48 turns of 20-mm-diameter wire on a mean diameter of 150 mm, and $G = 42 \text{ GN/m}^2$. Compute the maximum shearing stress in each spring using Eq. (3–9).   *Ans.*   For bronze, Max. $\tau = 9.93 \text{ MN/m}^2$

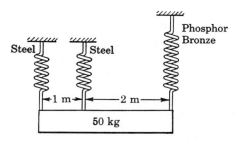

Figure P–355.

SUMMARY

The discussion of torsion in this chapter is limited to circular sections, solid or hollow. The shearing stress varies directly with the radial distance from the center of the cross section and is expressed by

$$\tau = \frac{T\rho}{J} \tag{3-2}$$

The maximum shearing stress in solid shafts of diameter d becomes

$$\tau = \frac{16T}{\pi d^3} \tag{3-2b}$$

In hollow shafts of external diameter D and internal diameter d, it is

$$\tau = \frac{16TD}{\pi(D^4 - d^4)} \tag{3-2c}$$

The angular deformation in a length L is expressed in radians by

$$\theta = \frac{TL}{JG} \tag{3-1}$$

which is converted to degrees by multiplying by $180/\pi = 57.3$. Equation (3–1) is useful not only in determining angular deformations, but also as a basis for solving statically indeterminate problems in torsion.

The relation between the torque T and power $\mathscr{P}$ transmitted by a shaft rotating at a frequency f is

$$T = \frac{\mathscr{P}}{2\pi f} \tag{3-3}$$

The study of flanged bolt couplings (Art. 3–3) is in effect the torsion formula reduced to a finite number of elements subjected to shear.

The existence of longitudinal shear stress (Art. 3–4) induced by torsional shear stress served to prove that the shear flow q is constant along the cross section of any thin-walled tube (Art. 3–5). In terms of the area A enclosed by the center line of the tube wall, its value is

$$q = \frac{T}{2A} \tag{3-7}$$

from which the average shearing stress across any thickness t is

$$\tau = \frac{q}{t} = \frac{T}{2At} \tag{3-8}$$

In close-coiled helical springs (Art. 3–6), the maximum shearing stress is expressed fairly accurately by

$$\tau = \frac{16PR}{\pi d^3}\left(1 + \frac{d}{4R}\right) \tag{3-9}$$

and more exactly by

$$\tau = \frac{16PR}{\pi d^3}\left(\frac{4m-1}{4m-4} + \frac{0.615}{m}\right) \tag{3-10}$$

where $m = 2R/d$.

The elongation of the spring generally neglects the effect of direct shearing deformation and is given by

$$\delta = \frac{64PR^3 n}{Gd^4} \tag{3-11}$$

Shear and Moment in Beams

4-1 INTRODUCTION

The basic problem in strength of materials is to determine the relations between the stresses and deformations caused by loads applied to any structure. In axial or torsional loadings, we had little trouble in applying the stress and deformation relations because in the majority of cases the loading either remains constant over the entire structure or is distributed in definite amounts to the component parts.

The study of bending loads, however, is complicated by the fact that the loading effects vary from section to section of the beam. These loading effects take the form of a shearing force and a bending moment, sometimes referred to as *shear* and *moment*. These terms will be defined in the next article. It will be shown in Chapter 5 that two kinds of stress act over the transverse section of a beam: (1) a bending stress, which varies directly with the bending moment, and (2) a shearing stress, which varies directly with the shear. As a preliminary to the study of

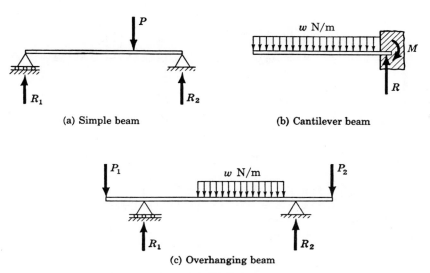

(a) Simple beam   (b) Cantilever beam

(c) Overhanging beam

Figure 4–1.   Statically determinate beams.

stresses in beams, therefore, this chapter is concerned with the variation in shear and bending moment in beams subjected to various combinations of loadings under different conditions of support, particularly the determination of the maximum values of shear and moment. Beam deflections will be discussed in Chapter 6.

Methods of supporting some types of beams are shown in Fig. 4–1. A simple beam is supported by a hinged reaction at one end and a roller support at the other, but is not otherwise restrained. A cantilever beam is supported at one end only, with a suitable restraint to prevent rotation of that end. An overhanging beam is supported by a hinge and a roller reaction, with either or both ends extending beyond the supports. These beams are all statically determinate; their reactions can be determined directly from the equations of static equilibrium.

Other methods of supporting beams are shown in Fig. 4–2. The propped beam, the fixed-ended or restrained beam, and the continuous beam all have at least one more reactive element than is absolutely necessary to support them. Such beams are statically indeterminate; the presence of excess supports requires the use of additional equations obtained from considering the elastic deformations of the beam. Their solution is discussed in Chapters 7 and 8.

A *concentrated load* is one that acts over so small a distance that it can be assumed to act at a point, as in Fig. 4–1a. In contrast, a *distributed load* acts over a considerable length of the beam. It may be distributed uniformly over the entire length, as in Fig. 4–1b, or over part of the length as in Fig. 4–1c. Distributed loads may also be uniformly varying or nonuniform. In a uniformly varying or triangular

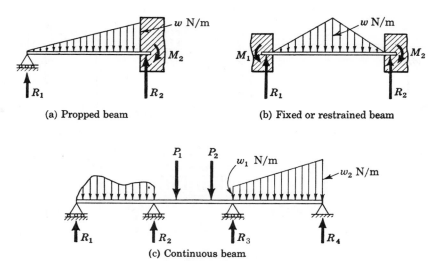

(a) Propped beam   (b) Fixed or restrained beam

(c) Continuous beam

Figure 4–2. Statically indeterminate beams.

load, the intensity of loading increases or decreases at a constant rate, as in Fig. 4–2a and 4–2b; this condition might result from water pressure acting on the face of a dam or from the dumping of a pile of sand. The trapezoidal loading in the right segment of Fig. 4–2c is a combination of a uniform and a uniformly varying load. The loading may also be nonuniform, as in the left segment of Fig. 4–2c; this may result from the haphazard piling of sand bags.

4–2 SHEAR AND MOMENT

Figure 4–3a shows a simple beam that carries a concentrated load P and is held in equilibrium by the reactions R_1 and R_2. For the time being, neglect the mass of the beam itself and consider only the effect of the load P. Assume that a cutting plane a–a at a distance x from R_1 divides the beam into two segments. The free-body diagram of the left segment in Fig. 4–3b shows that the externally applied load is R_1. To maintain equilibrium in this segment of the beam, the fibers in the exploratory section a–a must supply the resisting forces necessary to satisfy the conditions of static equilibrium. In this case, the external load is vertical, so the condition $\Sigma X = 0$ (the X axis is horizontal) is automatically satisfied.

To satisfy $\Sigma Y = 0$, the vertical unbalance caused by R_1 requires the fibers in section a–a to create a resisting force. This is shown as V_r, and is called the resisting shearing force. For the loading shown, V_r is numerically equal to R_1; but if additional loads had been applied

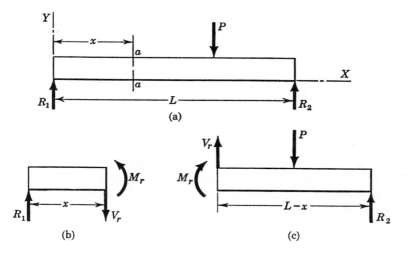

Figure 4–3. Equilibrium of segments to left and right of any exploratory section
a–a.

between R_1 and section *a–a* (as in Figs. 4–5 and 4–6), the net vertical
unbalance (which is equal but oppositely directed to the resisting
shearing force) would be found from the summation of their vertical
components. We define this net vertical unbalance as the shearing force
in the beam. It is denoted by V and may be determined from the
summation of the vertical components* of the external loads acting on
either side of the section. However, for the reason given in the footnote
on page 112, it is simpler to restrict this summation to the loads that act
on the segment to the *left* of the section. This definition of shearing
force (also called vertical shear or just shear) may be expressed mathematically as

$$V = (\Sigma Y)_L \qquad\qquad (4\text{–}1)$$

the subscript L emphasizing that the vertical summation includes only
the external loads acting on the beam segment to the left of the section
being considered.

The resisting shear V_r set up by the fibers in any section is always
equal but oppositely directed to the shearing force. V. In computing V,
upward acting forces or loads are considered as positive. This rule of
sign produces the effect shown in Fig. 4–4, in which a positive shearing
force tends to move the left segment upward with respect to the right,
and vice versa.

---

*The beam is assumed to be horizontal. With the beam in any other position,
the shearing force is computed from the summation of the components parallel to
the exploratory section.

Positive shear   Negative shear

Figure 4-4. Relative movements corresponding to signs of shearing force.

For complete equilibrium of the free-body diagram in Fig. 4–3b, the summation of moments must also balance. In this discussion, R_1 and V_r are equal, thereby producing a couple M that is equal to $R_1 x$ and is called the *bending moment* because it tends to bend the beam. The fibers in the exploratory section must create a numerically equal resisting moment, M_r, that acts as shown.* In most beams, the free-body diagram carries a number of loads, as shown in Fig. 4–5; hence a more complete definition of bending moment is necessary.

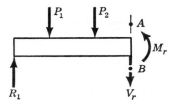

Figure 4-5.

Definition of bending moment

Bending moment is defined as the summation of moments about the centroidal axis of any selected section of all the loads acting either to the left or to the right side of the section, and is expressed mathematically as

$$M = (\Sigma M)_L = (\Sigma M)_R \tag{4-2}$$

the subscript L indicating that the bending moment is computed in terms of the loads acting to the left of the section, and the subscript R referring to loads to the right of the section.

Why the centroidal axis of the exploratory section must be chosen as the axis of bending moment may not be clear at this point; however, the reason is explained in Art. 5–2. Actually, in Fig. 4–5, where the

---

*Art. 4–3 shows that the bending moment, and hence the resisting moment, is always a couple.

loads are perpendicular to the beam, the axis of bending moment may be at point A, or B, or anywhere in the exploratory section, without changing the moment arms of the applied loads. But if the applied loads are inclined to the beam as shown in Fig. 4–6, the moment arms of the applied loads are unspecified unless the moment axis is at a definite location in the exploratory section. Such inclined loads cause combined axial and bending effects which are discussed in Art. 9–2.

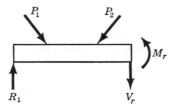

Figure 4–6.

Sign of bending moment

To many engineers, bending moment is positive if it produces bending of the beam concave upward, as in Fig. 4–7. We prefer to use an equivalent convention which states that *upward acting external forces cause positive bending moments with respect to any section; downward forces cause negative bending moments.* In so far as the left segment of a beam is concerned (Fig. 4–3b), this is equivalent to taking clockwise moments about the bending axis as positive, as indicated by the moment sense of R_1. With respect to the right segment of a beam (Fig. 4–3c), this convention means that the moment sense of the upward reaction R_2 is positive in a counterclockwise direction. This convention has the advantage of permitting bending moment to be computed, without any confusion in sign, in terms of the forces to either the left or the right of a section, depending on which requires the least arithmetical work. We never need think about whether a moment is clockwise or counterclockwise; upward acting forces always cause positive bending moments regardless of whether they act to the left or the right of the exploratory section.

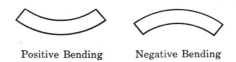

Positive Bending   Negative Bending

Figure 4–7. Curvatures corresponding to signs of bending moment.

The definition of shearing force and bending moment may be summarized mathematically as

$$V = (\Sigma Y)_L \qquad\qquad\qquad\qquad\qquad (4\text{--}1)$$

$$M = (\Sigma M)_L = (\Sigma M)_R \qquad\qquad\qquad (4\text{--}2)$$

in which positive effects are produced by upward forces and negative effects by downward forces. This rule of sign* will be used exclusively hereafter, and it will be further extended to give a positive sign to any quantity or expression in which such adjectives as "up" or "above" are used, and vice versa for negative signs. Remember that the subscripts L and R refer to the beam segment lying respectively to the left and right of the exploratory section.

ILLUSTRATIVE PROBLEM

401. Write shear and moment equations for the beam loaded as shown in Fig. 4–10a and sketch the shear and moment diagrams.

Solutions: Begin by computing the reactions. Applying $\Sigma M_{R_2} = 0$ gives $R_1 = 63$ kN, and $\Sigma M_{R_1} = 0$ yields $R_2 = 67$ kN. A check of these values is given by $\Sigma Y = 0$. The sections in the beam at which the loading conditions change are called *change of load points* and are designated by the letters A, B, C, and D.

If a section a–a is taken through the beam anywhere between A and B, the external loads on it appear as in Fig. 4–8. Applying the definitions of vertical shear and bending moment, and noting that they apply only to external loads, we obtain

$$[\,V = (\Sigma Y)_L\,] \qquad V_{AB} = (63 - 20x)\ \text{kN} \qquad\qquad (a)$$

$$[\,M = (\Sigma M)_L\,] \qquad M_{AB} = 63x - (20x)\frac{x}{2}$$

$$= (63x - 10x^2)\ \text{kN}\cdot\text{m} \qquad (b)$$

These equations are valid only for values of x between 0 and 5, that is, between points A and B. To obtain shear and moment equations between B and C, assume another exploratory section b–b taken anywhere between B and C. Note that the location of section b–b is still defined in terms of x as measured from the left end of the beam, although x now ranges between the limits of 5 and 10. The effects of the

---

*To avoid conflict with this rule, it is necessary to compute vertical shear in terms of the forces lying to the left of the exploratory section. If the forces to the *right* of the section were used, it would be necessary to take downward forces as positive so as to agree with the sign convention shown in Fig. 4–4.

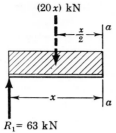

$(20x)$ kN

Figure 4-8.

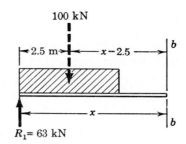

100 kN

2.5 m → ← $x - 2.5$

x

$R_1 = 63$ kN

Figure 4-9.

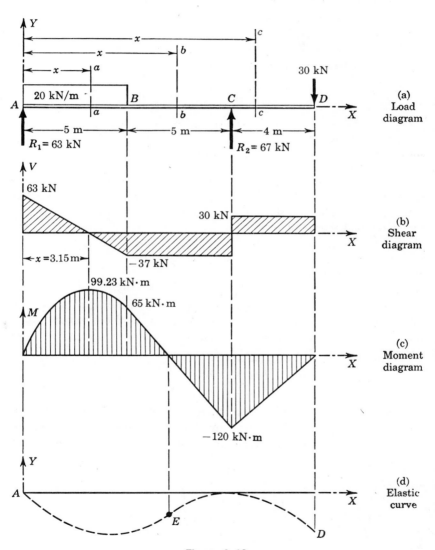

20 kN/m

A   B   C   D

—5 m—   —5 m—   —4 m—

$R_1 = 63$ kN $R_2 = 67$ kN

30 kN

(a)
Load
diagram

63 kN

$x = 3.15$ m

-37 kN

30 kN

(b)
Shear
diagram

99.23 kN·m

65 kN·m

-120 kN·m

(c)
Moment
diagram

A

E

D

(d)
Elastic
curve

Figure 4-10.

113

external forces on this section are determined by applying the definitions of shear and moment to Fig. 4–9.

$$[V = (\Sigma Y)_L] \qquad V_{BC} = 63 - 100 = -37 \text{ kN} \tag{c}$$

$$[M = (\Sigma M)_L] \qquad M_{BC} = 63x - 100(x - 2.5)$$
$$= (-37x + 250) \text{ kN} \cdot \text{m} \tag{d}$$

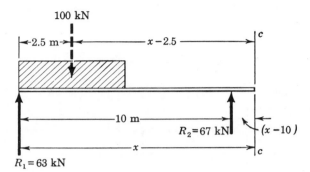

Figure 4–11.

The shear and moment equations for segment CD are obtained similarly by passing a section $c\text{–}c$ anywhere between C and D. The external loads acting on the beam to the left of this section are shown in Fig. 4–11, whence we obtain

$$[V = (\Sigma Y)_L] \qquad V_{CD} = 63 - 100 + 67 = +30 \text{ kN} \tag{e}$$
$$[M = (\Sigma M)_L] \qquad M_{CD} = 63x - 100(x - 2.5) + 67(x - 10)$$
$$= (30x - 420) \text{ kN} \cdot \text{m} \tag{f}$$

A simpler method of obtaining M_{CD} is to consider the forces lying to the right of section $c\text{–}c$ as shown in Fig. 4–12, from which noting that downward forces produce negative bending moment, we also obtain

$$[M = (\Sigma M)_R] \qquad M_{CD} = -30(14 - x)$$
$$= (30x - 420) \text{ kN} \cdot \text{m} \tag{f'}$$

Summarizing, we have computed V by considering only the external forces lying to the *left* of any exploratory section, whereas M may be computed by taking moments about the exploratory section caused

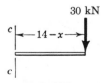

Figure 4–12.

by the external loads which lie *either* to the left or to the right of the section. We have been careful to assign plus signs to V and M caused by upward acting loads, and minus signs to V and M caused by downward acting loads. We shall be consistent in assigning a plus sign to any upward quantity and a minus sign to any quantity associated with the word "down" or its equivalent.

Note further that Figs. 4-8, 4-9, 4-11, and 4-12 have been used only for explanation; you will soon learn to visualize such diagrams directly from the original beam loading.

Shear and moment diagrams

Shear and moment diagrams are merely the graphical visualization of the shear and moment equations plotted on V-x and M-x axes, usually located below the loading diagram, as in parts (b) and (c) of Fig. 4-10.

The discontinuities in the shear diagram (Fig. 4-10b) are joined by vertical lines drawn up or down to represent the abrupt changes in shear caused respectively by upward or downward concentrated loads. Art. 4-4 (page 134) shows why this is correct. A final point to be observed at this time is that the highest and lowest points on the moment diagram (Fig. 4-10c) always correspond to sections of zero shear. This observation is also discussed in Art. 4-4, but it should be noted now that the value of x making M_{AB} maximum can be found by differentiating M_{AB} with respect to x and equating the result to zero. This result will be the shear equation V_{AB}. Thus we see that maximum moment corresponds to the section of zero shear.

Shear and moment at change of load points can be computed by substituting appropriate values of x in the foregoing V and M equations (a to f), but it is simpler and more direct to compute these numerical values by applying the fundamental definitions of V and M to specific sections. For example, the section of zero shear between A and B occurs because the downward force due to x meters of load applied at 20 kN/m must balance the vertical shear of 63 kN at A. Hence we have

$$63 = 20x \quad \text{or} \quad x = 3.15 \text{ m}$$

The moment at this section of zero shear is computed by taking moments of the forces to the left of the section. These forces consist of the upward reaction $R_1 = 63$ kN and the downward load of 63 kN caused by the length of the uniformly distributed load necessary to cause zero shear. From the definition of bending moment, we obtain

$$[M = (\Sigma M)_L]$$

at $x = 3.15$, $M = (63)(3.15) - 63\left(\dfrac{3.15}{2}\right) = 99.23 \text{ kN} \cdot \text{m}$

A final point of interest is brought out in Fig. 4–10d, which shows the shape taken by the beam under the given loading, assuming the beam to be quite flexible. The beam between A and E is concave up, and between E and D it is concave down. Since anything associated with *up* has a positive sign, it is not surprising that the moment diagram has positive values corresponding to the region AE, while for the portion ED, where the beam is concave down, the moment diagram has negative values. Sketching the shape of the beam therefore provides a check of the sign of bending moment.

At point E, where the beam changes its shape from concave up to concave down, we have what is called a *point of inflection*; it corresponds to the section of zero bending moment. Its position may be calculated by setting Eq. (*d*) equal to zero, which yields

$$\left[\, M_{BC} = 0 \,\right] \qquad -37x + 250 = 0; \quad x = 6.76 \text{ m}$$

ILLUSTRATIVE PROBLEM

402. Write the shear and moment equations for the cantilever beam carrying the uniformly varying load and concentrated load shown in Fig. 4–13. Also sketch the shear and moment diagrams.

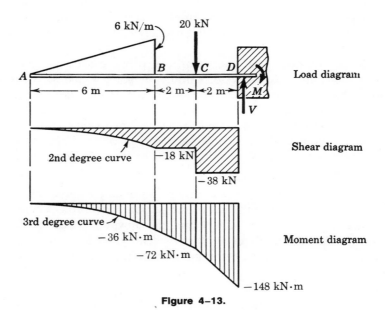

Figure 4–13.

Solution: Shear and moment calculations for a cantilever beam are simplified by drawing the load diagram with the restraining wall at the right end. Drawing diagrams similar to Fig. 4–8 or 4–9 where necessary, we find the shear and moment equations between each change of load position by applying Eqs. (4–1) and (4–2). For the region AB, in which x varies from 0 to 6, we have (see Fig. 4–14),

$$[V = (\Sigma Y)_L] \qquad V_{AB} = -\frac{x^2}{2} \text{ kN}$$

$$[M = (\Sigma M)_L] \qquad M_{AB} = -\frac{x^2}{2}\left(\frac{x}{3}\right) = -\frac{x^3}{6} \text{ kN·m}$$

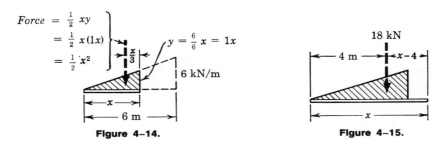

Figure 4–14.   Figure 4–15.

After passing B, the resultant force due to the triangular load (equal to its area) is constant at value of $\frac{1}{2}(6)(6) = 18$ kN acting through the centroid of the triangular load diagram at 4 m from A. For the region BC, in which x varies between 6 and 8, we therefore obtain (see Fig. 4–15)

$$[V = (\Sigma Y)_L] \qquad V_{BC} = -18 \text{ kN}$$
$$[M = (\Sigma M)_L] \qquad M_{BC} = -18(x - 4) = (-18x + 72) \text{ kN·m}$$

For a section between C and D (Fig. 4–16) in which x varies from 8 to 10, we obtain

$$[V = (\Sigma Y)_L] \qquad V_{CD} = -18 - 20 = -38 \text{ kN}$$
$$[M = (\Sigma M)_L] \qquad M_{CD} = -18(x - 4) - 20(x - 8)$$
$$= (-38x + 232) \text{ kN·m}$$

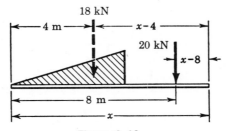

Figure 4–16.

The shear and moment equations are graphed as shown in the shear and moment diagrams in Fig. 4–13. At the wall D, the diagrams are returned to zero by means of the shear and moment reactions exerted by the wall. Observe that the maximum shear and the maximum bending moment always occur at the restrained end of a cantilever beam. An exception to this rule is possible if some of the loads are upward and the other loads downward. Note also that the moment equation M_{CD} is not necessary in computing the bending moment at D; this is found by a direct application of Eq. (4–2):

$$M = (\Sigma M)_L = -18(6) - 20(2) = -148 \text{ kN} \cdot \text{m}$$

PROBLEMS

Write shear and moment equations for the beams in the following problems. Also draw shear and moment diagrams, specifying values at all change of loading positions and at all points of zero shear. Neglect the mass of the beam in each problem.

403. Beam loaded as shown in Fig. P–403.

$$Ans. \quad V_{CD} = 20 \text{ kN}; \; M_{CD} = (20x - 140) \text{ kN} \cdot \text{m}$$

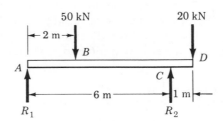

Figure P–403.

404. Beam loaded as shown in Fig. P–404.

$$Ans. \quad M_{CD} = (-4x + 28) \text{ kN} \cdot \text{m}$$

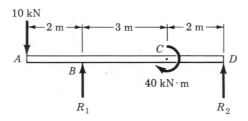

Figure P–404.

405.   Beam loaded as shown in Fig. P–405.

$$Ans. \quad M_{BC} = (-5x^2 + 44x + 60) \text{ kN} \cdot \text{m}$$

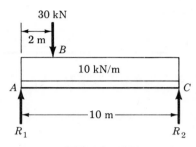

Figure P–405.

406.   Beam loaded as shown in Fig. P–406.

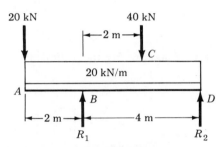

Figure P–406.

407.   Beam loaded as shown in Fig. P–407.

$$Ans. \quad \text{Max. } M = 57.6 \text{ kN} \cdot \text{m}$$

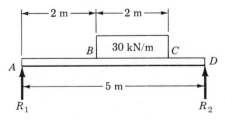

Figure P–407.

408. Beam loaded as shown in Fig. P–408.

Ans.   Max. $M = 83.33$ kN·m

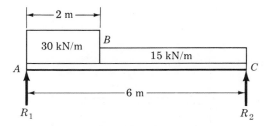

Figure P–408.

409. Cantilever beam loaded as shown in Fig. P–409.

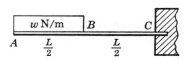

Figure P–409.

410. Cantilever beam carrying the uniformly varying load shown in Fig. P–410.

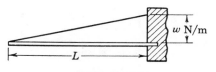

Figure P–410.

411. Cantilever beam carrying a distributed load varying from w N/m at the free end to zero at the wall, as shown in Fig. P–411.

Ans. $M = (wx^3/6L) - (wx^2/2)$

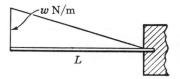

Figure P–411.

412.   Beam loaded as shown in Fig. P–412.

Ans.   Max. $M = 25$ kN·m

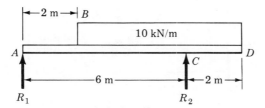

Figure P–412.

413.   Beam loaded as shown in Fig. P–413.

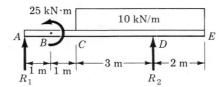

Figure P–413.

414.   Cantilever beam carrying the loads shown in Fig. P–414.

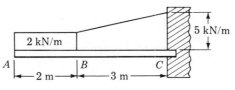

Figure P–414.

415.   Cantilever beam loaded as shown in Fig. P–415.

Ans. $M_{BC} = (-4x^2 + 20x - 40)$ kN·m

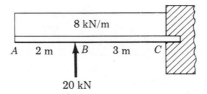

Figure P–415.

416. Beam carrying uniformly varying load shown in Fig. P–416.

Ans. Max. $M = wL^2/9\sqrt{3}$

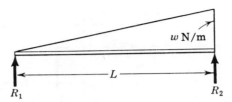

Figure P–416.

417. Beam carrying the triangular loading shown in Fig. P–417.

Ans. Max. $M = wL^2/12$

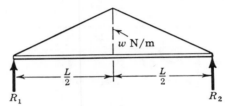

Figure P–417.

418. Cantilever beam loaded as shown in Fig. P–418.

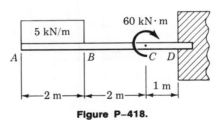

Figure P–418.

419. Beam loaded as shown in Fig. P–419.

Ans. Max. $M = 27.89$ kN·m

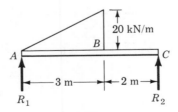

Figure P–419.

420.   A total distributed load of 60 kN supported by a uniformly distributed reaction as shown in Fig. P–420.

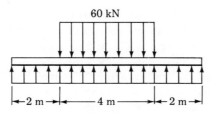

60 kN

|←2 m→|←   4 m   →|←2 m→|

Figure P–420.

421.   Write the shear and moment equations for the built-in circular bar shown in Fig. P–421 if (a) the load P is vertical as shown and (b) the load P is horizontal to the left.

   Ans.   (a) $V = -P \cos \theta$; $M = -PR \sin \theta$

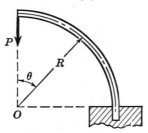

Figure P–421.

422.   Write the shear and moment equations for the semicircular arch shown in Fig. P–422 if (a) the load P is vertical as shown and (b) the load P is applied horizontally to the left at the top of the arch.

   Ans.   (a) $M_{AB} = \frac{1}{2} PR(1 - \cos \theta)$; $M_{BC} = \frac{1}{2} PR(1 + \cos \theta)$

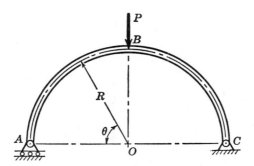

Figure P–422.

4–3   INTERPRETATION OF VERTICAL SHEAR
AND BENDING MOMENT

The beam in Fig. 4–17a carries a uniformly distributed load as well as concentrated loads. The external effects of each load acting to the left of section b–b are shown separately in parts (b), (c), and (d). In each of these figures, the effect of the applied load has been transferred to the exploratory section by adding a pair of equal but oppositely directed forces at that section which, as shown in the right side of the figures, reduce to a force at that section plus a couple. The moment of the couple is equal to the bending moment of the load. Hence, as is shown in the composite figure (e), the effect of the loads at one side of an exploratory section reduces to a system of forces whose vertical summation is the vertical shear and a system of couples whose algebraic summation is the bending moment.

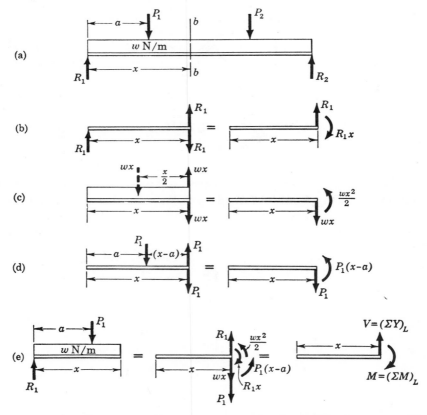

Figure 4–17. Shear and moment are resultant effects of loads acting to one side of exploratory section.

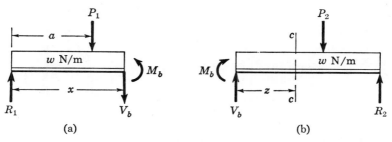

Figure 4–18. Free-body diagrams of segments of Fig. 4–17 in terms of shear and moment.

We may therefore conclude that the resultant effect of the forces at one side of an exploratory section reduces to a single force and a couple which are respectively the vertical shear and the bending moment at that section. Applying these forces to a free-body diagram of a beam segment produces equilibrium of that segment. Thus, in Fig. 4–18, the segments to the left and right of section $b–b$ in Fig. 4–17 are held in equilibrium by the shear and moment at section $b–b$. As far as the left segment is concerned, these equilibrants are resisting shear and resisting moment; but insofar as the right segment is concerned, they represent actual shear and bending moment. In other words, a beam may be cut at any section and the effect of the loads acting to one side of the section be replaced by the shear and moment at that section.

One application of this concept is that we can compute shear and moment at any section in terms of the shear and moment at another section. Thus in Fig. 4–18b, the moment at section $c–c$ will be

$$M_c = M_b + V_b z - \frac{wz^2}{2}$$

4-4   RELATIONS BETWEEN LOAD, SHEAR, AND MOMENT

In this article we shall discuss the relations existing between the loads, shears, and bending moments in any beam. These relations provide a method of constructing shear and moment diagrams without writing shear and moment equations. The relations are not independent of the basic definitions of shear and moment; instead they supplement them and are used in conjunction with them.

We begin by considering the beam in Fig. 4–19a, which is assumed to carry any general loading. The free-body diagram of a segment of this beam of length dx is shown magnified in Fig. 4–19b. As we saw in the preceding article, the effect of the loads to the left of this segment reduces to the shear V and the moment M, and the loads to the right of this segment produce the slightly different values of shear and

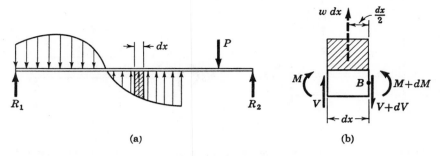

Figure 4–19.

moment $V + dV$ and $M + dM$. Although the loading is variable, it may be assumed constant at the intensity of w N/m over the small length dx, thereby producing the upward load $w\,dx$ which completes the free-body diagram.

Applying the conditions of static equilibrium to Fig. 4–19b, a summation of vertical forces yields

$$[\Sigma Y = 0] \qquad V + w\,dx - (V + dV) = 0$$

which reduces to

$$dV = w\,dx \tag{a}$$

From a moment summation about point B we have

$$[\Sigma M_B = 0] \qquad M + V\,dx + (w\,dx)\frac{dx}{2} - (M + dM) = 0$$

The third term in this equation is the square of a differential that is negligible in comparison with the other terms; hence the equation reduces to

$$dM = V\,dx \tag{b}$$

Integrating Eq. (a), we obtain

$$\int_{V_1}^{V_2} dV = \int_{x_1}^{x_2} w\,dx$$

in which the limits are the shear V_1 at position x_1 and the shear V_2 at position x_2. The left-hand term is easily integrable; it reduces to $V_2 - V_1$ and represents the change in shearing force between sections x_2 and x_1. We denote this change in shear as ΔV. In the right-hand term the product $w\,dx$ represents the area of an element of the load diagram like that shown shaded in Fig. 4–19a. Hence the definite integral $\int_{x_1}^{x_2} w\,dx$, which means the summation of such terms, represents the area under the load diagram between positions x_1 and x_2. Therefore the integration of Eq. (a) yields

$$V_2 - V_1 = \Delta V = (\text{Area})_{\text{load}} \tag{4–3}$$

Similarly the integration of Eq. (*b*) gives

$$\int_{M_1}^{M_2} dM = \int_{x_1}^{x_2} V \, dx$$

This reduces to

$$M_2 - M_1 = \Delta M = (\text{Area})_{\text{shear}} \qquad \qquad \textbf{(4-4)}$$

inasmuch as the product $V \, dx$ in the right-hand integral represents the area of an element under the shear diagram. Therefore the integral itself is equivalent to the area under the shear diagram between positions x_1 and x_2. Expressed in words, Eq. (4–4) shows that the change in bending moment ΔM between any two sections is equal to the area of the shear diagram for this interval.

Positive shearing forces are plotted upward from the X axis; hence positive shear areas are those which lie above the X axis and represent increases in the bending moment. The load diagram, however, is usually drawn with the loads on top of the beam because this is their natural position; as a consequence, the area of such downward acting loads is considered negative and represents *decreases* in the shearing force.

Equations (4–3) and (4–4) provide a convenient means of computing the changes in shear and moment and also the numerical values of shear and moment at any section, as will be demonstrated in the illustrative examples below. Of almost equal importance are the following variations of Eqs. (*a*) and (*b*), which enable us to sketch the shapes of the shear and moment diagrams:

$$w = \frac{dV}{dx} = \textbf{slope of shear diagram} \qquad \qquad \textbf{(4-5)}$$

$$V = \frac{dM}{dx} = \textbf{slope of moment diagram} \qquad \qquad \textbf{(4-6)}$$

As an application of these principles, consider the simply supported beam carrying the variable loading shown in Fig. 4–20a. Since positive slopes are directed up to the right and negative slopes are directed down to the right, i.e.,

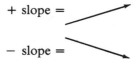

+ slope =

− slope =

we observe from Eq. (4–5) that the shear diagram in Fig. 4–20b must slope continuously down to the right. The inclination varies directly with the corresponding ordinate of the load diagram, being steepest where the load ordinate is maximum, and horizontal (or of zero slope) at the ends where the intensity of loading is zero.

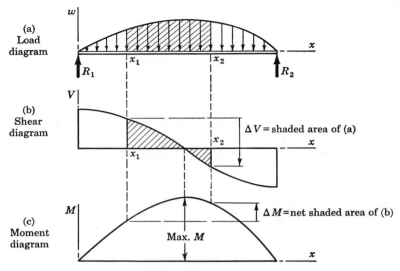

(a)
Load
diagram

(b)
Shear
diagram

(c)
Moment
diagram

Figure 4–20.   Relations between load, shear, and moment diagrams.

Similarly, by means of Eq. (4–6), the slope and shape of the moment diagram in Fig. 4–20c are determined by the corresponding ordinates of the shear diagram, which, being positive but decreasing in magnitude for the left portion, determine that the moment diagram slopes continuously up to the right with decreasing inclination. The slope becomes zero where the shear ordinate is zero. Note also that after the shear ordinates change sign and become increasingly larger negatively, the moment diagram slopes correspondingly more steeply down to the right. These conditions establish a maximum moment at the ordinate of zero shear.

The changes in shear (ΔV) and moment (ΔM) defined by Eqs. (4–3) and (4–4) are indicated in Figs. 4–20b and 4–20c. The shaded negative area of the load diagram determines ΔV to be negative or directed downward; in the shear diagram, the excess of positive over negative area in the region between x_1 and x_2 determines the positive change in moment ΔM, which is therefore directed upward.

A summary of the principles presented here and in Art. 4–2 suggests the following procedure for the construction of shear and moment diagrams:

1.   Compute the reactions.

2.   Compute values of shear at the change of load points, using either $V = (\Sigma Y)_L$ or $\Delta V = (\text{area})_{\text{load}}$.

3.   Sketch the shear diagram, determining the shape from Eq. (4–5); i.e., the intensity of the load ordinate equals the slope at the corresponding ordinate of the shear diagram.

4. Locate the points of zero shear.

5. Compute values of bending moment at the change of load points and at the points of zero shear, using either $M = (\Sigma M)_L = (\Sigma M)_R$ or $\Delta M = (\text{area})_{\text{shear}}$, whichever is more convenient.

6. Sketch the moment diagram through the ordinates of the bending moments computed in step 5. The shape of the diagram is determined from Eq. (4–6); i.e., the intensity of the shear ordinate equals the slope at the corresponding ordinate of the moment diagram.

ILLUSTRATIVE PROBLEMS

423. Using the semigraphical method described in this article, sketch shear and moment diagrams for the beam shown in Fig. 4–21, computing the values at all change of loading points and the maximum shear and maximum moment.

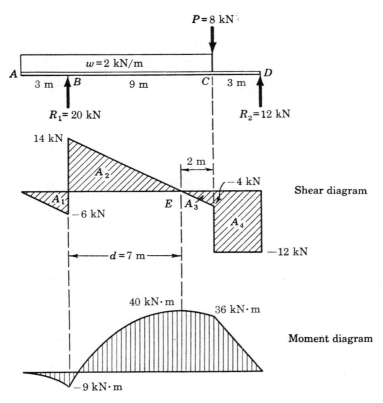

Figure 4–21. Load, shear, and moment diagrams.

Solution: The reactions are determined by equating to zero a moment summation about R_2 and then about R_1; this yields respectively $R_1 = 20$ kN and $R_2 = 12$ kN.

We next determine the values of vertical shear at the change of load positions. At A, the shear is zero. At the left of B, applying $V = (\Sigma Y)_L$ gives the shear as -6 kN caused by the downward resultant of the distributed load of 2 kN/m applied for 3 m. The same result may be obtained with Eq. (4–3), which indicates that the change in shear between A and B caused by the downward or negative uniformly distributed load is equal to the area of the load diagram in this interval, that is, $\Delta V = -2 \times 3 = -6$ kN. Hence the shear ordinate to the left of B has decreased 6 kN from the zero shear ordinate at A, to yield a net value of -6 kN. The concentrated load reaction at B causes the shear at B to increase abruptly by 20 kN to a net positive shear ordinate of 14 kN at the right of B.

Between B and C, the area of the load diagram is $-2 \times 9 = -18$, which by Eq. (4–3) represents the change in shear between B and C. The net shear ordinate at the left of C is therefore $V_C = V_B + \Delta V = 14 - 18 = -4$ kN. At C, the concentrated load of 8 kN changes the shear ordinate to -12 kN at the right of C. The shear ordinate stays constant at this value between C and D, since there is no load in this interval; at D, the upward reaction of 12 kN reduces the shear ordinate to zero.

The shape of the diagram connecting these shear ordinates is determined from Eq. (4–5), which shows that the slope is equal to the corresponding ordinates of the load diagram. Thus between A and B, the load intensity is constant and downward (or negative); hence the slope of the shear diagram in this interval is constant and down to the right. Similarly, between B and C, the load intensity is constant and negative; therefore the slope of the shear diagram here also is constant and down to the right. The slopes in the intervals AB and BC are parallel because they are each equal to the same load intensity. Finally between C and D, the intensity of loading is zero and the corresponding slope of the shear diagram is zero (a horizontal line).

We may conclude therefore that the shear diagram consists of straight horizontal lines for intervals in which the load intensity is zero, and of straight inclined lines for intervals of uniform load intensity.

The shear diagram passes through zero at B, where $x = 3$ m, and also at E. The position of E is determined from the fact that the shear at the right of B is 14 kN, which is reduced to zero in the interval BE at the rate of 2 kN/m. Hence $BE = d = 14/2 = 7$ m.

As a preliminary to computing the bending moments, we determine the areas of the shear diagram marked A_1, A_2, A_3, and A_4.

$$A_1 = \tfrac{1}{2}(3)(-6) = -9 \text{ kN·m}$$

$$A_2 = \tfrac{1}{2}(7)(14) = +49 \text{ kN} \cdot \text{m}$$
$$A_3 = \tfrac{1}{2}(2)(-4) = -4 \text{ kN} \cdot \text{m}$$
$$A_4 = 3(-12) = -36 \text{ kN} \cdot \text{m}$$

According to Eq. (4–4), the change in bending moment between any two sections equals the corresponding area of the shear diagram; hence, since the bending moment is zero at A (there are no loads to the left of A to cause a bending moment), the bending moment at B is given by A_1, or $M_B = -9$ kN·m.

Similarly, the bending moment at E is

$$M_E = M_B + \Delta M = A_1 + A_2 = -9 + 49 = +40 \text{ kN} \cdot \text{m}$$

The bending moment at C can also be computed as the sum of the areas A_1, A_2, and A_3, giving $M_C = 36$ kN·m; but small errors arising from neglecting sufficient significant figures in computing these areas may cause a cumulative error. Hence at sections near the right end of the beam, it is usually preferable to use the shear area to the right of such sections, or to apply the basic definition $M = (\Sigma M)_R$; whence in terms of the loads acting to the right of C we find $M_C = 12 \times 3 = 36$ kN·m. The correlation between this result and the area A_4 is evident if we observe that A_4 represents the amount by which the moment at C changes to become zero at D; that is,

$$M_D = M_C + \Delta M = M_C + A_4$$

or

$$0 = M_C - 36 \quad \text{and} \quad M_C = 36 \text{ kN} \cdot \text{m}$$

Whenever the change in bending moment between the ends of a beam is zero, as in this problem, the net area under the shear diagram is also zero; in other words, there must be as much positive as there is negative shear area. This provides a useful check on the accuracy of all intermediate values of bending moments computed from the area of a shear diagram.

After these bending moments are plotted as ordinates, the shape of the moment diagram connecting them is determined from Eq. (4–6); that is, the intensity of the shear ordinate equals the slope at the corresponding ordinate of the moment diagram. Thus as the shear ordinates between A and B change linearly from zero to -6 kN, the slopes at corresponding ordinates of the moment diagram change from horizontal or zero slope at A to increasingly steeper negative slopes as we pass from A to B, thereby producing a second-degree curve concave downward, as shown. In other words, the tangents to the moment curve have increasingly steeper slopes directed downward to the right.

The sudden change in shear at B from -6 to 14 kN causes the moment curve there to slope abruptly upward to the right; it becomes less steep, and eventually horizontal, as we move from B to E, because the corresponding positive shear ordinates decrease uniformly to zero. From E to C, however, the slope is again increasingly steeper and directed downward to the right as the corresponding shear ordinates change their sign to negative.

At C, the shear ordinate changes abruptly from -4 to -12 kN, at which it remains until D. The slope of the moment diagram is correspondingly abruptly steeper and constant, resulting in the straight line.

From the above discussion, it is apparent that an abrupt change in shear creates an abrupt change in the slope of the moment curve as at C, and that if the shear changes sign abruptly, there is a cusp in the moment diagram as at B. Also, where the shear diagram is constant because only concentrated loads are involved, as in the region CD, the moment diagram consists of straight lines, whereas for an interval in which the shear diagram varies uniformly because of a uniformly distributed load, the moment diagram is a parabolic arc with a vertical axis of symmetry at the section of zero shear, as at E.

424.   Without writing shear or moment equations, sketch shear and moment diagrams for the beam in Fig. 4–22, and compute values at all change of loading points and the values of maximum shear and maximum moment.

Solution: We begin by computing the reactions. Replacing the uniformly distributed reaction between C and D by its resultant R_2 and equating moments about R_1 to zero, we obtain $R_2 = 42$ kN. We then set moments about R_2 equal to zero, whence $R_1 = 24$ kN. Dividing R_2 by the length of 4 m, over which it is assumed to be uniformly distributed, gives the upward intensity of this reaction as 10.5 kN/m.

The shear diagram starts with an abrupt change in shear of 24 kN caused by the reaction R_1. Applying $\Delta V = \text{(area)}_{\text{load}}$ between A and B, the change in shear is equal to the area of the triangular load diagram, $\frac{1}{2} \times 9 \times (-12) = -54$ kN; this reduces the vertical shear at B to -30 kN. The shape of the shear diagram from A to B is determined from Eq. (4–5), which shows that the shape must vary from zero slope to increasingly steeper negative slopes corresponding to the increasingly greater downward intensity of the loading.

Between B and C, the intensity of loading is zero; hence, from Eq. (4–5), the slope of the corresponding portion of the shear diagram is zero, that is, horizontal.

From C to D, the loading is at a constant upward rate of 10.5 kN/m, creating a positive change of shear of 42 kN equal to the area of the load diagram in this interval, and a constant upward slope of the shear diagram as shown. This shear change of $+42$ kN added to the shear of -30 kN at C produces the shear of $+12$ kN at D. From D to

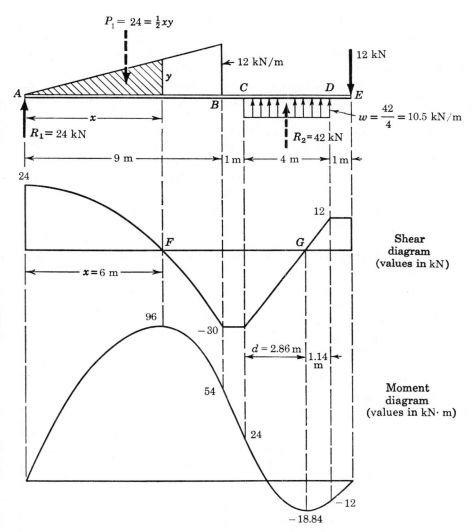

Figure 4–22. Load, shear, and moment diagrams.

E, the loading is zero, which means that the slope is horizontal. The concentrated load of 12 kN at E reduces the shear abruptly to zero.

Before we locate the positions of zero shear at F and G on the shear diagram, consider the effect of narrowing the distance over which the reaction R_2 is distributed. If points C and D are moved an equal amount toward each other, this narrowing will not change the magnitude or position of R_2. However, a reduction in the distance of CD will increase the intensity at which R_2 is distributed and cause a corresponding increase in the slope of the shear diagram in the narrowed interval

CD. For the extreme case in which the reaction is distributed over an infinitesimal width—that is, becomes a concentrated force—the intensity of loading is infinitely upward, and the corresponding slope of the shear diagram is vertically upward, as is the case at the reaction R_1. Similarly for a concentrated downward load at *E*, the intensity of loading is infinitely downward, and the corresponding slope of the shear diagram is vertically downward. *This explains why a concentrated load causes an abrupt change in the vertical shear.*

Let us now locate the sections of zero shear. The vertical shear of 24 kN at *A* is reduced to zero at *F* by the force P_1 due to the load diagram (shown shaded) applied over the interval *AF*. Evidently the magnitude of P_1 is also 24 kN and is equal to the area $\frac{1}{2}xy$, where *y* denotes the intensity of loading at the section *F*. Hence

$$24 = \tfrac{1}{2}xy \qquad\qquad (a)$$

Another relation between *x* and *y* is obtained from similar triangles in the load diagram:

$$\frac{y}{x} = \frac{12}{9} \quad \text{or} \quad y = \frac{12}{9}x \qquad\qquad (b)$$

which is substituted in Eq. (*a*) to obtain

$$24 = \frac{1}{2}x \cdot \frac{12}{9}x$$

from which

$$x^2 = \frac{24(18)}{12} = 36 \quad \text{and} \quad x = 6 \text{ m}$$

The section of zero shear at *G* is found from the fact that the upward reaction applied over the interval *CG* must total 30 kN in order to reduce the shear of -30 kN at *C* to zero at *G*. Since the reaction is distributed at 10.5 kN/m,

$$30 = 10.5d \quad \text{or} \quad d = 2.86 \text{ m}$$

or, in terms of *x* measured from the left end of the beam,

$$x = 10 + d = 12.86 \text{ m}$$

Bending Moment.   The moment at *F* where $x = 6$ m is found by applying the definition of bending moment.* Note that the shaded part of the triangular load applied between *A* and *F* totals 24 kN and acts at the centroid of the triangular area, that is, at $\frac{1}{3}$ of 6 m from *F*. Thus we

---

*It is better not to use Eq. (4–4) when the shear diagram is curved, since then the areas under the diagram are not too easily computed, especially in the interval *FB*. The computation of such areas is discussed in Chapter 6, especially on p. 231 and in Fig. 6–13. For the present, when the shear diagram is curved, as over the region *AB*, compute the bending moment by applying $(\Sigma M)_L$ or $(\Sigma M)_R$ rather than using the area of the shear diagram.

obtain

$$[M = (\Sigma M)_L]   \quad M_F = 24(6) - 24\left(\tfrac{6}{3}\right) = 96 \text{ kN} \cdot \text{m}$$

Similarly, the moment at B where $x = 9$ m is found to be

$$[M = (\Sigma M)_L]   \quad M_B = 24(9) - 54\left(\tfrac{9}{3}\right) = 54 \text{ kN} \cdot \text{m}$$

The moment at C can also be computed from this basic definition, but whenever the shear diagram consists of straight lines (either horizontal or inclined) it is usually simpler to apply Eq. (4–4); that is, the change in bending moment between any two sections equals the corresponding area of the shear diagram. For example, between B and C, the area of the shear diagram is a rectangle and equals $\Delta M = -30 \times 1 = -30$ kN·m. Hence, if the bending moment changes by -30 kN·m between B and C, the bending moment at C is

$$[M_C = M_B + \Delta M]   \quad M_C = 54 - 30 = 24 \text{ kN} \cdot \text{m}$$

The bending moment at D may be found from $M_D = (\Sigma M)_R = -12 \times 1 = -12$ kN·m; the negative sign comes from the downward load. However, to emphasize the significance of the shear diagram, note that the area of the shear diagram between D and E equals $12 \times 1 = 12$ kN·m. Since this change in moment is positive, it follows that to produce zero moment at the free end E, the moment at D must be -12 kN·m.

Similarly, the area of the shear diagram between G and D equals $\tfrac{1}{2} \times 12 \times 1.14 = 6.84$ kN·m. Since this is a positive increase in bending moment, the moment at G must be smaller by this amount than the moment at D, or

$$M_G = -12 - 6.84 = -18.84 \text{ kN} \cdot \text{m}$$

Sketching the bent beam, as in Fig. 4–10d in Problem 401, shows it to be concave downward at G and D, which is further verification of the negative bending moment signs at these sections.

Shape of the Moment Diagram.   After plotting the values of the bending moment, we consider next the shape of the moment curve connecting these points. Applying Eq. (4–6), we notice that since the shear ordinates are positive and decrease to zero as we move from A to F, the moment curve has correspondingly positive slopes (that is, directed upward to the right) that decrease to zero slope at F.

Between F and B, the vertical shear is increasingly negative, and hence the corresponding slopes of the moment curve become increasingly steeper downward to the right until the steepest slope is reached at section B. Between B and C, the shear stays constant; therefore the slope of the moment curve is constant, being represented by the straight line that joins sections B and C.

Between sections C and G, the shear is negative, becoming zero at G; hence the slope of the moment curve is negative and gradually reduces to zero (that is, horizontal) at G. Similarly the increasing positive shear between G and D results in an increasingly positive slope (directed upward to the right) up to D. Between D and E the slope remains constant because the vertical shear is constant between D and E.

The moment curve between C and D is a symmetrical parabola with its vertex at G, because at equal distances to either side of G the shear is numerically equal but of opposite sign, thereby producing equal slopes oppositely directed. The moment curve between A and B, however, is *not* symmetrical about F because the shear ordinates do not have equal values at equal distances on either side of F; here the moment curve is actually a third-degree parabola.

PROBLEMS

Without writing shear and moment equations, draw shear and moment diagrams for the beams specified in the following problems. Give numerical values at all change of loading positions and at all points of zero shear. (Note to instructor: Problems 403 to 420 may also be assigned for solution by the semigraphical method described in this article.)

425. Beam loaded as shown in Fig. P–425.

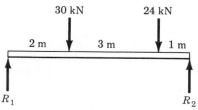

Figure P–425.

426. Cantilever beam acted upon by two forces and one couple as shown in Fig. P–426.

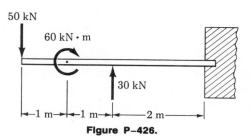

Figure P–426.

427. Beam loaded as shown in Fig. P–427.

Ans. Max. $M = 13.8$ kN·m

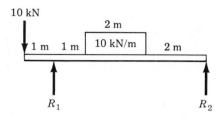

Figure P–427.

428. Beam loaded as shown in Fig. P–428.

Ans. Max. $M = -60$ kN·m

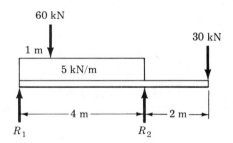

Figure P–428.

429. Beam loaded as shown in Fig. P–429.

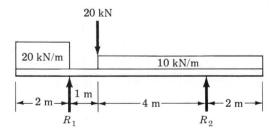

Figure P–429.

430.   In the overhanging beam shown in Fig. P–430, determine P so that the moment over each support equals the moment at midspan.

Ans. $P = 8.75$ kN

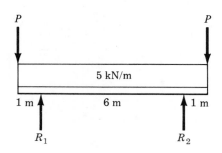

Figure P–430.

431.   Beam loaded as shown in Fig. P–431.

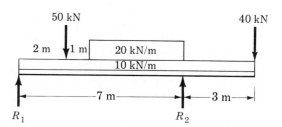

Figure P–431.

432.   A distributed load is supported by two distributed loads as shown in Fig. P–432.

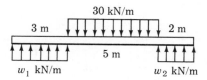

Figure P–432.

433. Overhanging beam loaded by a force and a couple as shown in Fig. P–433.   *Ans.* Max. $M = -160$ kN·m

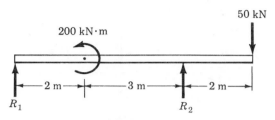

Figure P–433.

434. Beam loaded as shown in Fig. P–434.
                                        Ans. Max. $M = -36$ kN·m

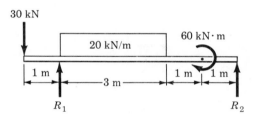

Figure P–434.

435. Beam loaded and supported as shown in Fig. P–435.

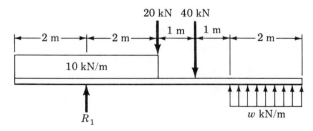

Figure P–435.

436.   Cantilever beam loaded as shown in Fig. P–436.

Ans.   Max. $M = 52.5$ kN·m

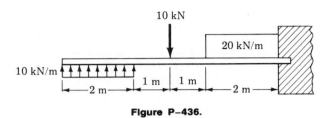

Figure P–436.

437.   Cantilever beam loaded as shown in Fig. P–437.

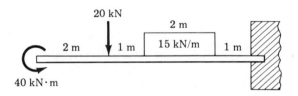

Figure P–437.

438.   A propped cantilever beam loaded as shown in Fig. P–438 consists of two segments joined by a frictionless hinge at which the bending moment is zero.

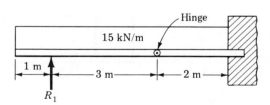

Figure P–438.

439. A beam supported on three reactions as shown in Fig. P–439 consists of two segments joined at a frictionless hinge at which the bending moment is zero.   *Ans.*   Max. $M = 57.6$ kN·m

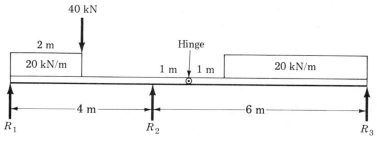

Figure P–439.

440. A frame $ABCD$, with rigid corners at B and C, supports the concentrated load as shown in Fig. P–440. (Draw shear and moment diagrams for each of the three parts of the frame.)

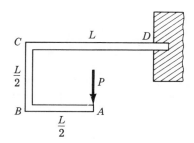

Figure P–440.

441. A beam $ABCD$ is supported by a hinge at A and a roller at D. It is subjected to the loads shown in Fig. P–441, which act at the

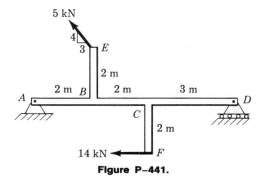

Figure P–441.

ends of the vertical members BE and CF. These vertical members are rigidly attached to the beam at B and C. (Draw shear and moment diagrams for the beam $ABCD$ only.)   *Ans.*   Max. $M = -22$ kN·m

442. Beam carrying the unformly varying load shown in Fig. P–422.   *Ans.*   Max. $M = wL^2/9\sqrt{3}$ at $x = L/\sqrt{3}$

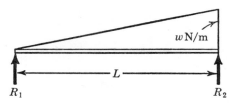

Figure P–442.

443. Beam carrying the triangular loads shown in Fig. P–443.
                                      Ans.   Max. $M = wL^2/12$

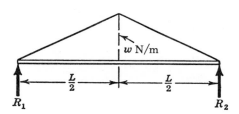

Figure P–443.

444. Beam loaded as shown in Fig. P–444.

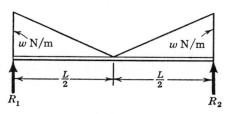

Figure P–444.

445.   Beam loaded as shown in Fig. P–445.

Ans.   Max. $M = -80\,\text{kN·m}$

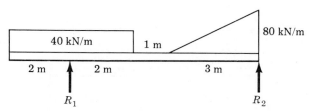

Figure P–445.

446.   Cantilever beam carrying the loads shown in Fig. P–446.

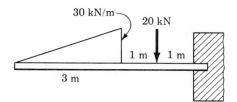

Figure P–446.

447.   Beam loaded as shown in Fig. P–447.

Ans.   Max. $M = -80\,\text{kN·m}$

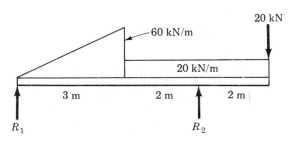

Figure P–447.

448.   Beam carrying the loads shown in Fig. P–448.

Ans.   Max. $M = 137.5 \text{ kN} \cdot \text{m}$

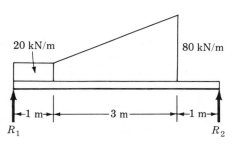

Figure P–448.

449.   A beam carrying the triangular load shown in Fig. P–449 is supported on a uniformly distributed reaction.

Ans.   Max. $M = -45 \text{ kN} \cdot \text{m}$

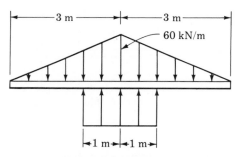

Figure P–449.

450.   Beam loaded and supported as shown in Fig. P–450.

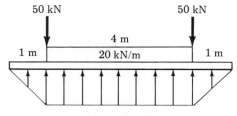

Figure P–450.

451. Beam loaded as shown in Fig. P–451.

Ans.   Max. $M = -45$ kN·m

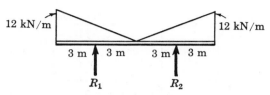

Figure P–451.

452. Beam loaded as shown in Fig. P–452.

Ans.   Max. $M = 49.5$ kN·m

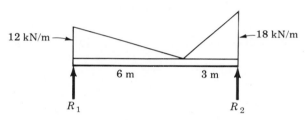

Figure P–452.

453. A uniformly varying load is supported on two distributed reactions as shown in Fig. P–453.

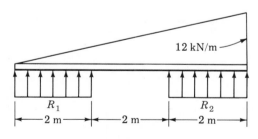

Figure P–453.

In the following problems, draw moment and load diagrams corresponding to the given shear diagrams where values are given in kilonewtons. Specify values at all change of load positions and at all points of zero shear.

454.   Shear diagram as shown in Fig. P–454.

                 Ans.   Max. $M = 30$ kN·m

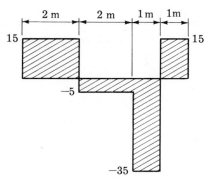

Figure P–454.

455.   Shear diagram as shown in Fig. P–455.

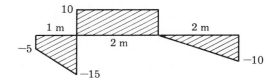

Figure P–455.

456.   Shear diagram as shown in Fig. P–456.

                 Ans.   Max. $M = -120$ kN·m

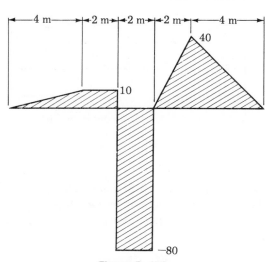

Figure P–456.

457. Shear diagram as shown in Fig. P–457.

Ans. M = 22.5 kN·m

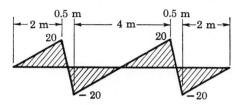

Figure P–457.

458. Shear diagram as shown in Fig. P–458.

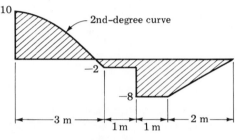

Figure P–458.

4–5 MOVING LOADS

A truck or other vehicle rolling across a beam or girder constitutes a system of concentrated loads at fixed distances from one another. For beams carrying only concentrated loads, the maximum bending moment occurs under one of the loads. Therefore, the problem here is to determine the bending moment under each load when each load is in a position to cause a maximum moment to occur under it. The largest of these various values is the maximum moment that governs the design of the beam.

In Figure 4–23, P_1, P_2, P_3, and P_4 represent a system of loads at fixed distances a, b, and c from one another; the loads move as a unit across the simply supported beam with span L. Let us locate the position of P_2 when the bending moment under this load is maximum. If we denote the resultant of the loads on the span by R and its position from P_2 by e, the value of the left reaction is

$$R_1 = \frac{R}{L}(L - e - x)$$

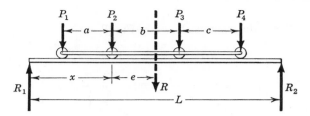

Figure 4-23. Moving loads.

The bending moment under P_2 is then

$$[M = (\Sigma M)_L] \qquad M_2 = \frac{R}{L}(L - e - x)(x) - P_1 a$$

To compute the value of x which will give maximum M_2, we set the derivative of M_2 with respect to x equal to zero:

$$\frac{dM_2}{dx} = \frac{R}{L}(L - e - 2x) = 0$$

from which

$$x = \frac{L}{2} - \frac{e}{2} \qquad\qquad (4\text{-}7)$$

This value of x is independent of the number of loads to the left of P_2, since the derivative of all terms of the form $P_1 a$ with respect to x will be zero.

Equation (4-7) may be expressed in terms of the following rule: *The bending moment under a particular load is a maximum when the center of the beam is midway between that load and the resultant of all loads then on the span.* With this rule we locate the position of each load when the moment at that load is a maximum and compute the value of each such maximum moment.

The maximum shearing force occurs at, and is equal to, the maximum reaction. The maximum reaction for a group of moving loads on a span occurs either at the left reaction, when the leftmost load is over that reaction, or at the right reaction, when the rightmost load is over it. In other words, the maximum reaction is the reaction to which the resultant load is nearest.

ILLUSTRATIVE PROBLEM

459. A truck and trailer combination having the axle loads shown in Fig. 4-24a rolls across the simply supported span of 12 m. Compute the maximum bending moment and the maximum shearing force.

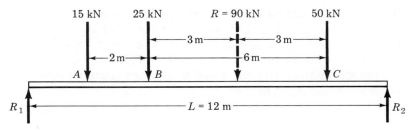

(a) Original loading

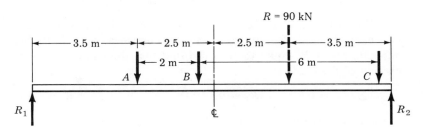

(b) Position of loads for maximum moment at A

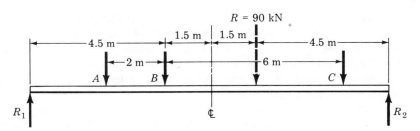

(c) Position of loads for maximum moment at B

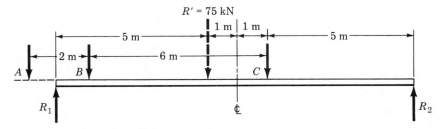

(d) Position of loads for maximum moment at C
with only B and C on span

Figure 4–24.   Moving loads.

Solution: The resultant of the three loads is $R = 90$ kN and is located as shown in Fig. 4–24a. The position of the loads that will cause the bending moment to be maximum under A is shown in Fig. 4–24b, in accordance with the rule expressed by Eq. (4–7) that the center line of the beam is midway between A and R. Taking moments about R_2 equal to zero, we find R_1 to be

$$\left[\Sigma M_{R_2} = 0\right] \qquad 12R_1 = 90(3.5) \qquad\qquad R_1 = 26.25 \text{ kN}$$

whence the bending moment at A is

$$\left[M = (\Sigma M)_L\right] \qquad M_A = 26.25(3.5) = 91.9 \text{ kN·m}$$

We next consider Fig. 4–24c, where the loads are so located that the center line of the beam is midway between B and R. Setting moments about R_2 equal to zero, we find that the value of R_1 for this position of the loads is

$$\left[\Sigma M_{R_2} = 0\right] \qquad 12R_1 = 90(4.5) \qquad\qquad R_1 = 33.75 \text{ kN}$$

whence the bending moment at B is

$$\left[M = (\Sigma M)_L\right] \qquad M_B = 33.75(4.5) - 15(2) = 122 \text{ kN·m}$$

If we now position the loads so that the center line of the beam is midway between C and R, in order to have the bending moment a maximum under C, we find that load A comes off the span, which is contrary to the assumption that all three loads are on the span. This indicates the possibility of a maximum bending moment under C when only loads B and C are on the span.

When only loads B and C are on the span, their resultant is $R' = 75$ kN at 2 m from C. This position of the loads to cause maximum bending moment under C is shown in Fig. 4–24d, in which the center line of the beam is midway between R' and C. Setting moments about R_1 equal to zero, we find R_2 for this condition to be

$$\left[\Sigma M_{R_1} = 0\right] \qquad 12R_2 = 75(5) \qquad\qquad R_2 = 31.25 \text{ kN}$$

whence the bending moment at C is computed to be

$$\left[M = (\Sigma M)_R\right] \qquad M_C = 31.25(5) = 156 \text{ kN·m}$$

It is left as an exercise for the reader to show that the maximum bending moments under A and B, when only loads A and B are on the span are, respectively, 96.3 kN·m and 105 kN·m, and that with only C on the span, the maximum moment occurs with C at midspan and equals 150 kN·m.

A comparison of the above results shows that the most dangerous bending moment is 156 kN·m, occurring under C when only loads B and C are on the span.

Maximum Shearing Force. If all three loads are on the span, the resultant load R is 3 m from R_2 when C is over R_2; it is 5 m from R_1 when A is over R_1. Evidently the maximum reaction, and consequently the maximum shearing force, is at R_2, since it is nearer the resultant load. By setting moments about R_1 equal to zero, the value of R_2 is found to be

$$\left[\Sigma M_{R_1} = 0\right] \qquad 12R_2 = 90(12 - 3) \qquad R_2 = \text{Max. } V = 67.5 \text{ kN}$$

We must also investigate the possibility of the maximum shearing force occurring when only loads B and C are on the span. The maximum reaction in this case will be at R_1, when B is over R_1 and the resultant load $R' = 75$ kN is 4 m from R_1. Its value will be $R_1 = \frac{75}{12}(12 - 4) = 50$ kN. The condition when only A and B are on the span need not be checked, because their resultant load of 40 kN is less than the reaction $R_2 = 67.5$ kN found above.

PROBLEMS

460. A truck with axle loads of 40 kN and 60 kN on a wheel base of 5 m rolls across a 10-m span. Compute the maximum bending moment and the maximum shearing force.

Ans. Max. $M = 160$ kN·m; Max. $V = 80$ kN

461. Repeat Problem 460 using axle loads of 30 kN and 50 kN on a wheel base of 4 m crossing an 8-m span.

462. A tractor with axle loads of 4 kN and 8 kN has a wheel base of 3 m. Compute the maximum moment and maximum shearing force when crossing a 6-m span.

Ans. Max. $M = 12.5$ kN·m; Max. $V = 10$ kN

463. Three equal wheel loads of 30 kN each, separated by 2 m between each load, roll as a unit across a 12-m span. Determine the maximum moment and maximum shear.

Ans. Max. $M = 210$ kN·m; Max. $V = 75$ kN

464. Three wheel loads roll as a unit across a 16-m span. The loads are $A = 10$ kN; $B = 20$ kN, 2 m to the right of A; and $C = 40$ kN, 4 m to the right of B. Determine the maximum moment and maximum shear in the simply supported span.

465. A truck and trailer combination crossing a 12-m span has axle loads of 10 kN, 20 kN, and 30 kN separated respectively by distances of 3 m and 5 m. Compute the maximum moment and maximum shear developed in the span.

Ans. Max. $M = 104$ kN·m; Max. $V = 45$ kN

SUMMARY

The fundamental definitons of shear and bending moment are expressed by

$$V = (\Sigma Y)_L \tag{4-1}$$

and

$$M = (\Sigma M)_L = (\Sigma M)_R \tag{4-2}$$

in which upward-acting forces or loads cause positive effects. The shearing force V should be computed only in terms of the forces to the left of the section being considered; the bending moment M may be computed in terms of the forces to either the left or the right of the section, depending on which requires less arithmetical work.

Relations between load, shear, and moment are given by

$$w = \frac{dV}{dx} \tag{4-5}$$

and

$$V = \frac{dM}{dx} \tag{4-6}$$

These relations are amplified in Art. 4–4 to provide a seimgraphical method of computing shear and moment which supplements Eqs. (4–1) and (4–2). We obtain

$$V_2 - V_1 = \Delta V = (\text{area})_{\text{load}} \tag{4-3}$$

and

$$M_2 - M_1 = \Delta M = (\text{area})_{\text{shear}} \tag{4-4}$$

which provide alternate methods of computing shear and moment. The relations (4–5) and (4–6), expressed in the form

intensity of load = corresponding slope of shear diagram

and

intensity of shear = corresponding slope of moment diagram

enable us to sketch the proper shapes of the shear and moment diagrams rapidly and correctly.

When systems of wheel loads move as a unit across a beam, the bending moment is a maximum under one of the loads. To determine the position of the loads when the moment is maximum under a particular load, the system of loads must be in such a position that the center line of the span is midway between that load and the resultant of all the loads then on the span. With the loads in this position, the reactions are computed and Eq. (4–2) is applied to compute the bending moment in the beam under the particular load.

5

Stresses in Beams

5-1 INTRODUCTION

In this chapter we derive the relations between the bending moment and the flexure stresses it causes, and between the vertical shear and the shearing stresses. In deriving these relations, the following assumptions are made:

1. Plane sections of the beam, originally plane, remain plane.

2. The material in the beam is homogeneous and obeys Hooke's law.

3. The moduli of elasticity for tension and compression are equal.

4. The beam is initially straight and of constant cross section.

5. The plane of loading must contain a principal axis of the beam cross section and the loads must be perpendicular to the longitudinal axis of the beam.

The application and limits of these assumptions will be discussed in succeeding articles as the reason for them becomes apparent.

5–2 DERIVATION OF FLEXURE FORMULA

The stresses caused by the bending moment are known as bending or *flexure stresses*, and the relation between these stresses and the bending moment is expressed by the *flexure formula*. The derivation of this relation follows the procedure developed in deriving the torsion formula (see Art. 3–2); that is, the elastic deformations plus Hooke's law determine the manner of stress variation, after which the conditions of equilibrium then establish the relation between stress and load.

Figure 5–1a shows two adjacent sections, *ab* and *cd*, separated by the distance *dx*. Because of the bending caused by load *P*, sections *ab* and *cd* rotate relative to each other by the amount *dθ*, as shown in Fig. 5–1b, but remain straight and undistorted in accordance with assumption 1 of the preceding article.

Fiber *ac* at the top is shortened, and fiber *bd* at the bottom is lengthened. Somewhere between them is located fiber *ef*, whose length is unchanged. Drawing the line *c'd'* through *f* parallel to *ab* shows that fiber *ac* is shortened an amount *cc'* and is in compression, and that fiber *bd* is lengthened by an amount *d'd* and is in tension.

The plane containing fibers like *ef* is called the *neutral surface* because such fibers remain unchanged in length and hence carry no stress. It will be shown shortly that this neutral surface contains the centroids of all transverse sections.

Consider now the deformation of a typical fiber *gh* located *y* units from the neutral surface. Its elongation *hk* is the arc of a circle of radius *y* subtended by the angle *dθ* and is given by

$$\delta = hk = y\,d\theta$$

The strain is found by dividing the deformation by the original length *ef*

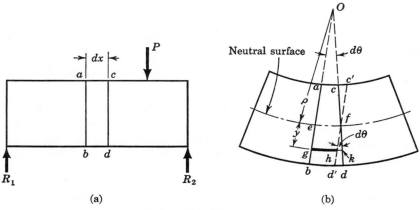

Figure 5–1. Deformations.

(a)   (b)

of the fiber:

$$\epsilon = \frac{\delta}{L} = \frac{y\,d\theta}{ef}$$

If we denote the radius of curvature of the neutral surface by ρ, the curved length ef is equal to $\rho\,d\theta$; whence the strain becomes

$$\epsilon = \frac{y\,d\theta}{\rho\,d\theta} = \frac{y}{\rho}$$

Assuming that the material is homogeneous and obeys Hooke's law (assumption 2), the stress in fiber gh is given by

$$\sigma = E\epsilon = \left(\frac{E}{\rho}\right)y \qquad\qquad (a)$$

Equation (a) indicates that the stress in any fiber varies directly with its location y from the neutral surface, since it is assumed that the modulus of elasticity E is equal in tension and compression (assumption 3) and the radius of curvature ρ of the neutral surface is independent of the location y of the fiber. However, the stresses must not exceed the proportional limit, for this would invalidate Hooke's law on which this stress variation is based.

To complete the derivation of the flexure formula, we apply the conditions of equilibrium. As we saw in Art. 4–3, the external loads that act to one side of an exploratory section are balanced by the resisting shear V_r and the resisting moment M_r. To create this balance, a typical element in the exploratory section is subjected to the forces shown in the pictorial sketch* in Fig. 5–2. The line of intersection between the neutral surface and the transverse exploratory section is called the *neutral axis*, abbreviated NA.

To satisfy the conditions that the external loads have no X components (assumption 5), we must have

$$[\Sigma X = 0] \qquad \int \sigma_x\,dA = 0$$

where σ_x is equivalent to σ in Eq. (a). On replacing σ_x by Ey/ρ, this becomes

$$\frac{E}{\rho}\int y\,dA = 0$$

The constant ratio E/ρ is written outside the integral sign. Since $y\,dA$ is the moment of the differential area dA about the neutral axis, the integral $\int y\,dA$ is the total moment of area. Hence

$$\frac{E}{\rho}A\bar{y} = 0$$

*The cross section is drawn as rectangular only for ease of representation; it may have any shape.

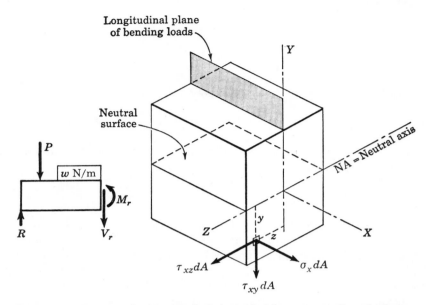

Figure 5–2. Forces acting on a typical element of the cross section of a beam.

However, since only $\bar{y}$ in this relation can be zero, we conclude that the distance from the neutral axis (which is the reference axis) to the centroid of the cross-sectional area must be zero; i.e., *the neutral axis must contain the centroid of the cross-sectional area.*

The condition that $\Sigma Y = 0$, resulting in $V = V_r$, leads to the shear stress formula, the derivation of which is postponed until later (Art. 5–7). It should be observed here that the resisting shear V_r is the summation of the shearing forces $\tau_{xy}\, dA$; that is, $V_r = \int \tau_{xy}\, dA$.

The condition $\Sigma Z = 0$ leads to $\int \tau_{xz}\, dA = 0$. Since the loading has no Z components, the system of shear forces $\tau_{xz}\, dA$ must be self-balancing. We examine this in greater detail in Art. 13–8, where the plane of loading may be offset from the XY plane but remains parallel to it. In those cases, the loading causes a moment about the X axis which is balanced by $\int y(\tau_{xz}\, dA) - \int z(\tau_{xy}\, dA)$ in order to satisfy the condition $\Sigma M_x = 0$. This condition is automatically satisfied for sections that are symmetrical about the Y axis because then the element under discussion has a symmetrically placed counterpart so the integrals are equal to zero. As a result, the plane of loading for sections symmetrical about the Y axis must coincide with the XY plane, or the beam will twist.

We consider next the condition $\Sigma M_y = 0$. The external loads have no moment about the Y axis nor do the internal forces $\tau_{xy}\, dA$ and $\tau_{xz}\, dA$. Therefore

$$\left[\Sigma M_y = 0\right] \qquad \int z(\sigma_x\, dA) = 0$$

Again replacing σ_x by Ey/ρ, we have

$$\frac{E}{\rho} \int zy \, dA = 0$$

The integral $\int zy \, dA$ is the product of inertia P_{zy}, which is zero only if Y or Z is an axis of symmetry or a principal axis. This is the justification for assumption 5.

The final condition of equilibrium, $\Sigma M_z = 0$, requires that the bending moment be balanced by the resisting moment; that is, $M = M_r$. The resisting moment about the neutral axis of a typical element being $y(\sigma_x \, dA)$, this condition requires that

$$M = \int y(\sigma_x \, dA)$$

which, by replacing σ_x by Ey/ρ from Eq. (a), becomes

$$M = \frac{E}{\rho} \int y^2 \, dA$$

Since $\int y^2 \, dA$ is defined as I, the moment of inertia* of the area about a reference axis, which here is the neutral axis (equivalent to the centroidal axis), we finally obtain

$$M = \frac{EI}{\rho} \qquad\qquad (b)$$

Observe now that it was necessary in Art. 4–2 to specify the centroidal axis of the exploratory section as the axis about which bending moment is computed in order to obtain a common axis for computing and equating M and M_r.

The usual form of writing Eq. (b) is

$$\frac{1}{\rho} = \frac{M}{EI} \qquad\qquad (5\text{–}1)$$

which we shall use in Arts. 6–2 and 6–3 as the basis for determining deflections in beams. Because curvature is equal to the reciprocal of the radius of curvature, Eq. (5–1) indicates that curvature is directly proportional to bending moment, an observation which we have already used (page 116) in checking the sign of bending moment with the shape of the deflected beam–positive curvature, which is concave upward, correlating with positive bending moment, and vice versa.

*A complete discussion of moment of inertia is given in Appendix A.

Equating the ratio E/ρ from Eq. (5–1) with its value from Eq. (a), we have

$$\frac{E}{\rho} = \frac{M}{I} = \frac{\sigma}{y}$$

which leads directly to the flexure formula*

$$\sigma = \frac{My}{I} \tag{5–2}$$

This formula indicates that the flexure stress in any section varies directly with the distance of the section from the neutral axis. In a more common form of the flexure formula y is replaced by the distance c, which is defined as the distance from the neutral axis to the remotest element. With this change, the maximum flexure stress in any section is given by

$$\text{Max. } \sigma = \frac{Mc}{I} \tag{5–2a}$$

If I/c is called the *section modulus* and denoted by S, another common variation of the flexure formula is

$$\text{Max. } \sigma = \frac{M}{I/c} = \frac{M}{S} \tag{5–2b}$$

This variation is useful for beams of constant cross section, as it shows that maximum flexure stress occurs at the section of maximum bending moment. Various values of section modulus for common cross sections are listed in Table 5–1.

An interesting analysis, similar to that we shall use later in analyzing reinforced concrete beams (Art. 10–4), is to consider the variation in flexure stress over a rectangular cross section, as shown in Fig. 5–3.

Because the horizontal summation of forces over a section must be zero, the total compressive force C in the upper half of the cross section is equal to the total tensile force T in the lower half. Thus the resisting moment M_r consists of the couple composed of the equal, oppositely directed forces C and T. The value of each of these forces is equal to the product of the average stress multiplied by the area. Therefore, since the average stress in a linear stress distribution is one-half the maximum stress,

$$T = C = (\sigma_{\text{ave.}})(\text{Area}) = \left(\frac{1}{2}\sigma\right)\left(b\frac{h}{2}\right)$$

---

*Note the similarity between the torsion formula $\tau = T\rho/J$ and the flexure formula $\sigma = My/I$. This similarity makes them easy to remember.

TABLE 5-1.   Section Moduli

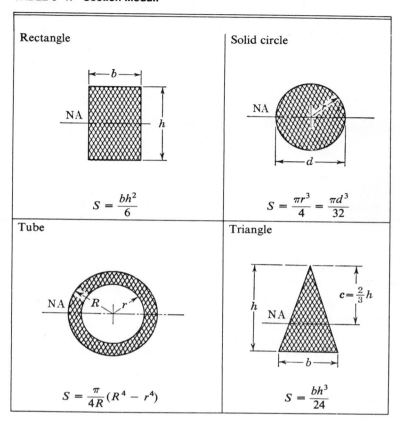

Rectangle

$$S = \frac{bh^2}{6}$$

Solid circle

$$S = \frac{\pi r^3}{4} = \frac{\pi d^3}{32}$$

Tube

$$S = \frac{\pi}{4R}(R^4 - r^4)$$

Triangle

$$c = \frac{2}{3}h$$

$$S = \frac{bh^3}{24}$$

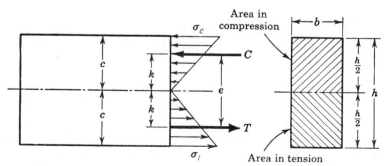

Figure 5-3.   Resisting moment is equivalent to the couple created by the resultant compressive and tensile forces.

The forces C and T act through the centroid of the triangular load distribution at a distance k from NA. Since $k = \frac{2}{3}c = \frac{2}{3}(h/2)$, the moment arm of the resisting couple is $e = 2k = \frac{2}{3}h$. Equating bending moment to resisting moment, we have

$$M = M_r = Ce = Te$$

$$M = \left(\frac{1}{2}\sigma\right)\left(b\frac{h}{2}\right)\left(\frac{2}{3}h\right) = \sigma\frac{bh^2}{6}$$

which agrees with Eq. (5–2b) for a rectangular section.

Modulus of rupture

Equation (5–2a) may be used to compute the flexure stress in a beam loaded to rupture in a testing machine. Because the proportional limit of the material is then exceeded, the stress determined in this manner is not a true stress; nevertheless, the fictitious stress so obtained is called the modulus of rupture. It is used to compare the ultimate strengths of beams of various sizes and materials.

ILLUSTRATIVE PROBLEMS

501.   A beam 150 mm wide by 250 mm deep supports the loads shown in Fig. 5–4. Determine the maximum flexural stress.

Solution: We begin by computing the maximum bending moment. The shear diagram shows that zero shear occurs at $x = 2$ m. Using the area of this diagram to compute the bending moment, we have at $x = 2$ m,

$$[\Delta M = (\text{area})_V] \qquad \text{Max. } M = \left(\frac{14 + 2}{2}\right)(2) = 16 \text{ kN} \cdot \text{m}$$

It is unnecessary to draw the moment diagram.

We now apply the flexure formula, being careful to use consistent units for the various quantities. From Table 5–1 we find that the section modulus is $S = bh^2/6$, so

$$\left[\sigma = \frac{M}{S} = \frac{6M}{bh^2}\right] \qquad \text{Max. } \sigma = \frac{6(16 \times 10^3)}{(0.150)(0.250)^2}$$

$$= 10.24 \text{ MPa} \qquad Ans.$$

502.   A timber beam 100 mm wide by 300 mm high and 8 m long carries the loading shown in Fig. 5–5. If the maximum flexural stress is 9 MPa, for what maximum value of w will the shear be zero under P, and what is the value of P?

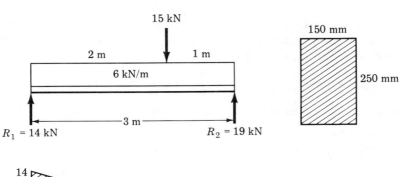

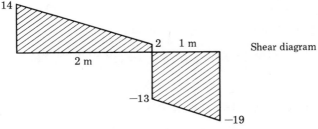

Shear diagram

Figure 5–4.

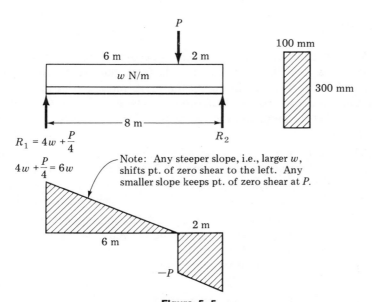

Note: Any steeper slope, i.e., larger w, shifts pt. of zero shear to the left. Any smaller slope keeps pt. of zero shear at P.

Figure 5–5.

Solution: To satisfy the given conditions, the shear diagram must appear as shown. The maximum value of w to reduce the shear to zero P is determined from Eq. (4–3):

$$[\Delta V = (\text{area})_{\text{load}}] \qquad 4w + \frac{P}{4} = 6w$$

which determines the following relation between P and w:

$$P = 8w \qquad\qquad (a)$$

The maximum bending moment occurs under P and is

$$[\Delta M = (\text{area})_V] \qquad \text{Max. } M = \tfrac{1}{2}(6)(6w) = 18w \text{ N} \cdot \text{m}$$

Applying the flexure formula, we obtain

$$\left[M = \sigma\frac{I}{c} = \sigma\frac{bh^2}{6} \right] \qquad 18w = (9 \times 10^6)\frac{(0.100)(0.300)^2}{6}$$

$$w = 750 \text{ N/m} \qquad Ans.$$

whence from relation (a), the value of P is

$$P = 8w = 8(750) = 6000 \text{ N} \qquad Ans.$$

PROBLEMS

503. A cantilever beam, 60 mm wide by 200 mm high and 6 m long, carries a load that varies uniformly from zero at the free end to 1000 N/m at the wall. (a) Compute the magnitude and location of the maximum flexural stress. (b) Determine the type and magnitude of the stress in a fiber 40 mm from the top of the beam at a section 3 m from the free end.   *Ans.* $(b) \sigma = 1.13$ MPa

504. A simply supported beam, 60 mm wide by 100 mm high and 4 m long, is subjected to a concentrated load of 800 N at a point 1 m from one of the supports. Determine the maximum fiber stress and the stress in a fiber located 10 mm from the top of the beam at midspan.

505. A high-strength steel band saw, 20 mm wide by 0.80 mm thick, runs over pulleys 600 mm in diameter. What maximum flexural stress is developed? What minimum diameter pulleys can be used without exceeding a flexural stress of 400 MPa? Assume $E = 200$ GPa.   *Ans.* $\sigma = 267$ MPa

506. A flat steel bar, 25 mm wide by 6 mm thick and 1 m long, is bent by couples applied at the ends so that the midpoint deflection is 20 mm. Compute the maximum stress in the bar and the magnitude of the couples. Use $E = 200$ GN/m².   *Ans.* $\sigma = 95.8$ MPa; $M = 14.4$ N $\cdot$ m

507.   In a laboratory test of a beam loaded by end couples, the fibers at layer AB in Fig. P–507 are found to increase 30×10^{-3} mm while those at CD decrease 90×10^{-3} mm in the 200–mm–gage length. Using $E = 100$ GPa, determine the flexural stress in the top and bottom fibers.

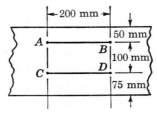

Figure P–507.

508.   Determine the minimum width b of the beam shown in Fig. P–508 if the flexural stress is not to exceed 10 MPa.

Ans. $b = 75.0$ mm

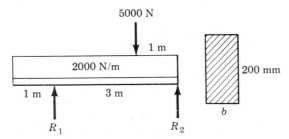

Figure P–508.

509.   A box section used in aircraft is constructed of tubes connected by thin webs as shown in Fig. P–509. Each tube has a cross-sectional area of 130 mm². If the average stress in the tubes is not to exceed 70 MPa, determine the total uniformly distributed load that can be supported on a simple span 4 m long. Neglect the effects of the webs.

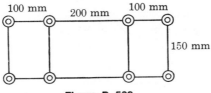

Figure P–509.

510. A 40-mm diameter bar is used as a simply supported beam 2 m long. Determine the largest uniformly distributed load which can be applied over the right half of the beam if the flexural stress is limited to 60 MN/m². *Ans.* $w = 1340$ N/m

511. A simply supported rectangular beam, 50 mm wide by 100 mm deep, carries a uniformly distributed load of 1200 N/m over its entire length. What is the maximum length of the beam if the flexural stress is limited to 20 MPa?

512. The circular bar 20 mm in diameter shown in Fig. P–512 is bent into a semicircle with a mean radius of 600 mm. If $P = 2000$ N and $F = 1000$ N, compute the maximum flexural stress developed in section a–a. Neglect the deformation of the bar.

Ans. $\sigma = 331$ MPa

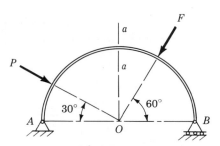

Figure P–512.

513. A rectangular steel beam, 50 mm wide by 80 mm deep, is loaded as shown in Fig. P–513. Determine the magnitude and location of the maximum flexural stress.

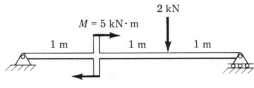

Figure P–513.

514. The right-angled frame shown in Fig. P–514 carries a uniformly distributed loading equivalent to 200 N for each horizontal projected meter of the frame; that is, the total load is 1000 N. Compute the maximum flexural stress at section a–a if the cross section is 50 mm square. *Ans.* $\sigma = 30.0$ MPa

515. Repeat Problem 514 to find the maximum flexural stress at section b–b.

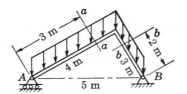

Figures P–514 and P–515.

516.   A rectangular steel bar, 20 mm wide by 40 mm high and 4 m long, is simply supported at its ends. If the density of steel is 7850 kg/m³, determine the maximum bending stress caused by the weight of the bar.

517.   A simply supported beam 4 m long is composed of two C230 × 30 channels riveted back to back. What uniformly distributed load can be carried, in addition to the weight of the beam, without exceeding a flexural stress of 140 MN/m² if (a) the webs are vertical and (b) the webs are horizontal. Refer to Appendix B for channel properties.

518.   A beam with a S380 × 74 section is simply supported at the ends. It supports a central concentrated load of 40 kN and a uniformly distributed load of 15 kN/m over its entire length, including the weight of the beam. Determine the maximum length of the beam if the flexural stress is not to exceed 140 MPa. Refer to Appendix B for properties of S shapes.

519.   A beam 10 m long is simply supported 1 m from each end. It is made of two C380 × 50 channels (see Appendix B) riveted back to back and used with the webs vertical. Determine the total uniformly distributed load that can be carried along its entire length without exceeding a flexural stress of 120 MPa.   *Ans.* $w = 22.0 \text{ kN/m}$

520.   A beam with a W200 × 27 section (see Appendix B) is used as a cantilever beam 6 m long. Find the maximum uniformly distributed load which can be applied over the entire length of the beam, in addition to the weight of the beam, if the flexural stress is not to exceed 140 MN/m².

521.   Repeat Problem 520 using a 4-m cantilever beam with a W250 × 67 section.

522.   In the portion of a butt joint shown in Fig. P–522, a 28-mm rivet secures 14-mm cover plates to a 20-mm main plate. Assuming that the loads are uniformly distributed along the rivet, determine the maximum bending stress in the rivet.

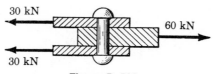

Figure P-522.

523. A square timber beam used as a railroad tie is supported by a uniformly distributed reaction and carries two uniformly distributed loads each totaling 48 kN as shown in Fig. P–523. Determine the size of the section if the maximum stress is limited to 8 MPa.

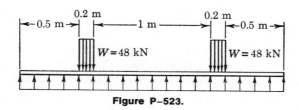

Figure P-523.

524. A wooden beam 150 mm wide by 300 mm deep is loaded as shown in Fig. P–524. If the maximum flexural stress is 8 MN/m², find the maximum values of w and P that can be applied simultaneously.

Ans. $w = 9$ kN/m; $P = 18$ kN

525. In Problem 524, if the load on the overhang is 10 kN/m and the overhang is x meters long, find the maximum values of P and x that can be used simultaneously.

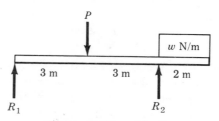

Figures P-524 and P-525.

526. A rectangular beam, 120 mm wide by 400 mm deep, is loaded as shown in Fig. P–526. If $w = 3$ kN/m, find P to cause a maximum bending stress of 10 MPa. *Ans.* $P = 32.5$ kN

527. Solve Problem 526 if $w = 6$ kN/m.

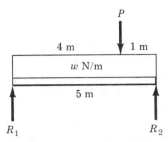

Figures P–526 and P–527.

5–3 ECONOMIC SECTIONS

In a beam having a rectangular or circular cross section, the fibers near the neutral axis are understressed compared with those at the top or bottom. The fact that a large portion of the cross section is thus understressed makes it inefficient for resisting flexure.

The flexure formula, $M = \sigma I / c$, shows that if the area of a beam of rectangular section (Fig. 5–6a) could be rearranged so as to keep the same overall depth but have the shape shown in Fig. 5–6b, the moment of inertia would be greatly increased, resulting in a greater resisting moment. Physically, the increase in resisting moment is due to more fibers being located at a greater distance from the NA, for such fibers carry a greater stress and have a larger moment arm about the NA to resist the applied bending moment. However, the section in Fig. 5–6b is not practicable; the two parts of it would collapse together. It is necessary to use some of the area to fix these parts in place relative to each other, as in Fig. 5–6c. We see later (page 197) that this web area transmits practically all the vertical shear, and we shall learn how to compute its dimensions.

Figure 5–6c represents a wide-flange beam (referred to as a W shape). This is one of the most efficient structural shapes manufactured

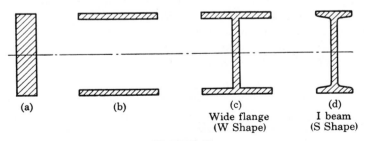

(a) (b) (c)
Wide flange
(W Shape)

(d)
I beam
(S Shape)

Figure 5–6.

because it not only provides great flexural strength with minimum weight of material but is highly efficient when used as a column (see Chapter 11). Another structural shape is the I beam (referred to as an S shape) in Fig. 5–6d; it preceded the wide flange and because it is not as efficient has been largely replaced by the wide flange beam. Properties of both these sections are given in Appendix B. A beam of either type is specified by stating its nominal depth in millimeters and its nominal mass per unit length in kilograms per meter. The designation W610 × 140, for example, indicates a wide-flange beam with nominal depth 610 mm and nominal mass per unit length of 140 kg/m. The tables in Appendix B indicate that the actual depth of this beam is 617 mm and its theoretical mass per unit length is 140.1 kg/m.* The structural tables give the dimensions and other properties of the cross-sectional area, such as moment of inertia (I), section modulus (S), and radius of gyration[†] (r) for each principal axis of the section.

In selecting a structural section to be used as a beam, it is obvious that the resisting moment $M_r = \sigma I / c = \sigma S$ must be equal to or greater than the applied bending moment M. This may be expressed as

$$S \geqslant \frac{M}{\sigma} \tag{5–3}$$

Equation (5–3) indicates that a beam must be selected whose section modulus is equal to or greater than the ratio of bending moment to allowable stress. Illustrative Problem 528 demonstrates the necessary procedure and cautions.

Lateral deflection of beams.

The compression flanges of beams tend to buckle horizontally sideways if the beam is too long. This buckling is a column effect. (Columns will be discussed in Chapter 11.) When this lateral deflection is prevented by the floor system or by bracing the compression flanges at proper intervals, the full allowable stresses may be used. Otherwise, the stresses should be reduced. Formulas for the reduction of the allowable stress are given in the specifications of the American Institute of Steel Construction. Hereafter, we assume that all beams are properly braced against lateral deflection.

---

*Many designs are based on the nominal mass per meter. However, to illustrate the use of the tables, we shall use the theoretical mass per meter.
†The use of r for radius of gyration conforms with American Institute of Steel Construction (AISC) notation. Be careful not to confuse this term with the r which is frequently used to denote the radius of a circle.

ILLUSTRATIVE PROBLEM

528. What is the lightest W shape beam that will support the load shown in Fig. 5–7 without exceeding a flexural stress of 120 MPa? Determine the actual stress in the beam. Assume the beam is properly braced against lateral deflection.

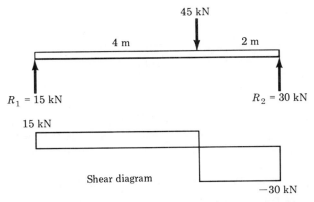

Figure 5–7. Live load. (Applied load exclusive of weight of beam.)

Solution: We begin by computing the reactions and sketching the shear diagram. The maximum moment occurs under the load and equals $15 \times 4 = 60$ kN · m. Applying Eq. (5–3), we have

$$\left[S \geqslant \frac{M}{\sigma} \right] \qquad S \geqslant \frac{60 \times 10^3}{120 \times 10^6} = 500 \times 10^{-6} \text{ m}^3$$

$$\geqslant 500 \times 10^3 \text{ mm}^3$$

Referring to the table of properties of W shapes (Appendix B) and starting at the bottom, we find that the first beam whose section modulus is greater than 500×10^3 mm³ is W200 × 52 with $S = 512 \times 10^3$ mm³. In the W250 group we find a W250 × 45 with $S = 534 \times 10^3$ mm³, which is also satisfactory, as well as being lighter. The W310 group lists a W310 × 39 beam with $S = 549 \times 10^3$ mm³. This one is the best one, because the lightest suitable beam in the remaining groups have a mass per meter which is greater than 38.7 kg/m, the mass per meter of the W310 × 39 beam.

The reader may wonder why more than one size beam is manufactured with approximately the same section modulus. The explanation is that although the lightest beam is the cheapest on the basis of weight alone, frequently headroom clearances require a beam of less depth than the lightest one.

The selection of the beam is not complete until a check calculation is made that includes the weight of the beam.* The beam's resisting moment M_B must be equal to or greater than the sum of the live load moment M_L caused by the applied loads and the dead load moment M_D caused by the dead weight of the beam:

$$M_B \geqslant M_L + M_D$$

Dividing each term of this equation by the stress σ gives

$$\frac{M_B}{\sigma} \geqslant \frac{M_L}{\sigma} + \frac{M_D}{\sigma}$$

whence, replacing M/σ by the section modulus S, we obtain the governing equation for design:

$$S_B \geqslant S_L + S_D$$

The weight of the beam in this example is 38.7 kg/m $\times$ 9.81 m/s^2 = 380 N/m, which is not sufficient to change the location of the maximum moment resulting from the combined live and dead loads. Hence, we compute the dead load moment M_D at $x = 4$ m (Fig. 5–8). From the definition of bending moment, $M = (\Sigma M)_R$, we have

$$M_D = (1.14)(2) - (0.380 \times 2)(1) = 1.52 \text{ kN} \cdot \text{m}$$

or, from the shaded area of the shear diagram,

$$M_D = \frac{(1.14 + 0.38)}{2}(2) = 1.52 \text{ kN} \cdot \text{m}$$

Therefore, the section modulus required to support the dead weight of the beam is

$$\left[S_D = \frac{M_D}{\sigma} \right] \qquad S_D = \frac{1.52 \times 10^3}{120 \times 10^6} = 12.7 \times 10^{-6} \text{ m}^3$$
$$= 12.7 \times 10^3 \text{ mm}^3$$

Applying the governing equation,

$$\left[S_B \geqslant S_L + S_D \right] \qquad 549 \times 10^3 > (500 + 12.7) \times 10^3$$

we see that a W310 $\times$ 39 beam is satisfactory.

---

*Frequently, the steel beam is encased in concrete for fireproofing or to form part of a concrete floor. The concrete may be assumed to extend at least 50 mm beyond the dimensions of the beam. For example, a W310 $\times$ 39 with a depth of 310 mm and a flange width of 165 mm will be encased in concrete having overall dimensions of 410 mm by 265 mm. Computed at a mass density of 2400 kg/m^3, the encasement adds an extra load of approximately 2.56 kN/m, which should be included in the weight of the beam. In subsequent problems, the weight of the encasement will be neglected. In actual practice, its weight is usually included in an estimate of the dead loads.

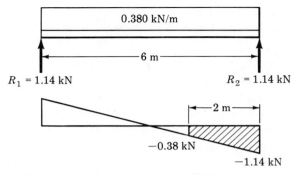

Figure 5–8.   Dead load. (Due to weight of beam.)

The actual beam stress is easily determined from the ratio of the beam modulus to the design modulus, viz.,

$$\left[M = \sigma' S_B = \sigma(S_L + S_D) \right]$$
$$\sigma'(549 \times 10^{-6}) = (120 \times 10^6)\left[(500 + 12.7) \times 10^{-6} \right]$$

from which

$$\sigma' = 112 \text{ MPa} \qquad Ans.$$

PROBLEMS

Assume that the beams in the following problems are properly braced against lateral deflection. Be sure to include the weight of the beam itself.

529.   A 10-m beam simply supported at the ends carries a uniformly distributed load of 16 kN/m over its entire length. What is the lightest W shape beam that will not exceed a flexural stress of 120 MPa? What is the actual stress in the beam selected?

Ans.   W610 × 82; 113 MPa

530.   Repeat Problem 529 if the distributed load is 12 kN/m and the length of the beam is 8 m.

531.   A concentrated load of 90 kN is applied at the center of a simply supported beam 8 m long. Select the lightest suitable W shape section using an allowable stress of 120 MN/m². *Ans.* W530 × 74

532.   Solve Problem 531 if the length of the beam is changed to 12 m.

533.   A beam simply supported on a 12-m span carries a uniformly distributed load of 30 kN/m over the middle 6 m. Using an allowable stress of 140 MPa, determine the lightest suitable W shape beam. What is the actual maximum stress in the selected beam?

Ans.   W610 × 125; 133 MPa

534. Repeat Problem 533 if the uniformly distributed load is changed to 80 kN/m.

535. A simply supported beam 16 m long carries a uniformly distributed load of 20 kN/m over the right half of the beam. Select the lightest suitable W shape beam if the allowable stress is 120 MN/m².

536. A simply supported beam 10 m long carries a uniformly distributed load of 20 kN/m over its entire length and a concentrated load of 40 kN at midspan. If the allowable stress is 120 MPa, determine the lightest W shape beam which can be used.

Ans.   W610 × 125

5–4 FLOOR FRAMING

Probably the most common structural use of beams is to provide support for the floors and frameworks of buildings. Figure 5–9 illustrates a typical detail in a home. The subfloor is supported by floor

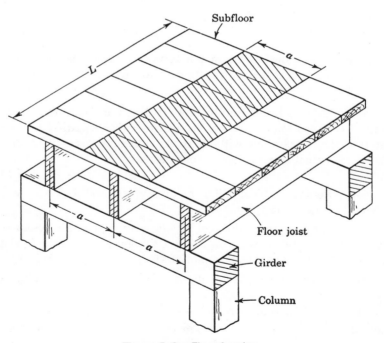

Figure 5–9. Floor framing.

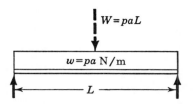

Figure 5–10. Loading on a floor joist.

joists (called floor beams in steel construction). The floor joists are assumed to act as simply supported beams. They are supported by heavier beams called girders, which in turn are supported by columns that transmit the loads to the foundation.

The floor load is specified as p N/m^2 and varies from 2.5 kN/m^2 for homes to as high as 25 kN/m^2 for industrial buildings. If the floor joists are L meters long and spaced a meters apart on centers, each joist is assumed to support the loading of an area aL m^2, shown shaded in Fig. 5–9. The loading on a typical floor joist therefore is the uniformly distributed load shown in Fig. 5–10. The total load W equals the load p N/m^2 acting over the area aL. This may be divided by the length L to give a loading per meter of $w = pa$ N/m.

In steel construction, the same general plan is followed except that the floor beams are usually riveted or bolted to the webs of the girders as shown in Fig. 1–12 (page 21). The figures in the following illustrative problem represent a small building and show how to construct the loading diagrams for the various beams.

ILLUSTRATIVE PROBLEM

537. Determine the loading diagrams for beams B-1, G-1, B-2, and G-2 for the building whose partial floor plan is shown in Fig. 5–11. The loading for each bay is indicated.

Solution: Beams supporting only floor loads are designated B-1, B-2, etc. Beams that support the reactions of floor beams are called girders and are denoted by G-1, G-2, etc. At beam B-1, the loading is 5 kN/m^2 uniformly distributed over a length of 4 m and a width of 2 m, resulting in the loading diagram shown in Fig. 5–12.

Beam G-1 is a girder used to support one end of beams B-1 and framed into beams B-2. It is loaded by the end reactions of beams B-1, as shown in Fig. 5–13.

Beam B-2 supports the end reaction of beam G-1, as well as half the loadings in the bays adjacent to it. Its loading diagram therefore is

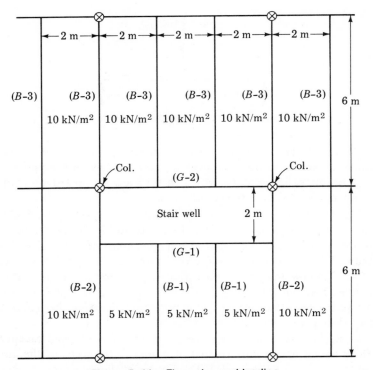

Figure 5–11. Floor plan and loading.

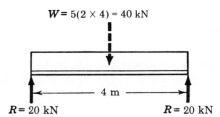

Figure 5–12. Beam (B-1).

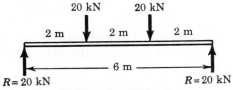

Figure 5–13. Girder (G-1).

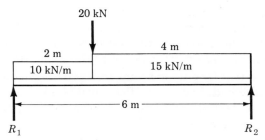

Figure 5–14. Beam (B-2).

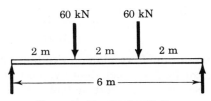

Figure 5–15. Girder (G-2).

as shown in Fig. 5–14. For the first 2 m, beam *B*-2 supports a total load of 10 kN/m² over a floor area 2 m × 1 m, equivalent to 20 kN applied at 10 kN/m. The reaction of beam *G*-1 is shown as a concentrated load of 20 kN. For the rest of the beam, the loading is 15 kN/m computed as the sum of the loadings per meter extending for 1 m into the 10-kN/m² bay and the 5-kN/m² bay.

The girder *G*-2 is loaded by the reactions of beams *B*-3 only, as shown in Fig. 5–15. Verify that the reaction of beam *B*-3 is 60 kN.

PROBLEMS

538. Floor joists 50 mm wide by 200 mm high, simply supported on a 4-m span, carry a floor loaded at 5 kN/m². Compute the center-line spacing between joists to develop a bending stress of 8 MPa. What safe floor load could be carried on a center-line spacing of 0.40 m?

Ans. 0.267 m; 3.34 kN/m²

539. Timbers 300 × 300 mm, spaced 0.90 m apart on centers, are driven into the ground and act as cantilever beams to back up the sheet piling of a coffer dam. What is the maximum safe height of water behind the dam if the density of water is 1000 kg/m³ and the bending stress is limited to 8 MN/m²? *Ans.* *h* = 2.90 m

540. Timbers 200 mm wide by 300 mm deep and 5 m long, simply supported at top and bottom, back up a dam restraining water

3 m deep. The density of water is $1000 \, \text{kg/m}^3$. (a) Determine the center-line spacing of the timbers if the flexural stress is 8 MPa. (b) Will this spacing be safe if the maximum flexural stress is limited to 12 MPa and the water reaches its maximum depth of 5 m?   *Ans.*   (a) 0.939 m

541.   The 6-m long floor beams in a certain building are simply supported at their ends and carry a floor load of $4 \, \text{kN/m}^2$. If the beams have W250 × 45 sections, determine the center-line spacing using an allowable flexural stress of 120 MPa.

542.   Select the lightest W shape sections which can be used for the beams and girders in Illustrative Problem 537 if the allowable stress is 120 MPa. Neglect the weights of the members.

Ans.   *B*-1: W250 × 18; *B*-2: W410 × 46;
G-1: W310 × 28; *G*-2: W410 × 60

543.   A portion of the floor plan of a certain building is shown in Fig. P–543. The total loading (including live and dead loads) in each bay is as shown. Select the lightest suitable W shape beams if the allowable flexural stress is 120 MPa. Assume the beams are adequately braced.

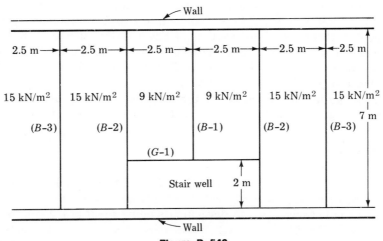

Figure P–543.

544.   Repeat Problem 543 if the $15 \, \text{kN/m}^2$ loading is changed to $24 \, \text{kN/m}^2$ and the $9 \, \text{kN/m}^2$ loading is changed to $12 \, \text{kN/m}^2$.

5–5   UNSYMMETRICAL BEAMS

All the beams discussed so far have been symmetrical with respect to the neutral axis. Because flexure stresses vary directly with distance from the neutral axis—which is the centroidal axis—such beam sections are desirable for materials that are equally strong in tension and compression. However, for materials relatively weak in tension and strong in compression, such as cast iron, it is desirable to use beams that are unsymmetrical with respect to the neutral axis. With such a cross section, the stronger fibers can be located at a greater distance from the neutral axis than the weaker fibers. The ideal treatment for such materials is to locate the centroidal or neutral axis in such a position that the ratio of the distances from it to the fibers in tension and in compression is exactly the same as the ratio of the allowable stresses in tension and in compression. The allowable stresses thus reach their permitted values simultaneously.

ILLUSTRATIVE PROBLEMS

545. A cast-iron beam carries a uniformly distributed load on a simple span. Compute the flange width b of the inverted T section (Fig. 5–16) so that the allowable stresses $\sigma_t = 30$ MPa and $\sigma_c = 90$ MPa reach their limits simultaneously.

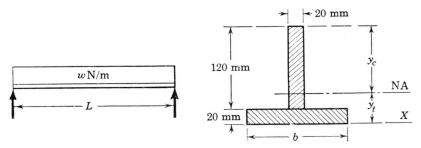

Figure 5–16.

Solution: The beam is bent concave upward so that the uppermost fibers are in compression and the lowermost fibers are in tension. As discussed in Art. 5–2, flexure stresses vary directly with their distance from the neutral axis. Therefore to cause σ_t and σ_c to reach their limits simultaneously, we have

$$\left[\frac{y_t}{y_c} = \frac{\sigma_t}{\sigma_c} \right] \qquad \frac{y_t}{y_c} = \frac{30}{90} = \frac{1}{3}$$

or

$$y_c = 3y_t \qquad\qquad\qquad\qquad\qquad (a)$$

Figure 5–16 shows that another relation between y_t and y_c is

$$y_t + y_c = 140 \text{ mm} \tag{b}$$

Solving relations (a) and (b), we obtain

$$y_t = 35 \text{ mm} \quad \text{and} \quad y_c = 105 \text{ mm}$$

Now consider the T section to consist of the two shaded rectangles. Since the neutral axis coincides with the centroidal axis, we take moments of areas with respect to an X axis through the base of the flange and obtain

$$[A\bar{y} = \Sigma ay]$$
$$(120 \times 20 + b \times 20)y_t = (120 \times 20)(20 + 60) + (b \times 20)(10)$$

In this is substituted the value $y_t = 35$, which gives

$$b = 216 \text{ mm} \quad \textit{Ans.}$$

546. Compute the maximum tensile and compressive stresses developed in the beam that is loaded and has the cross-sectional properties shown in Fig. 5–17.

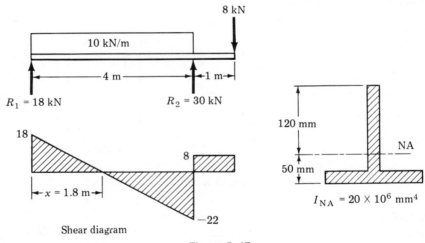

Figure 5–17.

Solution: Sections of zero shear are at $x = 1.8$ m and $x = 4$ m. The bending moments at these sections are $M_{1.8} = 16.2$ kN·m and $M_4 = -8$ kN·m. Check these values.

The positive moment at $x = 1.8$ m indicates curvature concave upward (see Art. 4–2); hence the upper fibers are in compression and the lower fibers are in tension. Applying Eq. (5–2) shows the flexure stresses to be

$$\left[\sigma = \frac{My}{I}\right] \quad \sigma_c = \frac{(16.2 \times 10^3)(0.120)}{20 \times 10^{-6}} = 97.2 \text{ MPa}$$

$$\sigma_t = \frac{(16.2 \times 10^3)(0.050)}{20 \times 10^{-6}} = 40.5 \text{ MPa}$$

Note that for the units to be consistent, M is expressed in $N \cdot m$, y in m, and I in m^4.

At $x = 4$ m, the negative bending moment is interpreted as curvature concave downward, so that the upper fibers are in tension and the lower ones in compression. Having thus interpreted the sign of the bending moment, we substitute the numerical value of the bending moment in Eq. (5–2) and obtain the following flexure stresses:

$$\left[\sigma = \frac{My}{I}\right] \quad \sigma_t = \frac{(8 \times 10^3)(0.120)}{20 \times 10^{-6}} = 48.0 \text{ MPa}$$

$$\sigma_c = \frac{(8 \times 10^3)(0.050)}{20 \times 10^{-6}} = 20.0 \text{ MPa}$$

Hence the maximum compressive stress is 97.2 MPa, occurring at $x = 1.8$ m, and the maximum tensile stress is 48.0 MPa, occurring at $x = 4$ m. In an unsymmetrical section having a reversal in curvature, the maximum stresses need not both occur at the section of maximum moment. The stresses at each section of zero shear must be investigated.

547.   The overhanging beam in Fig. 5–18 is made of cast iron, for which the allowable stresses are $\sigma_t = 40$ MPa and $\sigma_c = 100$ MPa. If the

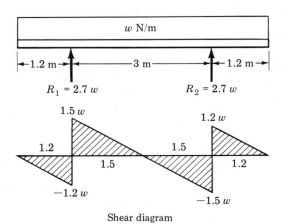

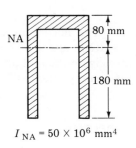

Shear diagram

Figure 5–18.

properties of the cross section are as shown, determine the maximum uniformly distributed load that can be supported.

Solution: At $x = 1.2$ m, the bending moment is $-0.72w$ N · m, the negative sign indicating tension in the upper fibers. Using Eq. (5–2), we find that the safe resisting moments in tension and compression are

$$\left[M_r = \frac{\sigma I}{y}\right] \quad M_t = \frac{(40 \times 10^6)(50 \times 10^{-6})}{0.080} = 25.0 \text{ kN · m}$$

$$M_c = \frac{(100 \times 10^6)(50 \times 10^{-6})}{0.180} = 27.8 \text{ kN · m}$$

Evidently tension governs, since the safe resisting moment is the lower value. Equating this to the bending moment, we have

$$[M = M_r] \quad 0.72w = 25.0 \times 10^3 \quad w = 34.7 \text{ kN/m}$$

Before concluding that this is the safe load, we must also investigate the other section of zero shear. At $x = 2.7$ m, $M = +0.405w$ N · m. Although this is lower than the moment at $x = 1.2$ m, the curvature is reversed, being concave upward and placing the upper fibers in compression and the lower ones in tension. From Eq. (5–2), the safe resisting moment is

$$\left[M_r = \frac{\sigma I}{y}\right] \quad M_c = \frac{(100 \times 10^6)(50 \times 10^{-6})}{0.080} = 62.5 \text{ kN · m}$$

$$M_t = \frac{(40 \times 10^6)(50 \times 10^{-6})}{0.180} = 11.1 \text{ kN · m}$$

Equating the lower resisting moment to the bending moment, we obtain

$$[M = M_r] \quad 0.405w = 11.1 \times 10^3 \quad w = 27.4 \text{ kN/m}$$

The maximum safe load is the lower of the values obtained at $x = 1.2$ m and $x = 2.7$ m, that is, 27.4 kN/m. Why is it unnecessary to investigate the section of zero shear at $x = 4.2$ m? Show that inverting the beam section will reduce the allowable load to 15.4 kN/m.

PROBLEMS

548.   The inverted T section of a 4-m simply supported beam has the properties shown in Fig. P–548. The beam carries a uniformly distributed load w over its entire length. Determine w if $\sigma_t \leqslant 30$ MN/m^2 and $\sigma_c \leqslant 70$ MN/m^2.   *Ans.* $w = 3750$ N/m

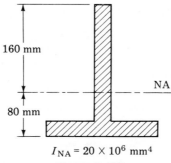

$I_{NA} = 20 \times 10^6$ mm^4

Figure P–548.

549. Determine the maximum tensile and compressive bending stresses developed in the beam shown in Fig. P–549. The cross section has the given properties.   *Ans.* $\sigma_t = 20.0$ MPa; $\sigma_c = 10.0$ MPa

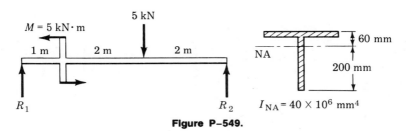

Figure P–549.

550. Find the maximum tensile and compressive flexure stresses for the cantilever beam shown in Fig. P–550.

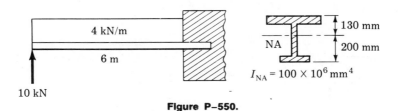

Figure P–550.

551. A beam with the cross section shown in Fig. P–551 is loaded in such a way that the maximum moments are $+1.5P$ N · m and $-2.2P$ N · m, where P is the applied load in newtons. Determine the maximum safe value of P if the working stresses are 30 MPa in tension and 70 MPa in compression.   *Ans.* $P = 16.0$ kN

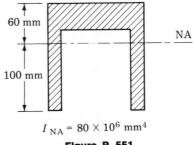

$I_{NA} = 80 \times 10^6 \text{ mm}^4$

Figure P–551.

552.   Resolve Problem 551 if the maximum moments are $+3.2P$ N · m and $-5.8P$ N · m.

553.   Determine the maximum safe value of W that can be carried by the beam shown in Fig. P–553 if $\sigma_t \leqslant 20$ MN/m^2 and $\sigma_c \leqslant 60$ MN/m^2.   *Ans.* $W = 3$ kN

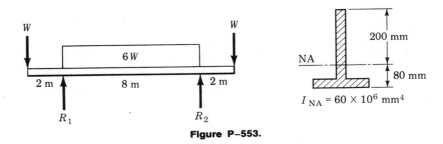

Figure P–553.

554.   What safe value of W can be applied to the beam loaded as shown in Fig. P–554 if $\sigma_t \leqslant 60$ MPa and $\sigma_c \leqslant 100$ MPa?

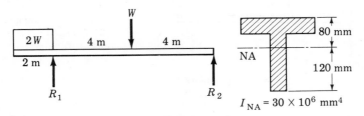

Figure P–554.

555.   A cast-iron beam carries the loads shown in Fig. P–555. If $\sigma_t \leqslant 20$ MN/m^2 and $\sigma_c \leqslant 80$ MN/m^2, compute the permissible limits of the overhang.   *Ans.* $x = 2.0$ to 2.5 m

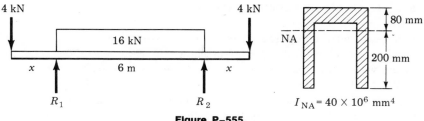

Figure P-555.

556. A T beam supports the three concentrated loads shown in Fig. P-556. Show that the NA is 70 mm from the top and that $I_{NA} = 15.52 \times 10^6$ mm⁴. Then use these values to determine the maximum value of P so that $\sigma_t \leqslant 30$ MPa and $\sigma_c \leqslant 70$ MPa.

Ans. $P = 1.41$ kN

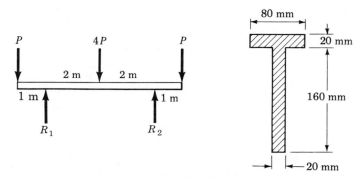

Figure P-556.

557. A cast-iron beam 10 m long and supported as shown in Fig. P-557 carries a uniformly distributed load of w N/m (including its own weight). The allowable stresses are $\sigma_t \leqslant 20$ MN/m² and $\sigma_c \leqslant 80$ MN/m². Determine the maximum safe value of w if $x = 1$ m.

Ans. $w = 1.92$ kN/m

558. In Problem 557, find the values of x and w so that w is a maximum.

Ans. $x = 1.59$ m; $w = 3.16$ kN/m

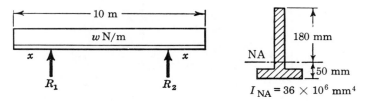

Figures P-557 and P-558.

5-6 ANALYSIS OF FLEXURE ACTION

If a beam were composed of many thin layers placed on each other, bending would produce the effect shown in Fig. 5–19. The separate layers would slide past each other and the total strength of the beam would be the sum of the strengths of the various layers. Such a built-up beam would be considerably weaker than a solid beam of equivalent dimensions. For a demonstration of this, flex a deck of playing cards between the fingers holding them rather loosely so that the cards can slide past one another as they are bent. Then grip the ends of them tightly, so that they cannot slip—thus approximating a solid section—and try to flex them. You will discover that considerably more effort is required.

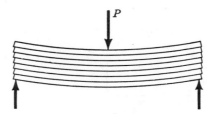

Figure 5–19. Sliding between layers of a built-up beam.

Figure 5–20a will aid in understanding this action. The figure is a pictorial representation of the flexural stress distribution over the portion to the left of the exploratory section $m–n$ of the solid beam in Fig. 5–20b.

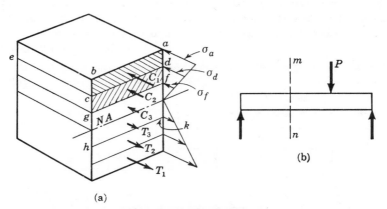

Figure 5–20. Distribution of compressive and tensile forces.

If we add the horizontal forces acting over the entire depth of the section, the compressive forces will exactly balance the tensile forces, as is required by the equilibrium condition $\Sigma X = 0$ (Art. 5–2). However, if we take a summation of horizontal forces over a partial depth of the section, say from the top elements a–b to those at c–d, the total compressive force C_1 over the area $abcd$ (equal to the area $abcd$ multiplied by the average of the stresses σ_a and σ_d) can be balanced only by a shear resistance developed at the horizontal layer dce. Such shear resistance is available in a solid beam but not in a built-up beam of unconnected layers.

If we extend the summation of horizontal forces down to layer fg, the resultant compressive force is increased by C_2, which is the average of the stresses σ_d and σ_f multiplied by the area $cdfg$. Thus a larger shear resistance must be developed over the horizontal layer at fg than at dce. Of course, the total compressive force C_1 plus C_2 acting over the area $abgf$ may also be computed as the average of the stresses σ_a and σ_f multiplied by the area $abgf$. However, the first method indicates the decreasing magnitude of the increase in the total compressive force as we descend by equal intervals from the top; i.e., although the total compressive force increases as we descend by equal intervals from the top, it does so by smaller increments.

This analysis shows that the maximum unbalanced horizontal force exists at the neutral axis. This unbalanced force decreases gradually to zero as the effects of layers below the neutral axis are included. This is so because the horizontal effect of the compressive forces is increasingly offset by the neutralizing effect of the tensile forces, until finally complete balance is attained and $\Sigma X = 0$ over the entire section.

This analysis also indicates that layers equidistant from the neutral axis, such as fg and hk, are subject to the same net horizontal unbalance, because in adding the horizontal forces from the top to these layers the equal compressive forces C_3 and T_3 cancel out. We conclude that equal shear resistances are developed at layers fg and hk. However, this requires that the areas from the neutral axis to the equidistant layers be symmetrical with respect to the neutral axis. The conclusion would not hold, for example, if the beam section were a triangle with its base horizontal.

PROBLEMS

559. A beam is composed of 6 planks, each 100 mm wide and 20 mm thick, piled loosely on each other to an overall dimension of 100 mm wide by 120 mm high. (a) Compare the strength of such a beam with that of a solid beam of equal overall dimensions. (b) What would be the ratio if the built-up beam consisted of 12 planks each 100 mm wide by 10 mm thick?   *Ans.*   (a) 1 to 6; (b) 1 to 12

560.   The wide-flange beam shown in Fig. P–560 is strengthened by riveting two cover plates 160 mm by 20 mm to the top and bottom flanges. If the maximum flexure stress is 110 MPa, compute the total force (a) in each cover plate and (b) in each flange. Neglect the weakening effect of the rivet holes.   *Ans.*   (a) 336 kN; (b) 304 kN

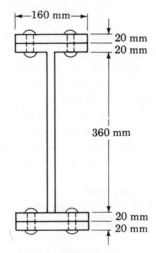

Figure P–560.

561.   A T section has the dimensions given in Fig. P–561. Show that the neutral axis is 60 mm below the top and that $I_{NA} = 26.67 \times 10^6$ mm^4. If the tensile stress at the bottom of the flange is 10 MN/m^2, determine (a) the total tensile force in the flange and (b) the total

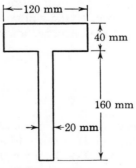

Figure P–561.

compressive force in the cross section. Also determine (c) the moment of the total compressive force and (d) the moment of the total tensile force about the NA. (e) How does the sum of (c) and (d) compare with the total applied bending moment as computed from the flexure formula?

      Ans.   (a) 96.0 kN; (b) 98.0 kN; (c) 9.15 kN · m; (d) 4.19 kN · m

562.   In any beam section having a maximum stress σ, show that the force on any partial shaded area A' in Fig. P–562 is given by $F = (\sigma/c)A'\bar{y}'$, where $\bar{y}'$ is the centroidal coordinate of A'. Also show that the moment of this force about the NA is

$$M' = \frac{\sigma}{c}I'$$

where I' is the moment of inertia of the shaded area about the NA.

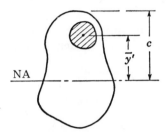

Figure P–562.

563.   A box beam is made from 50-mm by 150-mm pieces screwed together as shown in Fig. P–563. If the maximum flexure stress is 8 MPa, determine the force acting on the shaded piece and the moment of this force about the NA. (*Hint*: Use the results of Problem 562.)   *Ans.* $F = 45.0$ kN; $M = 3.50$ kN · m

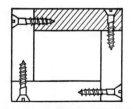

Figures P–563 and P–564.

564.   Resolve Problem 563 using one of the vertical pieces of wood instead of the shaded piece.

5-7 DERIVATION OF FORMULA FOR HORIZONTAL SHEARING STRESS

Consider two adjacent sections, (1) and (2), in a beam separated by the distance dx, as shown in Fig. 5–21, and let the shaded part between them be isolated as a free body. Figure 5–22 is a pictorial representation of this part, the beam from which it is taken being shown in dashed outline.

Assume the bending moment at section (2) to be larger than that at section (1), thus causing larger flexural stresses on section (2) than on section (1). Therefore, the resultant horizontal thrust H_2 caused by the compressive forces on section (2) will be greater than the resultant horizontal thrust H_1 on section (1). This difference between H_2 and H_1 can be balanced only by the resisting shear force dF acting on the bottom face of the free body, since no external force acts on the top or side faces of the free body.

Since $H_2 - H_1$ is the summation of the differences in thrusts $\sigma_2 \, dA$ and $\sigma_1 \, dA$ on the ends of all elements contained in the part shown

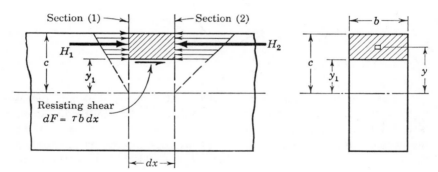

Figure 5–21.

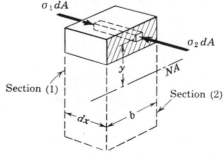

Figure 5–22.

in Fig. 5–22, a horizontal summation of forces gives

$$[\Sigma H = 0] \qquad dF = H_2 - H_1$$
$$= \int_{y_1}^{c} \sigma_2 \, dA - \int_{y_1}^{c} \sigma_1 \, dA$$

whence, replacing the flexural stress σ by its equivalent My/I, we obtain

$$dF = \frac{M_2}{I} \int_{y_1}^{c} y \, dA - \frac{M_1}{I} \int_{y_1}^{c} y \, dA = \frac{M_2 - M_1}{I} \int_{y_1}^{c} y \, dA$$

From Fig. 5–21 we note that $dF = \tau b \, dx$, where τ is the average shearing stress over the differential area of width b and length dx; also that $M_2 - M_1$ represents the differential change in bending moment dM in the distance dx; hence the above relation is rewritten as

$$\tau = \frac{dM}{Ib \, dx} \int_{y_1}^{c} y \, dA$$

From Art. 4–4 we recall that $dM/dx = V$, the vertical shear; so we obtain for the horizontal shearing stress,

$$\tau = \frac{V}{Ib} \int_{y_1}^{c} y \, dA = \frac{V}{Ib} A'\bar{y} = \frac{V}{Ib} Q \qquad (5\text{–}4)$$

We have replaced the integral $\int_{y_1}^{c} y \, dA$, which means the sum of the moments of the differential areas dA about the neutral axis, by its equivalent $A'\bar{y}$, where A' is the partial area of the section above the layer at which the shearing stress is being computed and $\bar{y}$ is the moment arm of this area with respect to the neutral axis; A' is the shaded area in the end view of Fig. 5–21. A variation of the product $A'\bar{y}$ is the symbol Q, which frequently is used to represent the static moment of area.

Shear flow

If the shearing stress τ is multiplied by the width b, we obtain a quantity q, known as shear flow, which represents the longitudinal force per unit length transmitted across the section at the level y_1. It is analogous to the shear flow discussed previously in the torsion of thin-walled tubes (see page 91). Using Eq. (5–4), we find that its value is given by

$$q = \tau b = \frac{V}{I} Q \qquad (5\text{–}4a)$$

One application of this relation is discussed in Art. 5–9; another is given in Illustrative Problem 1321 (page 541).

Relation between horizontal and vertical shearing stresses

Most students are surprised to find the term *vertical shear* (V) appearing in the formula for horizontal shearing stress (τ_h). However, as we shall show presently, a horizontal shearing stress is always accompanied by an equal vertical shearing stress. It is this vertical shearing stress τ_v, shown in Fig. 5–23, that forms the resisting vertical shear $V_r = \int \tau \, dA$ which balances the vertical shear V. Since it is not feasible to determine τ_v directly, we have resorted to deriving the numerically equal value of τ_h.

To prove the equivalence of τ_h and τ_v, consider their effect on a free-body diagram of a typical element in Fig. 5–23. A pictorial view of this element is shown in Fig. 5–24a; a front view, in Fig. 5–24b. For equilibrium of this element, the shearing stress τ_h on the bottom face requires an equal balancing shearing stress on the top face. The forces causing these shearing stresses (Fig. 5–23c) form a counterclockwise couple, which requires a clockwise couple to insure balance. The forces of this clockwise couple induce the shearing stresses τ_v on the vertical faces of the element as shown.

By taking moments about an axis through A (Fig. 5–24c), we obtain

$$[\Sigma M_A = 0] \qquad (\tau_h \, dx \, dz)dy - (\tau_v \, dy \, dz)dx = 0$$

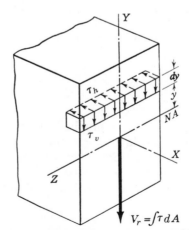

Figure 5–23. Horizontal and vertical shearing stresses.

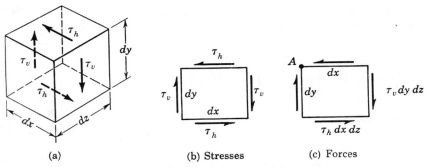

(a)   (b) Stresses   (c) Forces

Figure 5-24.   Shearing stresses on a typical element.

from which the constant product $dx\ dy\ dz$ is canceled to yield

$$\tau_h = \tau_v \tag{5-5}$$

We conclude therefore that a shearing stress acting on one face of an element is always accompanied by a numerically equal shearing stress acting on a perpendicular face.

Application to rectangular section

The distribution of shearing stresses in a rectangular section can be obtained by applying Eq. (5–4) to Fig. 5–25. For a layer at a distance y from the neutral axis, we have

$$\tau = \frac{V}{Ib}A'\bar{y} = \frac{V}{Ib}\left[b\left(\frac{h}{2} - y\right)\right]\left[y + \frac{1}{2}\left(\frac{h}{2} - y\right)\right]$$

which reduces to

$$\tau = \frac{V}{2I}\left(\frac{h^2}{4} - y^2\right)$$

This shows that the shearing stress is distributed parabolically across the depth of the section.

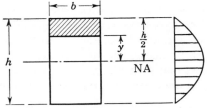

Figure 5-25.   Shearing stress is distributed parabolically across a rectangular section.

The maximum shearing stress occurs at the neutral axis and is found by substituting the dimensions of the rectangle in Eq. (5–4), as follows:

$$\tau = \frac{V}{Ib}A'\bar{y} = \frac{V}{(bh^3/12)b}\left(\frac{bh}{2}\right)\left(\frac{h}{4}\right)$$

which reduces to

$$\text{Max. } \tau = \frac{3}{2}\frac{V}{bh} = \frac{3}{2}\frac{V}{A} \tag{5–6}$$

This indicates that the maximum shearing stress in a rectangular section is 50% greater than the average shear stress.

Assumptions and limitations of formula

We have assumed, without saying so implicitly, that the shearing stress is uniform across the width of the cross section. Although this assumption does not hold rigorously, it is sufficiently accurate for sections in which the flexure forces are evenly distributed over a horizontal layer.

This condition is present in a rectangular section and in the wide-flange section shown in Fig. 5–26a, where the flexure forces on the vertical strips, both shaded and unshaded, are evenly distributed across any horizontal layer. But this condition does not exist in the triangular section in Fig. 5–26b, where the shearing stress is maximum at the left edge of the neutral axis, diminishing to zero at the right edge. Even here, however, Eq. (5–4) can be used to compute the *average* value of shearing stress across any layer. Another exception is a circular cross section (Fig. 5–26c). It can be shown that the stress at the edge of any layer must be tangent to the surface, as in the right half of the figure; but the direction of shearing stresses at interior points is unknown, although they are assumed to pass through a common center C as shown. The vertical components of these shearing stresses are usually assumed to be uniform across any layer, as in the left half of the figure, and are computed by means of Eq. (5–4). With this assumption, the maximum shearing stress across the neutral axis is $\frac{4}{3}(P/\pi r^2)$. A more elaborate study* shows that shearing stress actually varies at the neutral axis from $1.23P/\pi r^2$ at the edges to $1.38P/\pi r^2$ at the center.

*See S. Timoshenko and J. N. Goodier, *Theory of Elasticity*, 2nd ed., McGraw-Hill, New York, 1951, p. 321.

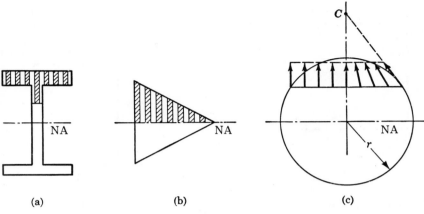

(a)   (b)   (c)

Figure 5–26.

ILLUSTRATIVE PROBLEMS

565. A simply supported beam 120 mm wide, 180 mm deep, and 6 m long carries a uniformly distributed load of 4 kN/m, as shown in Fig. 5–27. (a) Compute the shearing stress developed at horizontal layers 30 mm apart from top to bottom for a section 1 m from the left end. (b) Compute the maximum shearing stress developed in the beam.

Solution:

Part a. As shown on the shear diagram (Fig. 5–27a), the definition of vertical shear $V = (\Sigma Y)_L$ gives $V = 8$ kN at $x = 1$ m.

The moment of inertia about the neutral axis is

$$\left[I = \frac{bh^3}{12} \right] \qquad I_{NA} = \frac{120(180)^3}{12} = 58.32 \times 10^6 \text{ mm}^4$$

$$= 58.32 \times 10^{-6} \text{ m}^4$$

Applying Eq. (5–4) to a layer 30 mm from the top (Fig. 5–27b), we find that the shearing stress is

$$\left[\tau = \frac{V}{Ib} A'\bar{y} \right]$$

$$\tau = \frac{8000}{(58.32 \times 10^{-6})(0.120)}(0.120 \times 0.030)(0.075)$$

$$= 309 \text{ kPa}$$

Note that for consistency of units, the shear force is expressed in newtons, the moment of inertia in m^4 and the distances in m.

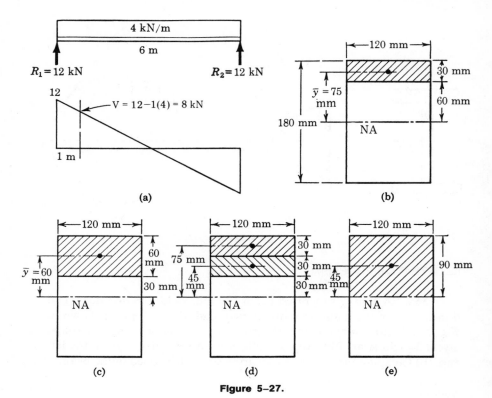

Figure 5–27.

At 60 mm from the top (Fig. 5–27c), the shearing stress is

$$\left[\tau = \frac{V}{Ib} A'\bar{y} \right]$$

$$\tau = \frac{8000}{(58.32 \times 10^{-6})(0.120)}(0.120 \times 0.060)(0.060)$$

$$= 494 \text{ kPa}$$

The shearing stress at 60 mm from the top can also be computed from Fig. 5–27d, in which the area A' is resolved into two strips 30 mm thick. Since a moment of area equals the sum of the moments of area of its parts (that is, $A'\bar{y} = \Sigma ay$), an identical result is obtained as follows:

$$\left[\tau = \frac{V}{Ib} \Sigma ay \right]$$

$$\tau = \frac{8000}{(58.32 \times 10^{-6})(0.120)}[(0.120 \times 0.030)(0.075)$$

$$+ (0.120 \times 0.030)(0.045)]$$

$$= 494 \text{ kPa}$$

Although this computation is admittedly more complex than the preceding one, it indicates the procedure to be followed when the area A' is more complex, as in the case of a wide-flange beam.

At the neutral axis, or at 90 mm from the top (Fig. 5–27e), the shearing stress is

$$\left[\tau = \frac{V}{Ib}A'\bar{y}\right]$$

$$\tau = \frac{8000}{(58.32 \times 10^{-6})(0.120)}(0.120 \times 0.090)(0.045)$$
$$= 555 \text{ kPa}$$

If desired, Eq. (5–6) may be used. As noted on page 192, this equation determines the maximum shearing stress on any rectangular section.

$$\left[\tau = \frac{3}{2}\frac{V}{bh}\right] \qquad \tau = \frac{3}{2}\frac{8000}{(0.120)(0.180)} = 555 \text{ kPa}$$

The shearing stress at the 120-mm layer and the 150-mm layer are determined similarly to be 494 and 309 kPa, respectively.

Note that equal values of τ are obtained at layers equidistant from the NA in any beam symmetrical about the neutral axis. Physically, this is true because, as was said on page 185, the compressive and tensile flexure forces between these layers cancel each other. Analytically it is true because the neutral axis is the centroidal axis, and hence the moment of area $A'\bar{y}$ computed for a partial area A' located above the NA equals that for a symmetrically placed area below the NA. Further, since the total moment of area is zero with respect to a centroidal axis, it follows that the moment of area about the NA of the area above any layer equals that of the area below that layer. Stated differently, in computing $A'\bar{y}$ we may use either the area above or that below any layer, depending upon which is easier to use.

Part b.   The maximum shearing stress occurs at the NA of the section of maximum shear. The shear diagram shows that maximum shear occurs at either end, and hence from Eq. (5–6) the maximum shearing stress is

$$\left[\tau = \frac{3}{2}\frac{V}{A}\right] \qquad \text{Max. } \tau = \frac{3}{2}\frac{12 \times 10^3}{(0.120 \times 0.180)} = 833 \text{ kPa} \qquad Ans.$$

566.   A beam has the wide-flange section shown in Fig. 5–28a. At a section where the vertical shear is $V = 70$ kN, compute (a) the maximum shearing stress and (b) the shearing stress at the junction of the flange and the web. (c) Plot the shearing stress distribution in the web, and determine the percentage of shear carried by the web alone.

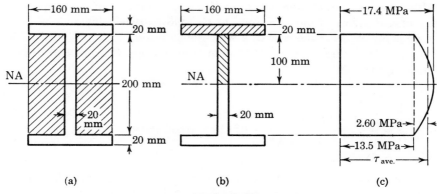

Figure 5–28.

Solution: The moment of inertia is found by resolving the section into a large rectangle from which is subtracted the two shaded rectangles. We obtain

$$\left[I = \Sigma \, \frac{bh^3}{12} \right]$$

$$I_{NA} = \frac{160(240)^3}{12} - 2 \left[\frac{70(200)^3}{12} \right]$$

$$= 91.0 \times 10^6 \, \text{mm}^4 = 91.0 \times 10^{-6} \, \text{m}^4$$

The maximum shearing stress occurs at the neutral axis. In applying Eq. (5–4), compute $A'\bar{y}$ as the sum of the moments of area of the rectangles shaded in Fig. 5–28b.

$$\left[\tau = \frac{V}{Ib} A'\bar{y} \right]$$

$$\text{Max. } \tau = \frac{70 \times 10^3}{(91.0 \times 10^{-6})(0.020)}$$
$$\times \left[(0.160 \times 0.020)(0.110) + (0.020 \times 0.100)(0.050) \right]$$
$$= 17.4 \, \text{MPa}$$

At the junction of the web and flange, there is a discontinuity in the shearing stress because the width $b = 160$ mm when computing the shearing stress in the flange whereas $b = 20$ mm when considering stresses in the web. Then, at the junction, the shearing stress *in the web* is

$$\left[\tau = \frac{V}{Ib} A'\bar{y} \right] \qquad \tau = \frac{70 \times 10^3}{(91.0 \times 10^{-6})(0.020)} (0.160 \times 0.020)(0.110)$$
$$= 13.5 \, \text{MPa}$$

Show that the shearing stress at the junction *in the flange* is 1.69 MPa.

The shearing stresses in the web vary parabolically from top to bottom, as shown in Fig. 5–28c. The average height of the parabolic segment is $\frac{2}{3}(17.4 - 13.5) = 2.60$ MPa. Therefore the average shear stress in the web is

$$\tau_{ave.} = 13.5 + 2.60 = 16.1 \text{ MPa}$$

The shearing force in the web is

$$[P = A\sigma_{ave.}] \qquad V_{web} = (0.200 \times 0.020)(16.1 \times 10^6) = 64.4 \text{ kN}$$

whence the percentage of shear carried by the web alone is

$$\%V_{web} = \frac{64.4}{70} \times 100 = 92.0\%$$

This shows that the flanges are almost ineffective in resisting the vertical shear. If it is assumed that the total vertical shear is carried by the web alone, the average shearing stress will be very close to the maximum stress as computed from Eq. (5–4). Thus

$$\left[\tau = \frac{V}{A_{web}}\right] \qquad \tau = \frac{70 \times 10^3}{(0.200)(0.020)} = 17.5 \text{ MPa}$$

This is very close to the computed maximum, 17.4 MPa.

This method gives results that closely approximate the actual maximum τ. In most design specifications or codes, however, the height of the web is not taken as the distance between flanges but is assumed to be the total depth of the beam. This procedure is not as accurate as the method just given, but lower allowable shearing stresses are usually specified in order to compensate.

PROBLEMS

567.   A timber beam 90 mm wide by 160 mm high is subjected to a vertical shear $V = 20$ kN. Determine the shearing stress developed at layers 20 mm apart from top to bottom of the section.

568.   Show that the shearing stress developed at the neutral axis of a beam with circular cross section is $\tau = \frac{4}{3}(V/\pi r^2)$. Assume that the shearing stress is uniformly distributed across the neutral axis.

569.   Show that the maximum shearing stress in a beam having a thin-walled tubular section of net area A is $\tau = 2V/A$.

570.   A simply supported beam 4 m long has the cross section shown in Fig. P–570. Determine the maximum uniformly distributed load which can be applied over the entire length of the beam if the shearing stress is limited to 1.2 MPa.   *Ans.* $w = 4.60$ kN/m

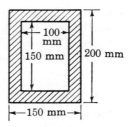

Figure P–570.

571. The T section shown in Fig. P–571 is the cross section of a beam formed by joining two rectangular pieces of wood together. The beam is subjected to a maximum shearing force of 60 kN. Show that the NA is 34 mm from the top and that $I_{NA} = 10.57 \times 10^6$ mm^4. Using these values, determine the shearing stress (a) at the neutral axis and (b) at the junction between the two pieces of wood.

Ans. (a) 3.28 MPa; (b) 3.18 MPa, 31.8 MPa

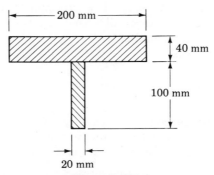

Figure P–571.

572. In Fig. P–572, if $P = 5$ kN, compute the shearing stress at horizontal layers 20 mm apart from bottom to top at the section of maximum V. The NA is 70 mm from the top and $I_{NA} = 15.52 \times 10^6$ mm^4.

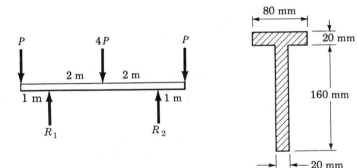

Figure P–572.

573. The cross section of a beam is an isosceles triangle with vertex uppermost, of altitude h and base b. If V is the vertical shear, show that the maximum shearing stress is $3V/bh$ located at the midpoint of the altitude.

574. In the beam section shown in Fig. P–574, prove that the maximum horizontal shearing stress occurs at a layer $h/8$ above or below the NA.

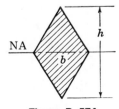

Figure P–574.

575. Determine the maximum and minimum shearing stress in the web of the wide-flange section in Fig. P–575 if $V = 100$ kN. Also compute the percentage of vertical shear carried only by the web of the beam.   *Ans.* Max. $\tau = 30.5$ MPa; Min. $\tau = 23.5$ MPa; 90.2%

576. Rework Problem 575 assuming that the web is 200 mm instead of 160 mm.

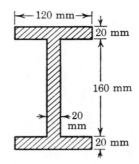

Figures P–575 and P–576.

577. A plywood beam is built up of 6-mm strips separated by blocks as shown in Fig. P–577. What shearing force V will cause a maximum shearing stress of 1.4 MPa?

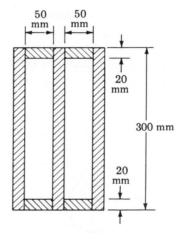

Figure P–577.

5–8   DESIGN FOR FLEXURE AND SHEAR

In this article we consider the determination of load capacity or the size of beam section that will satisfy allowable stresses in both flexure and shear. No principles are required beyond those already developed.

In heavily loaded short beams the design is usually governed by the shearing stress (which varies with V); but in longer beams the flexure stress generally governs because the bending moment varies with both load and length of beam. Shearing is more important in timber beams than in steel beams because of the low shearing strength of wood.

ILLUSTRATIVE PROBLEMS

578. A rectangular beam carries a distributed load of w N/m on a simply supported span of L meters. Determine the critical length at which the shearing stress τ and the flexure stress σ reach their allowable values simultaneously.

Solution: As shown in Fig. 5–29, max. $V = W/2$, where W is the total distributed load. The maximum load as limited by the allowable shearing stress is determined from Eq. (5–6):

$$\left[\text{Max. } \tau = \frac{3}{2}\frac{V}{bh} \right] \qquad \tau = \frac{3}{2} \cdot \frac{W/2}{bh} \qquad W = \frac{4}{3}bh\tau$$

Note that W is independent of the length.

At the point of zero shear, the maximum bending moment, computed from the area of the shear diagram, is

$$M = \frac{1}{2}\left(\frac{W}{2}\right)\left(\frac{L}{2}\right) = \frac{WL}{8} \text{ N} \cdot \text{m}$$

Substituting this value in the flexure formula, Eq. (5–2a), we obtain

$$\left[M = \frac{\sigma I}{c} = \frac{\sigma bh^2}{6} \right] \qquad \frac{WL}{8} = \frac{\sigma bh^2}{6}$$

Replacing W by its value in terms of the shear stress, we have

$$\left(\frac{4}{3}bh\tau\right)\left(\frac{L}{8}\right) = \frac{\sigma bh^2}{6}$$

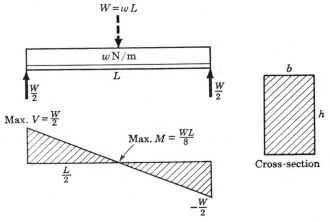

Figure 5–29.

which reduces to

$$L = \frac{\sigma h}{\tau}$$

For values larger than this critical length, flexure governs the design; for shorter values, shear governs.

579. A box beam supports the loads shown in Fig. 5–30. Compute the maximum value of P that will not exceed a flexural stress $\sigma = 8$ MPa or a shearing stress $\tau = 1.2$ MPa for sections between the supports.

Solution: We start by computing I for the net section, which is the difference between two rectangles. Hence

$$I = \Sigma \frac{bh^3}{12} = \frac{160(200)^3}{12} - \frac{120(160)^3}{12} = 65.7 \times 10^6 \text{ mm}^4$$
$$= 65.7 \times 10^{-6} \text{ m}^4$$

Determining the reactions from statics gives the shear diagram shown in Fig. 5–30. In terms of P, the maximum V is $-(\frac{1}{2}P + 2000)$. If the area of the cross section above the NA, where τ is a maximum, is resolved into the three rectangles shown, the static moment of the area, Q, is

$$[Q = \Sigma a \bar{y}] \qquad Q = (160 \times 20)(90) + 2(80 \times 20)(40)$$
$$= 416 \times 10^3 \text{ mm}^3 = 416 \times 10^{-6} \text{ m}^3$$

If desired, the area could also be resolved into the difference between the outer 100-mm by 160-mm rectangle and the inner 80-mm by 120-mm rectangle. This gives the same value of Q, viz.,

$$[Q = \Sigma a \bar{y}] \qquad Q = (160 \times 100)(50) - (120 \times 80)(40)$$
$$= 416 \times 10^3 \text{ mm}^3$$

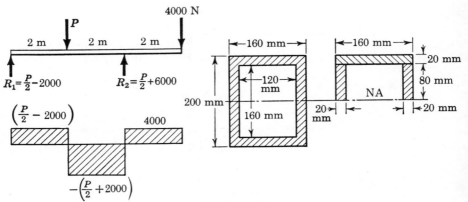

Figure 5–30.

We now substitute the absolute values of V and Q into Eq. (5–4) to obtain

$$\left[\tau = \frac{V}{Ib}Q\right] \quad 1.2 \times 10^6 = \frac{\frac{1}{2}P + 2000}{(65.7 \times 10^{-6})(0.040)}(416 \times 10^{-6})$$

from which

$$P = 11.2 \text{ kN}$$

The maximum moment between the supports in terms of P is at $x = 2$ m and has the value

$$M = \left(\tfrac{1}{2}P - 2000\right)(2) = (P - 4000) \text{ N} \cdot \text{m}$$

Applying the flexure formula, we have

$$\left[M = \frac{\sigma I}{c}\right] \quad P - 4000 = \frac{(8 \times 10^6)(65.7 \times 10^{-6})}{0.100}$$

$$P = 9.26 \text{ kN}$$

The maximum safe value of P is the smaller of the above values, namely, $P = 9.26$ kN.

PROBLEMS

580.   A rectangular beam of width b meters and height h meters carries a central concentrated load P on a simply supported span of length L meters. Express the maximum τ in terms of the maximum σ_f.

Ans. $\tau = \sigma_f h / 2L$

581.   A laminated beam is composed of three planks, each 150 mm by 60 mm, glued together to form a section 150 mm wide by 180 mm high. The allowable shear stress in the glue is 600 kPa, the allowable shear stress in the wood is 900 kPa, and the allowable flexure stress in the wood is 8 MPa. Determine the maximum uniformly distributed load which can be carried by the beam on a 2-m simple span.   *Ans.* $w = 12.2$ kN/m

582.   Find the cross-sectional dimensions of the smallest square beam which can be loaded as shown in Fig. P–582 if $\tau \leqslant 900$ kPa and $\sigma \leqslant 8$ MPa.

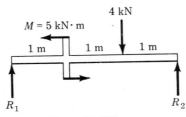

Figure P–582.

583.   A wide-flange section having the dimensions shown in Fig. P–583 supports a central concentrated load P on a simple span L meters long. Determine the ratio of the maximum flexure stress to the maximum shear stress.

584.   A built-up timber beam having the same cross section as that in Problem 583 is used to support a concentrated load P on a simply supported span 8 m long. Determine P and its location that would cause simultaneously a maximum flexural stress of 8 MPa and a maximum shearing stress of 1.2 MPa.

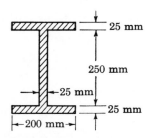

25 mm

250 mm

25 mm

25 mm

200 mm

Figures P–583 and P–584.

585.   A simply supported beam L meters long carries a uniformly distributed load of 16 kN/m over its entire length and has the cross section shown in Fig. P–585. Find L to cause a maximum flexural stress of 40 MPa. What maximum shearing stress is then developed?

Ans. $L = 1.77$ m; $\tau = 5.55$ MPa

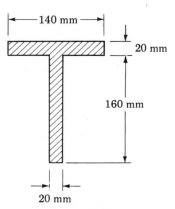

140 mm

20 mm

160 mm

20 mm

Figure P–585.

586.   A simply supported beam 6 m long carries a uniformly varying load which varies from zero at one end to w N/m at the other.

The beam section is the same as that in Fig. P–577. Find the maximum safe value of w if $\sigma_f \leqslant 10$ MPa and $\tau \leqslant 800$ kPa.

587.   The wide flange beam shown in Fig. P–587 supports the concentrated load W and a total uniformly distributed load of $2W$. Determine the maximum safe value of W if $\sigma_f \leqslant 10$ MPa and $\tau \leqslant 1.4$ MPa.   *Ans.* $W = 2.62$ kN

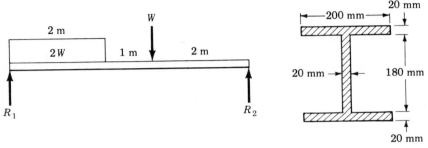

Figure P–587.

588.   The distributed load shown in Fig. P–588 is supported by a box beam with the given dimensions. Determine the maximum value of w that will not exceed a flexural stress of 14 MN/m² or a shearing stress of 1.2 MN/m².

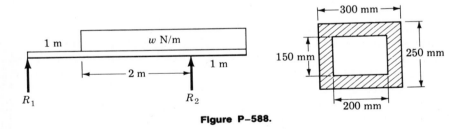

Figure P–588.

589.   A channel section carries two concentrated loads W and a total uniformly distributed load of $8W$, as shown in Fig. P–589. Verify that the NA is 50 mm above the bottom and that $I_{NA} = 15.96 \times 10^6$ mm⁴. Then use these values to determine the maximum value of W that will not exceed allowable stresses in tension of 30 MPa, in compression of 70 MPa, or in shear of 20 MPa.   *Ans.* $W = 3.19$ kN

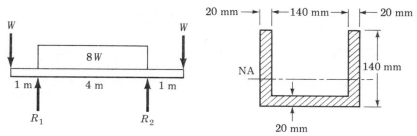

Figure P-589.

590. A rectangular beam, 150 mm wide by 250 mm high, carries a uniformly distributed load of 8 kN/m and a concentrated load P as shown in Fig. P–590. Determine the maximum safe value of P if $\sigma \leqslant$ 10 MPa and $\tau \leqslant$ 1.2 MPa.

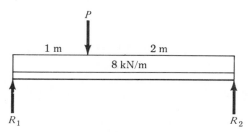

Figure P-590.

5-9   SPACING OF RIVETS OR BOLTS IN BUILT-UP BEAMS

In our analysis of flexure action (Art. 5–6) we showed that the various elements composing a built-up beam tend to slide past one another. We shall now consider the size and spacing of rivets or bolts in a built-up beam to resist this sliding action. The first step is to calculate the force to be resisted by such rivets.

Figure 5–31 shows a beam composed of three planks bolted together by two rows of bolts spaced e apart. Equation (5–4) gives the shearing stress at the contact surface between the two upper planks as

$$\tau = \frac{V}{Ib} Q$$

where Q is the static moment about the NA of the shaded area in the end view. Multiplying this shearing stress by the shaded area eb in the top view gives the force F to be resisted in a length e:

$$F = \tau(eb) = \frac{V}{Ib} Q(eb) = \frac{Ve}{I} Q$$

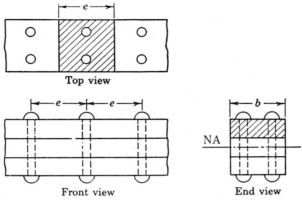

Top view

Front view   End view

Figure 5–31.

The same result can be obtained more directly by using the concept of shear flow, which is the longitudinal shearing force developed per unit length. Thus in the length e, Eq. (5–4a) determines the shear force to be

$$F = qe = \frac{VQ}{I} e$$

as before.

Friction being neglected, this force is resisted by the shearing or bearing strength R of the bolts, whichever is smaller. Equating R to F gives

$$R = \frac{Ve}{I} Q \qquad (5\text{–}7)$$

If the vertical shear varies in a beam, V is the average vertical shear in the interval e; but it is usually taken as the maximum V in this interval, especially in built-up steel girders where the length of the interval is taken as a panel length equal to the depth of the girder. In this case, Eq. (5–7) gives the rivet pitch in each panel length.

ILLUSTRATIVE PROBLEM

591.   A plate and angle girder is fabricated by attaching the short legs of four $125 \times 90 \times 13$ mm angles to a web plate 1100 mm by 10 mm to form a section 1120 mm deep, as shown in Fig. 5–32. The moment of inertia* about the NA is $I = 4140 \times 10^6$ mm⁴. At a section

---

*Current specifications of AISC call for no deduction for rivet holes in computing I, provided that the rivet hole area does not exceed 15% of the gross flange area. If it does, only the area in excess of 15% need be considered in modifying I to deduct for rivet holes.

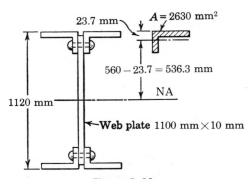

Figure 5–32.

where $V = 450$ kN, determine the spacing between 19-mm rivets that fasten the angles to the web plate. Use $\tau = 100$ MPa; in bearing, use $\sigma_b = 220$ MPa for rivets in single shear and $\sigma_b = 280$ MPa for rivets in double shear.

Solution: The rivets must resist the longitudinal force tending to slide the two flanges past the web. Hence it is the static moment of area of these two flange angles that must be used in Eq. (5–7). Referring to Fig. 5–32, we obtain

$$Q = 2(2630)(536.3) = 2820 \times 10^3 \text{ mm}^3 = 2820 \times 10^{-6} \text{ m}^3$$

The shearing resistance of a 19-mm rivet in double shear is

$$R_s = (A_s\tau)(2) = \frac{\pi}{4}(0.019)^2(100 \times 10^6)(2) = 56.7 \text{ kN}$$

The bearing resistance against the web plate is

$$R_b = (dt)\sigma_b = (0.019)(0.010)(280 \times 10^6) = 53.2 \text{ kN}$$

Using the lower of these values in Eq. (5–7), we get the required rivet pitch

$$e = \frac{RI}{VQ} = \frac{(53.2 \times 10^3)(4140 \times 10^{-6})}{(450 \times 10^3)(2820 \times 10^{-6})} = 0.174 \text{ m}$$
$$= 174 \text{ mm} \quad \textit{Ans.}$$

PROBLEMS

592.   A wide-flange section is formed by bolting together three planks, each 80 mm by 200 mm, arranged as shown in Fig. P–592. If each bolt can withstand a shearing force of 8 kN, determine the pitch if

the beam is loaded so as to cause a maximum shearing stress of
1.2 MPa.   *Ans.*   e = 98.2 mm

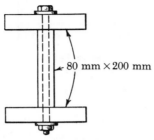

Figure P-592.

593.   A box beam, built up as shown in Fig. P-593. is secured by
screws spaced 100 mm apart. The beam supports a concentrated load *P*
at the third point of a simply supported span 3 m long. Determine the
maximum value of *P* that will not exceed a shearing stress of 800 kPa in
the beam or a total shearing force of 1200 N in the screws. What is the
maximum flexural stress in the beam?

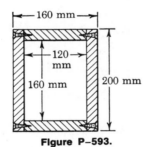

Figure P-593.

594.   A distributed load of *w* N/m is applied over the entire
length of a simply supported beam 4 m long. The beam section is that of
Problem 593, but used here so that the 160-mm dimension is vertical.
Determine the maximum value of *w* if $\sigma_f \leqslant 10$ MPa, $\tau \leqslant 800$ kPa, and
the screws have a shear strength of 800 N and a pitch of 50 mm.

Ans.   w = 2.05 kN/m

595.   A concentrated load *P* is carried at midspan of a simply
supported beam 6 m long. The beam is made of 50-mm by 150-mm
pieces of wood, screwed together as shown in Fig. P-595. If the
maximum flexural stress developed is 9 MN/m², find the pitch of the
screws if each screw can resist 800 N.

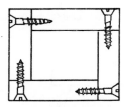

Figure P–595.

596. Three planks 100 mm by 150 mm, arranged as shown in Fig. P–596 and supported by bolts 0.4 m apart, are used to support a concentrated load P at the center of a simply supported span 6 m long. If P causes a maximum flexural stress of 12 MPa, determine the bolt diameters, assuming the shear between the planks is transmitted by friction only. The bolts are tightened to a tension of 140 MPa, and the coefficient of friction between the planks is 0.40.   *Ans.* $d = 19.1$ mm

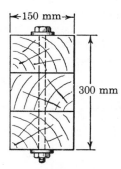

Figure P–596.

597. A plate and angle girder similar to that shown in Fig. 5–32 is fabricated by riveting the short legs of four 125 × 75 × 13 mm angles to a web plate 1000 mm by 10 mm to form a section 1020 mm deep. Cover plates, each 300 mm by 10 mm, are then riveted to the flange angles making the overall height 1040 mm. The moment of inertia of the entire section about the NA is $I = 4770 \times 10^6$ mm⁴. Using the allowable stresses specified in Illustrative Problem 591, determine the rivet pitch for 22-mm rivets attaching the angles to the web plate at a section where $V = 450$ kN.

598. As shown in Fig. P–598, two C380 × 60 channels are riveted together by pairs of 19-mm rivets spaced 200 mm apart along the length of the beam. What maximum vertical shear V can be applied to the section without exceeding the stresses given in Illustrative Problem 591?   *Ans.*   25.9 kN

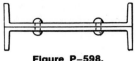

Figure P–598.

599. A beam is formed by riveting together two W250 × 73 sections as shown in Fig. P–599. It is used to support a uniformly distributed load of 30 kN/m (including the weight of the beam) on a simply supported span of 8 m. Compute the maximum flexural stress and the pitch between rivets that have a shearing strength of 26 kN.

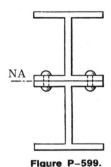

NA

Figure P–599.

SUMMARY

For homogeneous beams, originally straight, carrying transverse loads in the plane of symmetry, the bending moment creates flexural stresses expressed by

$$\sigma = \frac{My}{I} \tag{5--2}$$

The flexural stresses vary directly with their distance y from the neutral axis, which coincides with the centroidal axis of the cross section.

Maximum flexural stresses occur at the section of maximum bending moment at the extreme fibers of the section. The distance from the NA to the extreme fibers being denoted by c, the flexure formula becomes

$$\text{Max. } \sigma = \frac{Mc}{I} = \frac{M}{S} \tag{5--2a, b}$$

in which $S = I/c$ represents the section modulus of the beam. For geometric shapes, values of S are tabulated in Table 5–1 (page 159); for structural shapes, the values are given in Appendix B.

The vertical shear sets up numerically equal shearing stresses on longitudinal and transverse sections (Eq. 5–5, page 191), which are determined from

$$\tau = \frac{V}{Ib}A'\bar{y} = \frac{V}{Ib}Q \tag{5–4}$$

in which A' is the partial area of the cross section above a line drawn through the point at which the shearing stress is desired. $Q = A'\bar{y}$ is the static moment about the NA of this area (or of the area below this line).

Maximum shearing stresses occur at the section of maximum V and usually at the NA. For rectangular beams, the maximum shearing stress is

$$\text{Max. } \tau = \frac{3}{2}\frac{V}{bh} \tag{5–6}$$

In wide-flange beams, a very close approximation is

$$\text{Max. } \tau = \frac{V}{A_{\text{web}}}$$

where A_{web} is the web area between the flanges.

The rivet pitch in built-up beams is given by

$$e = \frac{RI}{VQ} \tag{5–7}$$

where R is the rivet resistance in the pitch length e, I is the moment of inertia of the gross section about the NA, V is the maximum vertical shear in the interval e, and Q is the moment of area about the NA of the elements whose sliding is resisted by the rivets.

Beam Deflections

6-1 INTRODUCTION

In this chapter we consider the rigidity of beams. Frequently the design of a beam is determined by its rigidity rather than by its strength. For example, in designing metalworking equipment for precision work, such as lathes, milling machines, or grinders, the deformations must be kept below the permissible tolerances of the work being machined. Again, floor beams carrying plastered ceilings beneath them are usually restricted to a maximum deflection of 1/360 of their length in order to avoid cracks in the plaster. One of the most important applications of beam deflections is to obtain equations with which, in combination with the conditions of static equilibrium, statically indeterminate beams can be analyzed. (See Chapters 7 and 8.)

Several methods are available for determining beam deflections. Although based on the same principles, they differ in technique and in their immediate objective. We shall consider first a modernization of the double-integration method which greatly broadens and simplifies its application. Another method, the area-moment method, is thought to be the most direct of any, especially when the deflection at a specific location is desired. After a preliminary discussion (Art. 6–4), it will be found to be not only simple, but extremely rapid to apply. A variation of it, which we will take up in Art. 8–7, is also rapid and easy to use.

Other methods are the conjugate-beam method and the method of superposition. The conjugate-beam method is a variation of the area-moment method but differs from it in technique. The method of superposition is not an independent method; it uses the deflection formulas for certain fundamental types of loadings to obtain results for loadings that consist of combinations of these fundamental types.

6–2 DOUBLE-INTEGRATION METHOD

The edge view of the neutral surface of a deflected beam is called the *elastic curve* of the beam. It is shown greatly exaggerated in Fig. 6–1. This article shows how to determine the equation of this curve, i.e., how to determine the vertical displacement y of any point in terms of its x coordinate.

Select the left end of the beam as the origin of an X axis directed along the original undeflected position of the beam, and a Y axis directed positive upward. The deflections are assumed to be so small that there is no appreciable difference between the original length of the beam and the projection of its deflected length. Consequently, the elastic curve is very flat and its slope at any point is very small. The value of the slope, $\tan \theta = dy/dx$, may therefore with only small error be set equal to θ; hence

$$\theta = \frac{dy}{dx} \qquad\qquad\qquad (a)$$

and

$$\frac{d\theta}{dx} = \frac{d^2y}{dx^2} \qquad\qquad\qquad (b)$$

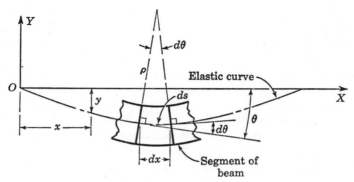

Figure 6–1. Elastic curve.

   If we now consider the variation in θ in a differential length ds caused by bending in the beam, it is evident that

$$ds = \rho\, d\theta \qquad\qquad\qquad (c)$$

where ρ is the radius of curvature over the arc length ds. Because the elastic curve is very flat, ds is practically equivalent to dx; so from Eqs. (c) and (b) we obtain

$$\frac{1}{\rho} = \frac{d\theta}{ds} \approx \frac{d\theta}{dx} \quad \text{or} \quad \frac{1}{\rho} = \frac{d^2y}{dx^2} \qquad\qquad (d)$$

   In deriving the flexure formula in Art. 5-2, we obtained on page 157 the relation

$$\frac{1}{\rho} = \frac{M}{EI} \qquad\qquad\qquad (5\text{-}1)$$

Equating the values of $1/\rho$ from Eqs. (d) and (5-1), we have

$$EI\frac{d^2y}{dx^2} = M \qquad\qquad\qquad \textbf{(6-1)}$$

This is known as the differential equation of the elastic curve of a beam. The product EI, called the *flexural rigidity* of the beam, is usually constant along the beam.

   The approximations we have made do not seriously invalidate Eq. (6-1), for if we replace $1/\rho$ by its exact value as found in any calculus text, we have, from Eq. (5-1),

$$\frac{\dfrac{d^2y}{dx^2}}{\left[1 + \left(\dfrac{dy}{dx}\right)^2\right]^{3/2}} = \frac{M}{EI}$$

Since dy/dx is very small, its square is negligible compared with unity, and hence we obtain

$$\frac{d^2y}{dx^2} = \frac{M}{EI}$$

which is the same as Eq. (6-1).

   If Eq. (6-1) is now integrated, assuming EI constant, we obtain

$$EI\frac{dy}{dx} = \int M\, dx + C_1 \qquad\qquad\qquad \textbf{(6-2)}$$

This is the slope equation specifying the slope or value of dy/dx at any point. Note that here M represents the moment equation expressed in terms of x, and C_1 is a constant to be evaluated from the given conditions of loading.

We now integrate Eq. (6–2) to obtain

$$EIy = \int \int M \, dx \, dx + C_1 x + C_2 \tag{6–3}$$

This is the required deflection equation of the elastic curve specifying the value of y for any value of x; C_2 is another constant of integration which must be evaluated from the given conditions of the beam and its loading.

If the loading conditions change along the beam, there is a corresponding change in the moment equation. This requires that a separate moment equation be written between each change of load point and that two integrations of Eq. (6–1) be made for each such moment equation. Evaluation of the constants introduced by each integration can become very involved. Fortunately, these complications can be avoided by writing a single moment equation in such a way that it becomes continuous for the entire length of the beam in spite of the discontinuity of loading.

For example, consider the beam shown in Fig. 6–2. Using the definition $M = (\Sigma M)_L$ discussed in Art. 4–2, we find that the moment equations between the change of load points are

$$M_{AB} = 480x \text{ N·m}$$
$$M_{BC} = \left[480x - 500(x - 2) \right] \text{ N·m}$$
$$M_{CD} = \left[480x - 500(x - 2) - \frac{450}{2}(x - 3)^2 \right] \text{ N·m}$$

Observe that the equation for M_{CD} will also be valid for both M_{AB} and M_{BC} provided that the terms $(x - 2)$ and $(x - 3)^2$ are neglected for values of x less than 2 and 3, respectively. In other words, the terms $(x - 2)$ and $(x - 3)^2$ are nonexistent for values of x for which the terms in parentheses are negative.

As a reminder of these restrictions, we adopt a notation in which the usual form of parentheses is replaced by pointed brackets, viz., $\langle \ \rangle$. With this change in notation, we obtain a single moment equation

$$M = \left(480x - 500\langle x - 2 \rangle - \frac{450}{2}\langle x - 3 \rangle^2 \right) \text{ N·m}$$

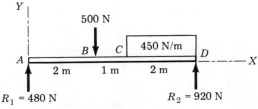

Figure 6–2.

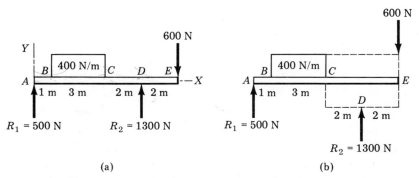

Figure 6–3.   Technique of establishing continuity of loading.

which is valid for the entire beam if we postulate that the terms between the pointed brackets do not exist for negative values; otherwise the term is to be treated like any ordinary expression.*

As another example, consider the beam in Fig. 6–3a. Here the distributed load extends only over the segment BC. We can create continuity, however, by assuming that the distributed load extends beyond C and adding an equal upward-distributed load to cancel its effect beyond C, as shown in Fig. 6–3b. The general moment equation, written for the last segment DE in our new notation using pointed brackets, is

$$M = \left(500x - \frac{400}{2}\langle x - 1\rangle^2 + \frac{400}{2}\langle x - 4\rangle^2 + 1300\langle x - 6\rangle\right) \text{N·m}$$

As before, we specify that the terms between the pointed brackets do not exist for negative values. Notice that all loadings are automatically included in the general moment equation by writing it for the last segment of the beam.

ILLUSTRATIVE PROBLEMS

601.   A concentrated load of 300 N is supported as shown in Fig. 6–4. Determine the equations of the elastic curve between each change of load point, and the maximum deflection in the beam.

*The justification for ignoring negative values of the terms in the pointed brackets depends on the fact that the general moment equation is written using the definition $M = (\Sigma M)_L$, which means that we consider the effects of loads lying only to the *left* of an exploratory section. A negative value of the terms in a pointed bracket indicates a loading that is to the *right* of an exploratory section, whereas a zero value merely indicates the start of a loading.

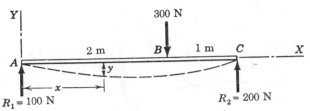

Figure 6–4.

Solution: Writing the general moment equation for the last segment *BC* of the beam, applying the differential equation of the elastic curve, and integrating twice, we obtain the following slope and deflection equations:

$$EI\frac{d^2y}{dx^2} = M = (100x - 300\langle x - 2\rangle)\, \text{N}\cdot\text{m} \qquad (a)$$

$$EI\frac{dy}{dx} = (50x^2 - 150\langle x - 2\rangle^2 + C_1)\, \text{N}\cdot\text{m}^2 \qquad (b)$$

$$EIy = \left(\frac{50}{3}x^3 - 50\langle x - 2\rangle^3 + C_1x + C_2\right)\text{N}\cdot\text{m}^3 \qquad (c)$$

To evaluate the two constants of integration which are physically equivalent to slope and deflection at the origin, we apply the following boundary conditions:

1. At *A* where $x = 0$, the deflection $y = 0$. Substituting these values in Eq. (*c*), we find that $C_2 = 0$. Remember that $\langle x - 2\rangle^3$ is to be ignored for negative values.

2. At the other support where $x = 3$, the deflection y is also zero. Knowing that $C_2 = 0$ and substituting these values in the deflection equation (*c*), we obtain

$$0 = \frac{50}{3}(3)^3 - 50(3 - 2)^3 + 3C_1 \quad \text{or} \quad C_1 = -133\, \text{N}\cdot\text{m}^2$$

Having thus evaluated the constants of integration, we return to Eqs. (*b*) and (*c*) to rewrite the slope and deflection equations in the conventional form shown in the tabulation on page 219.

Continuing the solution, we assume that the maximum deflection will occur in the segment *AB*. Its location may be found by differentiating Eq. (*e*) with respect to *x* and setting the derivative equal to zero, or, what amounts to the same thing, setting the slope equation (*d*) equal to zero and solving for the point of zero slope. We obtain

$$50x^2 - 133 = 0 \quad \text{or} \quad x = 1.63\, \text{m}$$

SEGMENT AB ($0 < x < 2$)
(d) $EI\dfrac{dy}{dx} = (50x^2 - 133)$ N·m^2
(e) $EIy = \left(\dfrac{50}{3}x^3 - 133x\right)$ N·m^3

SEGMENT BC ($2 < x < 3$)
(f) $EI\dfrac{dy}{dx} = [50x^2 - 150(x-2)^2 - 133]$ N· m^2
(g) $EIy = \left[\dfrac{50}{3}x^3 - 50(x-2)^3 - 133x\right]$ N·m^3

Since this value of x is valid for the segment AB, our assumption that the maximum deflection occurs in this region is confirmed. Hence, to determine the maximum deflection, we substitute $x = 1.63$ in Eq. (e), which yields

$$\text{Max. } EIy = -145 \text{ N·m}^3$$

The negative value obtained indicates that the deflection y is downward from the X axis. Frequently only the magnitude of the deflection, without regard to sign, is desired; this is denoted by δ, the use of y being reserved to indicate a directed value of deflection.

The unit of the product EIy is N·m^3. This follows from integrating Eq. (6–1) twice. With M in units of N · m, the first integration gives N·m^2 as the unit of the slope equation. A second integration results in N·m^3 as the unit of the deflection equation. For consistent units, E must be in units of N/m^2 and I in units of m^4. Then the deflection y will be determined in meters. For example, if $E = 10 \times 10^9$ N/m^2 and $I = 1.5 \times 10^6$ mm^4 = 1.5×10^{-6} m^4, the value of y is

$$(10 \times 10^9)(1.5 \times 10^{-6})y = -145$$

whence

$$y = -9.67 \times 10^{-3} \text{ m} = -9.67 \text{ mm}$$

602.   Find the value of EIy at the position midway between the supports and at the overhanging end for the beam shown in Fig. 6–5.

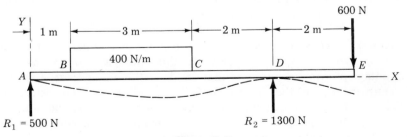

Figure 6–5.

Solution: This is the same beam for which we determined the general moment equation on page 217. Applying the differential equation of the elastic curve, and integrating twice, we obtain

$$EI\frac{d^2y}{dx^2} = M = \left(500x - \frac{400}{2}\langle x-1\rangle^2 + \frac{400}{2}\langle x-4\rangle^2\right.$$

$$\left. + 1300\langle x-6\rangle\right) \text{N·m}$$

$$EI\frac{dy}{dx} = \left(250x^2 - \frac{200}{3}\langle x-1\rangle^3 + \frac{200}{3}\langle x-4\rangle^3\right.$$

$$\left. + 650\langle x-6\rangle^2 + C_1\right) \text{N·m}^2$$

$$EIy = \left(\frac{250}{3}x^3 - \frac{50}{3}\langle x-1\rangle^4 + \frac{50}{3}\langle x-4\rangle^4\right.$$

$$\left. + \frac{650}{3}\langle x-6\rangle^3 + C_1x + C_2\right) \text{N·m}^3$$

To determine C_2, we note that $EIy = 0$ at $x = 0$, which gives $C_2 = 0$. Note that we ignore the negative terms in the pointed brackets. Next we use the condition that $EIy = 0$ at the right support where $x = 6$. This gives

$$0 = \frac{250}{3}(6)^3 - \frac{50}{3}(5)^4 + \frac{50}{3}(2)^4 + 6C_1 \quad \text{or} \quad C_1 = -1308 \text{ N·m}^2$$

Finally, to obtain the midspan deflection, we substitute $x = 3$ in the deflection equation for segment BC obtained by ignoring negative values of the bracketed terms $\langle x-4\rangle^4$ and $\langle x-6\rangle^3$. We obtain

$$EIy = \frac{250}{3}(3)^3 - \frac{50}{3}(2)^4 - 1308(3) = -1941 \text{ N·m}^3 \qquad Ans.$$

Also, at the overhanging end where $x = 8$, we have

$$EIy = \frac{250}{3}(8)^3 - \frac{50}{3}(7)^4 + \frac{50}{3}(4)^4 + \frac{650}{3}(2)^3 - 1308(8)$$

$$= -1814 \text{ N·m}^3 \qquad Ans.$$

603.   A simply supported beam carries the triangularly distributed load shown in Fig. 6–6a. Determine the deflection equation and the magnitude of the maximum deflection.

Solution: Because of symmetry, each reaction is one-half of the total load of $\frac{1}{2}wL$, or $R_1 = R_2 = \frac{1}{4}wL$. In this example, we take further advantage of symmetry to note that the deflection curve from A to B is the mirror image of that from C to B. The conditions of zero deflection at A and of zero slope at B do not require the use of a general moment equation. Only the moment equation for segment AB is needed, and this is easily found with the aid of Fig. 6–6b.

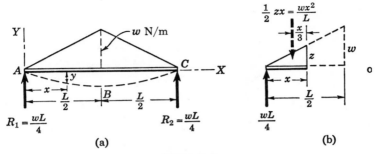

Figure 6–6.

Applying the differential equation of the elastic curve to segment AB and integrating twice, we obtain

$$EI\frac{d^2y}{dx^2} = M_{AB} = \frac{wL}{4}x - \frac{wx^2}{L}\cdot\frac{x}{3} \qquad (a)$$

$$EI\frac{dy}{dx} = \frac{wLx^2}{8} - \frac{wx^4}{12L} + C_1 \qquad (b)$$

$$EIy = \frac{wLx^3}{24} - \frac{wx^5}{60L} + C_1x + C_2 \qquad (c)$$

To evaluate the constants of integration, we note that at the support A, $y = 0$ at $x = 0$. Hence from Eq. (c), we have $C_2 = 0$. Also, because of symmetry, the slope $dy/dx = 0$ at midspan where $x = L/2$. Substituting these conditions in Eq. (b) yields

$$0 = \frac{wL}{8}\left(\frac{L}{2}\right)^2 - \frac{w}{12L}\left(\frac{L}{2}\right)^4 + C_1 \quad \text{or} \quad C_1 = -\frac{5wL^3}{192}$$

Hence the deflection equation from A to B (and also from C to B because of symmetry) becomes

$$EIy = \frac{wLx^3}{24} - \frac{wx^5}{60L} - \frac{5wL^3x}{192}$$

which reduces to

$$EIy = -\frac{wx}{960L}(25L^4 - 40L^2x^2 + 16x^4)$$

The maximum deflection at midspan, where $x = L/2$, is then found to be

$$EIy = -\frac{wL^4}{120} = -\frac{WL^3}{60}$$

where $W = \frac{1}{2}wL$ is the total load.

604.   Determine the equation of the elastic curve of a cantilever beam supporting a uniformly distributed load of w N/m over part of its length as shown in Fig. 6–7.

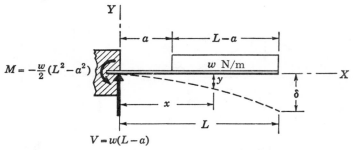

Figure 6–7.

Solution: At the wall, equilibrium conditions determine the shear and moment to be

$$V = w(L - a) \quad \text{and} \quad M = -w(L - a)\left(a + \frac{L - a}{2}\right)$$

$$= -\frac{w}{2}(L^2 - a^2)$$

In terms of the general moment equation, the differential equation of the elastic curve becomes

$$EI\frac{d^2y}{dx^2} = w(L - a)x - \frac{w}{2}(L^2 - a^2) - \frac{w}{2}\langle x - a \rangle^2 \qquad (a)$$

Integrating Eq. (a), we obtain the slope equation:

$$EI\frac{dy}{dx} = w(L - a)\frac{x^2}{2} - \frac{w}{2}(L^2 - a^2)x - \frac{w}{6}\langle x - a \rangle^3 + C_1 \quad (b)$$

However, the slope dy/dx is zero at $x = 0$, so $C_1 = 0$. We may now integrate the slope equation (with $C_1 = 0$) and obtain the deflection equation:

$$EIy = w(L - a)\frac{x^3}{6} - \frac{w}{4}(L^2 - a^2)x^2 - \frac{w}{24}\langle x - a \rangle^4 + C_2 \quad (c)$$

Since $y = 0$ at $x = 0$, we find that $C_2 = 0$ also. Observe that selecting the origin of axes at the perfectly restrained wall where the slope and deflection are zero makes the constants of integration also zero.

The value of the maximum deflection, which occurs at the free end, is denoted by δ. Evidently $\delta = -y$; so on substituting $x = L$ and simplifying, we obtain

$$EI\delta = \frac{w(L - a)}{8}\left(L^3 + L^2a + La^2 - \frac{a^3}{3}\right)$$

One important variation of this result occurs when $a = 0$. Then the entire length of the beam is uniformly loaded and the maximum deflection is given by

$$EI\delta = \frac{wL^4}{8} = \frac{WL^3}{8}$$

PROBLEMS

605. Determine the maximum deflection δ in a simply supported beam of length L carrying a concentrated load P at midspan.

Ans. $\delta = PL^3/48EI$

606. Determine the maximum deflection δ in a simply supported beam of length L carrying a uniformly distributed load of w N/m applied over its entire length.

Ans. $\delta = (5/384)(wL^4/EI) = (5/384)(WL^3/EI)$

607. Determine the maximum value of EIy for the cantilever beam loaded as shown in Fig. P–607. Take the origin at the wall.

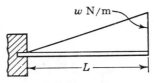

Figure P–607.

608. Find the equation of the elastic curve for the cantilever beam shown in Fig. P–608; it carries a load that varies from zero at the wall to w N/m at the free end. Take the origin at the wall.

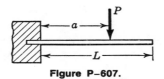

Figure P–608.

609. As shown in Fig. P–609, a simply supported beam carries two symmetrically placed concentrated loads. Compute the maximum deflection δ and compare one-half this result with the midspan δ to case 7, Table 6–2, page 270. Check your answer by letting $a = L/2$ and comparing it with the answer to Problem 605.

Ans. $\delta = (Pa/24EI)(3L^2 - 4a^2)$

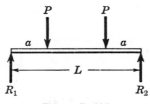

Figure P–609.

610.   The simply supported beam shown in Fig. P–610 carries a uniform load of w N/m symmetrically distributed over part of its length. Determine the maximum deflection δ and check your result by letting $a = 0$ and comparing with the answer to Problem 606.

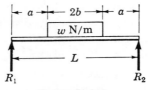

Figure P–610.

611.   Compute the value of $EI\delta$ at midspan for the beam loaded as shown in Fig. P–611. If $E = 10$ GN/m², what value of I is required to limit the midspan deflection to $1/360$ of the span?

Ans. $EI\delta = 500$ N·m³; $I = 4.50 \times 10^6$ mm⁴

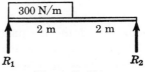

Figure P–611.

612.   Compute the midspan value of $EI\delta$ for the beam loaded as shown in Fig. P–612.   *Ans.* $EI\delta = 657$ N·m³

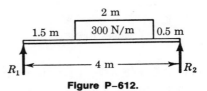

Figure P–612.

613.   Compute the value of EIy at the right end of the beam loaded as shown in Fig. P–613.   *Ans.* $EIy = 195$ N·m³

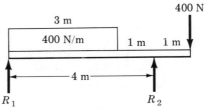

Figure P–613.

614. For the beam loaded as shown in Fig. P-614, (a) compute the slope of the elastic curve over the right support and (b) determine the maximum deflection between the supports.

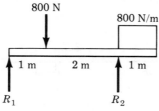

Figure P-614.

615. Compute the value of EIy midway between the supports for the overhanging beam shown in Fig. P-615.

Ans. $EIy = -4.66 \text{ kN} \cdot \text{m}^3$

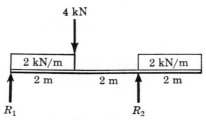

Figure P-615.

616. For the beam loaded as shown in Fig. P-616, determine (a) the deflection and slope under the load P and (b) the maximum deflection between the supports.

Ans.   (b) Max. $EIy = Pa^2b/9\sqrt{3}$

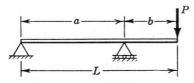

Figures P-616 and P-617.

617. Replace the load P by a clockwise couple M applied at the right end and determine the slope and deflection at the right end.

Ans. $EI \, dy/dx = -(M/3)(L + 2b)$; $EI\delta = (Mb/6)(2L + b)$

618.   A simply supported beam carries a couple M applied as shown in Fig. P–618. Determine the equation of the elastic curve and the deflection at the point of application of the couple. Then, letting $a = L$ and $a = 0$, compare your solution of the elastic curve with cases 11 and 12 in Table 6–2 on page 270.

Ans. $EIy = (Ma/3L)(L^2 - 3La + 2a^2)$

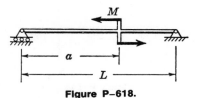

Figure P–618.

619.   Determine the midspan value of EIy for the beam loaded as shown in Fig. P–619. (*Hint*: Take advantage of symmetry to note that the slope is zero at midspan.)   *Ans.* $EIy = -3.33 \text{ kN·m}^3$

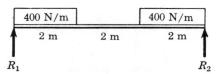

Figure P–619.

620.   Find the midspan deflection δ for the beam shown in Fig. P–620, carrying two triangularly distributed loads. (*Hint*: For convenience, select the origin of the axes at the midspan position of the elastic curve.)   *Ans.* $\delta = (9/1920)(wL^4/EI)$

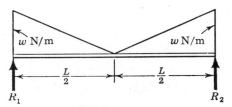

Figure P–620.

621.   Determine the value of $EI\delta$ midway between the supports for the beam shown in Fig. P–621. Check your result by letting $a = 0$ and comparing with Problem 606. (Apply the hint given in Problem 620.)

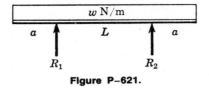

Figure P–621.

6–3   THEOREMS OF AREA-MOMENT METHOD

A useful and simple method of determining slopes and deflections in beams involves the area of the moment diagram and also the moment of that area—the *area-moment method*. We discuss first the two basic theorems of the method; then, after showing how to compute the area and moment of area of the moment diagram, we shall apply the method to several types of problems. The method is especially useful in directly determining the slope or deflection at a specified position. Depending as it does upon the geometry of the elastic curve, the area-moment method emphasizes the physical significance of slope and deflection.

The area-moment method is subject to the same limitations as the double-integration method; but in order to present it in its entirety as a completely independent alternative method, we repeat a small portion of the preceding article. Figure 6–8a shows a simple beam that supports any type of loading. The elastic curve is the edge view of the neutral surface and is shown, with greatly exaggerated deflections, in Fig. 6–8b; the moment diagram is assumed to be as in Fig. 6–8c.

As we saw in the derivation of the flexure formula, Art. 5–2, two adjacent plane sections of an originally straight beam will rotate through the angle $d\theta$ relative to each other. This is demonstrated in the enlarged detail of Fig. 6–8b, in which it is also apparent that the arc distance ds measured along the elastic curve between these two sections equals $\rho\,d\theta$, where ρ is the radius of curvature of the elastic curve at the given position. From Eq. (5–1) we have

$$\frac{1}{\rho} = \frac{M}{EI}$$

and since $ds = \rho\,d\theta$, we now write

$$\frac{1}{\rho} = \frac{M}{EI} = \frac{d\theta}{ds}$$

or

$$d\theta = \frac{M}{EI}\,ds \qquad\qquad (a)$$

In most practical cases the elastic curve is so flat that no serious error is made in assuming the length ds to equal its projection dx. With

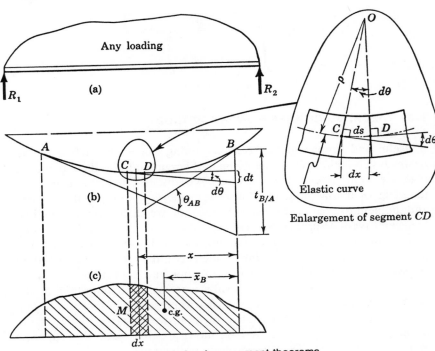

Figure 6–8. Area-moment theorems.

this assumption, we obtain

$$d\theta = \frac{M}{EI} dx \qquad (b)$$

It is evident that tangents drawn to the elastic curve at C and D in Fig. 6–8b are separated by the same angle $d\theta$ by which sections OC and OD (in the enlarged detail) rotate relative to each other. Hence the change in slope between tangents drawn to the elastic curve at any two points A and B will equal the sum of such small angles:

$$\theta_{AB} = \int_{\theta_A}^{\theta_B} d\theta = \frac{1}{EI} \int_{x_A}^{x_B} M \, dx \qquad (c)$$

Note also in Fig. 6–8b that the distance from B on the elastic curve (measured perpendicular to the original position of the beam) that will intersect a tangent drawn to this curve at any other point A is the sum of the intecepts dt created by tangents to the curve at adjacent points. Each of these intercepts may be considered as the arc of a circle of radius x subtended by the angle $d\theta$:

$$dt = x \, d\theta$$

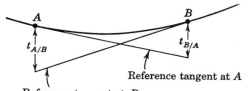

Figure 6–9.   Inequality of $t_{A/B}$ and $t_{B/A}$.

Hence

$$t_{B/A} = \int dt = \int x \, d\theta$$

Replacing $d\theta$ by the value in Eq. (b), we obtain

$$t_{B/A} = \frac{1}{EI} \int_{x_A}^{x_B} x(M \, dx) \qquad (d)$$

The length $t_{B/A}$ is known as the deviation of B from a tangent drawn at A, or as the tangential deviation of B with respect to A. The subscript indicates that the deviation is measured from B relative to a reference tangent drawn at A. Figure 6–9 illustrates the difference between the deviation $t_{B/A}$ of B from a reference tangent at A, and the deviation $t_{A/B}$ of A from a reference tangent at B. In general, such deviations are unequal.

The geometric significance of Eqs. (c) and (d) gives rise to the two basic theorems of the area-moment method. From the moment diagram in Fig. 6–8c we see that $M \, dx$ is the area of the shaded element located a distance x from the ordinate through B. Since $\int M \, dx$ means a summation of such elements, Eq. (c) may be expressed as

$$\theta_{AB} = \frac{1}{EI} (\text{area})_{AB} \qquad (6\text{–}4)$$

This is the algebraic expression of Theorem I, which is stated as follows:

Theorem I: The change in slope between tangents drawn to the elastic curve at any two points A and B is equal to the product of $1/EI$ multiplied by the area of the moment diagram between these two points.

Figure 6–8c shows that the expression $x(M \, dx)$ which appears under the integral sign in Eq. (d) is the moment of area of the shaded element about the ordinate at B. Hence the geometric significance of the integral $\int x(M \, dx)$ is that the integral is equivalent to the moment of area about the ordinate at B of that part of the moment diagram between A and B. Thus we obtain the algebraic form of Theorem II:

$$t_{B/A} = \frac{1}{EI}(\text{area})_{BA} \cdot \bar{x}_B \qquad (6-5)$$

This is stated more formally as:

Theorem II: The deviation of any point B relative to a tangent drawn to the elastic curve at any other point A, in a direction perpendicular to the original position of the beam, is equal to the product of $1/EI$ multiplied by the moment of area about B of that part of the moment diagram between points A and B.

The product EI is called *flexural rigidity*. Note that we have tacitly assumed that E and I remain constant throughout the length of the beam; this is usually the case. If they vary, however, they cannot be written outside the integral sign, and the manner of their variation with x must be known. Such variations are usually taken care of by dividing each moment ordinate by EI to obtain an M/EI diagram which is used in place of the moment diagram in the foregoing theorems.

In the two theorems, $(\text{area})_{AB}$ is the area of the moment diagram between points A and B, and $\bar{x}_B$ is the moment arm of this area measured from B. When the area of the moment diagram is composed of several parts (this is explained in Art. 6–4), the expression $(\text{area})_{AB} \cdot \bar{x}_B$ includes the moment of area of all such parts. The moment of area is always taken about an ordinate through the point at which the deviation is being computed. An automatic method of using the correct axis for moments is to give $\bar{x}$ the same subscript, for example, B (meaning that moment arms are to be measured from this point), as appears in the numerator of the subscript to t (i.e., B/A).

One rule of sign is very important: The deviation at any point is *positive* if the point lies above the reference tangent from which the deviation is measured, and *negative* if the point lies below the reference tangent. Positive and negative deviations are shown in Fig. 6–10. Conversely, a computed positive value for deviation means that the point must lie above the reference tangent.

(a) Positive deviation; B located above reference tangent

(b) Negative deviation; B located below reference tangent

Figure 6–10.   Signs of deviations.

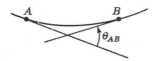 

(a) Positive change of slope; θ_{AB} (b) Negative change of slope; θ_{AB}
    is counterclockwise from left   is clockwise from left tangent
    tangent

Figure 6–11.   Signs of change of slope.

Another rule of sign that concerns slopes is shown in Fig. 6–11. A positive value for the change in slope θ_{AB} means that the tangent at the rightmost point B is measured in a counterclockwise direction from the tangent at the leftmost point, and vice versa.

6–4   MOMENT DIAGRAMS BY PARTS

In order to apply the theorems of the area-moment method, we should be able to compute easily and accurately the area under any part of a moment diagram, and also the moment of such an area about any axis. A method of doing this from calculus is to integrate the two expressions $\int M \, dx$ and $\int x(M \, dx)$ between proper limits, noting that the bending moment M must be expressed as a function of x.

Our purpose here, however, is to discuss a method of dividing moment diagrams into parts whose areas and centroids are known; this permits simple numerical calculations to replace integrations. The first step is to learn how to draw moment effects of each separate loading (hereafter called *moment diagram by parts*) instead of a *conventional* moment diagram.

The construction of moment diagrams by parts depends on two basic principles:

1. The resultant bending moment at any section caused by any load system is the algebraic sum of the bending moments at that section caused by each load acting separately. This statement is expressed algebraically by

$$M = (\Sigma M)_L = (\Sigma M)_R \tag{4–2}$$

where $(\Sigma M)_L$ indicates the sum of the moments caused by all the forces to the left of the section, and $(\Sigma M)_R$ is the sum of the moments caused by all the forces to the right of the section.

2. The moment effect of any single specified loading is always some variation of the general equation

$$y = kx^n \tag{a}$$

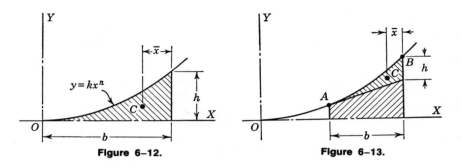

Figure 6–12.   **Figure 6–13.**

The graph of this equation is shown in Fig. 6–12. The shaded area and the location of its centroid are easily shown by calculus to be

$$\text{area} = \frac{1}{n+1} \cdot bh \qquad\qquad (b)$$

$$\bar{x} = \frac{1}{n+2} \cdot b \qquad\qquad (c)$$

where b is the base and h is the height.

In computing the area under the curve between positions like A and B in Fig. 6–13, Eqs. (b) and (c) refer to the shaded area between the curve, the *ordinate* at B and the *tangent* at A. To this area must be added, of course, the shaded trapezoidal area between the tangent and the X axis.

Table 6–1 demonstrates the truth of the second basic principle stated above, viz., that the moment effect of any load is some variation of the equation $y = kx^n$. This table gives data on four cantilever beams, each loaded differently with increasingly complex loads.

Note that a cantilever loaded by a couple C has a moment equation of the type $y = kx^n$ in which $k = -C$ and n is zero, viz., $M = -Cx^0$. In other words, a couple type of loading produces a moment equation of zero degree. Similarly, a concentrated load produces a moment equation of the first degree; a uniform load produces a moment equation of the second degree, etc.

In the column headed "Area," the area of the moment diagram is expressed in terms of a factor multiplied by the general base distance b and the maximum height h of the moment diagram. The position of the centroid of each moment diagram from the maximum ordinate of the diagram is defined as a factor of the base distance. These factors or coefficients, which increase very simply, are obtained from Eqs. (b) and (c) by assigning to n the value of the degree of each equation, i.e., $n = 0$, $n = 1$, $n = 2$, etc.

TABLE 6-1. Cantilever Loadings

TYPE OF LOADING	CANTILEVER BEAM	MOMENT EQUATION (moment at any section x)	DEGREE OF MOMENT EQUATION	MOMENT DIAGRAM	AREA	$\bar{x}$
Couple		$M = -C$	Zero (i.e., $M = -Cx^0$)	$b=L$, $h=-C$	$\frac{1}{1}\,bh$	$\frac{1}{2}\,b$
Concentrated		$M = -Px$	1st	$b=L$, $h=-PL$	$\frac{1}{2}\,bh$	$\frac{1}{3}\,b$
Uniformly distributed		$M = -\dfrac{w}{2}x^2$	2nd	$b=L$, $h=-\dfrac{wL^2}{2}$	$\frac{1}{3}\,bh$	$\frac{1}{4}\,b$
Uniformly varying		$M = -\dfrac{w}{6L}x^3$	3rd	$b=L$, $h=-\dfrac{wL^2}{6}$	$\frac{1}{4}\,bh$	$\frac{1}{5}\,b$

An example will illustrate how Table 6–1 is used to draw moment diagrams by parts. The simple beam in Fig. 6–14 is 3 m long and supports a uniformly distributed load of 300 N/m over the right 2 m of the span.

At any section a–a between A and B, the moment effect defined by $M = (\Sigma M)_L$ is caused only by R_1. Also, at any section b–b between B and C, the moment effect will be due to R_1 and to the portion of the uniformly distributed load included between B and b–b. Note that defining the bending moment in terms of the forces to the left of the section means that the uniformly distributed load has no moment effect on segment AB. Actually, the moment effect of R_1 at any section of the beam is equivalent to the cantilever loading at (a), whereas the moment effect of the uniform loading on any section of the beam is equivalent to the cantilever loading at (b).

By referring to Table 6–1, we can plot the moment diagrams of beams (a) and (b) on a common base line (the line of zero moment), as shown at (c) in the figure. That the algebraic sum of the shaded areas of (c) will yield the resultant or conventional moment diagram is evident from the fact that the moment at any section of the original beam is equal to the sum of the moments at that section caused by the individ-

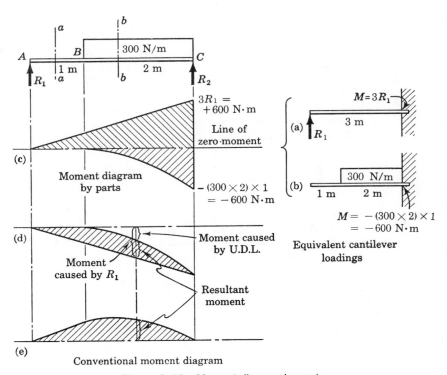

Figure 6–14. Moment diagram by parts.

ual loads (basic principle 1, above). Hence, if the triangular area is revolved about the line of zero moment as an axis, we obtain diagram (d). The shaded area of diagram (d) is evidently equal to the area of the conventional moment diagram (e) obtained by plotting the resultant moment at any section as ordinates to a horizontal base line. Hence the conventional moment diagram may be replaced by an equivalent moment diagram constructed of parts, as in Fig. 6–14c, whose areas and centroids can be easily computed from the data in Table 6–1.

To compute the moment of area of the moment diagram, we observe that the moment of the area of the conventional moment diagram (Fig. 6–14e) is equivalent to the sum of the moments of area of its parts, as drawn in Fig. 6–14c. Hence, noting that each such area is the product of a coefficient listed in Table 6–1 multiplied by the dimensions of the circumscribing rectangle, and that its centroidal location is the product of a similar coefficient times the base length of this rectangle, we obtain as the moment of area with respect to the right end C

$$\left[(\text{area})_{AC} \cdot \bar{x}_C = \Sigma ax \right]$$

$$(\text{area})_{AC} \cdot \bar{x}_C = \left(\frac{3 \times 600}{2} \right)\left(\frac{1}{3} \times 3 \right) - \left(\frac{2 \times 600}{3} \right)\left(\frac{1}{4} \times 2 \right)$$

$$= 700 \text{ N·m}^3 \qquad Ans.$$

The symbol $\bar{x}_C$ means that moment arms are to be measured from C. From Theorem II of the area-moment method, this result represents the product $EIt_{C/A}$, where $t_{C/A}$ is the deviation of point C from a tangent drawn to the elastic curve at point A.

One final observation: It is usually unnecessary to compute the beam reactions. Since the conventional bending moment at C must be zero (because no loads act to the right of C to cause bending moment), the moment of the reaction R_1 at C (equal to $3R_1$) must balance the bending moment of -600 N·m at C caused by the loading.

By now the reader may realize that the technique of drawing moment diagrams by parts is really a graphical interpretation of the method of writing a general moment equation described on page 217. Thus for the beam shown in Fig. 6–15, the general moment equation is

$$M = R_1 x + M_1 \langle x - a_1 \rangle^0 - \frac{w_1}{2} \langle x - a_2 \rangle^2 - \frac{w_2}{6b} \langle x - a_3 \rangle^3$$

If each of the terms in pointed brackets is replaced by a new variable that starts at the position where each loading begins, we see that each term in the general moment equation is the moment equation of a cantilever type given in Table 6–1. Thus replacing $\langle x - a_1 \rangle$ by u, $\langle x - a_2 \rangle$ by v, and $\langle x - a_3 \rangle$ by z, we obtain

$$M = R_1 x + M_1 u^0 - \frac{w_1}{2} v^2 - \frac{w_2}{6b} z^3$$

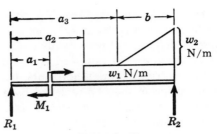

Figure 6–15.

ILLUSTRATIVE PROBLEMS

622. For the beam shown in Fig. 6–16a, compute the moment of area of the moment diagram about the left end.

Solution: From statics, the reactions are computed to be $R_1 = 300$ N and $R_2 = 600$ N. The moment diagram by parts is constructed by applying the definition $M = (\Sigma M)_L$ to segments AB and BC, and the definition $M = (\Sigma M)_R$ to segment CD. Observe that it is simpler to compute the bending moment for an exploratory section in segment CD by taking moments of forces to the right of the section rather than to the left. Computing bending moment by applying either $M = (\Sigma M)_L$ or

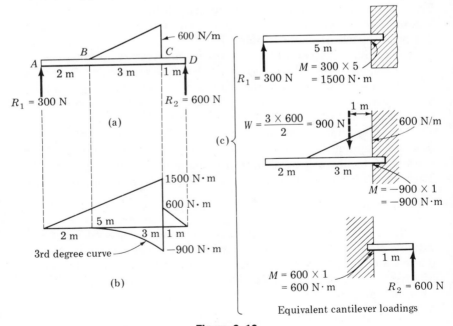

Figure 6–16.

$M = (\Sigma M)_R$ will generally indicate the simplest manner of drawing the moment diagram by parts.

The equivalent cantilever loadings are shown in Fig. 6–16c. The bending moment diagram in Fig. 6–16b is checked by the fact that the moment at C in terms of the forces to the left of C is $1500 - 900 = 600$ N·m, which equals, as it should, the moment at C expressed in terms of the forces acting to the right of C.

The moment of the area of the M diagram about the left end A is now computed as equal to the sum of the moments of area of its parts. Referring to Table 6–1 for the coefficients of area and of centroidal distances, we obtain:

$$\left[(\text{area})_{AD} \cdot \bar{x}_A = \Sigma ax \right]$$

$$(\text{area})_{AD} \cdot \bar{x}_A = \left(\frac{1500 \times 5}{2} \right)\left(\frac{2}{3} \times 5 \right) + \left(\frac{600 \times 1}{2} \right)\left(5 + \frac{1}{3} \times 1 \right)$$
$$- \left(\frac{900 \times 3}{4} \right)\left(2 + \frac{4}{5} \times 3 \right)$$

$$= 11.13 \text{ kN·m}^3 \qquad Ans.$$

What value of EIt does this result represent?

623. For the overhanging beam shown in Fig. 6–17a, compute the moment of area about C of the moment diagram included between the supports at A and C.

Solution: At any section between A and C, the conventional bending moment is computed more easily by applying $M = (\Sigma M)_L$, whereas between C and D it is simpler to apply $M = (\Sigma M)_R$. The moment diagram by parts shown in Fig. 6–17b is therefore constructed by combining the cantilever loadings in Fig. 6–17c.

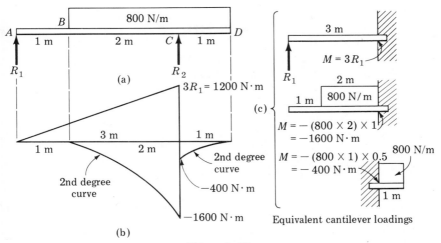

Figure 6–17.

In this problem, the value of the reactions need not be computed. The moment at C caused by R_1 is found from the fact that the bending moment at C of all forces to the left of C must equal the bending moment of all forces to the right of C, which is in accord with the fundamental definition $M = (\Sigma M)_L = (\Sigma M)_R$. In other words, $3R_1 - 1600 = -400$; hence $3R_1 = 1200$ N·m.

We obtain the moment of area of the M diagram between A and C about C by applying

$$\left[(\text{area})_{AC} \cdot \bar{x}_C = \Sigma ax \right]$$
$$(\text{area})_{AC} \cdot \bar{x}_C = \left(\frac{3 \times 1200}{2} \right)\left(\frac{3}{3} \right) - \left(\frac{2 \times 1600}{3} \right)\left(\frac{2}{4} \right)$$
$$= 1270 \text{ N·m}^3 \qquad Ans.$$

PROBLEMS

For each of the beams in Problems 624 to 629, compute the moment of area of the M diagram between the reactions about both the left and the right reaction.

624. Beam loaded as shown in Fig. P–624.

$$Ans. \quad (\text{area})_{AB} \cdot \bar{x}_A = 2500 \text{ N} \cdot \text{m}^3$$

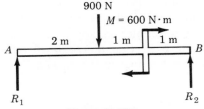

Figure P–624.

625. Beam loaded as shown in Fig. P–625. (*Hint:* Draw the moment diagram by parts from right to left.)

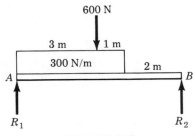

Figure P–625.

626.   Beam loaded as shown in Fig. P–626.

$Ans.$ $(area)_{AB} \cdot \bar{x}_B = 8.25 \text{ kN·m}^3$

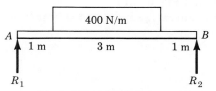

Figure P–626.

627.   Beam loaded as shown in Fig. P–627. (*Hint:*   Resolve the trapezoidal loading into a uniformly distributed load and a uniformly varying load.)

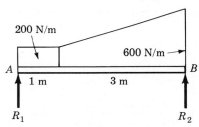

Figure P–627.

628.   Beam loaded with a uniformly varying load and a couple as shown in Fig. P–628. $Ans.$ $(area)_{AB} \cdot \bar{x}_A = 2.13 \text{ kN·m}^3$

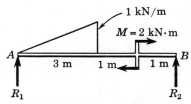

Figures P–628 and P–629.

629.   Solve Problem 628 if the sense of the couple is counterclockwise instead of clockwise as shown in Fig. P–628.

630. For the beam loaded as shown in Fig. P–630, compute the value of $(\text{area})_{AB} \cdot \bar{x}_A$. From this result determine whether the tangent drawn to the elastic curve at B slopes up or down to the right. (*Hint:* Refer to Eq. (6–5) and Fig. 6–10.)

$Ans.$ $(\text{area})_{AB} \cdot \bar{x}_A = -463 \text{ N} \cdot \text{m}^3$; slope is down to right

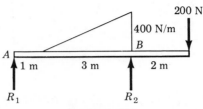

Figure P–630.

631. Determine the value of P for the beam loaded as shown in Fig. P–631 so that the moment of area about A of the M diagram between A and B will be zero. What is the physical significance of this result?

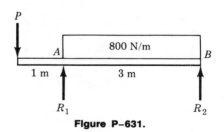

Figure P–631.

632. For the beam loaded as shown in Fig. P–632, compute the value of $(\text{area})_{AB} \cdot \bar{x}_A$. From this result, is the tangent drawn to the elastic curve at B directed up or down to the right? (*Hint:* Refer to Eq. (6–5) and Fig. 6–10.) $Ans.$ $(\text{area})_{AB} \cdot \bar{x}_A = 1.27 \text{ kN} \cdot \text{m}^3$; up to right

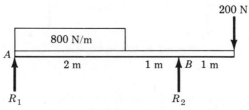

Figure P–632.

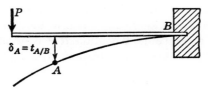

Figure 6–18.

6–5   DEFLECTION OF CANTILEVER BEAMS

It will be recalled that the tangential deviation at any point is the distance from the point on the elastic curve to a tangent drawn to the curve at some other point (Art. 6–3 and Fig. 6–8). As a consequence, the tangential deviation is generally not equal to the deflection. In cantilever beams, however, the wall is usually assumed to be perfectly fixed, and hence the tangent drawn to the elastic curve at the wall will be horizontal, as in Fig. 6–18. Therefore, if the tangential deviation at A is measured from a tangent drawn at B, the deviation $t_{A/B}$ will equal the deflection δ_A at A.

Several examples will illustrate how area-moment principles are used to determine slope and deflection in cantilever beams. Other types of beams are considered later.

ILLUSTRATIVE PROBLEMS

633.   For the cantilever beam in Fig. 6–19, it is assumed that $E = 12$ GN/m^2, $I = 10 \times 10^6$ mm^4. What value of P will cause a 20-mm deflection at the free end?

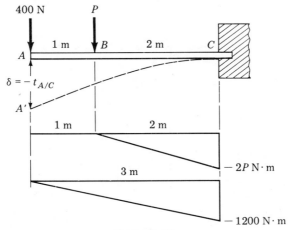

Figure 6–19.

Solution: The moment diagram by parts is drawn as shown. Evidently the deflection δ at A is numerically equal to the deviation of A from a tangent drawn at C. Since the deviation at A is negative because A lies below the tangent, we have from Theorem II,

$$t_{A/C} = \frac{1}{EI}(\text{area})_{AC} \cdot \bar{x}_A$$

$$-\delta = \frac{1}{EI}\left[-\left(\frac{2 \times 2P}{2}\right)\left(1 + \frac{2}{3} \times 2\right) - \left(\frac{3 \times 1200}{2}\right)\left(\frac{2}{3} \times 3\right) \right]$$

whence

$$EI\delta = (4.667P + 3600) \text{ N·m}^3$$

When substituting numerical values into the above result, it must be remembered that, for dimensional homogeneity, E must be expressed in N/m^2, I in m^4, and δ in m. Therefore, upon substitution, the above equation becomes

$$(12 \times 10^9)(10 \times 10^{-6})(20 \times 10^{-3}) = 4.667P + 3600$$

whence

$$P = -257 \text{ N} \qquad Ans.$$

The minus sign of P indicates that the direction of P must be opposite to that originally assumed; that is, P must act upward.

634. Compute the maximum slope and deflection for the cantilever beam shown in Fig. 6–20a which carries a load varying uniformly from zero at the wall to w N/m at the free end.

Solution: Although the given cantilever loading is not one of the types given in Table 6–1, it is easily transformed into them by replacing the given loading in Fig. 6–20a by those shown in Fig. 6–20c, i.e., superposing a downward uniformly distributed load and an upward uniformly

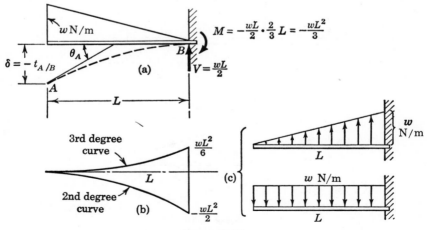

Figure 6–20.

varying load. Thus we obtain the parts of the moment diagram shown in Fig. 6–20b.

The elastic curve in Fig. 6–20a shows that the maximum slope and deflection occur at the free end A. The angle θ_A is clearly equal to the change in slope θ_{AB} measured between the tangents drawn to the elastic curve at A and B. Then by Theorem I we obtain

$$\theta_A = \theta_{AB} = \frac{1}{EI}(\text{area})_{AB} = \frac{1}{EI}\left[\frac{1}{4} \cdot \frac{wL^2}{6} \cdot L - \frac{1}{3} \cdot \frac{wL^2}{2} \cdot L\right]$$

$$= -\frac{wL^3}{8EI}$$

According to Fig. 6–11 on page 231 the minus sign means that the angle is measured in a clockwise sense from the left tangent at A to the right tangent at B; hence the slope at A is up to the right as shown.

The maximum deflection δ at A is numerically equal to $t_{A/B}$, which is the deviation of A from a tangent drawn at B. Since the deviation at A is negative because A lies below the tangent, we have from Theorem II

$$\left[t_{A/B} = \frac{1}{EI}(\text{area})_{AB} \cdot \bar{x}_A\right]$$

$$-\delta = \frac{1}{EI}\left[\left(\frac{1}{4} \cdot \frac{wL^2}{6} \cdot L\right)\left(\frac{4}{5}L\right) - \left(\frac{1}{3} \cdot \frac{wL^2}{2} \cdot L\right)\left(\frac{3}{4}L\right)\right]$$

which reduces to

$$\delta = \frac{11}{120}\frac{wL^4}{EI} \qquad Ans.$$

Remember that the symbol $\bar{x}_A$ means that the moment arms of the areas are to be measured from A.

635.   For the cantilever beam loaded as shown in Fig. 6–21a, compute the value of $EI\delta$ at A and at B.

Solution: The deflections at A and at B are numerically equal to the deviations of A and B from the horizontal tangent drawn to the elastic curve at C. Since A and B are both below this tangent, the deviations $t_{A/C}$ and $t_{B/C}$ are both negative. Applying Theorem II, we have

$$-\delta_A = t_{A/C} = \frac{1}{EI}(\text{area})_{AC} \cdot \bar{x}_A \qquad (a)$$

and

$$-\delta_B = t_{B/C} = \frac{1}{EI}(\text{area})_{BC} \cdot \bar{x}_B \qquad (b)$$

Observe that in Eq. (a) we shall use the M diagram between A and C, whereas in Eq. (b) we need only the M diagram between B and C.

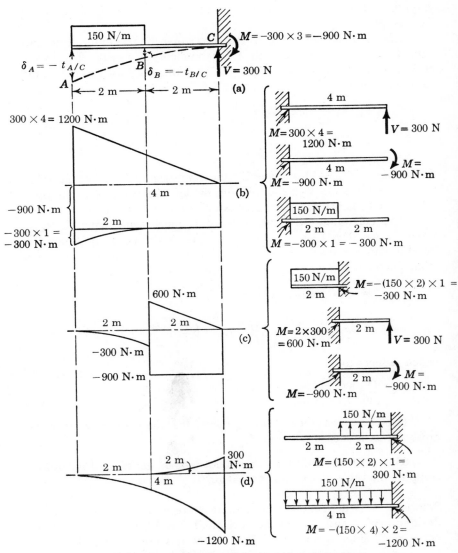

Figure 6–21. Variations of moment diagram by parts.

Let us examine first the various ways in which the M diagram by parts can be drawn so that the simplest diagram be used.

First, the M diagram in Fig. 6–21b is drawn by expressing the bending moment at any section in terms of the loads to the right of the section in accordance with the basic definition $M = (\Sigma M)_R$. The equivalent cantilever loadings are also shown.

Second, the M diagram can also be drawn by expressing the bending moment between A and B by the definition $M = (\Sigma M)_L$, and that between B and C by $M = (\Sigma M)_R$. This results in the M diagram and equivalent cantilever loadings shown in Fig. 6–21c.

Third, still another method is to replace the given loading by superposing the loadings in Fig. 6–21d. This gives what is probably the simplest M diagram by parts.

Although identical results will be obtained by using any of the aforementioned diagrams, we shall apply Eq. (a) to the third one (Fig. 6–21d). Noting that $\bar{x}_A$ means to take the moment of area about A, we obtain

$$- \delta_A = \frac{1}{EI}\left[\left(\frac{2 \times 300}{3}\right)\left(2 + \frac{3}{4} \times 2\right) - \left(\frac{4 \times 1200}{3}\right)\left(\frac{3}{4} \times 4\right)\right]$$

which reduces to

$$EI\delta_A = 4100 \text{ N·m}^3 \qquad Ans.$$

Since the deflection at B is expressed in terms of (area)$_{BC}$, either of the M diagrams in part (b) or (c) of Fig. 6–21 may be used, for they are identical for segment BC. Taking moments about B as indicated by $\bar{x}_B$ in Eq. (b), we obtain

$$- \delta_B = \frac{1}{EI}\left[\left(\frac{2 \times 600}{2}\right)\left(\frac{1}{3} \times 2\right) - (2 \times 900)\left(\frac{1}{2} \times 2\right)\right]$$

from which we have

$$- EI\delta_B = 400 - 1800 \quad \text{or} \quad EI\delta_B = 1400 \text{ N·m}^3 \qquad Ans.$$

Although not shown in fig. 6–21, a fourth way to draw the M diagram by parts from B to C is to use the concept discussed in Art. 4–3 on page 125: a beam may be cut at any section and the effect of the loads to one side of the section replaced by the shear and moment at the cut section. At B, the load causes a shear force of -300 N and a bending moment couple of -300 N·m whose effects from B to C produce a triangular and a rectangular moment diagram. The student may well sketch these diagrams and use them to compute δ_B as a check.

PROBLEMS

636. The cantilever beam shown in Fig. P–636 has a rectangular cross section 50 mm wide by h mm high. Find the height h if the maximum deflection is not to exceed 10 mm. Use $E = 10 \times 10^9$ N/m^2.

$Ans.$ $h = 328$ mm

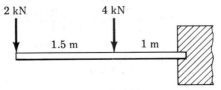

Figure P–636.

637.   For the beam loaded as shown in Fig. P–637, determine the deflection at 2 m from the wall. Use $E = 10 \times 10^9$ N/m^2 and $I = 20 \times 10^6$ mm^4.   *Ans.* $\delta = 22.7$ mm

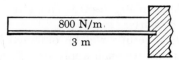

Figure P–637.

638.   For the cantilever beam shown in Fig. P–638, determine the value of $EI\delta$ at the left end. Is this deflection upward or downward?

Ans. $EI\delta = 6.67$ kN·m^3; upward

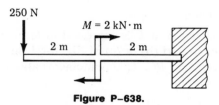

Figure P–638.

639.   A distributed load and an upward-concentrated load act as shown on the cantilever beam in Fig. P–639. Compute the amount the free end deflects upward or downward given that $E = 10 \times 10^9$ N/m^2 and $I = 60 \times 10^6$ mm^4.

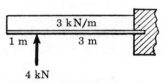

Figures P–639 and P–640.

640.   Compute the value of δ at the concentrated load in Problem 639. Is the deflection upward or downward?

Ans. $\delta = 46.9$ mm; downward

641.   For the cantilever beam shown in Fig. P–641, what value of P will cause zero deflection at A?   *Ans.* $P = 150$ N

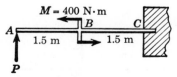

Figure P–641.

642. Find the maximum deflection for the cantilever beam loaded as shown in Fig. P–642 if the cross section is 50 mm wide by 150 mm high. Use $E = 69 \text{ GN/m}^2$. *Ans.* $\delta = 28.0$ mm

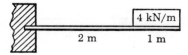

4 kN/m

2 m   1 m

Figure P–642.

643. Find the maximum value of $EI\delta$ for the cantilever beam shown in Fig. P–643. *Ans.* $EI\delta = (Pa^2/6)(3L - a)$

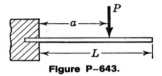

P

a

L

Figure P–643.

644. Determine the maximum deflection for the beam loaded as shown in Fig. P–644.

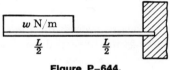

w N/m

$\dfrac{L}{2}$ $\dfrac{L}{2}$

Figure P–644.

645. Compute the deflection and slope at a section 2 m from the wall for the beam shown in Fig. P–645. Assume that $E = 10 \text{ GN/m}^2$ and $I = 30 \times 10^6 \text{ mm}^4$. *Ans.* $\delta = 16.4$ mm; $\theta = 0.739°$

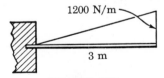

1200 N/m

3 m

Figure P–645.

646. For the beam shown in Fig. P–646, determine the value of I that will limit the maximum deflection to 20 mm. Use $E = 10 \times 10^9$ N/m^2.

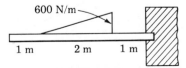

600 N/m

1 m   2 m   1 m

Figure P–646.

647. Find the maximum value of $EI\delta$ for the beam shown in Fig.
P–647.   *Ans.* $EI\delta = (121/1920)wL^4$

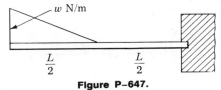

Figure P–647.

648. For the cantilever beam loaded as shown in Fig. P–648,
determine the deflection at any section x meters from the support.

Ans. $EI\delta = (wx^2/120L)(10L^3 - 10L^2x + 5Lx^2 - x^3)$

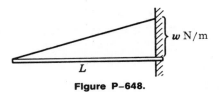

Figure P–648.

6–6   DEFLECTIONS IN SIMPLY SUPPORTED BEAMS

Deflections in cantilever beams were simplified by the fact that the
tangent to the elastic curve at the fixed end was known to be horizontal.
In simply supported beams, the position at which a tangent to the
elastic curve will be horizontal is usually unknown, and therefore a
different method must be used. This method may seem devious, but
actually it is simple and rapid. It is illustrated by Fig. 6–22, which shows
only the elastic curve of a simple beam. The loads and moment
diagrams have been omitted for clarity.

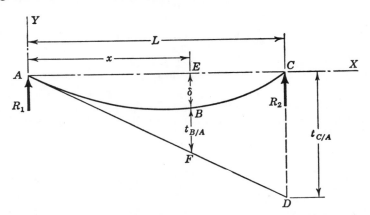

Figure 6–22. Geometry of area-moment method applied to simple beams.

The problem is to determine the value of the deflection δ at some position B. If a tangent to the elastic curve is drawn at A, the deviation $t_{B/A}$ at B from this tangent is evidently *not* the required deflection δ. However, the sum of δ and $t_{B/A}$ constitutes the distance EF; and if both EF and $t_{B/A}$ were known, δ could easily be found. Hence the distance EF must also be found. This is done by noting that the triangle AEF is similar to the triangle ACD, of which the leg CD equals the deviation $t_{C/A}$ of C from the reference tangent drawn at A.

The proper procedure to apply is obtained by reversing the steps of the above analysis into the following order:

1.   Compute $t_{C/A}$, using the relation

$$t_{C/A} = \frac{1}{EI}(\text{area})_{CA} \cdot \bar{x}_C$$

2.   From the relations between similar triangles, determine EF in terms of $t_{C/A}$. We obtain

$$EF = \frac{x}{L} \cdot t_{C/A}$$

3.   Compute $t_{B/A}$ from the relation

$$t_{B/A} = \frac{1}{EI}(\text{area})_{BA} \cdot \bar{x}_B$$

4.   Since EF is the sum of δ and $t_{B/A}$, the value of δ is given by

$$\delta = EF - t_{B/A}$$

As mentioned previously, this procedure may seem long, but actually it is rapid. Several examples will demonstrate the method. Only simple loadings are used, since this concentrates attention upon basic ideas. For more complex loadings the basic method is unchanged, the more complicated M diagram by parts being constructed as indicated in Art. 6–4.

ILLUSTRATIVE PROBLEMS

649.   The simple beam in Fig. 6–23 supports a concentrated load of 300 N at 2 m from the left support. Compute the value of $EI\delta$ at B, which is 1 m from the left support.

Solution: We begin by sketching the dashed outline of the elastic curve and drawing the M diagram by parts from left to right. Following

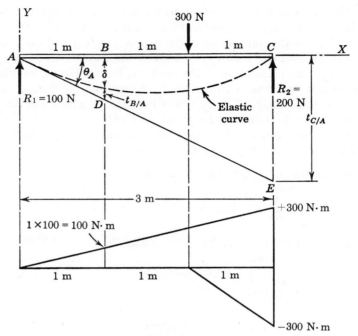

Figure 6–23.

the procedure just discussed, we first obtain $t_{C/A}$:

$$\left[t_{C/A} = \frac{1}{EI}(\text{area})_{CA} \cdot \bar{x}_C \right]$$

$$t_{C/A} = \frac{1}{EI}\left[\left(\frac{3 \times 300}{2} \right)\left(\frac{1}{3} \times 3 \right) - \left(\frac{1 \times 300}{2} \right)\left(\frac{1}{3} \times 1 \right) \right] = \frac{400}{EI}$$

$$(a)$$

Since triangle ABD is similar to triangle ACE,

$$BD = \frac{1}{3} \times t_{C/A} = \frac{400}{3EI} \tag{b}$$

The deviation $t_{B/A}$ is next obtained from

$$\left[t_{B/A} = \frac{1}{EI}(\text{area})_{BA} \cdot \bar{x}_B \right] \qquad t_{B/A} = \frac{1}{EI}\left[\left(\frac{1 \times 100}{2} \right)\left(\frac{1}{3} \times 1 \right) \right]$$

$$= \frac{100}{6EI} \tag{c}$$

Finally, the value of δ is given by

$$\left[\delta = BD = t_{B/A} \right] \qquad \delta = \frac{1}{EI}\left[\frac{400}{3} - \frac{100}{6} \right]$$

$$EI\delta = 116.7 \text{ N} \cdot \text{m}^3 \qquad Ans.$$

650.   Compute the slope of the elastic curve at the left reaction for the beam discussed in the preceding problem and shown in Fig. 6–23.

Solution: The slope of the elastic curve at A is given by tan θ_A, where θ_A is the angle between the horizontal and the tangent drawn to the elastic curve at A. (Remember that the deflections and slopes in beams are assumed to be very small compared to the length of the beam.) Hence, tan θ_A is practically equivalent to θ_A expressed in radians.

From Fig. 6–23, therefore, we obtain

$$\theta_A \approx \tan \theta_A = \frac{CE}{AC} = \frac{t_{C/A}}{AC}$$

whence, substituting the value $t_{C/A} = 400/EI$ from Eq. (*a*) in the preceding problem, we have

$$\theta_A = \frac{400/EI}{3} = \frac{400}{3EI} \qquad Ans.$$

651.   For the beam described in Illustrative Problem 649, locate the position of maximum deflection and compute the maximum $EI\delta$.

Solution: The required values can be determined by either of two methods. Both methods should be mastered because sometimes one is easier than the other.

Method I.   We begin by computing the deflection at any position B located x meters from the left reaction. Using the technique applied in Illustrative Problem 649, we obtain from Fig. 6–24,

$$\left[t_{B/A} = \frac{1}{EI} (\text{area})_{BA} \cdot \bar{x}_B \right] \quad t_{B/A} = \frac{1}{EI} \left[\left(\frac{1}{2} \cdot x \cdot 100\,x \right) \left(\frac{x}{3} \right) \right]$$

$$= \frac{100\,x^3}{6\,EI}$$

Using the value of $t_{C/A}$ given in Eq. (*a*) of that problem, we also have in Fig. 6–24,

$$\left[BD = \frac{x}{3} \cdot t_{C/A} \right] \quad BD = \frac{x}{3} \cdot \frac{400}{EI} = \frac{400x}{3EI}$$

The value of the deflection δ is the difference between BD and $t_{B/A}$. Hence the equation of the elastic curve, with deflections δ considered positive downward, is

$$[\delta = BD - t_{B/A}] \quad \delta = \frac{1}{EI} \left[\frac{400x}{3} - \frac{100x^3}{6} \right]$$

or

$$EI\delta = \frac{400x}{3} - \frac{100x^3}{6} \qquad (a)$$

This is valid for any position between the left reaction and the load

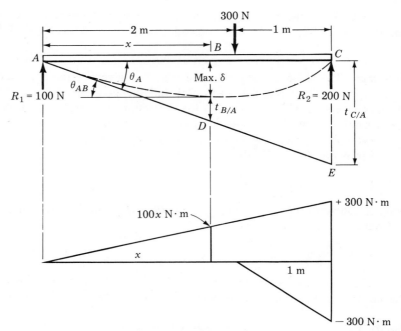

Figure 6–24.

(that is, between $x = 0$ and $x = 2$). The slope equation for this portion of the beam is found by differentiating the deflection equation, Eq. (a):

$$EI\frac{d\delta}{dx} = \frac{400}{3} - \frac{300x^2}{6} \qquad (b)$$

From the principle of maxima and minima developed in calculus, setting the first derivative of Eq. (a) equal to zero will determine the position of maximum deflection. This is equivalent to setting Eq. (b) equal to zero, thus determining the position of zero slope in the beam. We obtain

$$\frac{400}{3} - \frac{300x^2}{6} = 0 \quad \text{or} \quad x = 1.63 \text{ m} \qquad \textit{Ans.}$$

Substituting this value of x in Eq. (a) determines the maximum deflection.

$$\text{Max. } EI\delta = \frac{400}{3}(1.63) - \frac{100}{6}(1.63)^3 = 145 \text{ N·m}^3 \qquad \textit{Ans.}$$

It is instructive to compare the maximum value of $EI\delta$ with the value of $EI\delta$ at midspan. At midspan, $x = 1.5$ m, and substitution of this value of x in Eq. (a) gives

$$\text{Midspan } EI\delta = \frac{400}{3}(1.5) - \frac{100}{6}(1.5)^3 = 144 \text{ N·m}^3$$

This is about 1% less than the maximum value. This difference is so negligible in comparison with possible variations in the given data that for all practical purposes we may assume the midspan deflection to be equivalent to the actual maximum deflection. Indeed, it may be shown that, for a concentrated load located anywhere on a simple span, the maximum difference between midspan and maximum deflection is only 2.6%. In the next article we discuss a simplified method for computing midspan deflection.

Method II.   At the position of maximum deflection, the tangent to the elastic curve will be horizontal. As shown in Fig. 6–24, the change in slope between tangents at this position B and at A (i.e., θ_{AB}) is equal to the slope θ_A at A, since for small angles the radian measure and the tangent of the angle are practically equivalent, that is, $\theta_A = \tan \theta_A$.

From Theorem I of the area-moment method, we obtain

$$\left[\theta_{AB} = \frac{1}{EI} (\text{area})_{AB} \right] \qquad \theta_{AB} = \frac{1}{EI}\left(\frac{1}{2} \cdot x \cdot 100x \right) = \frac{50x^2}{EI}$$

which, on being equated to $\theta_A = 400/3EI$ obtained in Problem 650, gives

$$\frac{50x^2}{EI} = \frac{400}{3EI} \quad \text{or} \quad x = 1.63 \text{ m} \qquad Ans.$$

Computing the value of $EI\delta$ at this position gives maximum $EI\delta = 145$ N·m³, as was obtained with Method I.

652.   Determine the value of the deflection at D of the beam shown in Fig. 6–25a.

Solution: This problem brings out the importance of correctly interpreting the meaning of positive and negative deviation, particularly as it affects the geometry of the elastic curve. For illustrative purposes, this problem will be solved in two ways—first, by drawing the reference tangent at C; and second, by drawing the reference tangent at A.

Before drawing the reference tangent at C, we compute the deviation of A from this reference tangent. Correctly interpreting the sign of $t_{A/C}$ will indicate the direction in which the reference tangent slopes. The geometrically correct position of the reference tangent can then be drawn. Thus

$$t_{A/C} = \frac{1}{EI} (\text{area})_{AC} \cdot \bar{x}_A$$

$$= \frac{1}{EI}\left[\left(\frac{1}{2} \times 3 \times 900 \right)\left(\frac{2}{3} \times 3 \right) \right.$$

$$\left. - \left(\frac{1}{2} \times 2 \times 1000 \right)\left(1 + \frac{2}{3} \times 2 \right) \right] = \frac{367}{EI}$$

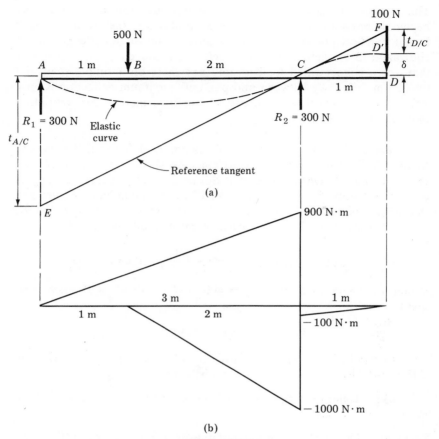

Figure 6–25.

The positive value of $t_{A/C}$ means that A on the elastic curve lies above the reference tangent at C. Hence the reference tangent at C slopes down to the left, as shown.

From the similar triangles ACE and CDF, we obtain

$$\frac{DF}{1} = \frac{t_{A/C}}{3} \quad \text{or} \quad DF = \frac{t_{A/C}}{3} = \frac{367}{3EI}$$

The deviation of D from the reference tangent at C is

$$\left[t_{D/C} = \frac{1}{EI}(\text{area})_{DC} \cdot \bar{x}_D \right] \qquad t_{D/C} = \frac{1}{EI}\left[-\left(\frac{1 \times 100}{2}\right)\left(\frac{2}{3} \times 1\right) \right]$$

$$= -\frac{100}{3EI}$$

The minus sign for $t_{D/C}$ means that D' on the elastic curve is below the reference tangent. Also, since DF is numerically greater than the absolute magnitude of $t_{D/C}$, it is now apparent that D is deflected upward from its original position. Therefore, the elastic curve between C and D is sketched as shown.

The deflection δ is obtained from

$$\delta = DF - |t_{D/C}| = \frac{367}{3EI} - \frac{100}{3EI} = \frac{89}{EI} \qquad \textit{Ans.}$$

If the reference tangent had been drawn at A and the elastic curve had been assumed to have the shape shown in Fig. 6–26, we would have obtained the following results:

$$t_{C/A} = \frac{1}{EI}(\text{area})_{CA} \cdot \bar{x}_C$$

$$= \frac{1}{EI}\left[\left(\frac{3 \times 900}{2}\right)\left(\frac{1}{3} \times 3\right) - \left(\frac{2 \times 1000}{2}\right)\left(\frac{1}{3} \times 2\right)\right] = \frac{683}{EI}$$

$$t_{D/A} = \frac{1}{EI}(\text{area})_{DA} \cdot \bar{x}_D$$

$$= \frac{1}{EI}\left[\left(\frac{3 \times 900}{2}\right)\left(1 + \frac{1}{3} \times 3\right) - \left(\frac{2 \times 1000}{2}\right)\left(1 + \frac{1}{3} \times 2\right)\right.$$

$$\left. - \left(\frac{1 \times 100}{2}\right)\left(\frac{2}{3} \times 1\right)\right]$$

$$= \frac{1000}{EI}$$

The similar triangles ACE and ADF give

$$\frac{DF}{4} = \frac{t_{C/A}}{3} \quad \text{or} \quad DF = \frac{4}{3}t_{C/A} = \frac{4}{3}\left(\frac{683}{EI}\right) = \frac{911}{EI}$$

Finally, using the elastic curve sketched in Fig. 6–26, we obtain the deflection

$$\delta = DF - t_{D/A} = \frac{1}{EI}(911 - 1000) = -\frac{89}{EI}$$

which, except for the minus sign, is the value obtained previously. Here the minus sign indicates that the deflection at D is opposite to the direction sketched in Fig. 6–26; i.e., it is directed upward as before. This is checked by the fact that $t_{D/A}$ is numerically larger than DF.

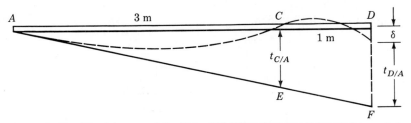

Figure 6–26. Elastic curve of Problem 652 with reference tangent drawn at A.

PROBLEMS

653. Compute the midspan value of $EI\delta$ for the beam shown in Fig. P–653. (*Hint:* Draw the M diagram by parts, starting from midspan toward the ends. Also take advantage of symmetry to note that the tangent drawn to the elastic curve at midspan is horizontal.)

Ans. $EI\delta = 3350 \text{ N} \cdot \text{m}^3$

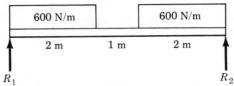

Figure P–653.

654. For the beam shown in Fig. P–654, find the value of $EI\delta$ at 1.0 m from R_2. (*Hint:* Draw the reference tangent to the elastic curve at R_2.)

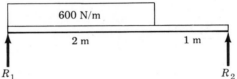

Figure P–654.

655. Find the value of $EI\delta$ under each concentrated load of the beam shown in Fig. P–655.   *Ans.* $608 \text{ N} \cdot \text{m}^3$; $850 \text{ N} \cdot \text{m}^3$

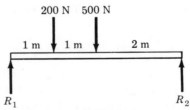

Figure P–655.

656. Find the value of $EI\delta$ at the point of application of the 100 N·m couple in Fig. P–656.   *Ans.* $EI\delta = 342 \text{ N} \cdot \text{m}^3$

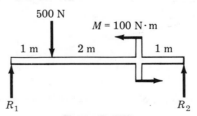

Figure P–656.

657.   Determine the midspan value of $EI\delta$ for the beam shown in Fig. P–657.

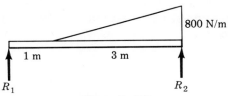

800 N/m

1 m   3 m

R_1 R_2

Figure P–657.

658.   For the beam shown in Fig. P–658, find the value of $EI\delta$ at the point of application of the couple.

$$Ans. \quad EI\delta = (Ma/3L)(L^2 - 3La + 2a^2)$$

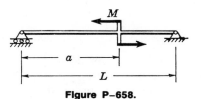

M

a

L

Figure P–658.

659.   A simple beam supports a concentrated load placed anywhere on the span, as shown in Fig. P–659. Measuring x from A, show that the maximum deflection occurs at

$$x = \sqrt{\frac{L^2 - b^2}{3}}$$

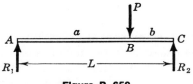

P

a b

A C

B

R_1 L R_2

Figure P–659.

660.   A simply supported beam is loaded by a couple M at its right end, as shown in Fig. P–660. Show that the maximum deflection occurs at $x = 0.577L$.

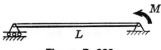

M

L

Figure P–660.

661. Compute the midspan deflection of the symmetrically loaded beam shown in Fig. P–661. Check your answer by letting $a = L/2$ and comparing with the answer to case 6 in Table 6–2 on page 270. Also compare one-half your answer with the midspan deflection of case 7 in Table 6–2.

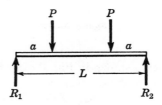

Figure P–661.

662. Determine the maximum deflection of the beam shown in Fig. P–662. Check your result by letting $a = L/2$ and comparing with case 8 in Table 6–2. Also use your result to check the answer to Problem 653.

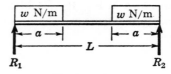

Figure P–662.

663. Determine the maximum deflection of the beam carrying a uniformly distributed load over the middle portion, as shown in Fig. P–663. Check your answer by letting $2b = L$ and comparing with case 8 in Table 6–2.   *Ans.* $EI\delta = (wb/24)(L^3 - 2Lb^2 + b^3)$

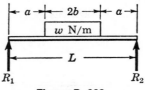

Figure P–663.

664. The middle half of the beam shown in Fig. P–664 has a moment of inertia twice that of the rest of the beam. Find the midspan deflection. (*Hint:* Convert the M diagram into an M/EI diagram.)

$$Ans. \quad \delta = \tfrac{3}{4}(Pa^3/EI)$$

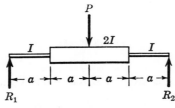

Figure P–664 and P–665.

665. Replace the concentrated load in Problem 664 by a uniformly distributed load of w N/m acting over the middle half of the beam. Find the maximum deflection.

666. Determine the value of $EI\delta$ at the right end of the overhanging beam shown in Fig. P–666. $Ans. \quad EI\delta = (wb^3/24)(4a + 3b)$

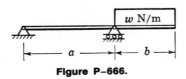

Figure P–666.

667. Determine the value of $EI\delta$ at the right end of the overhanging beam shown in Fig. P–667. Is the deflection up or down?

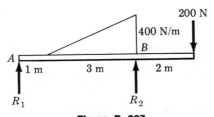

Figure P–667.

668. For the beam shown in Fig. P–668, compute the value of P that will cause the tangent to the elastic curve over support R_2 to be horizontal. $Ans. \quad P = 1350$ N

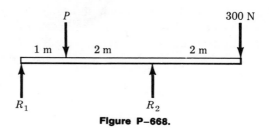

Figure P–668.

669.   Compute the value of $EI\delta$ at the left end of the beam shown in Fig. P–669.

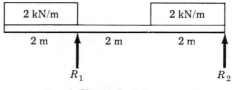

Figure P–669.

670.   Determine the value of $EI\delta$ at the left end of the overhanging beam shown in Fig. P–670.   *Ans.* $EI\delta = 428$ N·m³ down

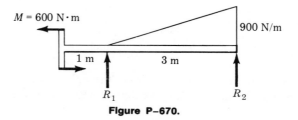

Figure P–670.

6–7 MIDSPAN DEFLECTIONS

In a symmetrically loaded simple beam, the tangent drawn to the elastic curve at midspan is horizontal and parallel to the unloaded beam. In such beams, the deviation at either support from the midspan tangent is equal to the midspan deflection.

For simple beams that are unsymmetrically loaded, the midspan deflection can be found as easily as for a symmetrically loaded beam. All that need be done is to add a symmetrically placed load for each load actually acting on the beam. The effect of this transformation to symmetry is to double the actual midspan deflection. In other words, the actual midspan deflection is equal to one-half the midspan deflection of the transformed symmetrically loaded beam. Note that there is so little difference between midspan deflection and the actual maximum deflection that practically the two values may be considered equivalent.

ILLUSTRATIVE PROBLEMS

671. A simply supported beam L meters long carries a load that varies uniformly from zero at the left end to w N/m at the right end, as shown in Fig. 6–27a. Determine the midspan deflection.

Solution: Create symmetry, as shown in Fig. 6–27b, by adding a load that varies uniformly from zero at the right end to w N/m at the left end. The result is a uniformly distributed load of w N/m over the entire span. The deviation at C from the midspan tangent drawn at B is equal to 2δ or twice the actual midspan deflection in Fig. 6–27a.

Since we are considering the deviation of C from the midspan tangent drawn at B, we need the moment diagram of only half the beam. This M diagram may be drawn by parts from C to B as shown in Fig. 6–27c or from B to C as shown in Fig. 6–27d. To facilitate understanding of Fig. 6–27d, the free-body diagram of the right half of the beam is also shown adjacent to it. Note that the midspan moment M is found from the fact that its effect at the right end must be equal and

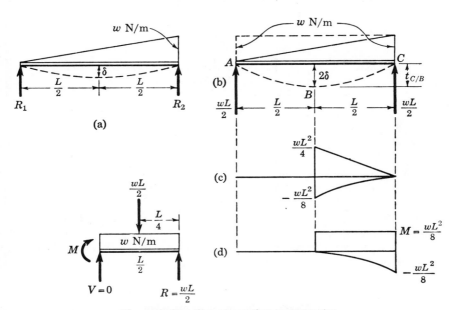

Figure 6–27. Transformation to symmetry.

opposite to that of the load in order to give a resultant bending moment
of zero at the end of the beam.

We shall apply the area-moment theorem to the M diagram in
part (d), leaving the student to verify that using part (c) yields the same
result. We obtain

$$\left[EIt_{C/B} = (\text{area})_{CB} \cdot \bar{x}_C \right]$$

$$2EI\delta = \left(\frac{wL^2}{8} \cdot \frac{L}{2} \right)\left(\frac{1}{2} \cdot \frac{L}{2} \right) - \left(\frac{1}{3} \cdot \frac{wL^2}{8} \cdot \frac{L}{2} \right)\left(\frac{1}{4} \cdot \frac{L}{2} \right) = \frac{5}{384}wL^4$$

whence

$$EI\delta = \frac{5}{768}wL^4 \quad Ans.$$

672.   Determine the midspan value of $EI\delta$ caused by the loads
shown in Fig. 6–28a.

Solution: The transformation to symmetry is shown in Fig. 6–28b.
Since we shall need the M diagram for only half of the beam, this is
shown for the left half in Fig. 6–28c, together with the corresponding

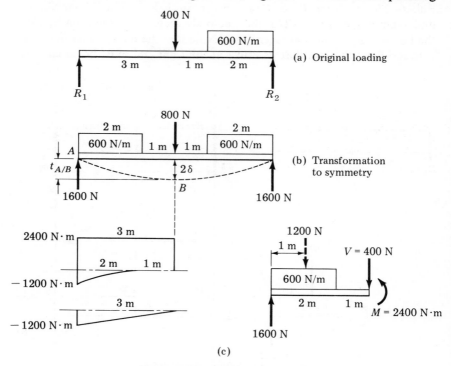

Figure 6–28.   Midspan deflection.

free-body diagram. Note that the midspan moment is found as described in the preceding problem. Also note that the reaction in the transformed beam is equal to one-half the symmetrically applied loads; that is, it is equivalent to the sum of the original loads.

The deviation $t_{A/B}$ of A from the horizontal tangent drawn at B is equal to twice the midspan deflection of the original loading. We obtain

$$\left[\, EIt_{A/B} = (\text{area})_{AB} \cdot \bar{x}_A \,\right]$$

$$2EI\delta = (2400 \times 3)\left(\frac{1}{2} \times 3\right) - \left(\frac{2 \times 1200}{3}\right)\left(\frac{1}{4} \times 2\right)$$

$$- \left(\frac{3 \times 1200}{2}\right)\left(\frac{1}{3} \times 3\right)$$

which reduces to

$$EI\delta = 4300 \text{ N·m}^3 \qquad Ans.$$

PROBLEMS

673. For the beam shown in Fig. P–673, show that the midspan deflection is $\delta = (Pb/48EI)(3L^2 - 4b^2)$.

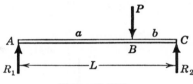

Figure P–673.

674. Find the deflection midway between the supports for the overhanging beam shown in Fig. P–674.

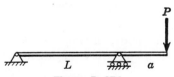

Figure P–674.

675. Repeat Problem 674 for the overhanging beam shown in Fig. P–675.   *Ans.* $\delta = wa^2L^2/32EI$

Figure P–675.

676. Determine the midspan deflection for the simply supported beam loaded by the couple shown in Fig. P–676.

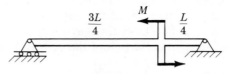

Figure P–676.

677. Determine the midspan deflection for the beam loaded as shown in Fig. P–677.

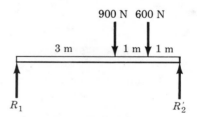

Figure P–677.

678. Determine the midspan value of $EI\delta$ for the beam shown in Fig. P–678.   *Ans.* $EI\delta = 3100 \ \text{N·m}^3$

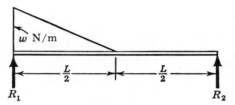

Figure P–678.

679. Determine the midspan value of $EI\delta$ for the beam shown in Fig. P–679 which carries a uniformly varying load over part of the span.
Ans. $EI\delta = 2940 \ \text{N·m}^3$

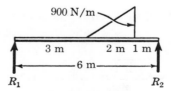

Figure P–679.

680.   Determine the midspan value of $EI\delta$ for the beam loaded as shown in Fig. P–680.

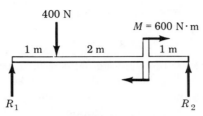

Figure P–680.

681.   Show that the midspan value of $EI\delta$ is $(wb/48)(L^3 - 2Lb^2 + b^3)$ for the beam in part (a) of Fig. P–681. Then use this result to find the midspan $EI\delta$ of the loading in part (b) by assuming the loading to extend over two separate intervals that start from midspan and adding the results.   *Ans.* $EI\delta = 9280 \text{ N·m}^3$

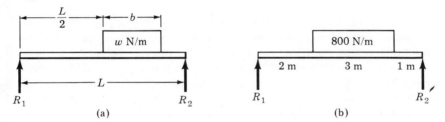

(a)   (b)

Figure P–681.

6–8   CONJUGATE-BEAM METHOD

Successive differentiation of the deflection equation discloses the following relations:

$$EIy = \text{deflection}$$

$$EI\frac{dy}{dx} = \text{slope}$$

$$EI\frac{d^2y}{dx^2} = \text{moment} = M$$

$$EI\frac{d^3y}{dx^3} = \text{shear} = V = \frac{dM}{dx}$$

$$EI\frac{d^4y}{dx^4} = \text{load} = \frac{dV}{dx} = \frac{d^2M}{dx^2}$$

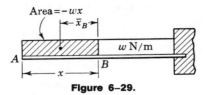

Figure 6–29.

It is evident that the relations among deflection, slope, and moment are the same as those among moment, shear, and load. This suggests that the area-moment method can be used to determine bending moment from the load diagram, just as deflection was obtained from the moment diagram. For example, in the load diagram in Fig. 6–29, the bending moment at B should equal (area of load diagram)$_{AB} \cdot \bar{x}_B$. That it does is seen from

$$(-wx)\left(\frac{1}{2}x\right) = -\frac{wx^2}{2}$$

Thus we could apply area-moment principles to compute bending moment, although this is impractical because better methods are available.

Nevertheless, the similarity of relations among load, shear, and moment, and among moment, slope, and deflection suggests that the relations among moment, slope, and deflection can be found by using the methods developed in Chapter 4 for computing shear and moment from load diagrams. We need merely assume that a beam is loaded, not with the actual loads, but with the M/EI diagram corresponding to these loads. Treating this M/EI diagram as a fictitious loading, we compute the shear and moment at any point caused by this loading. These fictitious shears and moments correspond to the actual slopes and deflections in the beam at corresponding points. This technique is known as the *conjugate-beam method* and sometimes as the method of elastic weights.

Applying the principles of shear and moment to a beam loaded with an M/EI diagram, we conclude that

1. **The actual slope = the fictitious shear**   **(6–6)**
2. **The actual deflection = the fictitious moment**   **(6–7)**

The method is especially useful for simply supported beams. For other beams, such as cantilevers or overhanging beams, artificial constraints must be applied; they are discussed later.

To evaluate the conjugate-beam method, let us compare it and the area-moment method when applied to a simple beam. Only in simply supported beams can the conjugate-beam method be applied directly without using artificial constraints.

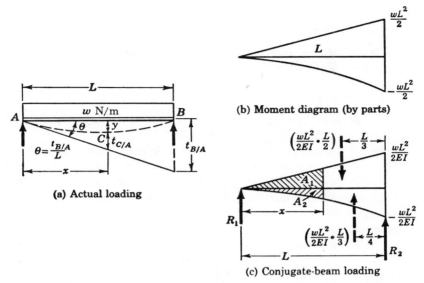

Figure 6–30. Comparison of conjugate-beam and area-moment methods.

Figure 6–30a shows a simply supported beam carrying a distributed load of w N/m. The moment diagram for this loading (drawn by parts in Fig. 6–30b) is multiplied by $1/EI$ and used as the conjugate-beam loading shown simply supported on the span L in Fig. 6–30c. The reaction R_1 of this conjugate beam is found by setting moments of the fictitious loads about B equal to zero. We obtain

$$[\Sigma M_B = 0] \qquad R_1 L = \left(\frac{wL^2}{2EI}\cdot\frac{L}{2}\right)\left(\frac{L}{3}\right) - \left(\frac{wL^2}{2EI}\cdot\frac{L}{3}\right)\left(\frac{L}{4}\right) \qquad (a)$$

The right-hand member of Eq. (a) will be recognized as $(1/EI)(\text{area})_{BA}$ $\cdot \bar{x}_B$, that is, $t_{B/A}$. Obviously, solving for R_1 is equivalent to $t_{B/A}/L$, which is the actual slope at A; this fact is evident from the geometry of the elastic curve in Fig. 6–30a. Nevertheless, this is confirmation of rule 1 of the conjugate-beam method: The fictitious shear equals the actual slope at the corresponding point in the actual beam.

To obtain the deflection at any point on the actual beam, we apply the definition of bending moment to the conjugate loads:

$$\text{Deflection } y = (\Sigma M)_L = R_1 x - A_1\frac{x}{3} + A_2\frac{x}{4}$$

$$= R_1 x - \left(A_1\frac{x}{3} - A_2\frac{x}{4}\right) \qquad (b)$$

However, in terms of the moment diagram in Fig. 6–30b, $[A_1(x/3) - A_2(x/4)]$ equals $(1/EI)(\text{area})_{CA} \cdot \bar{x}_C$, which equals $t_{C/A}$ on the elastic curve in Fig. 6–30a. Hence Eq. (*b*) may be rewritten as

$$y = R_1 x - t_{C/A} \qquad\qquad\qquad (c)$$

which, since $R_1 x = \theta x = (t_{B/A}/L)x$, is equivalent to the following area-moment relation:

$$y = t_{B/A}\left(\frac{x}{L}\right) - t_{C/A} \qquad\qquad (d)$$

This is the result previously obtained in Art. 6–6 for deflections in simple beams by the area-moment method.

     Thus the conjugate-beam method, which uses the fictitious shears and moments of an M/EI loading to determine actual slopes and deflections, involves precisely the same computations as the area-moment method but has the disadvantage of obscuring the physical significances of the computations. This disadvantage is even more pronounced when the method is used with cantilever and overhanging beams, where certain artificial constraints must be applied. Nevertheless, the conjugate-beam method offers an occasional advantage in certain routine work, in that it permits direct application of the definitions of shear and moment to the fictitious loading to find slope and deflection without any need of an elastic curve.

     Now for a word about the need for artificial constraints in certain cases. For the cantilever beam in Fig. 6–31a, the M/EI diagram appears as in Fig. 6–31b. This diagram cannot be applied directly as a fictitious load to a cantilever with the wall at the right end C because the fictitious shear and moment at B would be zero whereas the actual slope and deflection at B are not zero. Therefore, the diagram of fictitious loads must be modified, as in Fig. 6–31c, so that the fictitious

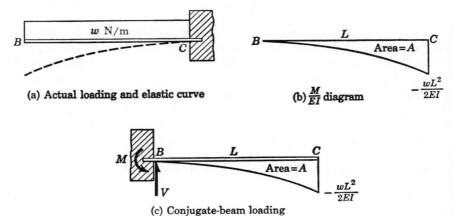

(a) Actual loading and elastic curve   (b) $\frac{M}{EI}$ diagram

(c) Conjugate-beam loading

Figure 6–31. Constraints required to solve cantilever beams by the conjugate-beam method.

loading will correspond to the actual slope and deflection at the free end.

The reason for supplying artificial constraints when solving cantilever problems should now be clear. To produce an actual zero slope at C in the original cantilever, the fictitious shear must be zero at C; therefore

$$[V = (\Sigma Y)_L] \qquad 0 = V - A$$

From this we see that the fictitious shear of the conjugate beam at B must equal the area A of the M/EI diagram. Also, to produce a zero fictitious moment at C, we must calculate the fictitious restraint M from

$$[M_C = (\Sigma M)_L] \qquad 0 = M + VL - A\frac{L}{4}$$

Only after the artificial constraints M and V have been found can the fictitious shear and moment (corresponding to actual slope and deflection) be computed. Hence, cantilever problems can be solved more simply and more directly by the area-moment method.

PROBLEMS

Problems 653 to 665 inclusive and cases 6 through 12 in Table 6–2 (page 270) may be assigned for solution by the conjugate-beam method.

6–9 DEFLECTIONS BY THE METHOD OF SUPERPOSITION

In a supplementary method of determining slopes and deflections, the results of a few simple loadings are used to obtain those for more complicated loadings. This procedure, called the *method of superposition*, determines the slope or deflection at any point in a beam as the resultant of the slopes and deflections at that point caused by each of the loads acting separately. The only restriction on this method is that the effect produced by each load must be independent of that produced by the other loads; i.e., each separate load must not cause an excessive change in the original shape or length of the beam.

The technique of superposition is advantageous primarily for loadings that combine the types in Table 6–2. For partially distributed loads, the method of superposition requires integration (see Illustrative Problem 683). In such cases, the double-integration method is preferable; if the deflection at only one specific location is desired, the area-moment method is generally best.

TABLE 6-2. Summary of Beam Loadings

CASE NO.	TYPE OF LOAD	MAX. MOMENT	SLOPE AT END	DEFLECTION EQUATION (y is positive downward)	MAXIMUM DEFLECTION
1		$M = -PL$	$\theta = \dfrac{PL^2}{2EI}$	$EIy = \dfrac{Px^2}{6}(3L - x)$	$\delta = \dfrac{PL^3}{3EI}$
2		$M = -Pa$	$\theta = \dfrac{Pa^2}{2EI}$	$EIy = \dfrac{Px^2}{6}(3a - x)$ for $0<x<a$ $EIy = \dfrac{Pa^2}{6}(3x - a)$ for $a<x<L$	$\delta = \dfrac{Pa^2}{6EI}(3L - a)$
3		$M = -\dfrac{wL^2}{2}$ $= -\dfrac{WL}{2}$	$\theta = \dfrac{wL^3}{6EI}$ $= \dfrac{WL^2}{6EI}$	$EIy = \dfrac{wx^2}{24}(6L^2 - 4Lx + x^2)$	$\delta = \dfrac{wL^4}{8EI} = \dfrac{WL^3}{8EI}$
4		$M = -\dfrac{wL^2}{6}$ $= -\dfrac{WL}{3}$	$\theta = \dfrac{wL^3}{24EI}$ $= \dfrac{WL^2}{12EI}$	$EIy = \dfrac{wx^2}{120L}(10L^3 - 10L^2x + 5Lx^2 - x^3)$	$\delta = \dfrac{wL^4}{30EI} = \dfrac{WL^3}{15EI}$
5		$M = -M$	$\theta = \dfrac{ML}{EI}$	$EIy = \dfrac{Mx^2}{2}$	$\delta = \dfrac{ML^2}{2EI}$
6		$M = \dfrac{PL}{4}$	$\theta_L = \theta_R = \dfrac{PL^2}{16EI}$	$EIy = \dfrac{Px}{12}\left(\dfrac{3}{4}L^2 - x^2\right)$ for $0<x<\dfrac{L}{2}$	$\delta = \dfrac{PL^3}{48EI}$

#	Diagram	M	θ	EIy	δ
7		$M = \dfrac{Pab}{L}$ at $x=a$	$\theta_L = \dfrac{Pb(L^2-b^2)}{6EIL}$ $\theta_R = \dfrac{Pa(L^2-a^2)}{6EIL}$	$EIy = \dfrac{Pbx}{6L}(L^2-x^2-b^2)$ for $0<x<a$ $EIy = \dfrac{Pb}{6L}\left[\dfrac{L}{b}(x-a)^3+(L^2-b^2)x-x^3\right]$ for $a<x<L$	$\delta = \dfrac{Pb(L^2-b^2)^{3/2}}{9\sqrt{3}\,EIL}$ at $x=\sqrt{\dfrac{L^2-b^2}{3}}$ At center (not max.) $\delta = \dfrac{Pb}{48EI}(3L^2-4b^2)$ when $a>b$
8		$M = \dfrac{wL^2}{8}$ $= \dfrac{WL}{8}$	$\theta_L = \theta_R = \dfrac{wL^3}{24EI}$	$EIy = \dfrac{wx}{24}(L^3-2Lx^2+x^3)$	$\delta = \dfrac{5wL^4}{384EI} = \dfrac{5WL^3}{384EI}$
9		$M = \dfrac{wL^2}{9\sqrt{3}}$ $= \dfrac{2WL}{9\sqrt{3}}$	$\theta_L = \dfrac{7wL^3}{360EI}$ $\theta_R = \dfrac{8wL^3}{360EI}$	$EIy = \dfrac{wx}{360L}(7L^4-10L^2x^2+3x^4)$	$\delta = \dfrac{2.5wL^4}{384EI} = \dfrac{5WL^3}{384EI}$ at $x=0.519L$
10		$M = \dfrac{wL^2}{12}$ $= \dfrac{WL}{6}$	$\theta_L = \theta_R = \dfrac{5wL^3}{192EI}$	$EIy = \dfrac{wx}{960L}(25L^4-40L^2x^2+16x^4)$ for $0<x<\dfrac{L}{2}$	$\delta = \dfrac{wL^4}{120EI} = \dfrac{WL^3}{60EI}$
11		$M = M$	$\theta_L = \dfrac{ML}{6EI}$ $\theta_R = \dfrac{ML}{3EI}$	$EIy = \dfrac{MLx}{6}\left(1-\dfrac{x^2}{L^2}\right)$	$\delta = \dfrac{ML^2}{9\sqrt{3}\,EI}$ at $x=\dfrac{L}{\sqrt{3}}$ At center (not max.) $\delta = \dfrac{ML^2}{16EI}$
12		$M = M$	$\theta_L = \dfrac{ML}{3EI}$ $\theta_R = \dfrac{ML}{6EI}$	$EIy = \dfrac{Mx}{6L}(L-x)(2L-x)$	$\delta = \dfrac{ML^2}{9\sqrt{3}\,EI}$ at $x=\left(L-\dfrac{L}{\sqrt{3}}\right)$ At center (not max.) $\delta = \dfrac{ML^2}{16EI}$

ILLUSTRATIVE PROBLEMS

682. Using the method of superposition, compute the midspan value of $EI\delta$ for the beam carrying two concentrated loads, shown in Fig. 6–32a.

Solution: From case 7 of Table 6–2, the midspan deflection of an eccentrically placed concentrated load is $EI\delta = (Pb/48)(3L^2 - 4b^2)$ where b is the smaller of the two segments into which the beam is divided by the load. Resolving the given loading into those shown in (b) and (c) in Fig. 6–32, we find that the midspan deflection of (a) is equal

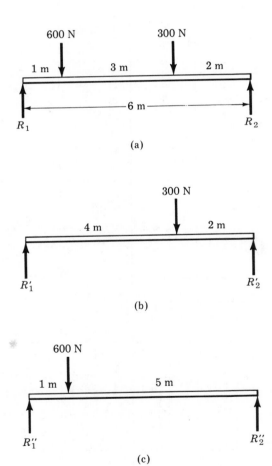

(a)

(b)

(c)

Figure 6–32.

to the sum of the midspan deflections of (b) and (c). Thus

$$EI\delta = \sum \frac{Pb}{48}(3L^2 - 4b^2)$$

$$= \frac{300(2)}{48}\left[3(6)^2 - 4(2)^2\right] + \frac{600(1)}{48}\left[3(6)^2 - 4(1)^2\right]$$

$$= 2450 \ \text{N} \cdot \text{m}^3 \qquad Ans.$$

683. A simply supported beam carries a uniformly distributed load over part of its length, as shown in Fig. 6–33. Compute the midspan value of $EI\delta$.

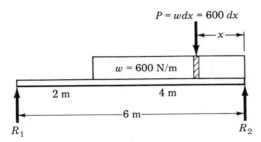

Figure 6–33.

Solution: The continuous load may be considered as a series of concentrated elemental loads, each of $P = w \, dx = 600 \, dx$ and located at a distance x from the end. At midspan, therefore, applying the result for case 7, we have

$$EI\delta = \sum \frac{Pb}{48}(3L^2 - 4b^2)$$

$$= \int_0^3 \frac{(600 \, dx)(x)}{48}\left[3(6)^2 - 4x^2\right] + \int_2^3 \frac{(600 \, dx)(x)}{48}\left[3(6)^2 - 4x^2\right]$$

$$= 5063 + 2562 = 7625 \ \text{N} \cdot \text{m}^3 \qquad Ans.$$

Two integrations are necessary—one from 0 to 3 for the right half of the beam, and the other from 2 to 3 for the left half. This is true because x, which replaces b, must be the length of the smaller segment into which P divides the beam.

If the given loading is divided into halves and each half is replaced by its resultant of 1200 N acting as shown in Fig. 6–34, the sum of the midspan deflections of these loads will be a good approximation of the

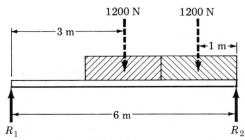

Figure 6–34.

actual midspan deflection. Applying the result for case 7 gives

$$EI\delta = \frac{1200(1)}{48}\left[3(6)^2 - 4(1)^2\right] + \frac{1200(3)}{48}\left[3(6)^2 - 4(3)^2\right]$$

$$= 8000 \text{ N·m}^3$$

A closer approximation could be obtained by subdividing the given loading into three or more parts. However, even with only two subdivisions, the result is only about 5% larger than the correct value, 7625 N·m³.

684. The overhanging beam in Fig. 6–35 carries a concentrated load P at its end. Determine the deflection under P.

Solution: The tangent to the elastic curve at R_2 is inclined at a very small angle θ with the horizontal, but the figure shows it greatly exaggerated for convenience in representation. Imagine the original position of the beam to coincide with this tangent and to be clamped in this inclined position at R_2. Application of the loads R_1 and P will produce the actual elastic curve. The deflections δ_1 and δ_2 produced by these loads are similar to those in case 1. Theoretically δ_1 (and δ_2 also) should be directed perpendicularly to the clamped tangent. However, this perpendicular distance $\delta_1 \cos\theta \approx \delta_1$ because θ is extremely small. In other words, there is no essential difference between a perpendicular to the tangent and the vertical deflection δ_1 shown.

From the geometry of the elastic curve, we have

$$\delta_1 = \theta a; \quad \text{hence} \quad \theta = \frac{\delta_1}{a}$$

Therefore the deflection at P is

$$y = \theta b + \delta_2 = \delta_1 \frac{b}{a} + \delta_2$$

whence, substituting $\delta = PL^3/3EI$ from case 1, we obtain

$$y = \frac{\dfrac{Pb}{a}(a^3)}{3EI} \cdot \frac{b}{a} + \frac{Pb^3}{3EI} = \frac{Pb^2}{3EI}(a + b) = \frac{Pb^2L}{3EI} \qquad Ans.$$

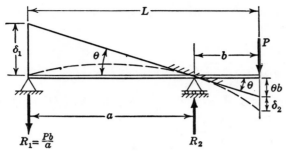

Figure 6-35.

Actually δ_1 and δ_2 are deviations from a tangent drawn at R_2, and this procedure duplicates that described in Illustrative Problem 652. The difference is that here we used a formula from Table 6-2 instead of computing deviations by using a moment diagram.

Alternate Solution: The rotation of the beam at the support R_2 may also be found by dividing the beam in Fig. 6-35 into the two parts shown in Fig. 6-36. The action of the overhang upon the portion between the supports is replaced by a shear force P and a moment $M = Pb$. The shear force is transmitted directly to the reaction R_2, whence the couple produces the effect of case 11. The rotation of the beam at R_2 is given by

$$\theta = \frac{ML}{3EI} = \frac{(Pb)a}{3EI}$$

The deflection at the end of the overhang may now be found as in the case of a cantilever beam (case 1) built in with an initial inclination θ. The total deflection is

$$y = \theta b + \delta_2 = \frac{Pba}{3EI} \cdot b + \frac{Pb^3}{3EI}$$

which, as before, reduces to

$$y = \frac{Pb^2}{3EI}(a + b) = \frac{Pb^2L}{3EI}$$

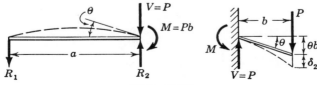

Figure 6-36.

The deflection equation for any point between the supports, as found from case 11, is

$$y = \frac{Pbax}{6EI}\left(1 - \frac{x^2}{a^2}\right)$$

and, in terms of x measured from R_2, the deflection equation for the overhanging portion is

$$y = \frac{Pba}{3EI}x + \frac{Px^2}{6EI}(3b - x)$$

685. Two cantilever beams, having the same cross section and made of the same material, jointly support a distributed load of w N/m as shown in Fig. 6–37. Determine the force P at the roller between them.

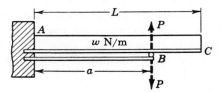

Figure 6–37.

Solution: The force P may be determined by the condition that at B both cantilevers have the same deflection. The deflection at B for the lower cantilever is found from case 1 to be

$$\delta = \frac{Pa^3}{3EI}$$

The upper cantilever is loaded with a combination of cases 3 and 2, the resultant deflection at B being

$$\delta = \frac{wa^2}{24EI}(6L^2 + a^2 - 4La) - \frac{Pa^3}{3EI}$$

Equating these deflections gives, for P,

$$P = \frac{w}{16a}(6L^2 + a^2 - 4La) \qquad Ans.$$

PROBLEMS

In solving the following problems, use Table 6–2.

686.   Determine the value of $EI\delta$ under each concentrated load in Fig. P–686.   *Ans.*   575 N·m³; 767 N·m³

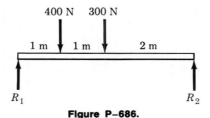

400 N   300 N

1 m   1 m   2 m

R_1 R_2

Figure P–686.

687.   Determine the midspan deflection of the beam shown in Fig. P–687 if $E = 10 \times 10^9 \text{ N/m}^2$ and $I = 20 \times 10^6 \text{ mm}^4$.

Ans.   25.8 mm

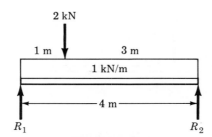

2 kN

1 m   3 m

1 kN/m

— 4 m —

R_1 R_2

Figure P–687.

688.   Determine the value of $EI\delta$ at the left end of the beam shown in Fig. P–688.

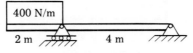

400 N/m

2 m   4 m

Figure P–688.

689. The beam shown in Fig. P–689 has a rectangular cross section 100 mm wide by 200 mm deep. Compute the value of P that will limit the midspan deflection to 40 mm. Use $E = 10 \times 10^9$ N/m^2.

Ans. $P = 7.07$ kN

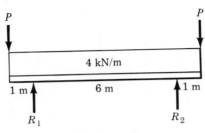

Figure P–689.

690. The beam shown in Fig. P–690 has a rectangular cross section 50 mm wide. Determine the proper depth d of the beam if the midspan deflection of the beam is not to exceed 20 mm and the flexural stress is limited to 10 MN/m^2. Use $E = 10$ GN/m^2.

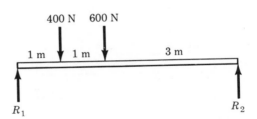

Figure P–690.

691. Determine the midspan deflection for the beam shown in Fig. P–691. (*Hint:* Apply case 7 and integrate.)

Ans. $EI\delta = (wa^2/48)(3L^2 - 2a^2)$

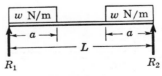

Figure P–691.

692. Determine the value of $EI\delta$ midway between the supports for the beam in Fig. P–692. (Hint: Combine case 11 and one-half of case 8.)

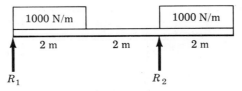

Figure P–692.

693. Determine the value of $EI\delta$ at the right end of the overhanging beam in Fig. P–693.   *Ans.* $EI\delta = 680$ N·m³, downward

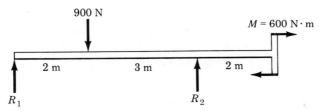

Figure P–693.

694. The frame shown in Fig. P–694 is of constant cross section and is perfectly restrained at its lower end. Compute the vertical deflection caused by the couple M.   *Ans.* $\delta = \dfrac{Ma}{EI}(b + \dfrac{a}{2})$

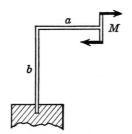

Figures P–694 and P–695.

695. Solve Problem 694 if the couple is replaced by a vertical downward load P.   *Ans.* $\delta = \dfrac{Pa^2}{EI}(b + \dfrac{a}{3})$

696. In Fig. P–696, determine the value of P for which the deflection under P will be zero.   *Ans.* $P = 400$ N

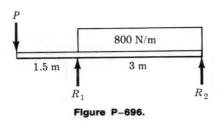

Figure P–696.

697. Two identical cantilever beams in contact at their ends support a distributed load over one of them as shown in Fig. P–697. Determine the restraining moment at each wall.

Ans. $M_A = -3wL^2/16;\ M_B = -5wL^2/16$

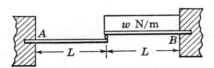

Figure P–697.

698. The beam in Fig. P–698 is supported at the left end by a spring which has a spring constant of 60 kN/m. For the beam, $E = 10 \times 10^9$ N/m^2 and $I = 60 \times 10^6$ mm^4. Compute the deflection of the spring.   *Ans.* $\delta = 13.6$ mm

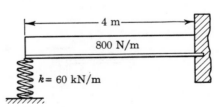

Figure P–698.

699. Two timber beams are mounted at right angles and in contact with each other at their midpoints. The upper beam A is 50 mm wide by 200 mm deep and simply supported on an 3-m span; the lower beam B is 80 mm wide by 200 mm deep and simply supported on a 4-m span. At their crossover point, they jointly support a load $P = 10$ kN. Determine the maximum flexural stress in the assembly.

SUMMARY

Starting with the relation $1/\rho = M/EI$ developed in Art. 5-2, two separate methods of determining slopes and deflections are discussed. The first one, the double-integration method, is primarily mathematical. Before the deflection at a particular point can be found, complete slope and deflection equations must be determined; this is easily accomplished by using the concept of a general moment equation developed on page 217. The constants of integration will become zero if the origin of axes is selected at a position where the slope and deflection are known to be zero, as at a perfectly restrained end or at the center of a symmetrically loaded beam.

The area-moment method is generally more direct than the double-integration method, especially when the deflection at a particular position is desired. Depending as it does upon the geometry of the elastic curve, the area-moment method emphasizes the physical significance of the computations. The two basic theorems of this method, developed in Art. 6-3, are summarized by

$$\theta_{AB} = \frac{1}{EI}(\text{area})_{AB} \qquad (6-4)$$

and

$$t_{B/A} = \frac{1}{EI}(\text{area})_{BA} \cdot \bar{x}_B \qquad (6-5)$$

The tangential deviation at any point is positive if that point lies above the reference tangent from which the deviation is measured; the converse is true for negative deviation. A positive value for the change in slope means that the tangent at the right point is rotated in a counterclockwise direction relative to the tangent at the left point.

The use of the area-moment theorems requires that the area and the moment of area of bending-moment diagrams be readily calculated. Therefore, beginning in Art. 6-4, and discussed more fully in succeeding articles, there is developed a method of drawing moment diagrams by parts (i.e., in terms of equivalent cantilever loadings) which is equivalent to, and replaces, the conventional moment diagram.

Deflections in cantilever beams (Art. 6-5) are easily found from the fact that the deflection at any point is equal to the deviation of that point from a reference tangent drawn at the fixed end.

The deflections in simply supported beams are determined by using a reference tangent to the elastic curve drawn at either reaction. The complete procedure is outlined in the four steps on page 249. Since midspan deflections are practically equivalent to maximum deflections, an easy method of finding them is developed in Art. 6-7.

The conjugate-beam method (Art. 6–8) shows how the definitions of shear and moment may be applied to a fictitious loading (in terms of the M/EI diagram of the original loading) to determine the slope and deflection at any point.

When beam loadings are combinations of the types listed in Table 6–2, deflections and slopes are obtained most easily by superposing the results listed there. See Art. 6–9 for further details.

Restrained
Beams

7-1 INTRODUCTION

Our study of simple stresses and torsion has shown that statically indeterminate problems require relations between the elastic deformations in addition to the equations of static equilibrium. Similarly, for our present study of indeterminate beams, additional relations must be found from the geometry of the elastic curves of the beams. Such relations are obtained from our study of the deflections in statically determinate beams.

Three techniques are discussed: (1) double-integration; (2) the method of superposition, which uses the general solutions in Tables 6–2 and 7–1; (3) the area-moment method, which deals directly with the shape of the elastic curve. As we shall see, sometimes the deflection at a particular position is required, sometimes a relation between the slopes at two positions, sometimes a combination of these concepts.

7-2 REDUNDANT SUPPORTS IN PROPPED
AND RESTRAINED BEAMS

A cantilever beam is supported by two reactive elements, the shear V and the moment M at the wall, as shown in Fig. 7–1a. Since these

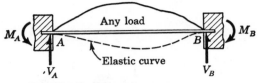

Figure 7-1. Determinate and indeterminate beams.

Figure 7-2. Perfectly restrained beam.

values are readily computed from $\Sigma Y = 0$ and $\Sigma M = 0$, the cantilever beam is statically determinate. Propping up the beam at some other position, as in Fig. 7-1b, introduces an additional reaction but does not increase the equations of static equilibrium, so the beam has one redundant support.

In other words, if any arbitrary value is assumed for R, values of V and M may be computed that will satisfy the equations of static equilibrium. Determination of the correct combination of R, V, and M therefore requires a condition in addition to those found from static equilibrium. Usually the most convenient condition is that the deflection under R is either zero or some known value. Another condition sometimes used, generally in the method of superposition, is that the slope at the wall is zero.

A beam restrained at both ends, as in Fig. 7-2, has four reactive elements. Because only two equations of static equilibrium are available —namely, $\Sigma Y = 0$ and $\Sigma M = 0$—the beam has two redundant supports. It is usually best to consider that the redundant supports are the shear and moment at the same wall, i.e., V_A and M_A; but sometimes, as we shall see in Art. 7-5, the two end moments are taken as the redundancies.

7-3   APPLICATION OF DOUBLE-INTEGRATION AND SUPERPOSITION METHODS

The double-integration method may be applied to propped or restrained beams in exactly the same manner as was described in Art. 6-2. In applying it, we select the origin of the reference axes at a fixed end where both the slope and deflection are zero, thereby causing the constants of integration to be zero. However, the general moment equation and its subsequent integrations will contain the unknown values of shear and moment at the fixed end. These are easily evaluated

TABLE 7–1 Slope and Deflection at Free End

$$EI\theta = \frac{ML}{n+1}; \quad EI\delta = \frac{ML^2}{n+2}; \quad M= \text{moment at wall}$$

LOADING	n	$EI\theta$	$EI\delta$
M, L	0	$\dfrac{ML}{1}$	$\dfrac{ML^2}{2}$
P, L	1	$\dfrac{ML}{2}$	$\dfrac{ML^2}{3}$
w N/m, L	2	$\dfrac{ML}{3}$	$\dfrac{ML^2}{4}$
w N/m, L	3	$\dfrac{ML}{4}$	$\dfrac{ML^2}{5}$

by substituting the boundary conditions at the other support into the slope and the deflection equations. The details are explained in the illustrative problems that follow.

The method of superposition may be applied by using the general results listed in Table 6–2, but it is more convenient to use the summary of slope and deflection at the free end of cantilevers under various loadings that is given in Table 7–1. In this table the results are expressed in terms of the moment M at the fixed end. Notice that the coefficients are exactly the same as those previously developed for area and for location of centroid in Table 6–1 (page 233). It is simple to verify that the general results are

$$EI\theta = \frac{ML}{n+1} \quad \text{and} \quad EI\delta = \frac{ML^2}{n+2}$$

where n is the degree of the moment curve.

ILLUSTRATIVE PROBLEMS

701. Solve for the reactive elements of the propped beam in Fig. 7–3a by two methods: (1) the method of superposition, considering R_A as the redundant support; (2) the double-integration method.

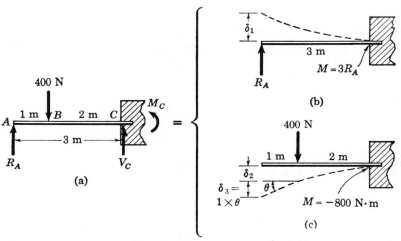

Figure 7–3.   Propped beam solved by superposition.

Solution:

Method of Superposition.   The propped beam may be duplicated by superposing the cantilever loadings shown in Fig. 7–3b and 7–3c so that the resultant deflection at A is zero; that is, $\delta_1 = \delta_2 + \delta_3$. Observe that δ_2 is the deflection under the 400-N load and that δ_3 is caused by rotation of the 1.0-m unloaded segment through the slope angle θ. Using Table 7–1, we combine the free end deflections of these loadings as follows to give the zero deflection at A in the original beam:

$$[\Sigma EI\delta = 0]\qquad \frac{(3R_A)(3)^2}{3} - \frac{(800)(2)^2}{3} - 1 \times \frac{(800)(2)}{2} = 0$$

from which we obtain

$$R_A = 207 \text{ N}$$

With R_A known, a vertical summation of forces determines V_C:

$$[\Sigma Y = 0]\qquad V_C + 207 - 400 = 0 \qquad V_C = 193 \text{ N}$$

From the definition of bending moment at C, which is equivalent to $\Sigma M_C = 0$ but simpler to apply, we obtain

$$[M = (\Sigma M)_L]\qquad M_C = 3(207) - 2(400) = -179 \text{ N·m}$$

An alternate method of superposition would be to let M_C be the redundant support. Its value could then be determined by replacing the given loading by a combination of cases 7 and 11 of Table 6–2 so that the sum of the right end slopes is zero. Check M_C using this procedure.

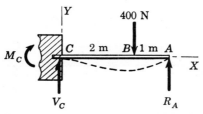

Figure 7-4. Solution by superposition.

Double-Integration Method. Returning to Fig. 7–3a, select the origin of axes at the fixed end C. For convenience, we redraw the beam as in Fig. 7–4 so that the fixed end is at the left. At this fixed end, the slope and deflection are both zero, so the constants of integration C_1 and C_2 will also be zero since these constants are physically equivalent to slope and deflection at the origin. Setting up the differential equation of the elastic curve in terms of the general moment equation, and integrating twice, we obtain

$$EI\frac{d^2y}{dx^2} = M_C + V_C x - 400\langle x - 2\rangle \qquad (a)$$

$$EI\frac{dy}{dx} = M_C x + \frac{V_C x^2}{2} - 200\langle x - 2\rangle^2 + C_1^{=0} \qquad (b)$$

$$EIy = \frac{M_C x^2}{2} + \frac{V_C x^3}{6} - \frac{200}{3}\langle x - 2\rangle^3 + C_2^{=0} \qquad (c)$$

To evaluate M_C and V_C, we note that the boundary conditions at A, where $x = 3$, are that the moment and the deflection are both zero. Hence substituting $x = 3$ in Eqs. (a) and (c), we obtain

$$M_C + 3V_C - 400(1) = 0$$

$$\frac{M_C}{2}(3)^2 + \frac{V_C}{6}(3)^3 - \frac{200}{3}(1)^3 = 0$$

Solving these simultaneously, we have, as before

$$V_C = 193\text{ N} \quad \text{and} \quad M_C = -179\text{ N·m} \qquad Ans.$$

One interesting aspect of this double-integration solution is that, if the given beam had been perfectly restrained at both ends, the redundants could be found by merely changing the boundary conditions at A to zero slope and zero deflection. Then by using Eqs. (b) and (c), in which to substitute $x = 3$, we would obtain $V_C = 104$ N and $M_C = -89$ N·m.

702. A beam 4 m long and perfectly restrained at the ends carries a uniformly distributed load over part of its length, as shown in Fig. 7–5. Use the double-integration method to compute the end shears and

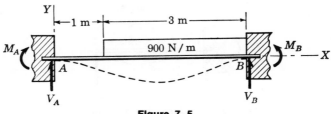

Figure 7–5.

end moments, and then check these results using the method of super-position.

Solution: The moment at A is shown acting in a positive or clockwise direction, although we know from the downward concavity of the elastic curve that this moment is actually negative or counterclockwise. By making this deliberate error in the direction of the vector quantity M_A, our solution will determine not only the correct numerical value of M_A but also specify its correct negative sign.

The elastic curve in Fig. 7–5 shows that the slope and deflection at A are both zero. Choosing the origin of axes at A will make the constants of integration C_1 and C_2 both zero since they represent respectively the slope and deflection at the origin. Setting up the differential equation of the elastic curve in terms of the general moment equation, and integrating twice, we obtain

$$EI\frac{d^2y}{dx^2} = M_A + V_A x - \frac{900}{2}\langle x - 1\rangle^2 \tag{a}$$

$$EI\frac{dy}{dx} = M_A x + \frac{V_A x^2}{2} - 150\langle x - 1\rangle^3 + C_1^{\nearrow = 0} \tag{b}$$

$$EIy = \frac{M_A x^2}{2} + \frac{V_A x^3}{6} - \frac{150}{4}\langle x - 1\rangle^4 + C_2^{\nearrow = 0} \tag{c}$$

To evaluate M_A and V_A, we note that at the other restrained end B, where $x = 4$, the slope and deflection are also zero. Hence substituting $x = 4$ in Eqs. (b) and (c), we obtain

$$4M_A + (4)^2\frac{V_A}{2} - 150(3)^3 = 0 \tag{d}$$

$$(4)^2\frac{M_A}{2} + (4)^3\frac{V_A}{6} - \frac{150}{4}(3)^4 = 0 \tag{e}$$

Solving these equations simultaneously yields

$$V_A = 949 \text{ N} \quad \text{and} \quad M_A = -886 \text{ N·m} \qquad Ans.$$

The negative sign for M_A indicates that the direction of M_A was incorrectly assumed, as was stated earlier. The moment is actually negative, and there should now be no confusion about sign in its subsequent use.

Having determined V_A and M_A, we apply the conditions of static equilibrium to the free-body diagram in Fig. 7–5 to compute the shear and moment at B. A vertical summation of forces gives

$$[\Sigma Y = 0] \qquad V_B + 949 - 900(3) = 0 \qquad V_B = 1750 \text{ N} \qquad Ans.$$

We can now determine M_B from the condition $\Sigma M_B = 0$, but it is simpler (and less confusing with respect to sign) to apply the definition of bending moment:

$$[M = (\Sigma M)_L] \qquad M_B = 4V_A + M_A - (900 \times 3)\left(\frac{3}{2}\right)$$
$$= 4(949) - 886 - 4050$$
$$= -1140 \text{ N·m} \qquad Ans.$$

As was shown in the preceding problem, this solution may also be adapted to a beam propped at B instead of fixed by changing the boundary conditions at B to zero moment and zero deflection. Then by substituting $x = 4$ in Eqs. (a) and (c), we could determine M_A and V_A if this beam were propped at B.

To apply the method of superposition, we replace the given loading repeated in Fig. 7–6a by the equivalent cantilever loadings in

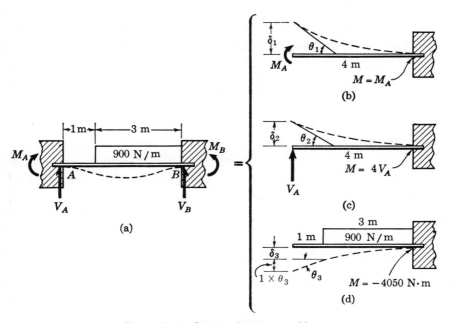

Figure 7–6. Solution by superposition.

Fig. 7–6b, c, and d. We now apply the conditions that $\theta_1 + \theta_2 - \theta_3 = 0$ and $\delta_1 + \delta_2 - \delta_3 - (1 \times \theta_3) = 0$.

Using Table 7–1, we then obtain

$$[\Sigma EI\theta = 0] \qquad M_A(4) + \frac{(4V_A)(4)}{2} - \frac{(4050)(3)}{3} = 0$$

and

$$[\Sigma EI\delta = 0] \qquad \frac{M_A(4)^2}{2} + \frac{(4V_A)(4)^2}{3} - \frac{(4050)(3)^2}{4}$$
$$- 1 \times \frac{(4050)(3)}{3} = 0$$

The solution of these equations yields, as before,

$$V_A = 949 \text{ N} \quad \text{and} \quad M_A = -886 \text{ N·m} \qquad Check$$

What would the answers be if the beam had been propped at A instead of being fixed?

PROBLEMS

In solving the following problems, use superposition or double-integration as directed by your instructor. Unless otherwise stated, the supports are assumed to remain at the same level. Additional problems may be selected from those in Art. 7–4.

703. For the propped beam shown in Fig. P–703, find R and draw the shear and moment diagrams. *Ans.* $R = (wb^3/8L^3)(4L - b)$

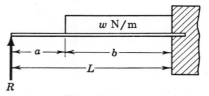

Figure P–703.

704. Compute the reaction R and sketch the shear and moment diagrams for the propped beam shown in Fig. P–704.

Ans. $R = wL/10$

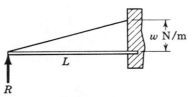

Figure P–704.

705. For the propped beam shown in Fig. P–705, determine the reaction R and sketch the shear and moment diagrams.

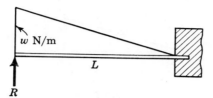

Figure P–705.

706. A couple M is applied at the propped end of the beam shown in Fig. P–706. Compute R at the propped end and also the wall restraining moment.   *Ans.* $R = 3M/2L$

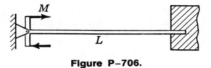

Figure P–706.

707. Determine the reaction R and sketch the shear and moment diagrams for the propped beam shown in Fig. P–707.

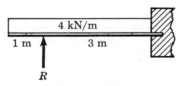

Figure P–707.

708. Determine the reaction R for the propped beam shown in Fig. P–708.

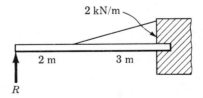

Figure P–708.

709. Determine the end moments for the restrained beam shown in Fig. P–709.   *Ans.* $M_A = -wL^2/30; M_B = -wL^2/20$

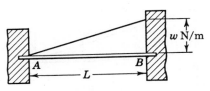

Figure P–709.

710. Solve for the end moments in the restrained beam loaded as shown in Fig. P–710.

Ans. $M_A = -Pab^2/L^2; M_C = -Pa^2b/L^2$

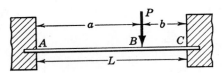

Figure P–710.

711. There is a small initial clearance Δ between the left end of the beam shown in Fig. P–711 and the roller support. Determine the reaction at the roller support after the uniformly distributed load is applied.

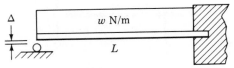

Figure P–711.

712. Compute the end moments for the restrained beam shown in Fig. P–712.   *Ans.* $M_A = -575$ N·m; $M_C = -1525$ N·m

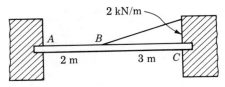

Figure P–712.

713. Determine the end moment and midspan value of $EI\delta$ for the restrained beam shown in Fig. P–713. (*Hint*: Because of symmetry, the end shears are equal and the slope is zero at midspan. Let the redundant be the moment at midspan.)

$$Ans. \quad M = -(2/3)Pa; \quad EI\delta = (5/24)Pa^3$$

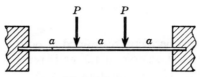

Figure P–713.

714. For the restrained beam shown in Fig. P–714, compute the end moment and maximum $EI\delta$. (*Hint*: Use equivalent cantilevers fixed at midspan and extending toward either end.)

$$Ans. \quad M = -2750 \text{ N} \cdot \text{m}$$

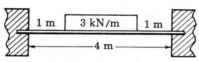

Figure P–714.

715. Determine the end moment and maximum $EI\delta$ for the restrained beam shown in Fig. P–715. (*Hint*: Let the redundants be the shear and moment at midspan. Also note that the midspan shear is zero. Why?)

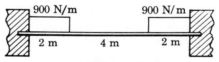

Figure P–715.

7–4 APPLICATION OF AREA-MOMENT METHOD

The method of superposition, described in the preceding article, requires the use of tabulated values of slopes and deflections as listed in Table 6–2 or 7–1. Instead of relying on such tabulations, it is usually preferable to use the area-moment theorems which develop the equations needed to determine the redundants directly from the moment diagram.

For propped beams we use the condition that the deviation of the support from the tangent drawn to the elastic curve at the wall is zero or some known value. Specific details are explained in Illustrative Problem

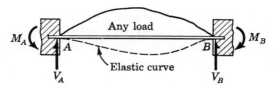

Figure 7-7.

716, below. For beams perfectly restrained at both ends, as in Fig. 7–7, the tangents to the elastic curve at the ends remain horizontal. Therefore, since there is no change in slope between the ends, $\theta_{AB} = 0$. In addition, if the ends A and B are at the same level, the deviation at B from a tangent drawn at A is zero; i.e. $t_{B/A} = 0$. Also, the deviation at A from a tangent drawn at B is zero, or $t_{A/B} = 0$. Applying the theorems of the area-moment method, we may put these conditions in the form:

$$EI\theta_{AB} = (\text{area})_{AB} = 0 \tag{a}$$

$$EIt_{B/A} = (\text{area})_{BA} \cdot \bar{x}_B = 0 \tag{b}$$

$$EIt_{A/B} = (\text{area})_{AB} \cdot \bar{x}_A = 0 \tag{c}$$

These three equations are not independent; any two may be used together with the equations of statics to determine the four reactive elements. As a rule, it is best to use Eq. (a) and either Eq. (b) or (c), depending on whether it is simpler to compute the moment of area of a particular moment diagram about the right or the left end. Which to choose will be apparent in the discussion of Illustrative Problem 717, below.

ILLUSTRATIVE PROBLEMS

716. Figure 7–8 shows a cantilever beam carrying a uniformly distributed load of w N/m. The beam is propped up at the free end by an unyielding support. Compute the reactions, and sketch the shear and moment diagrams.

Solution: The wall is assumed to provide absolute resistance to rotation of the beam, so that at B (Fig. 7–8a) a tangent to the elastic curve is horizontal and passes through A. Therefore, the deviation of A from the tangent at B is zero. Expressing this in terms of the moment diagram by parts shown in Fig. 7–8b, we obtain

$$\left[EIt_{A/B} = (\text{area})_{AB} \cdot \bar{x}_A = 0 \right]$$

$$\frac{(L)(R_A L)}{2}\left(\frac{2}{3}L\right) - \frac{(L)}{3}\left(\frac{wL^2}{2}\right)\left(\frac{3}{4}L\right) = 0$$

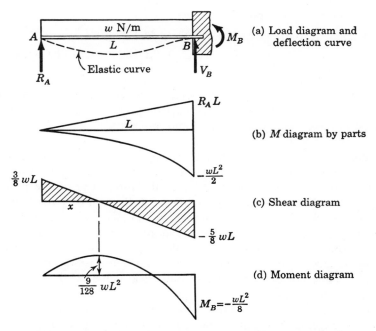

Figure 7-8. Propped beam solved by area-moment method.

or

$$R_A = \tfrac{3}{8}wL$$

A vertical summation of forces determines V at the wall:

$$[\Sigma Y = 0] \qquad R_A + V - wL = 0$$

whence, substituting the value of R_A, we obtain

$$V = wL - \tfrac{3}{8}wL = \tfrac{5}{8}wL$$

From the definition of bending moment at B, which is equivalent to $\Sigma M_B = 0$ but simpler to apply, we obtain

$$[M = (\Sigma M)_L] \qquad M_B = R_A L - \frac{wL^2}{2} = \left(\frac{3}{8}wL\right)L - \frac{wL^2}{2}$$

$$M_B = -\frac{wL^2}{8}$$

The shear diagram being drawn as in Fig. 7-8c, the location of zero shear is determined from

$$[V = (\Sigma Y)_L = 0] \qquad \tfrac{3}{8}wL - wx = 0 \qquad x = \tfrac{3}{8}L$$

whence the maximum positive moment is

$$[\Delta M = (\text{area})_V] \qquad M = \tfrac{1}{2}\left(\tfrac{3}{8}wL\right)\left(\tfrac{3}{8}L\right) = \tfrac{9}{128}wL^2$$

The moment diagram in Fig. 7–8d shows that the maximum moment is at the wall.

717. A beam 4 m long and perfectly restrained at the ends carries a uniformly distributed load over part of its length as shown in Fig. 7–9a. Compute the end shears and end moments.

Solution: This problem was solved on page 288 by double integration. Its solution here by the area-moment method shows the close similarity of the two methods and demonstrates the direct use of the moment diagram. We begin by drawing the moment diagram by parts from left to right, as in Fig. 7–9b. Although we know from the downward concavity of the elastic curve that the moment at A should be negative, we nevertheless show it as positive, that is, clockwise. Because of this deliberate error in the direction of the vector quantity M_A, our solution will determine not only the correct numerical value of M_A but also its correct negative sign.

The elastic curve in Fig. 7–9a shows that the change in slope between tangents drawn at A and B is zero. Applying the first theorem of the area-moment method, we obtain

$$[EI\theta_{AB} = (\text{area})_{AB} = 0] \qquad \frac{4(4V_A)}{2} + 4M_A - \frac{3(4050)}{3} = 0 \quad (a)$$

The deviation of B from a tangent drawn at A being zero, we obtain from the second theorem of area–moment:

$$[EIt_{B/A} = (\text{area})_{BA} \cdot \bar{x}_B = 0]$$

$$\frac{4(4V_A)}{2}\left(\frac{4}{3}\right) + 4M_A\left(\frac{4}{2}\right) - \frac{3(4050)}{3}\left(\frac{3}{4}\right) = 0 \qquad (b)$$

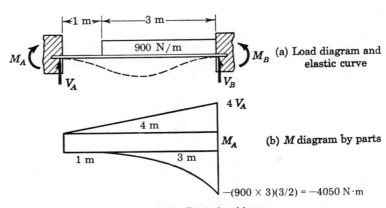

(a) Load diagram and elastic curve

(b) M diagram by parts

Figure 7–9. Restrained beam.

Solving Eqs. (*a*) and (*b*) simultaneously yields

$$V_A = 949 \text{ N} \quad \text{and} \quad M_A = -886 \text{ N·m} \quad Ans.$$

The negative sign for M_A confirms that the direction of M_A was incorrectly assumed. It actually is a negative moment, and there should now be no confusion regarding sign in its subsequent use.

The deviation at A from a tangent drawn at B is also zero, so that we could have used $EIt_{A/B} = (\text{area})_{AB} \cdot \bar{x}_A = 0$. A glance at the moment diagram (Fig. 7–9b) shows that it is simpler in this problem to take moments of area about B rather than about A. For this reason, we used $EIt_{B/A} = 0$ rather than $EIt_{A/B} = 0$.

Comparison with the solution by double integration on page 288 discloses that Eqs. (*d*) and (*e*) used there are identical with Eqs. (*a*) and (*b*) here. On the other hand, comparison with the superposition solution on page 290 discloses that its second equation is equivalent to $EIt_{A/B} = 0$, which we could have used here in place of $EIt_{B/A} = 0$.

Having determined V_A and M_A, we compute the shear and moment at B in the identical manner as was used on page 289. It is needless to repeat it here.

718.   Compute the end moment and maximum deflection for the symmetrically loaded and perfectly restrained beam shown in Fig. 7–10a.

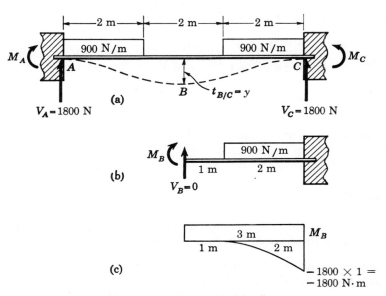

Figure 7–10.   Symmetrical loading.

Solution: Because of symmetry, the end shears are equal to each other, and each equals one-half the applied loads. The end moments also equal each other but are unknown. The simplest condition that determines the unknown end moment is that in a symmetrically loaded beam the tangent to the elastic curve at midspan is horizontal, and therefore the change in slope is zero between this tangent and a tangent drawn at either end.

In this problem, instead of solving directly for the end moment, it is simpler to solve first for the moment M_B at midspan and then apply the definition of bending moment to determine the end moment. We begin by drawing the free-body diagram of the segment BC as in Fig. 7–10b. Since the end shears each equal 1800 N, at midspan the shear $V_B = 0$; consequently the M diagram by parts is drawn as in Fig. 7–10c. Because there is no change in slope between the tangents at the midpoint B and the end C, we obtain

$$\left[EI\theta_{BC} = (\text{area})_{BC} = 0 \right] \qquad 3M_B - \frac{1800 \times 2}{3} = 0$$

$$M_B = 400 \text{ N·m}$$

Applying the definition of bending moment, we find the end moment to be

$$\left[M_C = (\Sigma M)_L \right] \qquad M_C = M_B - 1800 = 400 - 1800$$
$$= -1400 \text{ N·m} \qquad Ans.$$

The maximum deflection occurs at midspan and is numerically equal to the deviation of B from the horizontal tangent at C. Using the now known value of $M_B = 400$ N·m, we obtain

$$\left[EIt_{B/C} = (\text{area})_{BC} \cdot \bar{x}_B \right]$$
$$EIy = (400 \times 3)\left(\frac{3}{2}\right) - \left(\frac{1800 \times 2}{3}\right)\left(1 + \frac{3}{4} \times 2\right)$$
$$= -1200 \text{ N·m}^3 \qquad Ans.$$

The minus sign indicates that the deflection y is directed downward as shown.

PROBLEMS

Unless otherwise stated, assume unyielding supports. Additional problems may be selected from those in Art. 7–3.

719. For the propped beam shown in Fig. P–719, determine the reaction R and the midspan value of $EI\delta$.

Ans. $R = (5/16)P; \ EI\delta = (7/768)PL^3$

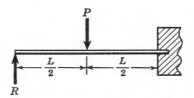

Figure P–719.

720. Compute the reaction R and sketch shear and moment diagrams for the propped beam shown in Fig. P–720.

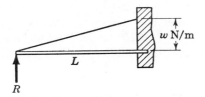

Figure P–720.

721. For the propped beam shown in Fig. P–721, determine the reaction R and sketch the shear and moment diagrams.

Ans. $R = (11/40)wL$

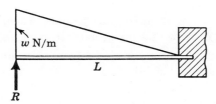

Figure P–721.

722. For the beam shown in Fig. P–722, compute the reaction R at the propped end and the moment at the wall. Check your results by letting $b = L$ and comparing with the results in Problem 706.

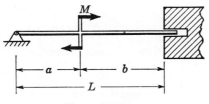

Figure P–722.

723. Find the reaction R and the moment at the wall for the propped beam shown in Fig. P–723.

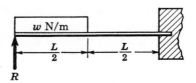

Figure P–723.

724. The beam shown in Fig. P–724 is only partially restrained at the wall so that, after the uniformly distributed load is applied, the slope at the wall is $wL^3/48EI$ upward to the right. If the supports remain at the same level, determine R. 　　　*Ans.*　$R = (7/16)wL$

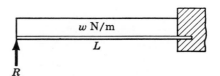

Figures P–724 and P–725.

725. If the support under the propped beam in Problem 724 settles an amount δ, show that the prop reaction decreases by $3EI\delta/L^3$.

726. A beam L meters long, perfectly restrained at both ends, supports a concentrated load P at midspan. Determine the end moment and maximum deflection.

727. Repeat Problem 726, assuming that the concentrated load is replaced by a uniformly distributed load of w N/m over the entire length. 　　　*Ans.*　$M = -wL^2/12$; $EI\delta = wL^4/384$

728. Determine the end moment and maximum deflection for a perfectly restrained beam loaded as shown in Fig. P–728.

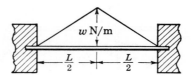

Figure P–728.

729.   For the restrained beam shown in Fig. P–729, compute the end moment and maximum $EI\delta$.

$$Ans. \quad M = -1830 \text{ N} \cdot \text{m}; \; EI\delta = 1080 \text{ N} \cdot \text{m}^3$$

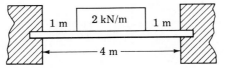

Figure P–729.

730.   Determine the end moment and maximum deflection for the perfectly restrained beam loaded as shown in Fig. P–730.

$$Ans. \quad M = -(wa^2/6L)(3L - 2a); \; EIy = -(wa^3/24)(L - a)$$

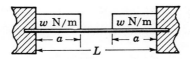

Figure P–730.

731.   The beam shown in Fig. P–731 is connected to a vertical rod. If the beam is horizontal at a certain temperature, determine the increase in stress in the rod if the temperature of the rod drops 50°C. Both the beam and the rod are made of steel with $E = 200 \times 10^9$ N/m². For the beam, use $I = 60 \times 10^6$ mm⁴.

$$Ans. \quad \sigma = 95.8 \text{ MN/m}^2$$

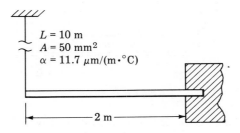

Figure P–731.

732. The midpoint of the steel beam in Fig. P–732 is connected to the vertical aluminum rod. Determine the maximum value of P if the stress in the rod is not to exceed 150 MN/m^2.

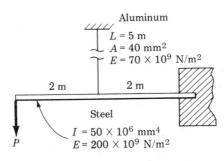

Aluminum
$L = 5$ m
$A = 40 \text{ mm}^2$
$E = 70 \times 10^9 \text{ N/m}^2$

2 m   2 m

Steel

$I = 50 \times 10^6 \text{ mm}^4$
$E = 200 \times 10^9 \text{ N/m}^2$

P

Figures P–732 and P–733.

733. The load P in Problem 732 is replaced by a counterclockwise couple M. Determine the maximum value of M if the stress in the vertical rod is not to exceed 100 MN/m^2.   *Ans.* $M = 41.1 \text{ kN·m}$

734. Determine the end moments for the restrained beam shown in Fig. P–734.

$$Ans. \quad M_A = -(5/192)wL^2; \ M_B = -(11/192)wL^2$$

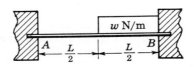

w N/m

$A \quad \dfrac{L}{2} \quad \dfrac{L}{2} \quad B$

Figure P–734.

735. The beam shown in Fig. P–735 is perfectly restrained at A but only partially restrained at B, where the slope is $wL^3/48EI$ directed up to the right. Solve for the end moments.

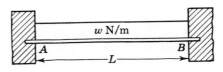

w N/m

$A \qquad\qquad\qquad B$

L

Figure P–735.

736. For the restrained beam shown in Fig. P–736, compute the end shears and end moments and sketch the shear and moment diagrams.

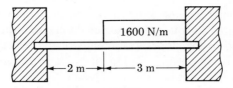

Figure P–736.

737. In the perfectly restrained beam shown in Fig. P–737, support B has settled a distance Δ below support A. Show that $M_B = -M_A = 6EI\Delta/L^2$.

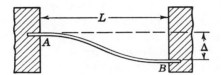

Figure P–737.

738. A perfectly restrained beam is loaded by a couple M applied where shown in Fig. P–738. Determine the end moments.

$$Ans. \quad M_A = \frac{Mb}{L}\left(\frac{3a}{L} - 1\right); \; M_B = -\frac{Ma}{L}\left(\frac{3b}{L} - 1\right)$$

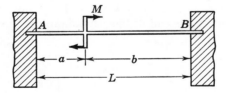

Figure P–738.

7–5 RESTRAINED BEAM EQUIVALENT TO SIMPLE BEAM WITH END MOMENTS

Usually the redundant elements in a restrained beam are most easily determined by applying the method which considers the redundant supports to be the shear and moment at one wall. However, it is sometimes desirable to treat the end moments as the redundant supports. To do this, the restrained beam is considered equivalent to a simple beam acted on not only by the given loading but also by end moments sufficient to rotate the ends of the beam until the slopes at the ends correspond to the slopes at the ends of the restrained beam. Thus the restrained beam in Fig. 7–11a may be considered equivalent to superposing the loadings in Figs. 7–11b and 7–11c.

It is evident that the unsymmetrical loading in Fig. 7–11b causes a greater slope θ_2 at the right end than the slope θ_1 at the left end. To cancel these end slopes by adding the loading in Fig. 7–11c, we must have $\theta_1 = \theta_1'$ and $\theta_2 = \theta_2'$, which requires that M_B be greater than M_A in order for θ_2' to be greater than θ_1'. In other words, *the larger end moment acts at the wall that is closer to the resultant of any single load.*

The difference between the end moments M_B and M_A is balanced by the couple $R'L$, consisting of the forces R' applied at the ends of the beam in Fig. 7–11c. By superposing the reactions in Figs. 7–11b and 7–11c, we obtain $V_A = R_1 - R'$ and $V_B = R_2 + R'$. If the loading were symmetrical, the end slopes θ_1 and θ_2 would be equal, which would require equal end moments M_A and M_B. In this case, there would be no couple reaction R', so the end shears would equal the end reactions of a similarly loaded simple beam. This conclusion agrees with the observation in Illustrative Problem 718 (page 297) on symmetrical loading.

To consider the end moments as the redundant supports, therefore, the moment diagram by parts for the beam in Fig. 7–11a will be drawn to correspond to the loadings in Figs. 7–11b and 7–11c and will appear as in Figs. 7–12a and 7–12b. Applying $EI\theta_{AB} = 0$ and $EIt_{B/A} = 0$, we can solve directly for M_A and M_B as the redundant supports, whence V_A and V_B are obtained by applying the equations of static equilibrium. This procedure is not generally as simple as treating the shear and moment at one end as the redundants.

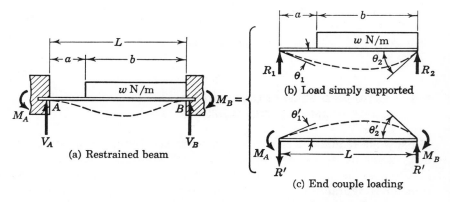

Figure 7–11.   Restrained beam resolved into simple beam loadings.

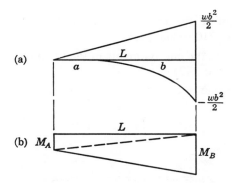

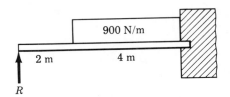

Figure 7–12.   Moment diagrams for beams in Fig. 7–11b and 7–11c.

PROBLEMS

Use the end moments as the redundants in solving these problems.

739.   Determine the wall moment in the propped beam described in Problem 705 (page 291).

740.   Solve for the wall moment in the propped beam shown in Fig. P–740.   *Ans.* $M = -3200 \, \text{N} \cdot \text{m}$

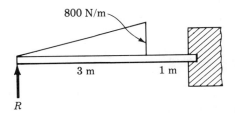

Figure P–740.

741.   Compute the moment at the restrained end of the propped beam shown in Fig. P–741.

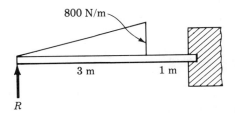

Figure P–741.

742. Determine the end moments in the restrained beam described in Problem 710 (page 292).

743. Compute the end moments in the restrained beam described in Problem 712 (page 292).

744. Determine the end moments in the perfectly restrained beam shown in Fig. P–744.

$Ans.$ $M_A = -1380\ \text{N}\cdot\text{m}$; $M_B = -1820\ \text{N}\cdot\text{m}$

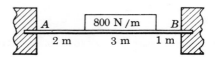

Figure P–744.

745. The restrained beam in Fig. P–745 carries a uniformly distributed load over part of the span and a couple. Compute the end moments. $Ans.$ $M_A = -2700\ \text{N}\cdot\text{m}$; $M_B = -2190\ \text{N}\cdot\text{m}$

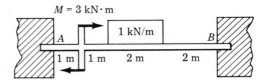

Figure P–745.

7–6 DESIGN OF RESTRAINED BEAMS

By applying the methods described in the preceding articles, we can determine general values for end moments and deflections in perfectly restrained beams that carry various general loadings. These values are summarized in Table 7–2. In the following problems we show how these general values are superposed to solve design problems involving various combinations of loads.

ILLUSTRATIVE PROBLEMS

746. Determine the section modulus required for a beam to support the loads shown in Fig. 7–13 without exceeding a flexure stress of 80 MN/m².

TABLE 7-2. Restrained Beam Loadings

CASE NO.	TYPE OF LOAD	END MOMENTS	VALUES OF EIy (y is positive downward)
1		$M_A = -\dfrac{Pab^2}{L^2}$ $M_B = -\dfrac{Pa^2b}{L^2}$	Midspan $EIy = \dfrac{Pb^2}{48}(3L-4b)$ Note: only for $a > b$
2		$M_A = M_B = -\dfrac{PL}{8}$	Max. $EIy = \dfrac{PL^3}{192}$
3		$M_A = M_B = -\dfrac{wL^2}{12} = -\dfrac{WL}{12}$	Max. $EIy = \dfrac{wL^4}{384} = \dfrac{WL^3}{384}$
4		$M_A = -\dfrac{5}{192}wL^2 = -\dfrac{5}{96}WL$ $M_B = -\dfrac{11}{192}wL^2 = -\dfrac{11}{96}WL$	Midspan $EIy = \dfrac{wL^4}{768} = \dfrac{WL^3}{384}$
5		$M_A = -\dfrac{wL^2}{30} = -\dfrac{WL}{15}$ $M_B = -\dfrac{wL^2}{20} = -\dfrac{WL}{10}$	Midspan $EIy = \dfrac{wL^4}{768} = \dfrac{WL^3}{384}$
6		$M_A = M_B = -\dfrac{5wL^2}{96} = -\dfrac{5WL}{48}$	Max. $EIy = \dfrac{7wL^4}{3840} = \dfrac{7WL^3}{1920}$
7		$M_A = \dfrac{Mb}{L}\left(\dfrac{3a}{L}-1\right)$ $M_B = -\dfrac{Ma}{L}\left(\dfrac{3b}{L}-1\right)$	
8		$M_A = -\dfrac{6EI\Delta}{L^2}$ $M_B = \dfrac{6EI\Delta}{L^2}$	

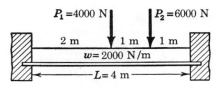

Figure 7–13.

Solution: The end moments due to the distributed load and the central load P_1 are equal because of symmetry. As discussed in Art. 7–5, the eccentric load P_2 causes a larger end moment at the nearer wall—the right wall, in this case. From Table 7–2, the maximum bending moment occurring at the right end is given by

$$\text{Max. } M = -\frac{wL^2}{12} - \frac{P_1 L}{8} - \frac{P_2 a^2 b}{L^2}$$

$$= -\frac{2000(4)^2}{12} - \frac{4000(4)}{8} - \frac{6000(3)^2(1)}{(4)^2}$$

$$\text{Max. } M = -8040 \text{ N} \cdot \text{m}$$

The negative sign of moment indicates a tensile stress at the top fibers. In an unsymmetrical section like a T beam, this would be important; but in a symmetrical section, only the numerical value of moment need be used. From the flexure formula, the section modulus required is therefore

$$\left[S = \frac{M}{\sigma} \right] \quad S = \frac{8040}{80 \times 10^6} = 100.5 \times 10^{-6} \text{ m}^3$$

$$= 100.5 \times 10^3 \text{ mm}^3 \quad \textit{Ans.}$$

747.   Select a suitable wide flange beam to support the loads shown in Fig. 7–14 without exceeding a flexural stress of 120 MPa. Compute the midspan deflection of this beam. Neglect the mass of the beam. Use $E = 200$ GPa.

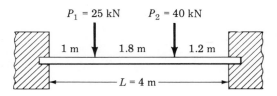

Figure 7–14.

Solution: Because of the unsymmetrical loading, the end at which the maximum moment occurs is not evident and computations for moments at each end must therefore be made. At the left end, we obtain

$$M_L = -\frac{P_1 ab^2}{L^2} - \frac{P_2 ab^2}{L^2}$$

$$= -\frac{(25)(1)(3)^2}{(4)^2} - \frac{(40)(2.8)(1.2)^2}{(4)^2} = -24.14 \text{ kN} \cdot \text{m}$$

At the right end, the moment is

$$M_R = -\frac{P_1 a^2 b}{L^2} - \frac{P_2 a^2 b}{L^2}$$

$$= -\frac{(25)(1)^2(3)}{(4)^2} - \frac{(40)(2.8)^2(1.2)}{(4)^2} = -28.21 \text{ kN} \cdot \text{m}$$

Substituting the larger numerical value of end moment into the flexure formula, we find that the required section modulus is

$$\left[S = \frac{M}{\sigma} \right] \quad S = \frac{28.21 \times 10^3}{120 \times 10^6} = 235 \times 10^{-6} \text{ m}^3 = 235 \times 10^3 \text{ mm}^3$$

A suitable beam is a W200 × 27 with $S = 249 \times 10^3$ mm^3 and $I = 25.8 \times 10^6$ mm^4 = 25.8×10^{-6} m^4. Table 7–2 gives the midspan deflection* as

$$EIy = \sum \frac{Pb^2}{48}(3L - 4b)$$

$$EIy = \frac{(25)(1)^2}{48}[3(4) - 4(1)] + \frac{(40)(1.2)^2}{48}[3(4) - 4(1.2)]$$

$$EIy = 12.81 \text{ kN} \cdot \text{m}^3$$

Substituting the numerical values for E and I, we obtain

$$(200 \times 10^9)(25.8 \times 10^{-6})y = (12.81 \times 10^3)$$

from which

$$y = 2.48 \times 10^{-3} \text{ m} = 2.48 \text{ mm} \qquad \textit{Ans.}$$

PROBLEMS

748.   A restrained beam 6 m long supports a concentrated load of 30 kN at 2 m from the left end and another concentrated load of 50 kN at 1.5 m from the right end. Select the lightest wide flange beam that will support these loads without exceeding a flexural stress of 120 MPa.

*In computing midspan deflection for a concentrated load, the term *b* is the smaller of the two segments into which the load divides the length of the beam.

Neglect the mass of the beam. Compute the midspan deflection of this beam if $E = 200 \text{ GN/m}^2$.   *Ans.*   W360 × 33; $\delta = 3.21$ mm

749.   A timber beam 150 mm wide by 300 mm deep and 6 m long is perfectly restrained at both ends. It supports a uniformly distributed load of 4 kN/m over its entire length and a concentrated load P at 2.5 m from the left end. Determine P so as not to exceed a flexural stress of 10 MN/m² or a midspan deflection of 1/360 of the span. Assume that $E = 10 \text{ GN/m}^2$.

750.   A W200 × 36 steel beam 5 m long is perfectly restrained at both ends. It carries a concentrated load of 20 kN at 1 m from the left end and another concentrated load of 30 kN at 2 m from the right end. Compute the maximum flexural stress and the midspan deflection. Neglect the mass of the beam. Use $E = 200$ GPa.

751.   A timber beam with a square cross section supports the loads shown in Fig. P–751. Determine the cross-sectional dimensions if the allowable flexural stress is 10 MN/m². What is the maximum shearing stress developed in the beam?

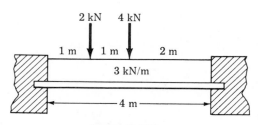

Figure P–751.

752.   Using Table 7–2, check the values of end moment and midspan deflection for the restrained beam in Problem 713 (page 293).

753.   A timber beam 100 mm wide by 150 mm deep supports the loads shown in Fig. P–753. Determine the maximum shearing stress developed.

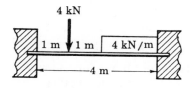

Figures P–753 and P–754.

754.   In Problem 753, compute the maximum flexural stress if the right end settles 20 mm relative to the left end. Assume that the beam is perfectly restrained against rotation at its ends and that $E = 10 \times 10^9$ N/m^2.   *Ans.* $\sigma = 16.1 \text{ MN/m}^2$

755.   An S130 $\times$ 22 steel beam 4 m long carries a load varying uniformly from zero at the left end to 15 kN/m at the right end. The beam is perfectly restrained against rotation at its ends, but the right end settles 10 mm relative to the left end. Determine the ratio of the maximum flexural stress to the flexural stress if no settlement had occurred. Use $E = 200$ GPa.   *Ans.*   1.06

SUMMARY

The principles of beam deflections studied in Chapter 6 are applied here to obtain additional equations which can be combined with the equations of static equilibrium to solve problems involving statically indeterminate beams.

In propped beams, we generally use the fact that the deflection under the redundant support is zero (if the support does not settle) or some known value (if the support does settle). For beams perfectly restrained at the ends, the elastic curve is such that there is no change in slope between the ends and the deflection of one end relative to the other end is zero.

The method of superposition is usually the easiest way to determine the redundant support in propped beams. It is also the best method of determining the end moments in restrained beams subjected to loadings of the types listed in Table 7–2.

With the double-integration method or the area-moment method, we generally take the shear and moment at one end as the redundant supports. Either method is as simple to apply as the other. The double-integration method is essentially mathematical and automatically determines deflections as well as redundant supports. The area-moment method, by emphasizing the geometric relations between the elastic curve and the moment diagram, is perhaps more direct in obtaining the equations that determine the redundant supports, but additional work is needed if deflections are required.

In Art. 7–5, the reduction of a restrained beam to the combination of a simply supported beam carrying the given loading and another simply supported beam subjected to end couples is valuable in visualizing which restrained end carries the larger moment. This concept is particularly advantageous in applying the area-moment method to continuous beams, as will be explained in the next chapter. In addition, it provides the basis for a rapid method of drawing the shear diagram, also explained in the next chapter.

8

Continuous Beams

8-1 INTRODUCTION

In this chapter we consider beams that are continuous over two or more spans, thereby having one or more redundant supports. It is possible to determine these redundancies by applying the deflection relations developed in Chapter 6, but a more convenient method is to consider the unknown bending moments at the supports of the beam as the redundancies. After these bending moments are found, it is comparatively simple to determine the reactions, as we shall show in Art. 8–5.

We present two methods of solving for the moments at the supports, either or both of which may be studied, depending on the available time. In the first method it is necessary to find a general relation between the bending moments at any three sections in a beam. This relation is known as the *three-moment equation* and is easily derived by applying the area-moment theorems. We shall show how this equation is used to determine deflections as well as redundancies in any type of beam. Actually, the three-moment equation can be used to solve all the problems in Chapters 6 and 7; however, in some instances, it is best used in combination with the area-moment or the double-integration method. Such combinations of techniques will also be discussed.

312

An alternate method of solving continuous beams is *moment distribution*, which is described in Art. 8–8. This method is completely independent of the three-moment equation, but Art. 8–5 for determining shear diagrams is common to both. Before the moment-distribution method may be applied, however, each span of the continuous beam must be assumed to be perfectly restrained at the supports and the fixed end moments must be computed. Usually the span loadings are such that the fixed end moments are readily obtained by superposition of the general results tabulated in Table 7–2 on page 307. However, this table does not list results for distributed loads over part of a span; for these loadings the three-moment equation is preferable if a more general list is not available.

8–2   GENERALIZED FORM OF THE THREE-MOMENT EQUATION

A portion of a beam that is loaded and supported in any manner is shown in Fig. 8–1a. At any three points 1, 2, and 3, pass cutting sections and replace the effects of the loads to the left or right of these sections by the proper values of vertical shear and bending moment. Thus the beam segments between points 1 and 2 and between points 2 and 3 (hereafter referred to as spans 1 and 2, respectively) may be isolated by means of the free-body diagrams in Fig. 8–1b. The lengths of the spans (or segments) are L_1 and L_2, and the bending moments at points 1, 2, and 3 are M_1, M_2, M_3; the vertical shears at these points are V_1, V_{-2}

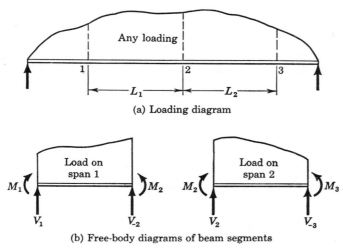

(a) Loading diagram

(b) Free-body diagrams of beam segments

Figure 8–1.   General loading on any beam.

(just to the left of point 2), V_2 (just to the right of point 2), and V_{-3} just to the left of point 3.

The technique discussed in Art. 7–5 enables us to resolve the free-body diagrams of the beam segments into simply supported spans that carry the actual beam loading, and spans loaded only by the bending moments and held in equilibrium by the couple reactions R_1' on span 1 and by R_2' on span 2. This equivalent loading is shown in Figs. 8–2a and 8–2b, respectively. When these loadings are superposed, they produce the free-body diagrams in Fig. 8–1b. Hence the vertical shears at points 1, 2, and 3 are equal to the algebraic sum of the simple beam reaction and couple reaction at these respective points.

In this manner, the moment diagram of each original beam segment is resolved into the moment diagram of the loads assumed to be carried on a simply supported span and the trapezoidal moment diagram caused by the bending moments in the original beam at the selected points 1, 2, and 3. These diagrams are shown in Figs. 8–2c and 8–2d, respectively.

For clarity, the elastic curve of the beam has been drawn separately in Fig. 8–3. The deflection of the curve is greatly exaggerated in order to show the geometric relations. Note that points 1, 2, and 3 lie on it.

A tangent drawn to the elastic curve at point 2 determines the tangential deviations $t_{1/2}$ at point 1 and $t_{3/2}$ at point 3. Another line

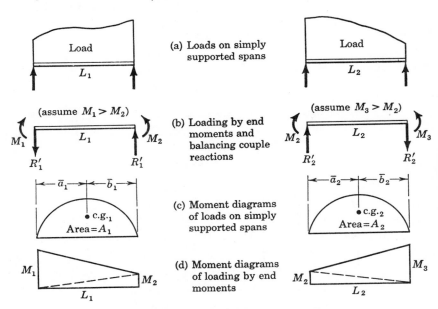

Figure 8–2.   Analysis of original loading.

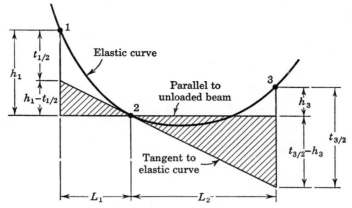

Figure 8-3. Elastic curve of any beam.

drawn through point 2 parallel to the initial position of the unloaded beam (which has been assumed horizontal for convenience) determines the heights of points 1 and 3 above point 2 to be h_1 and h_3. There are formed the shaded similar triangles having the bases L_1 and L_2 and the altitudes $(h_1 - t_{1/2})$ and $(t_{3/2} - h_3)$.

From the proportions between similar triangles, it is evident that

$$\frac{h_1 - t_{1/2}}{L_1} = \frac{t_{3/2} - h_3}{L_2}$$

which reduces to

$$\frac{t_{1/2}}{L_1} + \frac{t_{3/2}}{L_2} = \frac{h_1}{L_1} + \frac{h_3}{L_2} \qquad (a)$$

The values of the tangential deviations are found from

$$t_{1/2} = \frac{1}{EI}(\text{area})_{1-2} \cdot \bar{x}_1$$

and

$$t_{3/2} = \frac{1}{EI}(\text{area})_{3-2} \cdot \bar{x}_3$$

where $(\text{area})_{1-2} \cdot \bar{x}_1$ is the moment of area about point 1 of the moment diagram between points 1 and 2. As was said previously, this moment diagram has been resolved into the area A_1 (see Fig. 8-2c) and the two triangular areas into which the trapezoidal diagram for the end moments is divided (see Fig. 8-2d). Likewise, $(\text{area})_{3-2} \cdot \bar{x}_3$ is the moment about point 3 of the area of the moment diagram between points 2 and 3, as represented by area A_2 and the trapezoidal diagram for the end moments which has been subdivided into two triangles.

We can therefore express the tangential deviation $t_{1/2}$ at 1 from a tangent to the elastic curve drawn at 2 as

$$t_{1/2} = \frac{1}{EI}\left[A_1\bar{a}_1 + \frac{1}{2}M_1L_1 \times \frac{1}{3}L_1 + \frac{1}{2}M_2L_1 \times \frac{2}{3}L_1 \right] \qquad (b)$$

and the tangential deviation $t_{3/2}$ at 3 from the *same* tangent drawn at 2 as

$$t_{3/2} = \frac{1}{EI}\left[A_2\bar{b}_2 + \frac{1}{2}M_2L_2 \times \frac{2}{3}L_2 + \frac{1}{2}M_3L_2 \times \frac{1}{3}L_2 \right] \qquad (c)$$

Substituting these values of $t_{1/2}$ and $t_{3/2}$ in Eq. (a) gives

$$M_1L_1 + 2M_2(L_1 + L_2) + M_3L_2 + \frac{6A_1\bar{a}_1}{L_1} + \frac{6A_2\bar{b}_2}{L_2}$$
$$= 6EI\left(\frac{h_1}{L_1} + \frac{h_3}{L_2} \right) \qquad (8\text{-}1)$$

This equation expresses a general relation among moments at any three points in a beam, and hence is known as the *three-moment equation*.

When points 1, 2, and 3 are on the same level in the deflected beam, the heights h_1 and h_3 in Fig. 8–3 become zero and so does the right-hand term in Eq. (8–1). This is the usual condition in which the three-moment equation is applied. The three points selected in applying the equation to continuous beams are the points at the supports (usually assumed as rigid or else as settling the same amount); the equation is used to determine the bending moments in the beam over the supports.

If the three-moment equation is used for deflections, two of the points are selected over supports and the third is chosen at the point whose deflection is desired. Evidently the moments at the three points must first be known in order to compute deflections. We shall expand this application of the three-moment equation in Art. 8–7.

Rules of sign

Equation (8–1) was derived under the assumption that the bending moments at the selected points were positive and that points 1 and 3 were above point 2. Hence, heights h_1 and h_3 must be considered positive when measured upward from point 2. If the moment at any point is actually negative, the negative sign must be used when substituting its value in Eq. (8–1). Conversely, if an unknown moment is actually negative at any point, Eq. (8–1) will give a negative value for that moment; in other words, the sign of the moment at that point is automatically opposite to the positive value assumed in the derivation of the three-moment equation.

8–3   FACTORS FOR THE THREE-MOMENT EQUATION

The usefulness of the three-moment equation depends on the ease with which the expressions $6A\bar{a}/L$ and $6A\bar{b}/L$ in it can be found. As was said earlier, these expressions refer to the moment of area of the moment diagram resulting from carrying the applied loads on a simple span of the same length as the equivalent beam segment. The general expressions in Table 8–1 were obtained by the following procedure.

Case 3: Uniformly varying load

The loading over a span L in a continuous beam varies uniformly over the span. If this loading is assumed to be supported on a simple span, the moment diagram is drawn by parts from left to right, as in Fig. 8–4. The moment of area of this moment diagram about the right end is given by

$$A\bar{b} = \frac{1}{2}\left(\frac{wL^2}{6}\cdot L\right)\left(\frac{1}{3}L\right) - \frac{1}{4}\left(\frac{wL^2}{6}\cdot L\right)\left(\frac{1}{5}L\right)$$

$$= \frac{wL^4}{6}\left(\frac{1}{6} - \frac{1}{20}\right) = \frac{7}{360}wL^4$$

Multiplying this by $6/L$, we obtain the following general value for this type of loading:

$$\frac{6A\bar{b}}{L} = \frac{7}{60}wL^3 \qquad Ans.$$

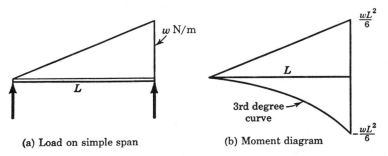

(a) Load on simple span   (b) Moment diagram

Figure 8–4.   Uniformly varying load.

Special loadings

For cases not listed in Table 8–1, or if the table is not available, the following example may be helpful.

Assume a continuous beam loaded as in Fig. 8–5; we wish to evaluate $6A_2\bar{b}_2/L_2$ for span 2. Take the loading on span 2 as if it were simply supported on a 4-m span, and draw the moment diagram by

TABLE 8–1.   Values of $6A\bar{a}/L$ and $6A\bar{b}/L$

CASE NO.	TYPE OF LOADING ON SPAN	$\dfrac{6A\bar{a}}{L}$	$\dfrac{6A\bar{b}}{L}$
1		$\dfrac{Pa}{L}(L^2 - a^2)$	$\dfrac{Pb}{L}(L^2 - b^2)$
2		$\dfrac{wL^3}{4} = \dfrac{WL^2}{4}$	$\dfrac{wL^3}{4} = \dfrac{WL^2}{4}$
3		$\dfrac{8}{60}wL^3 = \dfrac{8}{30}WL^2$	$\dfrac{7}{60}wL^3 = \dfrac{7}{30}WL^2$
4		$\dfrac{7}{60}wL^3 = \dfrac{7}{30}WL^2$	$\dfrac{8}{60}wL^3 = \dfrac{8}{30}WL^2$
5		$\dfrac{w}{4L}[b^2(2L^2 - b^2) - a^2(2L^2 - a^2)]$	$\dfrac{w}{4L}[d^2(2L^2 - d^2) - c^2(2L^2 -$
6		$\dfrac{5}{32}wL^3 = \dfrac{5}{16}WL^2$	$\dfrac{5}{32}wL^3 = \dfrac{5}{16}WL^2$
7		$-\dfrac{M}{L}(3a^2 - L^2)$	$+\dfrac{M}{L}(3b^2 - L^2)$

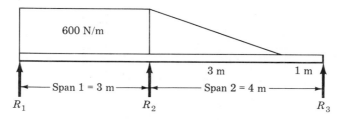

(a) Continuous beam

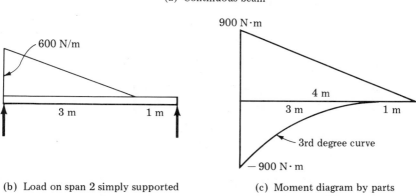

(b) Load on span 2 simply supported   (c) Moment diagram by parts

Figure 8–5. Evaluation of $6A_2\bar{b}_2/L_2$ for special loading.

parts from right to left (this is more convenient here). Since $6A\bar{b}/L$ means multiplying $6/L$ by the moment of area of the moment diagram, moments being taken about the right end, we have

$$\frac{6A_2\bar{b}_2}{L_2} = \frac{6}{4}\left[\left(\frac{900 \times 4}{2}\right)\left(\frac{2}{3} \times 4\right) - \left(\frac{900 \times 3}{4}\right)\left(1 + \frac{4}{5} \times 3\right)\right]$$

$$= \frac{6}{4}(4800 - 2295) = 3758 \text{ N} \cdot \text{m}^2 \qquad Ans.$$

If it had been necessary to evaluate $6A_2\bar{a}_2/L_2$ for span 2, the symbol $\bar{a}_2$ would have told us to take the moment of area about the left end, since, as Fig. 8–2c shows, the symbols $\bar{a}$ and $\bar{b}$ refer to moment arms measured respectively from the left and right ends of a span.

PROBLEMS

If the span loadings on a continuous beam reduce to the simply supported loads shown in each of the following problems, evaluate the factors $6A\bar{a}/L$ and $6A\bar{b}/L$.

801. See Fig. P–801. Check your result by letting $a = L/2$ and comparing with case 2 of Table 8–1.   *Ans.* $\frac{1}{2}wa^2(3L - 2a)$

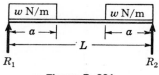

Figure P–801.

802. See Fig. P–802. When $b = L/2$, how does your result compare with case 2 of Table 8–1?

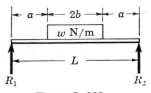

Figure P–802.

803. See Fig. P–803.   *Ans.* $\frac{5}{32}wL^3$

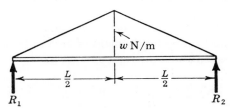

Figure P–803.

804. See Fig. P–804. Check your result by subtracting the answer for Problem 803 from case 2 of Table 8–1.

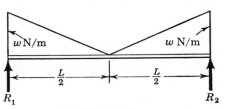

Figure P–804.

805.   See Fig. P–805. The roller support may resist upward or downward reaction.

  Ans. $6A\bar{a}/L = -(M/L)(3a^2 - L^2); \; 6A\bar{b}/L = (M/L)(3b^2 - L^2)$

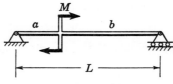

Figure P–805.

806.   See Fig. P–806.

  Ans. $6A\bar{a}/L = 5436 \text{ N} \cdot \text{m}^2; \; 6A\bar{b}/L = 4014 \text{ N} \cdot \text{m}^2$

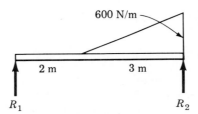

Figure P–806.

807.   See Fig. P–807. Solve by combining the results for Problems 805 and 806.

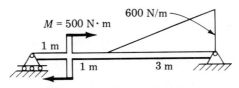

Figures P–807 and P–808.

808.   Solve Problem 807 if the couple is applied in a counterclockwise sense.

809.   See Fig. P–809.   *Ans.* $6A\bar{a}/L = 6648 \text{ N} \cdot \text{m}^2$

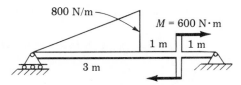

Figures P-809 and P-810.

810. Solve Problem 809 if the couple is applied in a counter-clockwise sense.

8-4 APPLICATION OF THE THREE-MOMENT EQUATION

We now see how the three-moment equation may be applied to determine the moments over the supports in various types of continuous beams. Later articles will show how these moments are used to determine the reactions of continuous beams and will describe a speedy method of drawing shear and moment diagrams.

ILLUSTRATIVE PROBLEMS

811. For the continuous beam in Fig. 8-6, determine the values of the moments over the supports. The supports are assumed to be rigid or, what amounts to the same thing, to have equal deformations. This assumption applies to all problems unless stated otherwise.

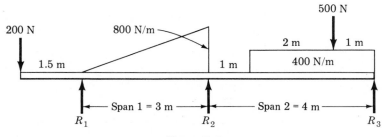

Figure 8-6.

Solution: Apply the three-moment equation to points over the supports. Since the supports remain at the same level, heights h_1 and h_3 are zero, and the equation reduces to

$$M_1L_1 + 2M_2(L_1 + L_2) + M_3L_2 + \frac{6A_1\bar{a}_1}{L_1} + \frac{6A_2\bar{b}_2}{L_2} = 0 \qquad (a)$$

We begin by noting that the bending moment at support 1, caused by the load to the left of R_1, is $M_1 = -200 \times 1.5 = -300$ N·m, whereas over support R_3 the moment M_3 is zero since no loads act to the right of R_3. Observe that the negative sign of M_1 must be retained when substituting its value in the three-moment equation.*

The factors in Eq. (a) are found by using Table 8–1. The load on span 1 is that of case 3; hence we have

$$\frac{6A_1\bar{a}_1}{L_1} = \frac{8}{60}wL^3 = \frac{8}{60}(800)(3)^3 = 2880 \text{ N} \cdot \text{m}^2 \qquad (b)$$

For span 2, the factor is found by adding the results listed for cases 1 and 5; hence we obtain

$$\frac{6A_2\bar{b}_2}{L_2} = \frac{Pb}{L}(L^2 - b^2) + \frac{wd^2}{4L}(2L^2 - d^2)$$

$$= \frac{(500)(1)}{4}\left[(4)^2 - (1)^2\right] + \frac{400(3)^2}{4(4)}\left[2(4)^2 - (3)^2\right]$$

$$= 1875 + 5175 = 7050 \text{ N} \cdot \text{m}^2 \qquad (c)$$

Substituting these results in Eq. (a) determines M_2, which is now the only unknown. We obtain

$$-300(3) + 2M_2(3 + 4) + 2880 + 7050 = 0$$

from which

$$M_2 = -\frac{9030}{14} = -645 \text{ N} \cdot \text{m} \qquad Ans.$$

812.   Determine the moments over the supports in the continuous beam shown in Fig. 8–8.

---

*A common error involves applying the three-moment equation between the overhang and span 1 as shown in Fig. 8–7, thus completely forgetting that the 200-N load causes the overhang to deflect downward an unknown distance h. Under these conditions, the right-hand part of the general three-moment equation contains the unknown h and *is not zero*. However, after M_1 and M_2 have been found, the general three-moment equation may be applied between spans 0 and 1 to solve for the deflection h. This procedure will be discussed in Illustrative Problem 858 on p. 342.

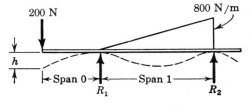

Figure 8-7.

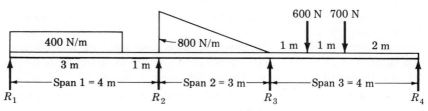

Figure 8–8.

Preliminary: Writing the three-moment equation between spans 1 and 2, and between spans 2 and 3, we have

$$M_1L_1 + 2M_2(L_1 + L_2) + M_3L_2 + \frac{6A_1\bar{a}_1}{L_1} + \frac{6A_2\bar{b}_2}{L_2} = 0 \qquad (a)$$

$$M_2L_2 + 2M_3(L_2 + L_3) + M_4L_3 + \frac{6A_2\bar{a}_2}{L_2} + \frac{6A_3\bar{b}_3}{L_3} = 0 \qquad (b)$$

From the definition of bending moment, both M_1 and M_4 are zero. Hence Eqs. (a) and (b) are a pair of simultaneous equations in M_2 and M_3, which can be solved when the values of $6A\bar{a}/L$ and $6A\bar{b}/L$ for the given loadings are known. Using Table 8–1, we compute these values as follows:

$$\frac{6A_1\bar{a}_1}{L_1} = \frac{wb^2}{4L}(2L^2 - b^2) = \frac{400(3)^2}{4(4)}\left[2(4)^2 - (3)^2\right] = 5175 \text{ N} \cdot \text{m}^2$$

$$\frac{6A_2\bar{b}_2}{L_2} = \frac{8}{60}\,wL^3 = \frac{8}{60}(800)(3)^3 = 2880 \text{ N} \cdot \text{m}^2$$

$$\frac{6A_2\bar{a}_2}{L_2} = \frac{7}{60}\,wL^3 = \frac{7}{60}(800)(3)^3 = 2520 \text{ N} \cdot \text{m}^2$$

$$\frac{6A_3\bar{b}_3}{L_3} = \sum \frac{Pb}{L}(L^2 - b^2)$$

$$= \frac{600(3)}{4}\left[(4)^2 - (3)^2\right] + \frac{700(2)}{4}\left[(4)^2 - (2)^2\right]$$

$$= 3150 + 4200 = 7350 \text{ N} \cdot \text{m}^2$$

Solution: The values just computed are substituted in Eqs. (a) and (b):

$$2M_2(4 + 3) + 3M_3 + 5175 + 2880 = 0$$
$$3M_2 + 2M_3(3 + 4) + 2520 + 7350 = 0$$

or

$$14M_2 + 3M_3 + 8055 = 0 \qquad (c)$$
$$3M_2 + 14M_3 + 9870 = 0 \qquad (d)$$

Solving Eqs. (c) and (d) simultaneously for the two unknowns M_2 and M_3 gives

$$M_2 = -445 \text{ N} \cdot \text{m} \quad \text{and} \quad M_3 = -610 \text{ N} \cdot \text{m} \qquad Ans.$$

PROBLEMS

Unless otherwise stated, the continuous beams in the following problems are supported on rigid foundations which are at the same level. In each problem, determine the bending moments in the beam over the supports.

813.   See Fig. P–813.

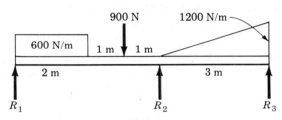

Figure P–813.

814.   See Fig. P–814.   *Ans.* $M_2 = -262 \text{ N} \cdot \text{m}$

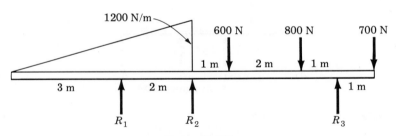

Figure P–814.

815.   Determine the lengths of the overhangs in Fig. P–815 so that the moments over the supports will be equal.   *Ans.* $x = L/\sqrt{6}$

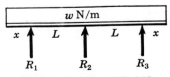

Figures P–815 and P–816.

816. Solve Problem 815 if one span is three-fourths the length of the other span.

817. See Fig. P–817.

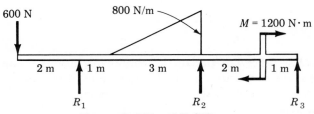

Figures P–817 and P–818.

818. In Problem 817, determine the changed value of the applied couple that will cause M_2 to become zero.

Ans. $M = 105 \text{ N} \cdot \text{m}$ clockwise

819. See Fig. P–819.

$$Ans. \quad M_2 = -\frac{Pa(L^2 - a^2)}{L^2} \cdot \frac{2(\alpha + \beta)}{4(1 + \alpha)(\alpha + \beta) - \alpha^2}$$

$$M_3 = +\frac{Pa(L^2 - a^2)}{L^2} \cdot \frac{\alpha}{4(1 + \alpha)(\alpha + \beta) - \alpha^2}$$

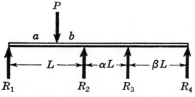

Figures P–819 and P–820.

820. Solve Problem 819 if the concentrated load is replaced by a uniformly distributed load of w N/m over the first span.

$$Ans. \quad M_2 = -\frac{wL^2}{4} \cdot \frac{2(\alpha + \beta)}{4(1 + \alpha)(\alpha + \beta) - \alpha^2}$$

$$M_3 = +\frac{wL^2}{4} \cdot \frac{\alpha}{4(1 + \alpha)(\alpha + \beta) - \alpha^2}$$

821. See Fig. P–821.

$$Ans. \quad M_2 = -\frac{3PL}{8} \cdot \frac{1 + 2\beta}{4(1 + \alpha)(1 + \beta) - 1}$$

$$M_3 = -\frac{3PL}{8} \cdot \frac{1 + 2\alpha}{4(1 + \alpha)(1 + \beta) - 1}$$

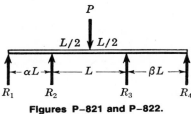

Figures P–821 and P–822.

822.   Solve Problem 821 if the concentrated load is replaced by a uniformly distributed load of w N/m over the middle span.

$$Ans. \quad M_2 = -\frac{wL^2}{4} \cdot \frac{1 + 2\beta}{4(1 + \alpha)(1 + \beta) - 1}$$

$$M_3 = -\frac{wL^2}{4} \cdot \frac{1 + 2\alpha}{4(1 + \alpha)(1 + \beta) - 1}$$

823.   A continuous beam simply supported over three 4-m spans carries a concentrated load of 2 kN at the center of the first span, a concentrated load of 3 kN at the center of the third span, and a uniformly distributed load of 900 N/m over the middle span. Solve for the moments over the supports and check your answers using the results obtained for Problems 819 and 822.

824.   The first span of a simply supported continuous beam is 4 m long, the second span is 2 m long, and the third span is 4 m long. Over the first span there is a uniformly distributed load of 2 kN/m, and over the third span there is a uniformly distributed load of 4 kN/m. At the midpoint of the second span, there is a concentrated load of 10 kN. Solve for the moments over the supports and check your answers, using Problems 820 and 821.

$$Ans. \quad M_2 = -2900 \text{ N} \cdot \text{m}; \quad M_3 = -6100 \text{ N} \cdot \text{m}$$

825.   See Fig. P–825.

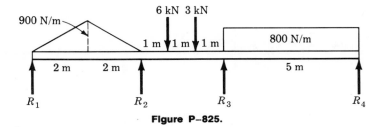

Figure P–825.

826. See Fig. P–826.

Ans. $M_2 = -1690 \text{ N} \cdot \text{m};\ M_3 = -3230 \text{ N} \cdot \text{m}$

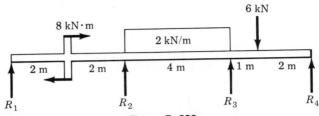

Figure P–826.

827. See Fig. P–827.

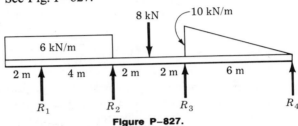

Figure P–827.

8–5 REACTIONS OF CONTINUOUS BEAMS; SHEAR DIAGRAMS

The major reason for computing the reactions of continuous beams is to be able to draw the shear diagram. Two methods of computing reactions are available: in one, reactions are computed by using the definition of bending moment; in the other, the reaction is divided into parts from which the shear diagram can be drawn easily. The second method is preferred for reasons which will be given later. In both methods the moments over the supports must first be determined.

As an example of the first method, consider the beam in Fig. 8–9 whose moments over the supports were found in Illustrative Problem 812 to be $M_2 = -445 \text{ N} \cdot \text{m}$ and $M_3 = -610 \text{ N} \cdot \text{m}$. Applying the definition of bending moment, we express M_2 in terms of the moments

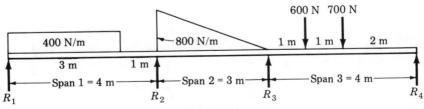

Figure 8–9.

about R_2 of all loads to the left of R_2 and obtain

$$[M_2 = (\Sigma M)_L] \qquad M_2 = -445 = 4R_1 - (400 \times 3) \times 2.5$$

whence

$$R_1 = 639 \text{ N} \qquad Ans.$$

To determine R_2, we apply the definition of M_3 to moments about R_3 of all loads to the left of R_3, as follows:

$$[M_3 = (\Sigma M)_L]$$

$$-610 = 7R_1 - (400 \times 3) \times 5.5 + 3R_2 - \left(\frac{800 \times 3}{2}\right) \times \frac{2}{3} \times 3$$

Substituting in this relation the now known value of $R_1 = 639$ N, we find that

$$R_2 = 1306 \text{ N} \qquad Ans.$$

The value of R_4 is also obtained from the value of M_3 by expressing M_3 in terms of the moments about R_3 of all loads to the right of R_3:

$$[M_3 = (\Sigma M)_R]$$
$$-610 = 4R_4 - 700 \times 2 - 600 \times 1$$

whence

$$R_4 = 348 \text{ N} \qquad Ans.$$

The value of R_3 can now be found by taking a vertical summation of all forces acting on the entire beam. This gives

$$[\Sigma Y = 0]$$

$$R_1 + R_2 + R_3 + R_4 = 400 \times 3 + \frac{800 \times 3}{2} + 600 + 700$$

$$639 + 1306 + R_3 + 348 = 1200 + 1200 + 600 + 700$$

whence

$$R_3 = 1407 \text{ N} \qquad Ans.$$

It is evident that this method carries through any numerical error and is also tedious if there are more than three spans to the beam. An alternate method eliminates both of these objections and presents the results in a form suitable for drawing the shear diagram rapidly. This alternate method depends on isolating each span and determining the supporting shear forces at the end of each one.

In Art. 8–2 and Fig. 8–1 we saw that any span can be isolated as a free body by applying to it the proper values of end moments and shears. The isolated span can then be resolved into a simply supported beam carrying the given loads, plus another beam loaded only by the end moments and couple reactions. Span 2 of Fig. 8–9 is thus resolved

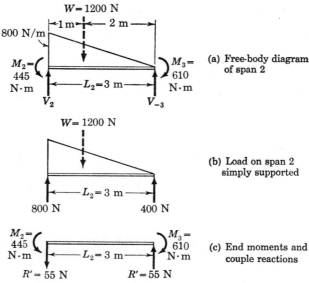

Figure 8-10. Component loadings on span 2.

into its component parts in Fig. 8-10. Because the end moments M_2 and M_3 are negative, they act as shown, and their absolute magnitudes may be used. The term V_2 denotes the vertical shear in the beam to the right of R_2, and V_3 is numerically equivalent to the vertical shear in the beam to the left of R_3. The minus sign in the subscript of V_{-3} indicates that it acts opposite to the actual vertical shear in order to create equilibrium in span 2.

Since parts (b) and (c) of Fig. 8-10 are superposed to form part (a), it follows that the actual end shears V_2 and V_{-3} are the algebraic sum of the equivalent simple beam reactions and the couple reactions R'. In this example, M_3 is numerically larger than M_2; hence there is an unbalanced clockwise couple acting on part (c) of magnitude $M_3 - M_2$. It can be balanced only by the counterclockwise moment caused by the couple reactions R' acting at the supports and having a moment arm equal to the length L_2 of the span. Evidently the numerical value of R' is given by

$$R'L_2 = M_3 - M_2$$

or

$$R' = \frac{M_3 - M_2}{L_2} = \frac{610 - 445}{3} = 55 \text{ N}$$

The couple reaction R' acts upward at the larger moment M_3 and downward at the smaller moment M_2. In the algebraic summation of reactions referred to previously, we take the upward direction as positive and the downward direction as negative.

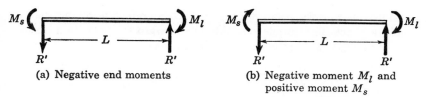

| (a) Negative end moments | (b) Negative moment M_l and positive moment M_s |

Figure 8–11. Couple reaction acts upward at the end having the larger negative moment.

Generalizing this discussion, we may state that the couple reaction R' on any span is given by

$$R' = \frac{M_l - M_s}{L} \qquad (8\text{–}2)$$

where M_l is the larger absolute value of end moment on the span, M_s is the smaller absolute value of end moment, and L is the length of the span. As a rule, the couple reaction R' acts upward at the end of the span having the larger absolute value, and downward at the other end. This assumes that negative moments act over the supports. If one support moment is actually positive in sign, the negative moment is taken as the larger absolute value and its numerical value is used as M_l in Eq. (8–2). The proof of these statements is shown by the free-body diagrams in Fig. 8–11.

A convenient way of arranging the values of simple beam and couple reactions is shown in Fig. 8–12a. The couple reactions were

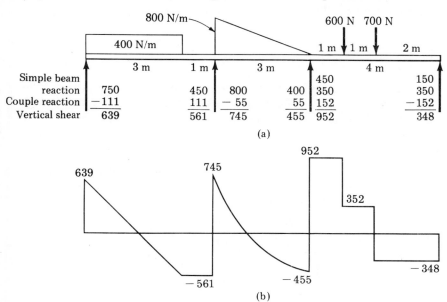

Figure 8–12. A second method of computing reactions and drawing shear diagram.

computed as follows, the numerical subscripts referring to the span on which they act. The couple reaction acts upward (or is given a positive sign) at the end having the larger numerical negative end moment.

$$\left[R' = \frac{M_l - M_s}{L} \right] \qquad R_1' = \frac{445 - 0}{4} = 111 \text{ N}$$

$$R_2' = \frac{610 - 445}{3} = 55 \text{ N}$$

$$R_3' = \frac{610 - 0}{4} = 152 \text{ N}$$

The shear diagram in Fig. 8–12b may now easily be plotted. Remember that the values of vertical shear in Fig. 8–12a which act to the left of the supports are equal but opposite to the actual vertical shear in the beam. This accounts for the minus values in the shear diagram. If the values of the reaction are desired, they may be obtained by adding the vertical shears acting at the reaction. Thus

$R_1 = 639$ N
$R_2 = 561 + 745 = 1306$ N
$R_3 = 455 + 952 = 1407$ N
$R_4 = 348$ N

which agree with the values determined by the first method.

PROBLEMS

In the following problems, determine the reactions and sketch the shear diagrams. Then compute the values of maximum vertical shear V and maximum positive bending moment M. In solving the problems, use the moments determined in the reference problems unless otherwise instructed.

828. A continuous beam carries a uniform load over two equal spans as shown in Fig. P–828.

Ans. $M_2 = - wL^2/8; \ R_1 = R_3 = \frac{3}{8}wL; \ R_2 = \frac{5}{4}wL$

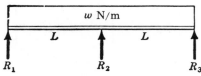

Figure P–828.

829.   A uniform load is carried over three equal spans as shown in Fig. P–829.

Ans. $M_2 = M_3 = -\dfrac{wL^2}{10}$; $R_1 = R_4 = 0.4wL$; $R_2 = R_3 = 1.1wL$

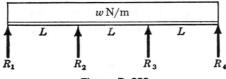

Figure P–829.

830.   Refer to Problem 814.

831.   Refer to Problem 817 for which $M_2 = +\,156$ N $\cdot$ m.
Ans. $R_2 = 109$ N; max. $+M = 452$ N $\cdot$ m

832.   Refer to Problem 824.

833.   Refer to Problem 825 for which $M_2 = -\,2.04$ kN$\cdot$m and $M_3 = -\,2.81$ kN $\cdot$ m.

834.   Refer to Problem 826.
Ans. $R_2 = 6.04$ kN; $R_3 = 9.46$ kN; max. $+M = 3.15$ kN $\cdot$ m

835.   Refer to Problem 827 for which $M_2 = -\,1.895$ kN $\cdot$ m and $M_3 = -\,16.42$ kN $\cdot$ m.

836.   For the continuous beam loaded as shown in Fig. P–815 on page 325, determine the length x of the overhangs that will cause equal reactions.   *Ans.* $x = 0.44L$

8–6   CONTINUOUS BEAMS WITH FIXED ENDS

For continuous beams with fixed ends, assume the fixed end to be equivalent to an imaginary span with an imaginary loading. The three-moment equation, when applied to the beam, includes this imaginary span; however, all the terms that refer to the imaginary span have zero values.

The foregoing statement is easily proved by using the last span of a continuous beam as shown in Fig. 8–13a. The moment M_1 at V_1 is due to the loads on the beam that lie to the left of V_1. The right end at B is assumed to be perfectly fixed, that is, a tangent to the elastic curve at B will be perfectly horizontal. The effect of the perfectly fixed end may be duplicated by adding the reflection of the loads (that is, assuming that the wall at B is a mirror), as shown in Fig. 8–13b. Because of the symmetry of loading thus obtained, the tangent to the elastic curve at B will be horizontal, which is the effect given by a perfectly fixed end.

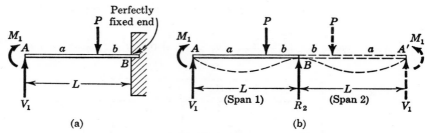

Figure 8-13. A fixed end is equivalent to an imaginary span.

Applying the three-moment equation to spans 1 and 2 of Fig. 8–13b gives

$$M_1L_1 + 2M_2(L_1 + L_2) + M_3L_2 + \frac{6A_1\bar{a}_1}{L_1} + \frac{6A_2\bar{b}_2}{L_2} = 0 \qquad (a)$$

whence, on substituting values corresponding to those in the figure, we obtain

$$M_1L + 2M_2(L + L) + M_1L + \frac{6A_1\bar{a}_1}{L_1} + \frac{6A_1\bar{a}_1}{L_1} = 0$$

or

$$2M_1L + 4M_2L + 2 \times \frac{6A_1\bar{a}_1}{L_1} = 0$$

Dividing by 2, we obtain

$$M_1L + 2M_2L + \frac{6A_1\bar{a}_1}{L_1} = 0 \qquad (b)$$

This would have been obtained from Eq. (a) at once if zero had been substituted for all the terms referring to the imaginary span (span 2 in this example). The principle that a fixed end is equivalent to an imaginary span has thus been proved and will now be applied to several examples.

ILLUSTRATIVE PROBLEMS

837. Find the moments over the supports for the propped beam in Fig. 8–14. The right end is assumed to be perfectly fixed.

Solution: This problem can be solved by the basic area-moment method, but it can be solved more easily and more quickly by considering the fixed end to be equivalent to an imaginary span.

The three-moment equation is applied between spans 1 and 2, whose supports are at the same level, so we obtain

$$M_1L_1 + 2M_2(L_1 + L_2) + M_3L_2 + \frac{6A_1\bar{a}_1}{L_1} + \frac{6A_2\bar{b}_2}{L_2} = 0$$

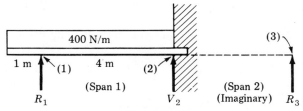

Figure 8–14.

The moment at R_1 due to the overhang is

$$M_1 = -(400 \times 1) \times \tfrac{1}{2} = -200\,\text{N}\cdot\text{m}$$

and zero is substituted for all terms referring to span 2. Table 8–1 shows that $6A_1\bar{a}_1/L_1 = wL^3/4$, so the three-moment equation reduces to

$$-200 \times 4 + 2M_2 \times 4 + \frac{400 \times 4^3}{4} = 0$$

from which

$$M_2 = -700\,\text{N}\cdot\text{m} \qquad Ans.$$

838.   Find the moments over the supports for the continuous beam in Fig. 8–15. Both ends of the beam are assumed to be perfectly fixed.

Solution: The perfectly fixed ends are considered equivalent to the imaginary spans 0 and 3. Writing the three-moment equation for spans 0 and 1, for spans 1 and 2, and for spans 2 and 3, we have

$$M_0L_0 + 2M_1(L_0 + L_1) + M_2L_1 + \frac{6A_0\bar{a}_0}{L_0} + \frac{6A_1\bar{b}_1}{L_1} = 0 \qquad (a)$$

$$M_1L_1 + 2M_2(L_1 + L_2) + M_3L_2 + \frac{6A_1\bar{a}_1}{L_1} + \frac{6A_2\bar{b}_2}{L_2} = 0 \qquad (b)$$

$$M_2L_2 + 2M_3(L_2 + L_3) + M_4L_3 + \frac{6A_2\bar{a}_2}{L_2} + \frac{6A_3\bar{b}_3}{L_3} = 0 \qquad (c)$$

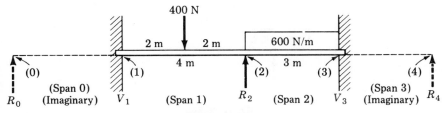

Figure 8–15.

In applying these equations, we neglect any terms referring to the imaginary spans.

Using Table 8–1, we next compute the following values:

$$\frac{6A_1\bar{b}_1}{L_1} = \frac{Pb}{L}(L^2 - b^2) = \frac{400 \times 2}{4}(16 - 4) = 2400 \text{ N} \cdot \text{m}^2$$

$$\frac{6A_1\bar{a}_1}{L_1} = \frac{Pa}{L}(L^2 - a^2) = \frac{400 \times 2}{4}(16 - 4) = 2400 \text{ N} \cdot \text{m}^2$$

$$\frac{6A_2\bar{a}_2}{L_2} = \frac{6A_2\bar{b}_2}{L_2} = \frac{wL^3}{4} = \frac{600 \times 3^3}{4} = 4050 \text{ N} \cdot \text{m}^2$$

Substituting these values in the three-moment equations gives

From Eq. (a):	$8M_1 + 4M_2 + 2400 = 0$	(d)
From Eq. (b):	$4M_1 + 14M_2 + 3M_3 + 6450 = 0$	(e)
From Eq. (c):	$3M_2 + 6M_3 + 4050 = 0$	(f)

Solving Eqs. (d), (e), and (f) simultaneously gives

$$M_1 = -147 \text{ N} \cdot \text{m}, \quad M_2 = -307 \text{ N} \cdot \text{m}, \quad M_3 = -522 \text{ N} \cdot \text{m} \quad Ans.$$

PROBLEMS

In the following problems, the ends of the beams are assumed to be perfectly fixed by the walls against rotation. All supports are assumed to remain at the same level.

839.   Determine the prop reaction for the beam in Fig. P–839.

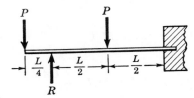

Figure P–839.

840.   For the propped beam shown in Fig. P–840, determine the prop reaction and the maximum positive bending moment.

$$Ans. \quad R = 2.96 \text{ kN}; \ M = 696 \text{ N} \cdot \text{m}$$

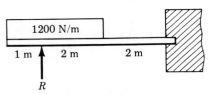

Figure P–840.

841.   Determine the wall moment and prop reaction for the beam shown in Fig. P–841.   *Ans.* $M = -1.35 \text{ kN} \cdot \text{m}; \ R = 2.33 \text{ kN}$

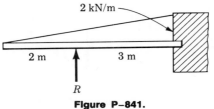

Figure P–841.

842.   For the propped beam shown in Fig. P–842, determine the wall moment and the reaction of the prop support.

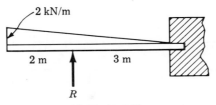

Figure P–842.

843.   For the propped beam shown in Fig. P–843, determine the wall moment and the prop reaction.

Ans. $M = -1566 \text{ N} \cdot \text{m}; \ R = 359 \text{ N}$

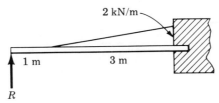

Figure P–843.

844.   In the propped beam shown in Fig. P–844, determine the prop reaction.

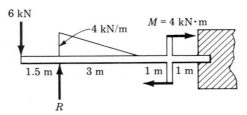

Figure P–844.

845. Compute the moments over the supports for the beam shown in Fig. P–845 and then draw the shear diagram.

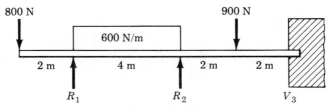

Figure P–845.

846. Sketch the shear diagram for the continuous beam shown in Fig. P–846.   *Ans.* $M_2 = -2.52$ kN·m; $M_3 = +733$ N·m

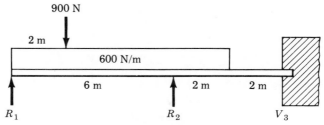

Figure P–846.

847. Compute the moments over the supports and sketch the shear diagram for the continuous beam shown in Fig. P–847.

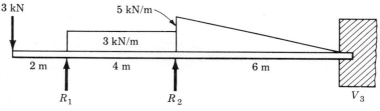

Figure P–847.

848. Determine the support moments and reactions for the beam shown in Fig. P–848.

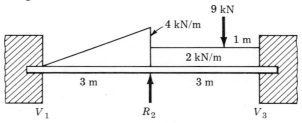

Figure P–848.

849.   Find the moments over the supports for the beam shown in Fig. P–849.

Ans. $M_1 = -300$ N·m; $M_2 = -1500$ N·m; $M_3 = -2700$ N·m

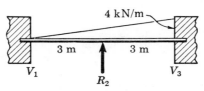

Figure P–849.

850.   Determine the moments over the supports for the beam loaded as shown in Fig. P–850.

$$Ans. \quad M_1 = -\frac{wL^2}{8} \cdot \frac{2 + 3\alpha}{3 + 3\alpha}; \quad M_2 = -\frac{wL^2}{8} \cdot \frac{2}{3 + 3\alpha};$$

$$M_3 = +\frac{wL^2}{8} \cdot \frac{1}{3 + 3\alpha}$$

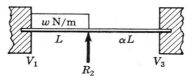

Figures P–850 and P–851.

851.   Replace the distributed load in Problem 850 by a concentrated load P at midspan and solve for the moments over the supports. $$Ans. \quad M_1 = -\frac{3PL}{16} \cdot \frac{2 + 3\alpha}{3 + 3\alpha}; \quad M_2 = -\frac{3PL}{16} \cdot \frac{2}{3 + 3\alpha};$$

$$M_3 = +\frac{3PL}{16} \cdot \frac{1}{3 + 3\alpha}$$

852.   Use the results of Problems 850 and 851 to check the answers to Illustrative Problem 838.

853.   For the continuous beam shown in Fig. P–853, determine the moments over the supports. Also draw the shear diagram and compute the maximum positive bending moment. (*Hint:* Take advantage of symmetry.)

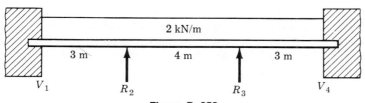

Figure P–853.

854.   Solve for the moments over the supports in the beam loaded as shown in Fig. P–854.

$$Ans. \quad M_1 = M_4 = + \frac{wL^2}{12} \cdot \frac{1}{2 + \alpha}; \quad M_2 = M_3 = - \frac{wL^2}{12} \cdot \frac{2}{2 + \alpha}$$

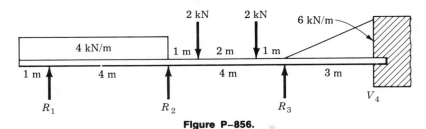

Figures P–854 and P–855.

855.   If the distributed load in Problem 854 is replaced by a concentrated load P at midspan, determine the moments over the supports.

856.   For the beam shown in Fig. P–856, determine the moments over the supports. Also draw the shear diagram and compute the maximum positive bending moment.

$$Ans. \quad M_2 = - 4460 \text{ N} \cdot \text{m}; \quad M_3 = - 661 \text{ N} \cdot \text{m}; \quad M_4 = - 3270 \text{ N} \cdot \text{m}$$

Figure P–856.

8–7   DEFLECTIONS DETERMINED BY THE THREE-MOMENT EQUATION

Before discussing the use of the general three-moment equation to find deflections let us review a few facts. The three-moment equation determines the relation among the moments at *any* three points in *any* beam. These three points determine two segments of the beam, and the terms $6A_1\bar{a}_1/L_1$ and $6A_2\bar{b}_2/L_2$ of the three-moment equation refer to the moment diagram resulting from the loads acting on these segments. The loads are assumed to be simply supported on spans that are as long as the segments. Heights h_1 and h_3 refer to the heights of points 1 and 3 relative to point 2 (see Fig. 8–3, page 315); the heights are considered positive if above point 2 and negative if below it.

The general method for determining deflections by means of the three-moment equation is to select points 1, 2, and 3 so that either (or both) of the heights h_1 and h_3 is equal to the desired deflection. This occurs when two of the points are selected over supports and the third is chosen at the location where the deflection is to be determined. The values of the moments at points 1, 2, and 3 must first be known or computed. This method will now be illustrated.

ILLUSTRATIVE PROBLEMS

857. Use the three-moment equation to determine the value of $EI\delta$ at 1 m from the left support of the beam loaded as shown in Fig. 8–16.

Solution: The exaggerated position of the deflection curve is shown by the dashed line. Selecting point 2 at the position of the desired deflection and points 1 and 3 at the reactions will make h_1 and h_3 each equal to the desired deflection. Also, since 1 and 3 are above the horizontal line through 2, heights h_1 and h_3 will be positive. The segments into which the beam is divided by the three points are designated as span 1 and span 2.

The general three-moment equation is

$$M_1L_1 + 2M_2(L_1 + L_2) + M_3L_2 + \frac{6A_1\bar{a}_1}{L_1} + \frac{6A_2\bar{b}_2}{L_2}$$
$$= 6EI\left(\frac{h_1}{L_1} + \frac{h_3}{L_2}\right)$$

Since span 1 in Fig. 8–16 is unloaded, $6A_1\bar{a}_1/L_1$ is zero. For span 2, case 5 of Table 8–1 gives

$$\frac{6A_2\bar{b}_2}{L_2} = \frac{wd^2}{4L}(2L^2 - d^2) = \frac{900(2)^2}{4(3)}(2 \times 9 - 4) = 4200 \text{ N} \cdot \text{m}^2$$

By taking moments about R_2, we determine that $R_1 = 450$ N. Having found R_1, the bending moment at point 2 is $M_2 = 450 \times 1 = 450$ N · m. Since $M_1 = M_3 = 0$ and $h_1 = h_3 = \delta$, substituting the above

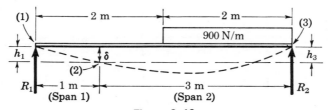

Figure 8–16.

values in the three-moment equation yields

$$0 + 2(450)(1 + 3) + 0 + 0 + 4200 = 6EI\left(\frac{\delta}{1} + \frac{\delta}{3}\right)$$

which reduces to

$$EI\delta = 975 \text{ N} \cdot \text{m}^3 \qquad Ans.$$

858. Determine the value of $EI\delta$ under the 200-N load of the continuous beam shown in Fig. 8–17.

Solution: This beam is the one for which the support moments were determined in Illustrative Problem 811. Here we select points 0, 1, and 2, as shown, between which to write the three-moment equation. Point 0 on the elastic curve is below point 1; hence $h_0 = -\delta$ and $h_2 = 0$. Using the results of Illustrative Problem 811, we find that $M_0 = 0$, $M_1 = -300 \text{ N} \cdot \text{m}$, and $M_2 = -645 \text{ N} \cdot \text{m}$.

Applying the three-moment equation, we write

$$M_0 L_0 + 2M_1(L_0 + L_1) + M_2 L_1 + \frac{6A_0\bar{a}_0}{L_0} + \frac{6A_1\bar{b}_1}{L_1}$$

$$= 6EI\left(\frac{h_0}{L_0} + \frac{h_2}{L_1}\right)$$

In this instance span 0 is unloaded and hence $6A_0\bar{a}_0/L_0$ is zero. Case 3 of Table 8–1 gives for span 1

$$\frac{6A_1\bar{b}_1}{L_1} = \frac{7}{60}wL^3 = \frac{7}{60} \times 800 \times 27 = 2520 \text{ N} \cdot \text{m}^2$$

Substituting these values in the three-moment equation, with careful note of the minus signs of M_1 and M_2, we obtain

$$2(-300)(1.5 + 3) + (-645) \times 3 + 2520 = 6EI\left(\frac{-\delta}{1.5} + 0\right)$$

whence

$$EI\delta = 529 \text{ N} \cdot \text{m}^3 \qquad Ans.$$

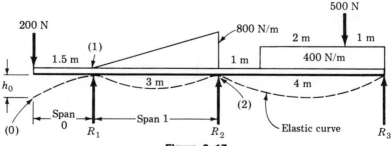

Figure 8–17.

The positive value of the result indicates that the deflection is downward as assumed.

PROBLEMS

859. Determine the value of $EI\delta$ under P in Fig. P–859. What is the result if P is replaced by a clockwise couple M?

$$Ans. \quad EI\delta = \frac{PLb^2}{3}; \; EI\delta = \frac{Mb(2L + b)}{6}$$

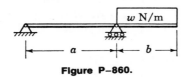

Figure P–859.

860. Determine the value of $EI\delta$ at the end of the overhang and midway between the supports for the beam shown in Fig. P–860.

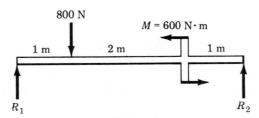

Figure P–860.

861. For the beam shown in Fig. P–861, determine the value of $EI\delta$ at 1 m and 3 m from the left support. *Ans.* $900 \text{ N} \cdot \text{m}^3$; $767 \text{ N} \cdot \text{m}^3$

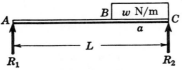

Figure P–861.

862. Determine the value of $EI\delta$ at B for the beam shown in Fig. P–862.

$$Ans. \quad EI\delta = \frac{wa^3}{24L}(4L - 3a)(L - a)$$

P–862.

Figure P–862.

863.   For the beam shown in Fig. P–863, determine the value of $EI\delta$ midway between the supports and at the left end.

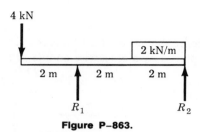

Figure P–863.

864.   A 6-m beam, simply supported at 1 m from each end, carries a uniformly distributed load of 800 N/m over its entire length. Compute the value of $EI\delta$ at the middle and at the ends.

865.   For the beam shown in Fig. P–865, compute the value of $EI\delta$ at $x = 3$ m and at the end of the overhang.

Ans.   At overhang, $EI\delta = 813$ N · m³ down

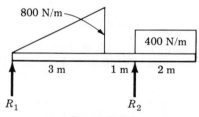

Figure P–865.

866.   Determine the midspan value of $EI\delta$ for the beam shown in Fig. P–866.

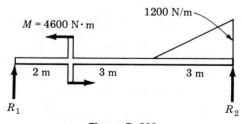

Figure P–866.

867.   For the beam in Fig. P–867, compute the value of P that will cause a zero deflection under P.   *Ans.* $P = 489$ N

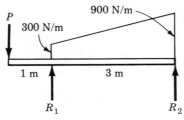

Figure P–867.

868.   Determine the values of $EI\delta$ at midspan and at the ends of the beam loaded as shown in Fig. P–868.

Ans.   At ends, $EI\delta = 10.9$ kN $\cdot$ m^3 up

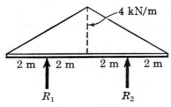

Figure P–868.

869.   Find the value of $EI\delta$ at the center of the first span of the continuous beam in Fig. P–869 if it is known that $M_2 = -2040$ N $\cdot$ m and $M_3 = -2810$ N $\cdot$ m.   *Ans.* $EI\delta = 120$ N $\cdot$ m^3 up

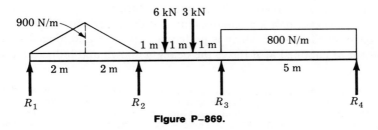

Figure P–869.

870. Compute the value of $EI\delta$ at the overhanging end of the beam in Fig. P–870 if it is known that the wall moment is $+1100\ \text{N}\cdot\text{m}$.

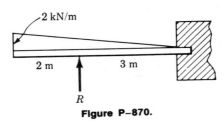

Figure P–870.

871. The continuous beam in Fig. P–871 is supported at its left end by a spring whose constant is $50\ \text{kN/m}$. For the beam, $E = 10 \times 10^9\ \text{N/m}^2$ and $I = 40 \times 10^6\ \text{mm}^4$. Compute the deflection of the spring.

                                Ans. $\delta = 79.4\ \text{mm}$

872. Repeat Problem 871 assuming that the loadings on the spans are interchanged.

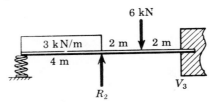

Figures P–871 and P–872.

8–8   MOMENT DISTRIBUTION

Modern techniques of designing continuous structures are based on a method of successive approximations popularized by Hardy Cross.* This method, which is widely known as the *moment-distribution method,* is applicable to all types of rigid-frame analysis. Its application to continuous beams will serve to introduce this powerful tool of the structural engineer.

    Several preliminary concepts are necessary. The first, the *carry-over moment,* is defined as the moment induced at the fixed end of a beam by the action of a moment applied at the other end. Thus consider the beam in Fig. 8–18a which is perfectly fixed at A and hinged at B. A moment M_B applied at B flexes the beam as shown and induces the wall

*See Cross's papers, Continuity as a factor in reinforced concrete design. *Proc. A.C.I.* pp. 669–711 (1929); Simplified rigid frame design. *Proc. A.C.I.* **26,** 170–183 (1930); Analysis of continuous frames by distributing fixed end moments. *Trans. A.S.C.E.* **96,** 1–156 (1932). See also H. Cross and N. D. Morgan, *Continuous Frames of Reinforced Concrete,* Wiley, New York, 1932.

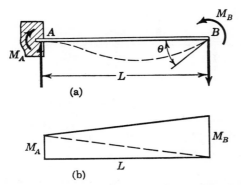

Figure 8-18. Carry-over moment and beam stiffness.

moment M_A. The moment diagram, which is drawn as described in Art 7-5, is shown in Fig. 8-18b. Although M_A is actually negative (because of the downward curvature of the elastic curve at A), it is convenient to consider it positive as shown; consequently the solution will determine not only its absolute value but also the correct sign (negative).

The deviation at B from a reference tangent drawn at A is zero because of the perfect wall constraint at A. Hence

$$\left[EIt_{B/A} = (\text{area})_{BA} \cdot \bar{x}_B \right] \quad 0 = \left(\tfrac{1}{2} M_A L \right) \left(\tfrac{2}{3} L \right) + \left(\tfrac{1}{2} M_B L \right) \left(\tfrac{1}{3} L \right)$$

whence

$$M_A = -\tfrac{1}{2} M_B \tag{8-3}$$

This result means that a moment applied at the hinged end B "carries over" to the fixed end A a moment that is half the amount and of opposite sign.

A second concept needed for the moment-distribution method is *beam stiffness*. Beam stiffness is the moment required at the simply supported end of a beam to produce unit rotation of that end, the other end being rigidly fixed. Note that this definition implies no relative linear displacement of the two ends of the beam.

The slope at B in Fig. 8-18a is found from the first theorem of the area-moment method and is expressed in terms of the moment diagram in Fig. 8-18b:

$$\left[EI\theta_{AB} = (\text{area})_{AB} \right] \quad EI\theta = \tfrac{1}{2} M_A L + \tfrac{1}{2} M_B L$$

Replacing M_A by $-\tfrac{1}{2} M_B$ from Eq. (8-3) gives

$$M_B = \frac{4EI\theta}{L}$$

As was said above, the value of M_B when θ equals 1 radian is known as the *beam stiffness*. It varies with the ratio I/L as well as with E. It is

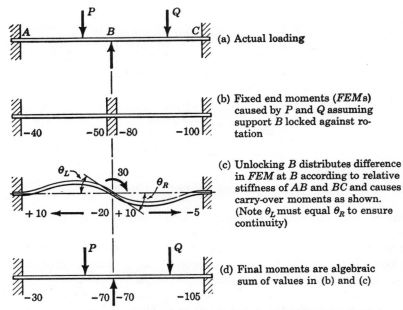

Figure 8–19.   Qualitative description of moment-distribution procedure.

denoted by the symbol K; hence

$$\text{absolute } K = \frac{4EI}{L} \qquad (8\text{–}4)$$

However, in many structures E remains constant, so only a relative measure of resistance to rotation is required. This may be called *relative beam stiffness* and is expressed by

$$\text{relative } K = \frac{I}{L} \qquad (8\text{–}5)$$

We are now ready to describe qualitatively the moment-distribution procedure.

The continuous beam in Fig. 8–19a is perfectly restrained at A and C and simply supported at B. Assume that at B the beam is temporarily locked or rigidified against the rotation caused by the loads P and Q. Under these conditions, segments AB and BC will act as fixed ended beams subjected to the fixed end moments caused by loads P and Q. These fixed end moments (hereafter abbreviated as FEM) are assumed to have the values shown in Fig. 8–19b.

If the support at B is now released or unlocked, the difference in FEM between sections to the left and right of B creates an unbalanced moment of 30 N · m which causes the beam at B to rotate, as shown in Fig. 8–19c, until the moments at B are balanced. Obviously, the moment to the left of B will be increased by some amount, say 20 N·m; and to the right of B the moment will be decreased by the remaining 10 N·m of the 30 N·m difference between the FEM at B in Fig. 8–19b. Thus the unbalanced moment is distributed at the unlocked support. The rotation of B caused by these distributed moments induces, at A and C, carry-over moments of half the amount and of opposite sign. These carry-over moments are indicated by the arrows in Fig. 8–19c.

The ratio of distribution of the unbalanced moment at B is fixed by the fact that the two beams must rotate through the same angle at B. This means that the unbalanced moment must be distributed in the ratio of the stiffness factors of the adjacent beams. The ratio of distribution to any beam is called a *distribution factor*, DF, and is defined by

$$DF = \frac{K}{\Sigma K} \tag{8–6}$$

where K is the stiffness factor for that beam and ΣK is the sum of the stiffness factors for adjacent beams. If the beams are of the same material (as is generally the case), only relative K need be used. Further, if they are of the same cross section, relative K (that is, beam stiffness) is inversely proportional to the length (see Eq. 8–5). In distributing the FEMs the object is to secure balance at the unlocked support.

The final moments in Fig. 8–19d are obtained by superposition of the FEMs in Fig. 8–19b and of the distributed moments and carry-over moments in Fig. 8–19c.

Sign Convention

In the preceding discussion, signs were based on conventional bending moments. This convention requires the carry-over moment to be of opposite sign and often leads to confusion regarding the magnitude and sign of the unbalanced moment to be distributed, especially when more than two members frame into a common joint. Computational accuracy is increased and confusion eliminated by using an alternate sign convention based upon the sense of rotation of the end moments.

With this convention, counterclockwise moments acting on the beam are considered to be positive, and clockwise end moments are negative. As a result of this alternate convention, several minor modifications occur. The first is that carry-over moments are of the *same sign*. The second is that in distributing the unbalanced moments at each support, the distributed moments are of the *same sign* and are so applied

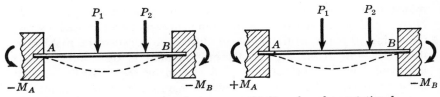

(a) Signs based on conventional restraining moments   (b) Signs based on rotational sense of restraining moments

Figure 8–20. Difference between sign conventions.

as to make the algebraic sum of the moments at a support or joint equal to zero. Finally, in a fixed-ended beam carrying downward loads as in Fig. 8–20, the fixed end moments will be positive at the left end and negative at the right end.

Sometimes the alternate sign convention is based upon the rotation of the end joint rather than the member. Then a moment tending to rotate a joint clockwise is considered positive. However, since the action of the member upon the joint is equal but opposite to that of the joint upon the member, this convention based on joint rotation is exactly equivalent to that of moment rotation on the member.

The moment-distribution method may be summarized in the following steps:

1. Assume that all supports are fixed or locked and compute *fixed end moments* for each span considered separate from every other span. Table 7–2 (page 307) will be helpful in computing these FEMs.

2. Unlock each support and *distribute* the unbalanced moment at each one to each adjacent span by means of Eq. (8–6). Then relock each support.

3. After distributing the unbalanced moment to each adjacent span, *carry over* one-half this amount, with the *same* sign, to the other end of each span.

This completes one cycle of distribution. Steps 2 and 3 must be repeated because of the new unbalance caused by the carry-over moments. Such repetitions are made until the carry-over moments become zero or negligibly small. The process may be stopped when any distribution is completed, the accuracy of the final results depending on the number of cycles. As a rule, no more than four cycles are necessary, since the unbalance caused by the carry-over moments usually decreases rapidly to zero.

The following illustrative problems show the method of recording results, as well as some suggested modifications or shortcuts.

ILLUSTRATIVE PROBLEMS

873. The continuous beam of constant cross section and material shown in Fig. 8–21 is perfectly restrained at the ends. Compute the moments over the supports.

Solution: Although I is not specified, it is convenient to take I as a common multiple of the span lengths, that is, 12 units. Then the values of relative stiffness $K = I/L$ are shown, and the distribution factors (DF) are computed from Eq. (8–6) and also listed.

Assuming all supports locked, and using Table 7–2 (page 307), we compute values for the fixed end moments (FEM):

$$\text{Span } AB: \quad M_A = -\frac{Pab^2}{L^2} = -\frac{4500(4)(2)^2}{(6)^2} = -2000 \text{ N} \cdot \text{m}$$

$$M_B = -\frac{Pa^2b}{L^2} = -\frac{4500(4)^2(2)}{(6)^2} = -4000 \text{ N} \cdot \text{m}$$

$$\text{Span } BC: \quad M_B = M_C = -\frac{wL^2}{12} = -\frac{6000(4)^2}{12} = -8000 \text{ N} \cdot \text{m}$$

Using the rotation sign convention, these FEMs are inserted in the table in Fig. 8–21 with a "plus" sign at the left end of each span and a "minus" sign at the right end.

With B unlocked, the unbalanced moment is the numerical difference between the FEM at B, or $8000 - 4000 = 4000$ N·m. Using

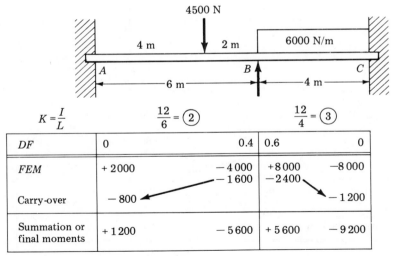

$K = \dfrac{I}{L}$		$\dfrac{12}{6} = ②$	$\dfrac{12}{4} = ③$	
DF	0	0.4	0.6	0
FEM	+ 2000	− 4 000	+8 000	−8 000
		− 1 600	−2 400	
Carry-over	− 800			− 1 200
Summation or final moments	+ 1 200	− 5 600	+ 5 600	− 9 200

Figure 8–21.

the values of DF, we distribute part of this unbalanced moment to the left of B as $0.4(4000) = 1600$ N $\cdot$ m and the remainder to the right of B as $0.6(4000) = 2400$ N $\cdot$ m. As indicated in the table in Fig. 8–21, the signs of these distributed moments are chosen to be minus so that the sum of the moments at B will become zero. One-half the values of these distributed moments are now carried over with the same sign. Thus -1600 applied to the left of B is carried over as -800 to A, and -2400 applied to the right of B is carried over as -1200 to C.

Since A and C are locked or fixed and are specified as remaining so, they absorb these carry-over moments and the distribution is completed. The final values of the moment at each support are obtained by algebraic summations of each vertical column, giving the results shown. If desired, the moments in the final summation are easily converted back to conventional bending moments by merely changing the sign at the left end of each span.

874. The continuous beam in Fig. 8–22 carries the same loads as the beam in Illustrative Problem 873, but the ends at A and C are simply supported. Compute the support moments.

Solution: Values of K and DF are computed and listed as in Illustrative Problem 873. Assuming all supports locked, the FEMs are also

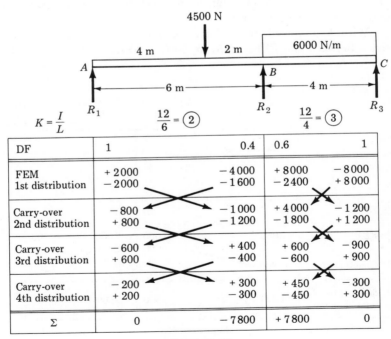

DF	1		0.4	0.6		1
FEM 1st distribution	$+2000$ -2000		-4000 -1600	$+8000$ -2400		-8000 $+8000$
Carry-over 2nd distribution	-800 $+800$		-1000 -1200	$+4000$ -1800		-1200 $+1200$
Carry-over 3rd distribution	-600 $+600$		$+400$ -400	$+600$ -600		-900 $+900$
Carry-over 4th distribution	-200 $+200$		$+300$ -300	$+450$ -450		-300 $+300$
Σ	0		-7800	$+7800$		0

Figure 8–22.

computed and listed, with "plus" signs at the left end of each span and "minus" signs at the right end.

All the supports are now unlocked, which restores the beam to the specified conditions at each support. The unbalanced moment at each support set up by unlocking the supports must now be distributed. At B the distribution is as described in Illustrative Problem 873; but releasing

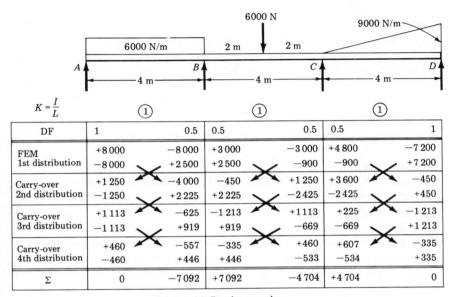

Solution (a) Regular procedure

$K=\dfrac{I}{L}$	①		①		①	
DF	1	0.5	0.5	0.5	0.5	1
FEM 1st distribution	+8 000 −8 000	−8 000 +2 500	+3 000 +2 500	−3 000 −900	+4 800 −900	−7 200 +7 200
Carry-over 2nd distribution	+1 250 −1 250	−4 000 +2 225	−450 +2 225	+1 250 −2 425	+3 600 −2 425	−450 +450
Carry-over 3rd distribution	+1 113 −1 113	−625 +919	−1 213 +919	+1 113 −669	+225 −669	−1 213 +1 213
Carry-over 4th distribution	+460 −460	−557 +446	−335 +446	+460 −533	+607 −534	−335 +335
Σ	0	−7 092	+7 092	−4 704	+4 704	0

Solution (b) Shortcut procedure

$K=\dfrac{I}{L}$	$\frac{3}{4}\times1=\left(\frac{3}{4}\right)$		①		$\frac{3}{4}\times1=\left(\frac{3}{4}\right)$	
DF	1	$\frac{3}{7}$	$\frac{4}{7}$	$\frac{4}{7}$	$\frac{3}{7}$	1
FEM Release A & D	+8 000 −8 000	−8 000 → −4 000	+3 000	−3 000	+4 800 +3 600	−7 200 ← +7 200
Adjusted FEM 1st distribution	0	−12 000 +3 857	+3 000 +5 143	−3 000 −3 086	+8 400 −2 314	0
Carry-over 2nd distribution		+661	−1 543 +882	+2 572 −1 470	−1 102	
Carry-over 3rd distribution		+315	−735 +420	+441 −252	−189	
Carry-over 4th distribution		+54	−126 +72	+210 −120	−90	
Σ	0	−7 113	+7 113	−4 705	+4 705	0

Figure 8–23.

A and *C* is equivalent to adding moments of -2000 and $+8000$, respectively, so as to cause a final moment of zero at these free ends. When this distribution is completed and all the supports are relocked, the distributed moments cause the carry-over effects indicated by the arrows, and this again introduces unbalance at the locked supports. However, this unbalance is appreciably smaller.

We continue to unlock each support, distribute the unbalanced moment, and relock each support, thus completing another cycle of distribution, until the carry-over moments become negligibly small or, as here in cycle 3, until the sum of the carry-over moment and the distributed moment is zero on each side of a support. Further cycles of distribution will then produce no effect, as cycle 4 shows. Note that the analysis must end with a distribution and not with a carry-over.

875. Apply the moment-distribution method to the continuous beam of three spans with free ends shown in Fig. 8–23.

Solution: There are two solutions. The first, in (a), involves the same procedure that was described in Illustrative Problem 874 for free ends on two spans. It is inconvenient to treat a free end as fixed, carry moment over to it, and then release it again. It is simpler to use a modification in which the free end, initially assumed fixed, is released only once and has no moment carried over to it for further distribution. To understand this modification, we shall show how the moment that is distributed to the left of *B* in the first distribution of solution (a) is carried through the computations. This moment, denoted by *M* in Fig. 8–24, carries over to *A* as $+\frac{1}{2}M$ if *A* is locked. If *A* is a free end, releasing it causes the distributed moment $-\frac{1}{2}M$ at *A*, which then carries over to *B* the moment $\frac{1}{2}(-\frac{1}{2}M) = -\frac{1}{4}M$. A summation of these values gives zero at *A* (which is now freely hinged) and $\frac{3}{4}M$ at *B*.

If the initial distribution at *B* had been modified so that only $\frac{3}{4}M$ were distributed to *B*, these results would have been obtained directly, with no moment carried over to *A*. In other words, if the stiffness factor

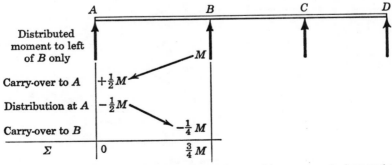

Figure 8–24. Modification of distribution at *B* to avoid carry-over to free end *A*.

for AB is multiplied by $\frac{3}{4}$, the distribution at B is modified so that no moment need be carried over to the free end.

This shortcut is applied in the second solution (b) shown in Fig. 8–23. As a start, reduce A and D to free ends by releasing them, applying the balancing moments of -8000 at A and $+7200$ at D, and carrying over half these amounts with the same signs to B and C as

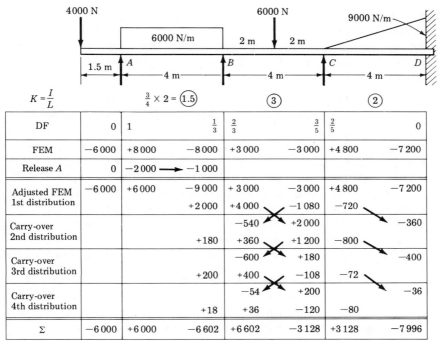

$$K = \frac{I}{L} \qquad \frac{3}{4} \times 2 = \boxed{1.5} \qquad \boxed{3} \qquad \boxed{2}$$

DF	0	1	$\frac{1}{3}$	$\frac{2}{3}$	$\frac{3}{5}$	$\frac{2}{5}$	0
FEM	−6 000	+8 000	−8 000	+3 000	−3 000	+4 800	−7 200
Release A		0	−2 000 ⟶ −1 000				
Adjusted FEM 1st distribution	−6 000	+6 000	−9 000	+3 000	−3 000	+4 800	−7 200
			+2 000	+4 000	−1 080	−720	
Carry-over 2nd distribution			−540	+2 000			−360
			+180	+360	+1 200	−800	
Carry-over 3rd distribution			−600	+180			−400
			+200	+400	−108	−72	
Carry-over 4th distribution			−54	+200			−36
			+18	+36	−120	−80	
Σ	−6 000	+6 000	−6 602	+6 602	−3 128	+3 128	−7 996

Solution (a) Regular procedure

DF	0	1	$\frac{1}{3}$	$\frac{2}{3}$	$\frac{3}{5}$	$\frac{2}{5}$	0
FEM 1st distribution	−6 000	+8 000	−8 000	+3 000	−3 000	+4 800	−7200
		0	−2 000	+1 667	+3 333	−1 080	−720
Carry-over 2nd distribution			−1 000	−540	+1 667		−360
			+513	+1 027	−1 000	−667	
Carry-over 3rd distribution				−500	+514		−334
			+167	+333	−308	−206	
Carry-over 4th distribution				−154	+167		−103
			+51	+103	−100	−67	
Σ	−6 000	+6 000	−6 602	+6 602	−3 140	+3 140	−7 997

Solution (b) Alternate procedure

Figure 8–25.

shown. Now distribute the FEMs at B and C, using the distribution factors obtained by modifying the stiffness of AB and CD as described above. Note that no moment is carried over to the free ends under these conditions. Note also that we are approaching the final result more rapidly in the fourth distribution here than in the corresponding fourth distribution in the first solution.

876. The loading in Illustrative Problem 875 is modified in Fig. 8–25 by adding an overhang at A and fixing the end at D. The moments of inertia of segments AB and CD are each equal to 20 units, but that of BC is increased to 30 units. The relative stiffnesses are therefore $K_{AB} = 2$, $K_{BC} = 3$, and $K_{CD} = 2$. Compute the moments over the supports.

Solution: This problem illustrates two additional concepts: (1) The cross section may vary from segment to segment and is taken into account by computing the I/L ratio for each segment.* (2) The overhanging end offers no resistance to rotation. Hence, when A is released, the unbalance of $2000 \ \text{N} \cdot \text{m}$ must be distributed as zero to the left of A and as $-2000 \ \text{N} \cdot \text{m}$ to the right of A. When $-1000 \ \text{N} \cdot \text{m}$ is carried over to B, the adjusted fixed end moments are as shown below the first double line. Also, to avoid carrying moments back to A, we multiply K_{AB} by $\frac{3}{4}$, and the shortcut procedure described in Illustrative Problem 875 is then applied, as shown in Fig. 8–25. Note that the fixed end at D absorbs the carry-over moments from C but does not transfer any back to C.

An alternate distribution preferred by some engineers gives substantially the same results and is also shown in Fig. 8–25. Here, instead of starting with an adjusted FEM at A and B, the unbalanced moment at all supports is first distributed at each support and then carried over as shown.

PROBLEMS

By means of the moment-distribution method, solve for the moments over the supports in the continuous beams referred to below.

877. See Problem 814 (page 325).

878. See Problem 826.

879. See Problem 827.

880. See Problem 845 (page 338).

881. See Problem 846.

---

*More complex cases, in which the cross section varies along the segment, are treated by H. Cross and N. D. Morgan, *Continuous Frames of Reinforced Concrete*, Wiley, New York, 1932.

882.   See Problem 849.

883.   See Problem 853.

884.   See Problem 856.

885.   Solve Problem 856 if the moment of inertia varies from span to span so that the relative stiffness for span 1 is 2, that for span 2 is 1.5, and that for span 3 is 1.

886.   Solve for the support moments in Problem 825 (page 327) if the ends are perfectly fixed instead of simply supported.
Ans. $M_1 = -73$ N·m; $M_2 = -2100$ N·m; $M_3 = -2380$ N·m;
$$M_4 = -1310 \text{ N} \cdot \text{m}$$

SUMMARY

The general form of the three-moment equation is

$$M_1 L_1 + 2M_2(L_1 + L_2) + M_3 L_2 + \frac{6A_1 \bar{a}_1}{L_1} + \frac{6A_2 \bar{b}_2}{L_2}$$

$$= 6EI\left(\frac{h_1}{L_1} + \frac{h_3}{L_2}\right) \quad (8-1)$$

The factors in this equation are tabulated for various span loadings in Table 8–1 on page 318.

For continuous beams whose supports are on the same level, the terms h_1 and h_3 reduce to zero and the three-moment equation readily determines the moments at the supports. If one end of a continuous beam is fixed, it may be treated as though it were an imaginary span.

When determining deflections, the three-moment equation is written between three points, 1, 2, and 3, so that either (or both) of the heights h_1 and h_3 is equal to the desired deflection. Generally, two of the points are at the supports, whereas the third point is at the location of desired deflection. The values of the moments at these three points must first be known or computed.

An alternate method of determining support moments in continuous beams (as well as in more general frames) is moment distribution. To apply it, each span is first considered to be fixed at the supports, and the fixed end moments are computed from the general values listed in Table 7–2 on page 307. Each support is then released, and the resulting unbalanced moments are distributed at each support and also carried over to adjacent supports in accordance with the details given in Art. 8–8. As explained in Art. 8–5, once the support moments have been found, it is a simple matter to determine the shear diagram, from which, as explained in Chapter 4, the maximum shear and bending moment can be found.

Combined Stresses

9-1 INTRODUCTION

In preceding chapters we studied three basic types of loading: axial, torsional, and flexural. Each of these types was discussed on the assumption that only one of these loadings was acting on a structure at a time. The present chapter is concerned with cases in which two or more of these loadings act simultaneously upon a structure. The three basic types of loading and the corresponding stress formula may be summarized as follows:

Axial loading $\qquad \sigma_a = \dfrac{P}{A}$

Torsional loading $\qquad \tau = \dfrac{T\rho}{J}$

Flexural loading $\qquad \sigma_f = \dfrac{My}{I}$

There are four possible combinations of these loadings: (1) axial and flexural; (2) axial and torsional; (3) torsional and flexural; and (4) axial, torsional, and flexural, acting simultaneously. We shall consider

358

the axial and flexural combination first because it combines only normal stresses and is therefore the simplest. All the others combine shearing and normal stresses and require a preliminary discussion (see Arts. 9–4 to 9–7) before they can be considered.

9–2   COMBINED AXIAL AND FLEXURAL LOADS

The simply supported beam in Fig. 9–1a carries a concentrated load Q. The supports are hinged to the beam at its centroidal surface. At point A, the flexural stress $\sigma_f = My/I$. It is tensile and is directed normal to the surface of the cross section, as shown. The force exerted on the element at A is $\sigma_f \, dA$.

If the same beam supported in the same way is loaded only with an axial load P (Fig. 9–1b), the axial stresses are uniformly distributed across any transverse section (Art. 1–3). Their magnitude is $\sigma_a = P/A$; they are tensile and directed normal to the cross section. The force exerted on the element at A is $\sigma_a \, dA$.

If both loads act simultaneously on the beam (Fig. 9–1c), the resultant stress at A is equal to the superposition of the two separate effects. Thus, the resultant force at A is the vector sum of the collinear forces $\sigma_a \, dA$ and $\sigma_f \, dA$. Dividing this by the area dA gives the resultant stress $\sigma = \sigma_a + \sigma_f$ directed normal to the cross section.

Similarly, at a point B in the same section, also at a distance y from the neutral axis but above it, the resultant stress is the difference between the axial and flexural stresses. If tensile stress is denoted by a positive sign and compressive stress by a negative sign, the resultant stress at any point of the beam is given by the algebraic sum of the axial and flexural stresses at that point:

$$\sigma = \sigma_a \pm \sigma_f$$

or

$$\sigma = \overset{\oplus}{\underset{\ominus}{}} \frac{P}{A} \pm \frac{My}{I} \qquad (9\text{–}1)$$

Note that the axial stress may be compressive; this is the reason for the circled $\oplus$ and $\ominus$ signs before P/A. The circling of these signs is a reminder that the axial stress is uniform and of the same type all over a cross section whereas the magnitude and type of the flexural stress vary with position.

In Eq. (9–1) we used the method of superposition. One note of caution is necessary, as the following makes clear: Fig. 9–2 shows, in exaggerated form, the flexing effect of Q. If P is tensile, as in Fig. 9–2a, the bending moment of P at any section, i.e., $P\delta$, tends to reduce the bending moment due to Q and hence slightly reduces the flexural stress.

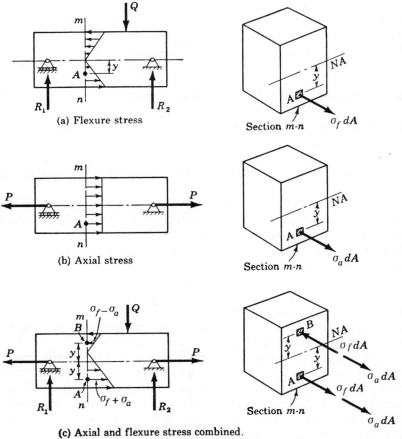

(a) Flexure stress

Section m-n $\sigma_f \, dA$

(b) Axial stress

Section m-n $\sigma_a \, dA$

(c) Axial and flexure stress combined.
Note shift in position of line of zero stress.

Figure 9–1.

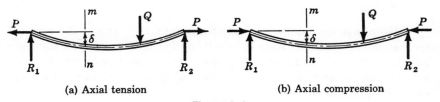

(a) Axial tension

(b) Axial compression

Figure 9–2.

The opposite effect occurs when the axial load is compressive, as in Fig. 9–2b, where the additional bending moment $P\delta$ slightly increases the flexural stress. In other words, the values given by Eq. (9–1) are slightly high when P is tensile and slightly low when P is compressive. These effects are negligible in the case of most structural members, which are usually so stiff that stresses produced by bending moments like $P\delta$ can be neglected. But in long slender members or columns, the effect is significant and more exact methods must be used. (Columns are discussed in Chapter 11.)

ILLUSTRATIVE PROBLEM

901. A cantilever beam (Fig. 9–3) has the profile shown so that it will provide sufficient clearances for large pulleys mounted on the line shaft it supports. The reaction of the line shaft is a load $P = 25$ kN. Determine the resultant normal stresses at A and B at the wall.

Solution: We begin by determining the bending moment of P. This is computed in terms of its components $P_x = 20$ kN and $P_y = 15$ kN by taking moments about the centroidal axis of section AB:

$$\left[M = (\Sigma M_{cg})_R \right]$$
$$M = -(15 \times 10^3)(0.450) + (20 \times 10^3)(0.150) = -3750 \text{ N} \cdot \text{m}$$

Since P_y acts down, its moment effect is negative (Art. 4–2) and the opposite moment effect of P_x must be positive. The negative sign of the bending moment at AB indicates that the beam curvature at section A–B is concave downward (Art. 4–2), thereby causing tension at A and compression at B. Having thus interpreted the sign of the bending moment, we use only its absolute value in applying Eq. (9–1).

However, it may not as yet be obvious that the axial tensile effect is due solely to P_x. Use the principles of mechanics to convert the given

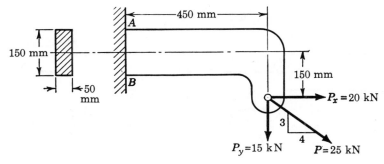

Figure 9–3.

load into either of the equivalent loadings in Fig. 9–4. It is evident from the principle of transmissibility that the entire moment effect is due to P_y in Fig. 9–4b and therefore the axial effect is caused by P_x alone. Or we may add a pair of collinear forces each equal to P_x, as in Fig. 9–4c, thereby reducing the system to that shown in Fig. 9–4d. Once again we see that the axial effect is caused by P_x, since the bending moment, which consists of $-0.450P_y$ plus the couple $0.150P_x$, is equivalent to the bending moment as computed above.

We are now ready to compute the resultant stresses by applying Eq. (9–1). At A we obtain

$$\left[\sigma = \frac{P}{A} + \left(\frac{Mc}{I} = \frac{6M}{bh^2} \right) \right]$$

$$\sigma_A = \frac{20 \times 10^3}{(0.050)(0.150)} + \frac{6(3750)}{(0.050)(0.150)^2}$$

$$= (2.67 \times 10^6) + (20.00 \times 10^6) = 22.67 \text{ MPa} \qquad Ans.$$

At B, where the flexural stress is compressive, we have

$$\left[\sigma = \frac{P}{A} - \left(\frac{Mc}{I} = \frac{6M}{bh^2} \right) \right]$$

$$\sigma_B = \frac{20 \times 10^3}{(0.50)(0.150)} - \frac{6(3750)}{(0.050)(0.150)^2}$$

$$= (2.67 \times 10^6) - (20.00 \times 10^6) = -17.33 \text{ MPa} \qquad Ans.$$

The signs indicate that the stress is tensile at A and compressive at B.

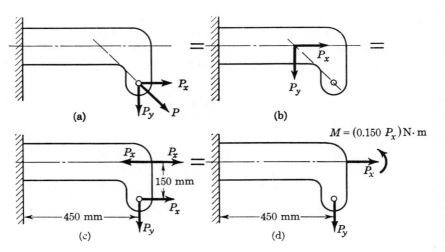

Figure 9–4.

PROBLEMS

902. Compare the maximum stress in a bent rod 10 mm square, where the load P is 10 mm off center as shown in Fig. P–902, with the maximum stress were the rod straight and the load applied axially. (*Note*: This problem illustrates why lateral deflection in columns is so dangerous.)   *Ans.*   7 to 1

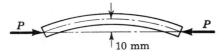

Figure P–902.

903. A cast iron link is 40 mm wide by 200 mm high by 500 mm long. The allowable stresses are 40 MN/m² in tension and 80 MN/m² in compression. Compute the largest compressive load that can be applied to the ends of the link along a longitudinal axis that is located 150 mm above the bottom of the link.

904. To avoid interference, a link in a certain machine is designed so that its cross-sectional area is reduced one-half at section $A–B$, as shown in Fig. P–904. Compute the maximum tensile stress developed across section $A–B$ if (a) the cross section of the link is 160 mm square and (b) the cross section is a circle 160 mm in diameter.
                                    Ans.   (a) 75.0 MPa; (b) 85.3 MPa

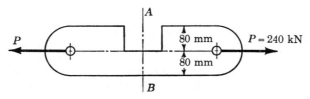

Figure P–904.

905. A wooden beam 100 mm by 200 mm, supported as shown in Fig. P–905, carries a load P. What is the largest safe value of P if the maximum stress is not to exceed 10 MPa?

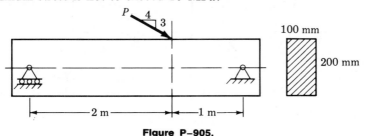

Figure P–905.

906.   The bent steel bar shown in Fig. P–906 is 200 mm square. Determine the normal stresses at A and B.

Ans. $\sigma_A = -29.2$ MPa; $\sigma_B = 4.2$ MPa

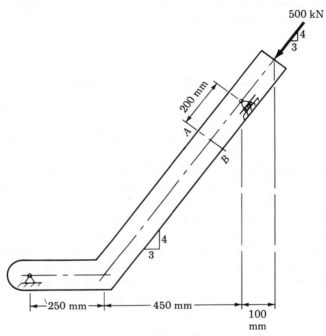

Figure P–906.

907.   Determine the largest load P that can be supported by the platform of the cast-iron bracket shown in Fig. P–907 if $\sigma_t \leqslant 30$ MN/m^2 and $\sigma_c \leqslant 70$ MN/m^2.   *Ans.* $P = 32.9$ kN

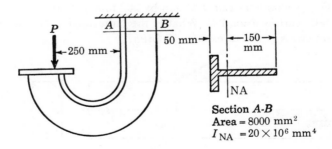

Section A-B
Area = 8000 mm^2
$I_{NA} = 20 \times 10^6$ mm^4

Figure P–907.

908.   A punch press has the cast-steel frame shown in Fig. P–908. Determine the largest force P that can be exerted at the jaws of the punch without exceeding a normal stress of 120 MPa at section A–B. The properties of the area are as shown where 1–1 is the centroidal axis.

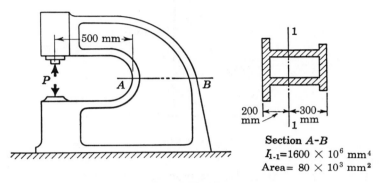

Section A-B
$I_{1\text{-}1} = 1600 \times 10^6$ mm^4
Area $= 80 \times 10^3$ mm^2

Figure P–908.

909.   A rectangular beam, 100 mm wide by 400 mm deep, is pinned at A, supported by a cable CD, and carries a load P, as shown in Fig. P–909. Determine the maximum value of P that will not exceed a normal stress of 120 MPa. Neglect the possibility of buckling.

Ans. $P = 457$ kN

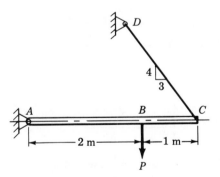

Figure P–909.

910.   The inclined beam in Fig. P–910 is supported by a pin at A and a roller at C. The cross section is 100 mm by 300 mm. Determine the maximum compressive stress developed in the beam.

Ans. $\sigma_c = 70.7$ MPa

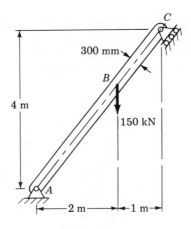

Figure P–910.

911. If $P = 100$ kN for the bracket shown in Fig. P–911, compute the maximum tensile and compressive stresses developed at section $A–B$.

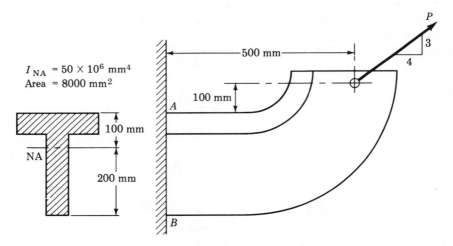

Figures P–911 and P–912.

912. Determine the maximum safe load P that can be applied to the bracket in Fig. P–911 if the allowable stresses at section $A–B$ are $\sigma_t \leq 8$ MPa and $\sigma_c \leq 12$ MPa.   *Ans.* $P = 12.1$ kN

913. Compute the stresses at A and B for the link loaded as shown in Fig. P–913.   *Ans.* $\sigma_A = -14.1$ MPa; $\sigma_B = 5.10$ MPa

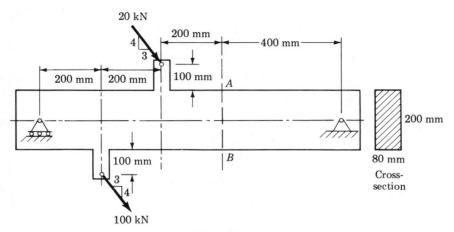

Figure P–913.

914. A timber beam AD, 100 mm thick by 300 mm high and loaded as shown in Fig. P–914, is pinned at its lower end and supported by a horizontal cable CE. Determine the maximum compressive stress in the beam.   *Ans.* $\sigma = 14.3$ MPa

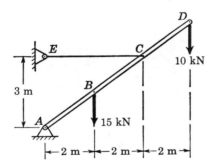

Figure P–914.

915. A concrete dam has the profile shown in Fig. P–915. If the density of concrete is 2400 kg/m^3 and that of water is 1000 kg/m^3, determine the maximum compressive stress on section m–n if the depth of the water behind the dam is $h = 15$ m.

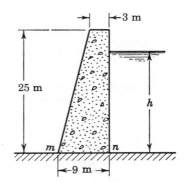

Figure P–915.

916. For the pin-connected frame shown in Fig. P–916, determine the maximum normal stress in member *BD* if its cross section is 100 mm wide by 400 mm deep. Neglect the weights of the members.

Ans. $\sigma = 25.4$ MPa

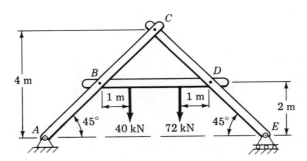

Figure P–916.

917. The structure shown in Fig. P–917 is hinged to fixed supports at *A* and *E*. Compute the maximum compressive stress developed in bar *BDE* if its cross section is 200 mm square. Neglect the weights of the members.

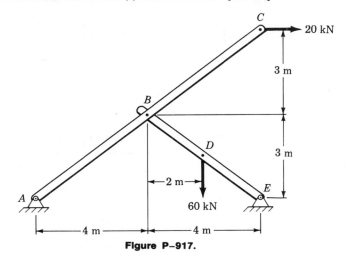

Figure P–917.

9–3   KERN OF A SECTION; LOADS APPLIED
OFF AXES OF SYMMETRY

A special case of combined axial and flexural loads is illustrated in Fig.
9–5a, in which a short strut* carries a compressive load P applied with
an eccentricity e along one of the principal axes† of the section. The
addition of a pair of forces P_1 and P_2, each of magnitude P and acting
at the centroid of the section, causes the equivalent loading shown in
Fig. 9–5b. The stresses across any typical section m–n are the result of
the superposition of the direct compressive stress ($\sigma_a = P/A$) in Fig.
9–5c and the flexural stress ($\sigma_f = Mc/I = (Pe)c/I$) in Fig. 9–5d. If the
maximum flexural stress is larger than the direct compressive stress,
the resultant stress appears as in Fig. 9–5e. The point of zero stress N is
the new location of the neutral axis and is easily found by computing
the distance a at which the tensile flexural stress equals the direct
compressive stress:

$$\frac{P}{A} = \frac{My}{I} = \frac{(Pe)a}{I}$$

whence

$$a = \frac{I}{Ae} \tag{9–2}$$

*A short strut is one whose length is no more than ten times its smallest
lateral dimension; the flexural deflection is so small that its effect can be neglected.
The eccentric loading of long bars is discussed in Art. 11–6.
  †The principal axes are the axes of maximum and minimum moments of
inertia.

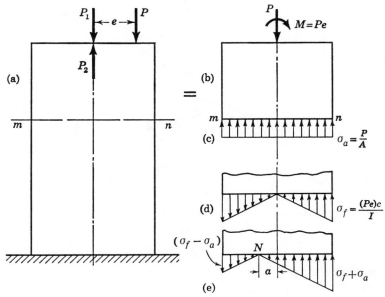

Figure 9–5.

It is evident that there will be no tensile stress anywhere over the section if the direct compressive stress equals or exceeds the maximum flexural stress. Thus, for a rectangular section of dimensions b and h, with P applied at an eccentricity e (Fig. 9–6), we obtain

$$\frac{P}{A} = \frac{Mc}{I} = \frac{Pe\left(\dfrac{h}{2}\right)}{\dfrac{bh^3}{12}}$$

The maximum eccentricity to avoid tension is thus

$$e = \frac{h}{6} \tag{9–3}$$

This formula is the basis of the well-known rule that, in designing masonry or other structures weak in tension, the resultant load should fall in the middle third of the section.

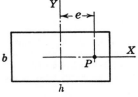

Figure 9–6.

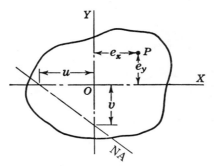

Figure 9–7.

We now consider the general case* in which the load P is applied at any point with respect to the principal axes X and Y, as in Fig. 9–7. If e_x and e_y represent the eccentricities of P, the moments of P with respect to the X and Y axes are respectively Pe_y and Pe_x. By superposition, the stress at any point of the cross section whose coordinates are x and y is

$$\sigma = -\frac{P}{A} - \frac{(Pe_x)x}{I_y} - \frac{(Pe_y)y}{I_x} \tag{9-4}$$

To determine the neutral axis or line of zero stress in the cross section, we set $\sigma = 0$. Using $I_y = Ar_y^2$ and $I_x = Ar_x^2$, where r_y and r_x are respectively the radii of gyration† relative to the Y and X axes, we obtain

$$0 = 1 + \frac{e_x}{r_y^2}x + \frac{e_y}{r_x^2}y \tag{a}$$

The intercepts u and v of the neutral axis with the X and Y axes respectively are found by substituting first $y = 0$ and then $x = 0$ in Eq. (a). This gives

$$u = -\frac{r_y^2}{e_x} \quad \text{and} \quad v = -\frac{r_x^2}{e_y} \tag{b}$$

The neutral axis passes through the quadrant which is opposite to that containing P, and in general it is *not* perpendicular to the direction OP. For example, the stress distribution on a rectangular section caused by a load P not on the principal axes (Fig. 9–8a) is shown in Fig. 9–8b. If the stresses at A, B, and C are computed, the intersection of the

---

*Actually, this is an application of unsymmetrical bending, which is discussed at length in Art. 13–9.

†Here we are using r to denote radius of gyration to conform to AISC notation. Be careful not to confuse this r with the r that is frequently used to denote the radius of a circle.

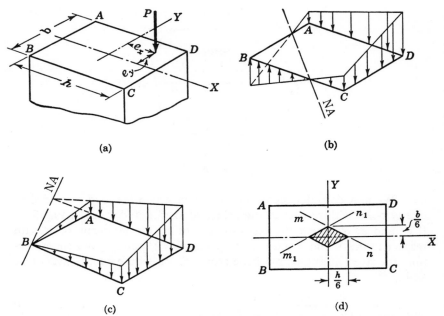

(a)

(b)

(c)

(d)

Figure 9–8. Neutral axis for load P eccentrically applied and kern of rectangular section.

neutral axis with AB and BC (or their extensions) can be easily determined by proportion.

Let us now determine the coordinates e_x and e_y of the load P, for which the neutral axis passes through the corner B, as in Fig. 9–8c. Substituting $\sigma = 0$, $x = -h/2$, and $y = -b/2$ in Eq. (9–4), we obtain

$$0 = -\frac{P}{bh} + \frac{(Pe_x)\left(\dfrac{h}{2}\right)}{\dfrac{bh^3}{12}} + \frac{(Pe_y)\left(\dfrac{b}{2}\right)}{\dfrac{hb^3}{12}}$$

or

$$\frac{e_x}{h/6} + \frac{e_y}{b/6} = 1 \tag{c}$$

This is the equation of the straight line mn in Fig. 9–8d; it intersects the X and Y axes at $h/6$ and $b/6$, respectively. This line is the locus of points of application of P, for which corner B has zero stress. Any compressive load above and to the right of this line causes tension at B. Similarly the line m_1n_1 is the locus of loads that cause zero stress in corner C. Continuing this procedure indicates that evidently no corner and no part of the cross section will be in tension if the resultant compressive load lies on or within the diamond-shaped figure. This shaded area is known as the *kern* of the cross section.

Show that the kern of a circular section is a circle whose diameter is one-quarter the diameter of the section.

PROBLEMS

918. A compressive load $P = 80$ kN is applied, as shown in Fig. 9–8a, at a point 40 mm to the right and 60 mm above the centroid of a rectangular section for which $h = 400$ mm and $b = 200$ mm. Compute the stress at each corner and the location of the neutral axis. Illustrate your answers with a sketch similar to Fig. 9–8b.

919. From the data in Problem 918, what additional load applied at the centroid is necessary so that no tensile stress exists anywhere on the cross section?   *Ans.* 112 kN

920. A compressive load $P = 100$ kN is applied, as shown in Fig. 9–8a, at a point 70 mm to the left and 30 mm above the centroid of a rectangular section for which $h = 300$ mm and $b = 150$ mm. What additional load, acting normal to the cross section at its centroid, will eliminate tensile stress anywhere over the cross section? *Ans.* 160 kN

921. Calculate and sketch the kern of a W360 × 122 section.
        Ans. A diamond-shaped figure having coordinates on the
        X axis of ±30.9 mm and on the Y axis of ±130 mm

922. Calculate and sketch the kern of a W310 × 500 section.

9–4 VARIATION OF STRESS WITH INCLINATION OF ELEMENT

In Art. 1–2 we saw that the magnitude and type of stress depend on the inclination of an element. As a review of that discussion, consider that the body in Fig. 9–9a is acted upon by the given forces which are in

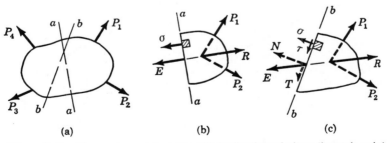

                (a)   (b)   (c)

Figure 9–9. Stress at a point varies with inclination of plane through point.

equilibrium. Pass two exploratory sections a–a and b–b through the body, section a–a perpendicular to the resultant R of P_1 and P_2 as in Fig. 9–9b, and section b–b inclined to R as in Fig. 9–9c. The element in Fig. 9–9b is subjected only to a normal stress, but the element in Fig. 9–9c is subjected to both normal and shearing stress caused by the N and T components of the equilibrant E. Thus at the same position in a stressed body (located here at the intersection of sections a–a and b–b), the stresses on an element vary with the orientation of the element.

In the following articles we discuss the manner in which these stresses vary with the orientation of the element. Our purpose will be to determine the orientation of the element on which the maximum normal stress exists, and its magnitude; also the orientation of the element on which maximum shearing stress exists, and its magnitude.

In general, it is not possible to compute directly the stresses that exist on any arbitrarily chosen surface. In beams, for example, the flexure formula determines stresses only on planes normal to the longitudinal axis of the beam. So also for torsional shearing stresses: the torsion formula determines shearing stresses only on sections normal to the longitudinal axis of a twisted bar. Thus, in a bar subjected to simultaneous bending and twisting as in Fig. 9–10, we can compute the flexural and shearing stresses only for elements in the position shown. But the discussion of Fig. 9–9 indicates that if the element in Fig. 9–10 were rotated about the axis shown, there would be a particular position at which maximum normal stress would exist.

There are two methods of determining this position of the element and computing the maximum stresses to which it is subjected: one is analytical; the other is graphical, based on *Mohr's circle*. The analytical

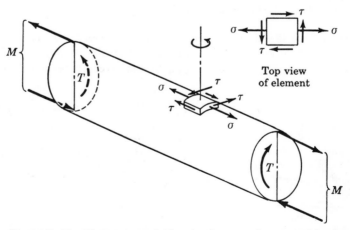

Figure 9–10. Stresses caused by simultaneous flexure and torsion.

discussion in Art. 9–6 is given primarily to demonstrate the construction and validity of Mohr's circle, which is discussed at length in Art. 9–7. (The formulas in Art. 9–6 should *not* be memorized.)

9–5   STRESS AT A POINT

The average stress over an area is obtained by dividing the force by the area over which it acts. If the average stress is constant over the area, the stress is said to be uniform. If the stress is not uniform, the stress at any point is found by permitting the area enclosing the point to approach zero as a limit. In other words, *stress at a point* really defines the uniform stress distributed over a differential area. In Fig. 9–11, for example, the normal stress in an X direction acting at the point whose coordinates are x, y and z means the uniform stress acting over the differential area $dy\ dz$.

When the stress at a point is defined by components acting in several directions, the stresses may be represented as acting on the differential element (i.e., volume) enclosing the point. For example, let the stresses at a point be σ_x, σ_y, and τ_{xy}; Fig. 9–12a shows these stresses as they act on the differential element enclosing the point. The element is usually represented by its front view, as in Fig. 9–12b. Note that there

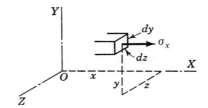

Figure 9–11.   Stress at a point.

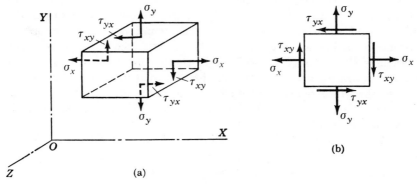

(a)

(b)

Figure 9–12.   Stress components.

is a shearing stress τ_{yx} acting on the Y face in the X direction. This is due to the fact that a shearing stress on any plane induces an equal shearing stress on a plane perpendicular to the first one. (See Art. 5–7, page 188.)

The notation used here defines a *normal* stress by means of a single subscript corresponding to the face on which it acts. A face takes the name of the axis normal to it, for example, the X face is perpendicular to the X axis. A *shearing* stress is denoted by a double subscript, the first letter corresponding to the face on which the shearing stress acts and the second indicating the direction in which it acts. Thus the shearing stress on the X face acting in the Y direction is denoted by τ_{xy}, and the shearing stress on the Y face acting in the X direction is denoted by τ_{yx}. Of course $\tau_{xy} = \tau_{yx}$, since the shearing stresses on perpendicular planes are equal.

In this and succeeding articles, we consider only plane stress in which the stresses act parallel to a single plane such as the XY plane. In a triaxial stress, the Z face of an element may be subject to the normal stress σ_z as well as to shearing stresses τ_{zx} and τ_{zy}. These shearing stresses then induce the numerically equal shearing stresses τ_{xz} and τ_{yz} which act, respectively, on the X and Y faces.

9–6   VARIATION OF STRESS AT A POINT: ANALYTICAL DERIVATION

The stress acting at a point is represented by the stresses acting on the faces of the element enclosing the point. As we saw in Art. 9–4, the stresses change with the inclination of the planes passing through that point; that is, the stresses on the faces of the element vary as the angular position of the element changes.

In determining these stress variations analytically, a plane is passed that cuts the original element into two parts and the conditions of equilibrium are applied to either part. Figure 9–13b shows the normal and shearing stress components acting on the plane whose normal N makes an angle θ with the X axis (see Fig. 9–13a). The triangular element in Fig. 9–13b is in equilibrium under the action of the forces arising from the stresses that act over its faces. The area of the inclined face being denoted by A, these forces are shown in the free-body diagram in Fig. 9–13c. The point diagram of these forces is shown in Fig. 9–13d.

Applying the conditions of equilibrium to axes chosen as in Fig. 9–13d, we obtain

$$[\Sigma N = 0] \quad A\sigma = (\sigma_x A \cos \theta) \cos \theta + (\sigma_y A \sin \theta) \sin \theta$$
$$- (\tau_{xy} A \cos \theta) \sin \theta - (\tau_{yx} A \sin \theta) \cos \theta \quad (a)$$

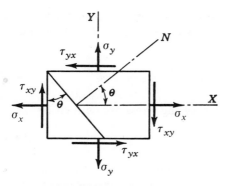

(a) Original state of stress

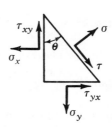

(b) Stresses acting on wedge

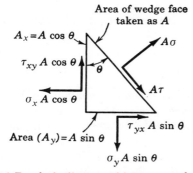

(c) Free-body diagram of forces on wedge

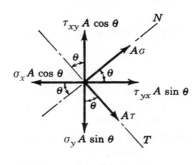

(d) Point diagram of forces

Figure 9–13. Variation of stress components.

and

$$[\Sigma T = 0] \qquad A\tau = (\sigma_x A \cos\theta)\sin\theta - (\sigma_y A \sin\theta)\cos\theta$$
$$+ (\tau_{xy} A \cos\theta)\cos\theta - (\tau_{yx} A \sin\theta)\sin\theta \qquad (b)$$

Since the common term A can be canceled and since τ_{yx} is numerically equal to τ_{xy}, we use the relations

$$\cos^2\theta = \frac{1 + \cos 2\theta}{2}, \quad \sin^2\theta = \frac{1 - \cos 2\theta}{2},$$

$$\sin\theta\cos\theta = \frac{1}{2}\sin 2\theta$$

to reduce Eqs. (a) and (b) to

$$\sigma = \frac{\sigma_x + \sigma_y}{2} + \frac{\sigma_x - \sigma_y}{2}\cos 2\theta - \tau_{xy}\sin 2\theta \qquad (9\text{–}5)$$

and

$$\tau = \frac{\sigma_x - \sigma_y}{2}\sin 2\theta + \tau_{xy}\cos 2\theta \qquad (9\text{–}6)$$

The planes defining maximum or minimum normal stresses are found by differentiating Eq. (9–5) with respect to θ and setting the derivative equal to zero, whence

$$\tan 2\theta = -\frac{2\tau_{xy}}{\sigma_x - \sigma_y} \qquad (9\text{–}7)$$

Similarly, the planes of maximum shearing stress are defined by

$$\tan 2\theta_s = \frac{\sigma_x - \sigma_y}{2\tau_{xy}} \qquad (9\text{–}8)$$

Equation (9–7) gives two values of 2θ that differ by 180°. Hence, the planes on which maximum and minimum normal stresses occur are 90° apart. Similarly, from Eq. (9–8) the planes on which the maximum shearing stress occurs are also found to be 90° apart.

The planes of zero shearing stress may be determined by setting τ equal to zero in Eq. (9–6); this gives

$$\tan 2\theta = -\frac{2\,\tau_{xy}}{\sigma_x - \sigma_y}$$

which is identical with Eq. (9–7). Hence *maximum and minimum normal stresses occur on planes of zero shearing stress*. The maximum and minimum normal stresses are called the *principal stresses*, sometimes referred to as the *p* and *q* stresses.

Equation (9–8) is also the negative reciprocal of Eq. (9–7). This means that the values of 2θ defined by Eqs. (9–7) and (9–8) differ by 90°. In other words, *the planes of maximum shearing stress are at 45° with the planes of principal stress*.

Substituting values of 2θ from Eqs. (9–7) and (9–8) respectively in Eqs. (9–5) and (9–6), we obtain the following expressions for maximum stresses:

$$(\sigma)_{\substack{\text{max.}\\ \text{min.}}} = \frac{\sigma_x + \sigma_y}{2} \pm \sqrt{\left(\frac{\sigma_x - \sigma_y}{2}\right)^2 + (\tau_{xy})^2} \qquad (9\text{–}9)$$

$$(\tau)_{\text{max.}} = \pm \sqrt{\left(\frac{\sigma_x - \sigma_y}{2}\right)^2 + (\tau_{xy})^2} \qquad (9\text{–}10)$$

If a similar analysis is made for a plane whose normal is perpendicular to N, the stress components on this perpendicular plane will be

$$\sigma' = \frac{\sigma_x + \sigma_y}{2} - \frac{\sigma_x - \sigma_y}{2}\cos 2\theta + \tau_{xy}\sin 2\theta \qquad (9\text{–}5a)$$

$$\tau' = \frac{\sigma_x - \sigma_y}{2}\sin 2\theta + \tau_{xy}\cos 2\theta \qquad (9\text{–}6a)$$

Adding Eqs. (9–5) and (9–5a) shows that the sum of the normal stresses on any two perpendicular planes is a constant equal to $\sigma_x + \sigma_y$. Also, comparison of Eqs. (9–6) and (9–6a) confirms the equivalence of shearing stress on perpendicular planes.

9–7 MOHR'S CIRCLE

The formulas developed in the preceding article may be used for any case of two-dimensional stress. A visual interpretation of them, devised by the German engineer Otto Mohr in 1882, eliminates the necessity for remembering them.* In this interpretation a circle is used; accordingly, the construction is called Mohr's circle. If this construction is plotted to scale, the results can be obtained graphically; usually, however, only a rough sketch is drawn, analytical results being obtained from it by following the rules given later.

We can easily show that Eqs. (9–5) and (9–6) define a circle by first rewriting them as follows:

$$\sigma - \frac{\sigma_x + \sigma_y}{2} = \frac{\sigma_x - \sigma_y}{2}\cos 2\theta - \tau_{xy}\sin 2\theta \qquad (a)$$

$$\tau = \frac{\sigma_x - \sigma_y}{2}\sin 2\theta + \tau_{xy}\cos 2\theta \qquad (b)$$

By squaring both these equations, adding the results, and simplifying, we obtain

$$\left(\sigma - \frac{\sigma_x + \sigma_y}{2}\right)^2 + \tau^2 = \left(\frac{\sigma_x - \sigma_y}{2}\right)^2 + (\tau_{xy})^2 \qquad (c)$$

Recall that σ_x, σ_y, and τ_{xy} are known constants defining the specified state of stress, whereas σ and τ are variables. Consequently $(\sigma_x + \sigma_y)/2$ is a constant, say C, and the right-hand member of Eq. (c) is another constant, say R. Using these substitutions, we transform Eq. (c) into

$$(\sigma - C)^2 + \tau^2 = R^2 \qquad (d)$$

---

*Equations (9–5) and (9–6), as well as the succeeding variations of them, are identical to the equations that express the variations in moments of inertia with respect to U and V axes inclined at an angle θ to the reference axes X and Y. Replacing normal stress by the moment of inertia I and the shearing stress by the product of inertia P, we obtain

$$I_u = \frac{I_x + I_y}{2} + \frac{I_x - I_y}{2}\cos 2\theta - P_{xy}\sin 2\theta$$

and

$$P_{uv} = \frac{I_x - I_y}{2}\sin 2\theta + P_{xy}\cos 2\theta$$

A Mohr's circle treatment of these equations is described in detail in Appendix A, p. 627.

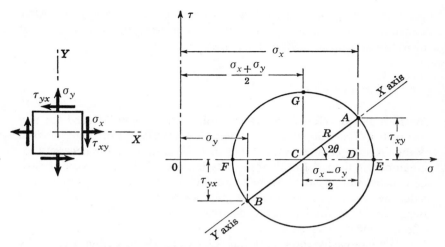

Figure 9–14. Mohr's circle for general state of stress.

This, being of the form $(x - C)^2 + y^2 = R^2$, is readily recognized as a circle of radius

$$R = \sqrt{\left(\frac{\sigma_x - \sigma_y}{2}\right)^2 + (\tau_{xy})^2}$$

whose center is offset rightward a distance

$$C = \frac{\sigma_x + \sigma_y}{2}$$

from the origin.

Figure 9–14 represents Mohr's circle for the state of stress which was analyzed in the preceding article. The center C is the average of the normal stresses, and the radius

$$R = \sqrt{\left(\frac{\sigma_x - \sigma_y}{2}\right)^2 + (\tau_{xy})^2}$$

is the hypotenuse of the right triangle CDA. How do the coordinates of points E, F, and G compare with the expressions derived in Eqs. (9–9) and (9–10)? We shall see that Mohr's circle is a graphic visualization of the stress variation given by Eqs. (9–5) and (9–6). The following rules summarize the construction of Mohr's circle.

Rules for applying Mohr's circle to combined stresses

1. On rectangular $\sigma - \tau$ axes, plot points having the coordinates (σ_x, τ_{xy}) and (σ_y, τ_{yx}). These points represent the normal and shearing stresses acting on the X and Y faces of an element for which the stresses

are known. In plotting these points, assume tension as plus, compression as minus, and shearing stress as plus when its moment about the center of the element is clockwise.*

2. Join the points just plotted by a straight line. This line is the diameter of a circle whose center is on the σ axis.

3. As different planes are passed through the selected point in a stressed body, the normal and shearing stress components on these planes are represented by the coordinates of points whose position shifts around the circumference of Mohr's circle.

4. The radius of the circle to any point on its circumference represents the axis directed normal to the plane whose stress components are given by the coordinates of that point.

5. The angle between the radii to selected points on Mohr's circle is twice the angle between the normals to the actual planes represented by these points, or to twice the space angularity between the planes so represented. The rotational sense of this angle corresponds to the rotational sense of the actual angle between the normals to the planes; i.e., if the N axis is actually at a counterclockwise angle θ from the X axis, then on Mohr's circle the N radius is laid off at a counterclockwise angle 2θ from the X radius.

ILLUSTRATIVE PROBLEMS

923. At a certain point in a stressed body, the principal stresses are $\sigma_x = 80$ MPa and $\sigma_y = -40$ MPa. Determine σ and τ on the planes whose normals are at $+30°$ and $+120°$ with the X axis. Show your results on a sketch of a differential element.

Solution: The given state of stress is shown in Fig. 9–15a. Following the rules given above, draw a set of rectangular axes and label them σ and τ as shown in Fig. 9–15b. (Note that, for convenience, the stresses are plotted in units of MPa.) Since the normal stress component on the X face is 80 MPa and the shear stress on that face is zero, these components are represented by point A which has the coordinates (80, 0). Similarly, the stress components on the Y face are represented by point B (-40, 0).

According to rule 2, the diameter of Mohr's circle is AB. Its center C, lying midway between A and B, is 20 MPa from the origin O. The

---

*This special rule of sign for shearing stress makes $\tau_{xy} = -\tau_{yx}$ in Mohr's circle. From here on, we shall use this rule to designate positive shearing stress. However, the mathematical theory of elasticity uses the convention that shearing stress is positive when directed in the positive coordinate direction on a positive face of an element, i.e., when acting upward on the right face or rightward on the upper face. This other rule makes $\tau_{xy} = \tau_{yx}$, which is convenient for mathematical work but confusing when applied to Mohr's circle.

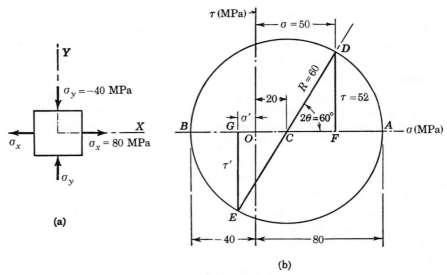

(a)

(b)

Figure 9–15.

radius of the circle is the distance $CA = 80 - 20 = 60$ MPa. From rule 4, the radius CA represents the X axis. In accordance with rules 4 and 5, point D represents the stress components on the face whose normal is inclined at $+30°$ to the X axis, and point E represents the stress components on the perpendicular face. Observe that positive angles on the circle are plotted in a counterclockwise direction from the X axis and are double the angles between actual planes.

From rule 3, the coordinates of point D represent the required stress components on the $30°$ face. From the geometry of Mohr's circle, these values are

$$\sigma = OF = OC + CF = 20 + 60 \cos 60° = 50 \text{ MPa}$$
$$\tau = DF = 60 \sin 60° = 52.0 \text{ MPa}$$

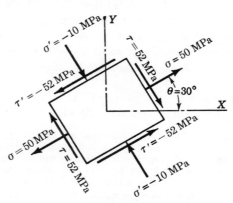

Figure 9–16.

On the perpendicular 120° face we have

$$\sigma' = OG = OC - CG = 20 - 60 \cos 60° = -10 \text{ MPa}$$
$$\tau' = GE = -60 \sin 60° = -52.0 \text{ MPa}$$

Both sets of the above stress components are shown on the differential element in Fig. 9–16. Observe the clockwise and counterclockwise moments of τ and τ', respectively, relative to the center of the element (see rule 1). Finally, note that a complete sketch of a differential element shows the stress components acting on all four faces of the element and the angle at which the element is inclined.

924. A state of stress is specified in Fig. 9–17a. Determine the normal and shearing stresses on (a) the principal planes, (b) the planes of maximum shearing stress, and (c) the planes whose normals are at $+36.8°$ and $+126.8°$ with the X axis. Show the results of parts a and b on complete sketches of differential elements.

Solution: Mohr's circle for the given state of stress is shown in Fig. 9–17b. The stresses on the X face are represented by point A, which has an abscissa of 32 and a negative ordinate of 20. τ_{xy} is considered negative because its moment sense is counterclockwise about the center of the element in Fig. 9–17a. The stresses on the Y face are given by point B, which has an abscissa of -10 (negative because σ_y is compressive) and an ordinate of $+20$ (positive because the moment sense of τ_{yx} is clockwise). Joining A and B gives the diameter of Mohr's circle, its center C being midway between the abscissae of A and B, or at 11 MPa from the origin O. Hence the radius R is computed from the right triangle whose sides are 21 and 20; the radius is 29.0 MPa.

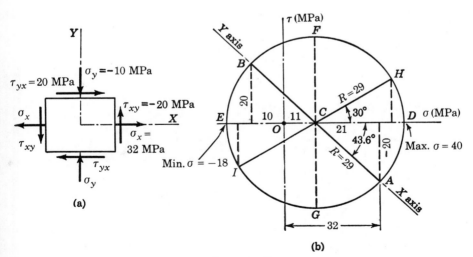

(a)

(b)

Figure 9–17.

The principal stresses are represented by points D and E, where the shearing stress coordinates are zero. From the geometry of the circle, we obtain

Max. $\sigma = OD = 11 + 29 = +40$ MPa
Min. $\sigma = OE = 11 - 29 = -18$ MPa

The radius to D makes a counterclockwise angle 2θ measured from the radius CA, which denotes the X axis. From the circle, we see that $\tan 2\theta = 20/21 = 0.952$, and hence $2\theta = 43.6°$ and $\theta = 21.8°$. The principal stresses and principal planes are as shown on the differential element in Fig. 9–18a. There is, of course, no shearing stress on the principal planes.

The stresses on the planes of maximum shearing stress are given by the coordinates of points F and G, the values being max. $\tau = 29$ MPa and min. $\tau = -29$ MPa; the normal stress on each plane is $+11$ MPa. The radius CF is 90° counterclockwise from CD, so the normal to the plane of maximum shearing stress is 45° counterclockwise from the maximum principal plane, or at $45° + 21.8° = 66.8°$ with the X axis. These results are shown on the element in Fig. 9–18b.

To complete the solution, the stresses on the plane whose normal is at $+36.8°$ with the X axis are represented by point H, located at the intersection of the radius CH with Mohr's circle (see rule 3). From rule 5, the angle between the normals to any two faces is laid off double size on the circle; hence, angle $ACH = 2 \times 36.8 = 73.6°$ and angle $HCD = 73.6 - 43.6 = 30°$. Therefore the coordinates of point H are

$\sigma = 11 + 29 \cos 30° = 36.1$ MPa
$\tau = 29 \sin 30° = 14.5$ MPa

The stresses on the plane whose normal is at $+126.8°$ with the X axis are represented by point I. Points H and I are 180° apart on the circle since the planes they represent are actually 90° apart. The coordinates of point I are

$\sigma' = 11 - 29 \cos 30° = -14.1$ MPa
$\tau' = -29 \sin 30° = -14.5$ MPa

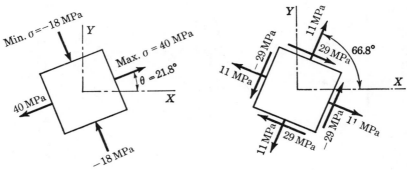

(a) Principal stresses   (b) Maximum shear stresses

Figure 9–18.

The student should show the above stress components on a complete sketch of a differential element.

PROBLEMS

925. Two wooden joists 50 mm × 100 mm are glued together along the joint AB as shown in Fig. P–925. Determine the normal stress and shearing stress in the glue if $P = 100$ kN.   *Ans.* $\sigma = 15$ MPa

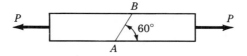

Figure P–925.

926. A short 50-mm-diameter circular bar is made of a material for which the maximum safe compressive stress is 80 MN/m² and the maximum safe shearing stress is 30 MN/m². Determine the maximum axial compressive force which can be safely applied to the bar.

927. An element is subjected to the principal stresses $\sigma_x = -50$ MPa and $\sigma_y = 30$ MPa. Compute the stress components on planes whose normals are at $+30°$ and $+120°$ with the X axis. Show your answers on a complete sketch of a differential element.

928. A small block is 50 mm long, 30 mm high, and 10 mm thick. The block is subjected to uniformly distributed tensile forces having the resultant values shown in Fig. P–928. Determine the stress components developed along the diagonal AB.
                    Ans. $\sigma = 83.9$ MPa; $\tau = -39.7$ MPa

929. Solve Problem 928 assuming that the 30 kN forces are reversed to act in compression.

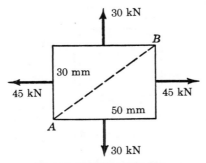

Figures P–928 and P–929.

930. A closed cylindrical tank is fabricated from 10-mm plate and subjected to an internal pressure of 1400 kPa. Determine the maximum diameter if the maximum shearing stress is limited to 30 MPa. [*Hint:* In a closed-end thin-walled cylindrical pressure vessel, the longitudinal stress is given by $pD/4t$ and the circumferential stress is $pD/2t$ (see Art. 1–6). Also, refer to Problem 941.]

931. For the state of stress shown in Fig. P–931, determine the principal stresses and the maximum shearing stress. Show all results on complete sketches of differential elements.

Ans.   Max. $\sigma = 90$ MPa at $\theta = -18.4°$

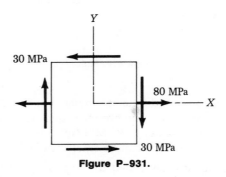

Figure P–931.

932. If a point is subjected to the state of stress shown in Fig. P–932, determine the principal stresses and the maximum shearing stress. Show all results on complete sketches of differential elements.

Ans.   Max. $\sigma = 58.1$ MPa at $\theta = +19.9°$;
Min. $\sigma = -98.1$ MPa at $\theta = +109.9°$

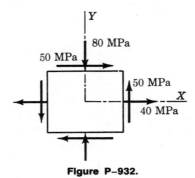

Figure P–932.

933. For the state of stress shown in Fig. P–933, determine normal and shearing stresses on the planes whose normals are at $+60°$ and $+150°$ with the X axis. Show these stresses on a sketch of the element. *Ans.* For $\theta = 60°$, $\sigma = 7.32$ MPa and $\tau = 10$ MPa

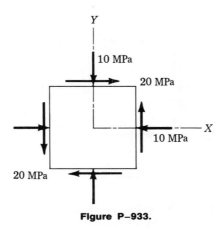

Figure P–933.

934. If an element is subjected to the state of stress shown in Fig. P–934, find the principal stresses and the maximum shearing stresses. Also determine the stress components on planes whose normals are at 45° and 135° with the X axis. Show all results on complete sketches of the appropriate elements.

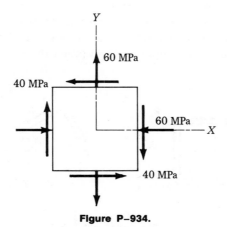

Figure P–934.

935. For the element shown in Fig. P–935, determine the values of σ_x and σ_y if the principal stresses are known to be 20 MPa and −80 MPa.

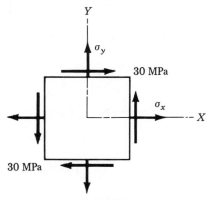

Figure P–935.

936. A tube with an external diameter of 150 mm is fabricated from 10-mm plate. The spiral weld used makes an angle of +30° with the longitudinal axis. Determine the maximum torque that can be applied if the shearing stress along the weld is limited to 30 MN/m².

Ans. $T = 17.3$ kN·m

937. A closed-end cylindrical pressure vessel having an outside diameter of 600 mm and fabricated from 10-mm steel plate is subjected to an internal pressure of 1400 kPa. Determine the normal and shearing stresses on the spiral weld used which makes an angle of +30° with the longitudinal axis.

938. The state of stress at a point is the result of two separate actions: one produces the pure shear of 35 MPa shown in part (a) of Fig. P–938, and the other produces the pure shear of 30 MPa shown in part (b). Find the resultant stress by rotating the state of stress in part

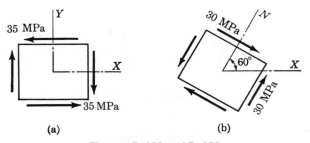

Figures P–938 and P–939.

(b) to coincide with that in part (a) so that the stresses can be superposed and added directly. Then determine the principal stresses and principal planes.   *Ans.*   Max. $\sigma = 32.9$ MPa at $\theta = -18.8°$

939. Solve Problem 938 assuming that the directions of the 30 MPa shearing stresses in Fig. P–938b are reversed.

940. The state of stress at a point is the result of three separate actions that produce the three states of stress shown in Fig. P–940. Determine the principal stresses and principal planes caused by the superposition of these three stress states.

Ans.   Max. $\sigma = 37.8$ MPa at $\theta = 26.2°$

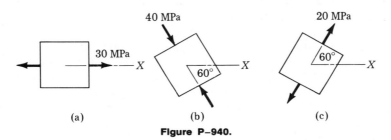

(a)   (b)   (c)

Figure P–940.

941. The principal stresses on an element are σ_x, σ_y, and σ_z. Assuming $\sigma_x > \sigma_y > \sigma_z$, show that the maximum shearing stress on *any* plane through the element is equal to $\frac{1}{2}(\sigma_x - \sigma_z)$. (*Hint:* Consider all orientations of the element by drawing three separate Mohr's circles, each representing a rotation about one of the three principal axes.)

942. A state of plane stress is defined by $\sigma_x = 20$ MN/m², $\sigma_y = 40$ MN/m², and $\tau_{xy} = 20$ MN/m². Determine the maximum shearing stress on *any* plane through the stressed point. (*Hint:* First find the principal stresses and then apply the results of Problem 941.)

Ans.   Max. $\tau = 26.2$ MN/m²

9-8   APPLICATIONS OF MOHR'S CIRCLE
TO COMBINED LOADINGS

The most important use of combined stresses is in the design of members subjected to combined loadings, or the determination of safe loads. Here Mohr's circle makes possible a visualization of conditions that is superior to mere analytical manipulation. The usual procedure is to consider an element on which the effect of the three fundamental loadings—axial, torsional, and flexural—can be computed. A study of Mohr's circle for this element indicates the design criteria. The illustrative problems at the end of this article are typical of the procedures involved.

Stress trajectories

An element on the surface of the cylinder in Fig. 9–19a is subjected to the indicated torsional shearing stress. Fig. 9–19b shows Mohr's circle for this state of stress. The radius OA specifies the X axis. The maximum tensile stress is denoted by point D, whose radius OD is 90° clockwise from OA. Hence the normal to the plane of maximum tensile stress is 45° clockwise from the X axis, as shown in Fig. 9–19a. The lines in Fig. 9–20 that follow the directions of the principal stresses are called *stress trajectories*. For torsion, they are 45° helices. If the material is weakest in tension, as is common for brittle materials, failure occurs along a 45° helix such as AB. This may be confirmed experimentally by twisting a piece of chalk until it breaks.

Further visualization of the tensile and compressive stresses induced by pure shear is provided in Fig. 9–21. The distorted appearance of the element $ABCD$, originally rectangular, indicates that the diagonal AC has been lengthened and BD shortened. These deformations agree with the directions of the tensile and compressive stresses previously obtained.

In beams, the directions of the principal stresses vary with the intensities of the flexural stress σ_f and the horizontal shearing stress τ. For example, at point A of the cantilever beam in Fig. 9–22, Mohr's circle shows that the direction of the principal compressive stress makes a clockwise angle θ with the X axis; the principal tensile stress is at right

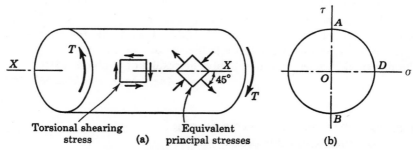

Torsional shearing stress **(a)** Equivalent principal stresses **(b)**

Figure 9–19. Cylinder subjected to torsion.

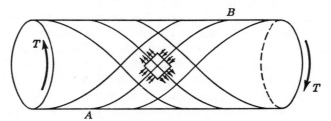

Figure 9–20. Stress trajectories for torsion.

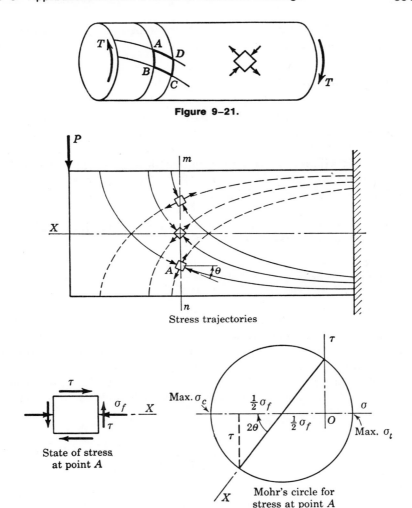

Figure 9–21.

Stress trajectories

State of stress
at point A

Mohr's circle for
stress at point A

Figure 9–22.

angles. The value of θ varies with the ratio τ/σ_f (i.e., $\tan 2\theta = 2\tau/\sigma_f$). At the extreme fibers m and n of the section through A, τ is zero and the principal stress directions are horizontal and vertical. At the neutral plane where σ_f is zero, the principal stresses are at 45° to the X axis.

The solid and dashed lines on the beam represent the stress trajectories. They consist of two systems of orthogonal curves whose tangents at each point are in the direction of the principal stresses at that point. The solid lines indicate the direction of the maximum compressive stresses, and the dashed lines indicate the direction of the maximum tensile stresses. Be careful not to confuse stress trajectories with lines of constant stress. *Stress trajectories are lines of principal stress direction but of variable stress intensity.*

ILLUSTRATIVE PROBLEMS

943. A shaft 100 mm in diameter that rotates at 30 r/s is subjected to bending loads that produce a maximum bending moment of 2500π N·m. Determine the torque that can also act simultaneously on the shaft without exceeding a resultant shearing stress $\tau = 80$ MPa or a resultant normal stress $\sigma = 100$ MPa. What is the maximum power that can be transmitted by the shaft?

Solution: The bending moment produces a maximum flexural stress at the top or bottom of the shaft. Its value is

$$\left[\sigma_f = \frac{Mc}{I} = \frac{4M}{\pi r^3} \right] \qquad \sigma_f = \frac{4(2500\pi)}{\pi(0.050)^3} = 80 \text{ MPa}$$

An applied torque T, as yet undetermined, produces a torsional shearing stress, maximum at the periphery of the shaft, which also acts on the element at the top or bottom of the shaft.* This state of stress is shown in Fig. 9–23a. Although the torsional shearing stress (denoted by τ_t to distinguish it from the maximum resultant shearing stress τ which the shaft can withstand) is not yet known, Mohr's circle can be drawn in terms of it, as in Fig. 9–23b.

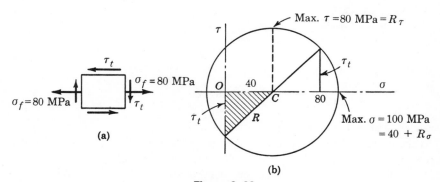

(a)

(b)

Figure 9–23.

---

*At the extremities of the neutral axis, the torsional shearing stress is added to the shearing stress caused by the vertical shear, thereby forming a resultant shearing stress. In a short heavily loaded shaft, this can be the maximum resultant shearing stress that limits the value of the torsional shearing stress. However, this possibility is ignored here.

To produce the maximum permitted resultant shearing stress, the radius of the circle must be $R_\tau = 80$ MPa. However, the radius that will produce the maximum permitted normal stress must satisfy the condition $\sigma = 100 = OC + R_\sigma = 40 + R_\sigma$, whence $R_\sigma = 60$ MPa.

It should be clear that the proper radius of Mohr's circle is R_τ or R_σ, whichever is smaller, so that the allowable values of max. τ or max. σ will not be exceeded. Having thus determined the proper radius, that is, $R = 60$ MPa, we compute, from the shaded triangle in Fig. 9-23b, the maximum torsional shearing stress τ_t that can be combined with the flexural stress. Hence

$$\tau_t^2 = R^2 - (40)^2 = (60)^2 - (40)^2$$

or

$$\tau_t = 44.7 \text{ MPa}$$

The torsion formula shows that the torque required to produce this torsional shearing stress is

$$\left[T = \frac{\tau J}{r} = \frac{\tau \pi r^3}{2} \right] \qquad T = \frac{(44.7 \times 10^6)\pi(0.050)^3}{2}$$
$$= 8780 \text{ N·m} \qquad Ans.$$

Finally, the maximum power that can be transmitted by the shaft is given by

$$\left[\mathscr{P} = 2\pi fT \right] \qquad \mathscr{P} = 2\pi(30)(8750) = 1650 \text{ kW} \qquad Ans.$$

944. A solid shaft is subjected to simultaneous twisting and bending due to a torque T and a maximum bending moment M. Express the maximum resultant shearing stress τ and the maximum resultant normal stress σ in terms of T, M, and the radius r of the shaft. Using these relations, determine the proper diameter of a solid shaft to carry simultaneously $T = 1200$ N·m and $M = 900$ N·m, if $\sigma \leqslant 100$ MPa and $\tau \leqslant 70$ MPa.

Solution: The simultaneous bending and twisting in this problem is commonly encountered in designing shafts. The formulas that will be developed are very useful, but their use should be limited to cases in which both M and T are known. *Under any other conditions, Mohr's circle should be used.*

The state of stress of an element subjected to simultaneous flexure and torsion is shown in Fig. 9-24a, and the equivalent Mohr's circle in Fig. 9-24b. The maximum resultant shearing stress τ is equal to the radius R, which, from the shaded right triangle, is

$$\text{Max. } \tau = R = \sqrt{\left(\tfrac{1}{2}\sigma_f\right)^2 + (\tau_t)^2} \qquad (a)$$

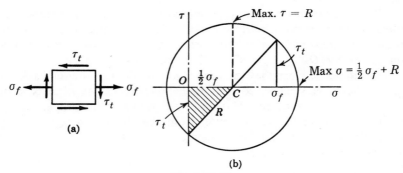

Figure 9–24.

The following variations of the flexure and torsion formulas as applied to a circular shaft are used:

$$\sigma_f = \frac{4M}{\pi r^3} \quad \text{and} \quad \tau = \frac{2T}{\pi r^3} \qquad (b)$$

Substituting these values in Eq. (a) yields

$$\text{Max. } \tau = \sqrt{\left(\frac{2M}{\pi r^3}\right)^2 + \left(\frac{2T}{\pi r^3}\right)^2}$$

which reduces to

$$\text{Max. } \tau = \frac{2}{\pi r^3} \sqrt{M^2 + T^2}$$

Letting $T_e = \sqrt{M^2 + T^2}$, we obtain finally

$$\textbf{Max. } \tau = \frac{2T_e}{\pi r^3} \qquad (9\text{--}11)$$

The similarity between Eq. (9–11) and the torsion formula in Eq. (b) suggests *equivalent torque* as a suitable name for T_e.

An equation for maximum resultant normal stress that is similar to the flexure formula but involves an *equivalent moment* M_e is obtained as follows: In Fig. 9–24b, the maximum and resultant normal stress is max. $\sigma = \frac{1}{2}\sigma_f + R$. Substituting $\sigma_f = 4M/\pi r^3$ and $R = 2T_e/\pi r^3$, we have

$$\text{Max. } \sigma = \frac{2}{\pi r^3}(M + T_e)$$

Multiplying and dividing the right side by 2 gives

$$\textbf{Max. } \sigma = \frac{4M_e}{\pi r^3} \qquad (9\text{--}12)$$

which is equivalent to the flexure formula in Eq. (*b*) if the equivalent moment M_e is expressed by

$$M_e = \tfrac{1}{2}(M + T_e)$$

It is not necessary to memorize Eqs. (9–11) and (9–12) because they are so similar to the torsion and flexure formulas. In using them, however, the following definitions of equivalent torque and equivalent moment must be remembered:

$$T_e = \sqrt{M^2 + T^2} \tag{9–13}$$
$$M_e = \tfrac{1}{2}(M + T_e) \tag{9–14}$$

For the numerical data given in this problem, the equivalent torque and equivalent moment are

$$T_e = \sqrt{M^2 + T^2} = \sqrt{(900)^2 + (1200)^2} = 1500 \text{ N·m}$$
$$M_e = \tfrac{1}{2}(M + T_e) = \tfrac{1}{2}(900 + 1500) = 1200 \text{ N·m}$$

The shaft radius required so that the maximum shearing stress will not be exceeded is found from Eq. (9–11):

$$\left[\tau = \frac{2T_e}{\pi r^3}\right]$$

$$70 \times 10^6 = \frac{2(1500)}{\pi r^3} \quad \text{or} \quad r = 23.9 \times 10^{-3} \text{ m} = 23.9 \text{ mm}$$

The radius that will avoid exceeding the maximum normal stress is, from Eq. (9–12),

$$\left[\sigma = \frac{4M_e}{\pi r^3}\right]$$

$$100 \times 10^6 = \frac{4(1200)}{\pi r^3} \quad \text{or} \quad r = 24.8 \times 10^{-3} \text{ m} = 24.8 \text{ mm}$$

The larger of these two radii will satisfy both stress conditions; hence the proper diameter is

$$d = 2 \times 24.8 = 49.6 \text{ mm} \qquad Ans.$$

945. Design a solid shaft to carry the loads shown in Fig. 9–25, if max. $\tau \leqslant 70$ MPa and max. $\sigma \leqslant 120$ MPa. The belt pulls on pulleys B and C are vertical, and those on pulley E are horizontal. Neglect the masses of the pulleys and shaft.

Solution: The given loading produces bending in both the vertical and horizontal planes. The bending moment diagrams in these planes are given in Fig. 9–25b, c. The resultant moment at any section is given by $M = \sqrt{M_h^2 + M_v^2}$. Therefore the resultant bending moments at

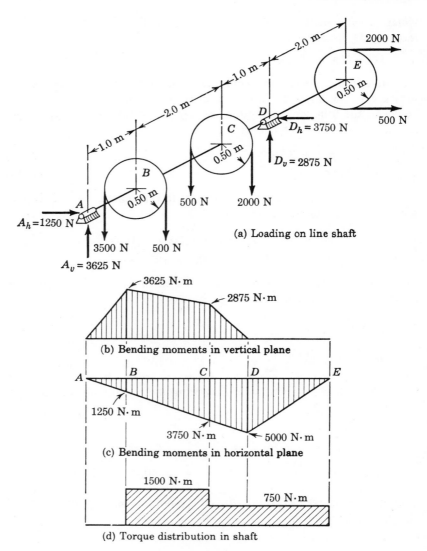

(a) Loading on line shaft

(b) Bending moments in vertical plane

(c) Bending moments in horizontal plane

(d) Torque distribution in shaft

Figure 9–25.

B, C, and D are $M_B = 3834$ N·m, $M_C = 4725$ N·m, and $M_D = 5000$ N·m. Combining these values with the torque distribution in the shaft (Fig. 9–25d) shows that the dangerous sections are at C and D.

Since both moment and torque values are known, it is advantageous to use the method in Illustrative Problem 944. Applying Eqs. (9–13) and (9–14), we find the equivalent torque and equivalent moment to be

at C: $T_e = \sqrt{M^2 + T^2} = \sqrt{(4725)^2 + (1500)^2} = 4957 \text{ N·m}$

$M_e = \frac{1}{2}(M + T_e) = \frac{1}{2}(4725 + 4957) = 4841 \text{ N·m}$

at D: $T_e = \sqrt{M^2 + T^2} = \sqrt{(5000)^2 + (750)^2} = 5056 \text{ N·m}$

$M_e = \frac{1}{2}(M + T_e) = \frac{1}{2}(5000 + 5056) = 5028 \text{ N·m}$

The larger value of T_e is applied to Eq. (9–11), and the larger value of M_e is applied to Eq. (9–12). In this problem, both max. T_e and max. M_e occur at D. Hence, we obtain

$$\left[\tau = \frac{2T_e}{\pi r^3} \right]$$

$70 \times 10^6 = \dfrac{2(5056)}{\pi r^3}$ $r = 35.8 \times 10^{-3} \text{ m} = 35.8 \text{ mm}$

$$\left[\sigma = \frac{4M_e}{\pi r^3} \right]$$

$120 \times 10^6 = \dfrac{4(5028)}{\pi r^3}$ $r = 37.7 \times 10^{-3} \text{ m} = 37.7 \text{ mm}$

The larger value of r determines the proper radius. Therefore the required diameter of the shaft is $d = 2 \times 37.7 = 75.4$ mm. In view of the standard sizes of shafting, a shaft of 80-mm diameter would be selected.

PROBLEMS

946. Explain why the stress trajectories in Fig. 9–22 tend to become horizontal as they approach the wall. Where are they exactly horizontal? What are the stress trajectories for axial tension or compression?

947. The solid shaft in a small hydraulic turbine is 100 mm in diameter and supports an axial compressive load of 140π kN. Determine the maximum power that be developed at 4 r/s without exceeding a maximum shearing stress of 70 MN/m^2 or a maximum normal stress of 90 MN/m^2.   *Ans.* 273 kW

948. A solid shaft 100 mm in diameter is subjected simultaneously to an axial compressive force of 600 kN and to a torque which twists the shaft through an angle of 1.5° in a length of 8 m. If $G = 80 \times 10^9$ N/m², compute the maximum normal and shearing stresses in the shaft.   *Ans.* $\tau = 40.4$ MN/m²

949.   A solid shaft 100 mm in diameter carries simultaneously an axial tensile load of 50π kN, a maximum bending moment of 2π kN·m, and a torque of 3π kN·m. Compute the maximum tensile, compressive, and shearing stresses produced in the shaft.

Ans. $\sigma_t = 106$ MPa; $\sigma_c = 74.8$ MPa; $\tau = 63.8$ MPa

950.   Repeat Problem 949 assuming that the axial load is changed to a compressive load of 40π kN.

951.   A solid shaft 80 mm in diameter is subjected to a maximum bending moment of 800π N·m and an axial tensile load of 40π kN. Find the maximum torque which can be safely applied if max. $\sigma \leqslant 100$ MN/m² and max. $\tau \leqslant 80$ MN/m².

952.   A solid circular shaft is used to transmit simultaneously a torque of 2600 N·m and a maximum bending moment of 2000 N·m. Determine the radius of the smallest shaft which can be used if max. $\sigma \leqslant 80$ MPa and max. $\tau \leqslant 60$ MPa.   *Ans.* $r = 34.8$ mm

953.   A shaft 80 mm in diameter carries a maximum bending moment of 3 kN·m. What torque can also be applied without exceeding a maximum shearing stress of 80 MN/m² or a maximum normal stress of 120 MN/m²?

954.   A closed-end cylindrical pressure vessel has an external diameter of 400 mm and a wall thickness of 20 mm. The vessel carries simultaneously an internal pressure of 4 MPa, a torque of 80 kN·m, and a bending moment of 20 kN·m. Determine the maximum tensile stress in the walls of the vessel. Neglect the possibility of buckling.

Ans.   Max. $\sigma_t = 53.9$ MPa

955.   A closed-end cylindrical pressure vessel has an external diameter of 300 mm and is fabricated from 10-mm steel plate. If the tank is subjected to an internal pressure of 6 MN/m², find the maximum torque which can also be applied if the normal stress in the walls of the vessel is limited to 100 MN/m². Neglect the possibility of buckling.

956.   Compute the principal stresses and maximum shearing stress at point A in Fig. P–956 at the section $x = 250$ mm. The beam is rectangular, 20 mm wide by 120 mm deep, and point A is 20 mm above the centerline of the beam. Assume the 50-kN load acts at the centroid of the cross section. Show answers on complete sketches of appropriate differential elements. (*Hint*: Be sure to include the shearing stress caused by the applied load.)

> *Ans.*   Max. $\sigma = 72.6$ MPa at $\theta = -12.9°$

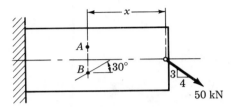

Figures P-956 and P-957.

957.   For the 20 mm by 120 mm beam described in Problem 956, determine the stress components on the 30° plane at point *B*. Assume that $x = 300$ mm and that *B* is 20 mm below the centerline of the beam. Show answers on a complete sketch of a differential element.

958.   A 50-mm diameter bracket, securely fastened to the wall as shown in Fig. P-958, carries the given horizontal and vertical loads. Find the principal stresses and maximum shearing stress at point *A* on the upper surface of the shaft.   *Ans.*   Max. $\sigma = 41.2$ MN/m^2

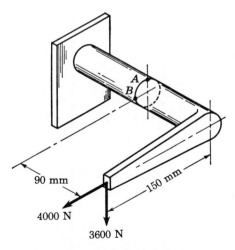

Figures P-958 and P-959.

959. Repeat Problem 958 to find the principal stresses and maximum shearing stress at point B on the front surface of the shaft.

960. A 100-mm diameter shaft carries an axial load P and a torque T, as shown in Fig. P–960. Determine the normal and shearing stresses on the spiral weld which makes an angle of 30° with the axis of the shaft.   *Ans.* $\sigma = 18.4$ MPa; $\tau = -16.5$ MPa

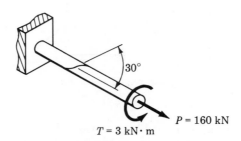

$T = 3$ kN·m

$P = 160$ kN

Figure P–960.

961. Twenty kilowatts are transmitted through a speed reducer. At one part of the machine, as shown in Fig. P–961, the pinion drives the gear A of shaft AB at 6 r/s. Determine the minimum diameter of shaft AB if max. $\sigma \leqslant 80$ MN/m² and max. $\tau \leqslant 60$ MN/m². Consider only torsional and bending stresses in the shaft.

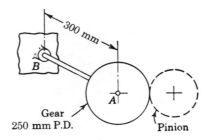

Figure P–961.

962. A line shaft 50 mm in diameter is subjected to the loads shown in Fig. P–962. The belt pulls on pulley A are horizontal, and those on pulley B are vertical. Calculate the maximum normal and shearing stresses developed in the shaft.

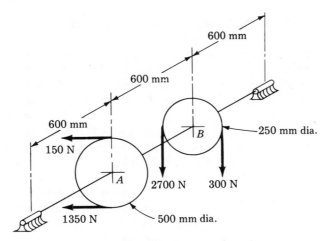

Figure P–962.

963.   Design a solid shaft to carry the loads shown in Fig. P–963 if max. $\tau \leqslant 60$ MPa and max. $\sigma \leqslant 80$ MPa. The belt pulls at A and C are horizontal, and those on pulley E are vertical.

Ans. $d = 68.4$ mm

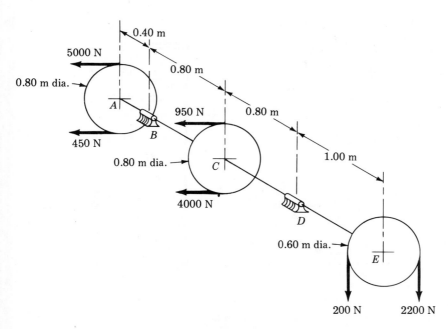

Figure P–963.

9-9   TRANSFORMATION OF STRAIN COMPONENTS

Most of the problems encountered in engineering design involve a combination of axial, torsional, and flexural loads applied to homogeneous materials of a prismatic shape. In such cases, the stresses may be computed as described in the preceding articles and maximum resultant stresses used as a criterion of design. Occasionally, however, irregularities in a structure, or conditions which violate the basic assumptions of the torsion or flexure theories, require us to resort to experimental methods of determining stresses. Since stress is a mathematical concept that represents the intensity of force on a unit area, it *cannot* be measured directly. Nevertheless, the stress–strain relations defined by Hooke's law permit us to estimate stresses from strains which *can* be measured. In this article we study the transformation of a given set of strains into principal strains. In the next article we consider the application of strain measurements and their conversion into stresses.

Consider now an element subjected to the general state of plane stress shown in Fig. 9–26a. The normal stresses (assumed to be tensile) elongate the element in the X and Y directions, and the shearing stresses distort the element through the shearing strain γ_{xy} as shown in Fig. 9–26b. The effect of these strains on any line element OA in Fig. 9–26a

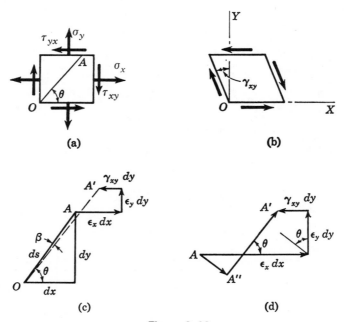

Figure 9–26.

is shown in Fig. 9–26c, where OA is elongated to OA' and also changes its angular position by a very small amount that we call β. The movement of A to A' is the vector sum of the elongation $\epsilon_x \, dx$ in the X direction, $\epsilon_y \, dy$ in the Y direction, and the shear distortion $\gamma_{xy} \, dy$ in the X direction. These movements of A are exaggerated in Fig. 9–26d to show how their vector sum may be resolved into two components parallel and perpendicular to OA. The parallel component $A''A'$ represents the increase in length of OA, whereas the perpendicular component AA'' causes the change β in the angular position of OA.

The magnitude of $A''A'$, found by projecting $\epsilon_x \, dx$, $\epsilon_y \, dy$, and $\gamma_{xy} \, dy$ upon the direction of OA, is

$$A''A' = \epsilon_x \, dx \cos \theta + \epsilon_y \, dy \sin \theta - \gamma_{xy} \, dy \cos \theta \qquad (a)$$

This increase in length of OA divided by its original length ds is defined as the strain ϵ_a in the direction OA:

$$\epsilon_a = \frac{A''A'}{ds} = \frac{\epsilon_x \, dx \cos \theta}{ds} + \frac{\epsilon_y \, dy \sin \theta}{ds} - \frac{\gamma_{xy} \, dy \cos \theta}{ds} \qquad (b)$$

But from Fig. 9–26c, $dx/ds = \cos \theta$ and $dy/ds = \sin \theta$. With these relations, Eq. (b) reduces to

$$\epsilon_a = \epsilon_x \cos^2 \theta + \epsilon_y \sin^2 \theta - \gamma_{xy} \sin \theta \cos \theta \qquad (c)$$

On substituting for $\cos^2 \theta$ and $\sin^2 \theta$ their equivalents in terms of 2θ, this becomes

$$\epsilon_a = \frac{\epsilon_x + \epsilon_y}{2} + \frac{\epsilon_x - \epsilon_y}{2} \cos 2\theta - \frac{1}{2}\gamma_{xy} \sin 2\theta \qquad (9\text{--}15)$$

The angular deviation of OA (i.e., β) is determined by dividing the perpendicular component AA'' by the original length ds of OA. Projecting the movement of point A upon the direction perpendicular to OA, we obtain, from Fig. 9–26d,

$$AA'' = \epsilon_x \, dx \sin \theta - \epsilon_y \, dy \cos \theta - \gamma_{xy} \, dy \sin \theta \qquad (d)$$

whence

$$\begin{aligned} \beta &= \frac{AA''}{ds} = \frac{\epsilon_x \, dx \sin \theta}{ds} - \frac{\epsilon_y \, dy \cos \theta}{ds} - \frac{\gamma_{xy} \, dy \sin \theta}{ds} \\ &= \epsilon_x \sin \theta \cos \theta - \epsilon_y \sin \theta \cos \theta - \gamma_{xy} \sin^2 \theta \end{aligned} \qquad (e)$$

For the line element at right angles to OA, the angular deviation β' may be found by substituting $\theta + 90°$ for θ. Since $\sin(\theta + 90°) = \cos \theta$ and $\cos(\theta + 90°) = -\sin \theta$, we obtain

$$\beta' = -\epsilon_x \sin \theta \cos \theta + \epsilon_y \sin \theta \cos \theta - \gamma_{xy} \cos^2 \theta \qquad (f)$$

Since β and β' rotate in opposite directions,* their absolute sum is equal to their algebraic difference. Hence the total change in the right angle between OA and its normal OB, which defines the shearing strain for an element located $\theta°$ from the $X-Y$ axes, becomes

$$\gamma_{ab} = \beta - \beta' = \epsilon_x(2 \sin \theta \cos \theta) - \epsilon_y(2 \sin \theta \cos \theta)$$
$$+ \gamma_{xy}(\cos^2 \theta - \sin^2 \theta)$$

which, in terms of the double angle 2θ, reduces to

$$\frac{1}{2}\gamma_{ab} = \frac{(\epsilon_x - \epsilon_y)}{2} \sin 2\theta + \frac{1}{2}\gamma_{xy} \cos 2\theta \qquad (9\text{--}16)$$

A comparison of Eqs. (9–15) and (9–16) with the normal and shearing stress transformation given by Eqs. (9–5) and (9–6) shows that they are identical in form. We conclude that the normal and shearing *strains* can be represented by a Mohr's circle for strain, constructed in the same manner as Mohr's circle for stress *except that half values of shearing strain are plotted* instead of shear stress.

In applying Mohr's circle for strain, we use the following rules of sign. Extensional strains are considered positive, compressive strains negative, and shearing strains positive when they *increase* the original right angle of an unstrained element.[†] A more general rule of sign for shearing strain is obtained by denoting the strain between two perpendicular directions OA and OB by γ_{ab}, the first subscript indicating the direction OA associated with the angle θ. The shearing strain is considered as positive if the directional line associated with the first subscript (OA in this case) moves clockwise relative to the other directional line, and vice versa. This rule of sign makes $\gamma_{ab} = -\gamma_{ba}$. This agrees with our convention for shearing stress, that $\tau_{xy} = -\tau_{yx}$, as well as for products of inertia, that $P_{xy} = -P_{yx}$.

The similarity in form between stress and strain components is further exemplified by the following statement: A Mohr's circle for strain can be transformed into a concentric Mohr's circle for stress by means of the scale transformations

$$R_\sigma = R_\epsilon \frac{E}{1 + \nu} \qquad (9\text{--}17)$$

$$(OC)_\sigma = (OC)_\epsilon \frac{E}{1 - \nu} \qquad (9\text{--}18)$$

*If β and β' were in the same direction, there would be no change in the right angle, i.e., no shearing strain, as is easily verified by adding β and β'.

[†]This is contrary to the convention adopted in the mathematical theory of elasticity where the *closing* of a right angle is taken as positive shearing strain. The convention adopted in this book not only achieves consistency between Mohr's circles of inertia, stress, and strain but avoids the artificiality of plotting values of positive shearing strain in a negative or downward direction.

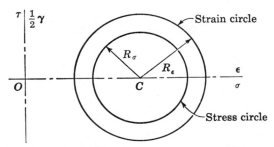

Figure 9–27.   Transformation of Mohr's circle of strains to Mohr's circle of stress.

in which R_σ and R_ϵ are respectively the radii of the stress and strain circles in Fig. 9–27 and $(OC)_\sigma$ and $(OC)_\epsilon$ are respectively the stress and strain coordinates of the centers of the concentric circles. Also E is the modulus of elasticity and ν is Poisson's ratio. The proof of these relations is called for in Problem 965. Their application is discussed in the following illustrative problem.

ILLUSTRATIVE PROBLEM

964.   In a body subjected to plane strain, there act at a certain point $\epsilon_x = 800 \times 10^{-6}$ m/m, $\epsilon_y = 200 \times 10^{-6}$ m/m, and $\gamma_{xy} = 600 \times 10^{-6}$ rad. Compute (a) the principal strains and the principal strain axes; also (b) the strain ϵ_a in a direction of 60° with the X axis, the strain ϵ_b perpendicular to ϵ_a, and the shearing strain γ_{ab}. Further, if $E = 200$ GPa and $\nu = 0.30$, determine the principal stresses and the normal and shearing stresses on the element rotated 60° from the X axis.

Solution: Mohr's circle for the given state of strain is shown in Fig. 9–28. The factor 10^{-6} being omitted, the coordinates of point A are $\epsilon_x = 800$ and $\frac{1}{2}\gamma_{xy} = 300$; and the coordinates of B are $\epsilon_y = 200$ and $\frac{1}{2}\gamma_{yx} = -300$. The X axis is represented by the radius CA, and the Y axis by the radius CB. The radius of the circle is computed to be 424, whence the maximum principal strain ϵ_1, denoted by point D, equals $500 + 424 = 924 \times 10^{-6}$. The minimum principal strain ϵ_2, denoted by point E, equals $500 - 424 = 76 \times 10^{-6}$. The angle between the maximum strain axis and the X axis is one-half the angle ACD, or 22.5°, in a clockwise direction from the X axis, as shown in Fig. 9–29.

To determine the strain ϵ_a, lay off the radius CF at twice 60°, or 120°, counterclockwise from the X axis denoted by CA. The strain coordinates of point F are $\epsilon_a = 500 - 424 \cos 15° = 90 \times 10^{-6}$, and $\frac{1}{2}\gamma_{ab} = 424 \sin 15° = 110$, whence $\gamma_{ab} = 220 \times 10^{-6}$ radians.

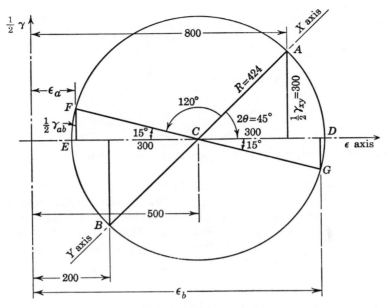

Figure 9–28.   Strain circle.

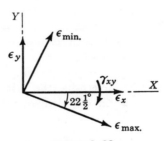

Figure 9–29.

Laying off the 90° angle between the directions of ϵ_a and ϵ_b to double scale locates G as diametrically opposite from F; hence $\epsilon_b = 500 + 424 \cos 15° = 910 \times 10^{-6}$.

Except for changing the symbols and plotting only half values of shearing strain, the procedure is the same as that given in Art. 9–7 describing Mohr's circle for stress.

The stress components may now be found by applying Hooke's law to the strain components, as will be shown later; but the most convenient method is to transform the strain circle into the stress circle. The only results needed from the strain circle are the radius and the location of the center. These are transformed into corresponding values for the stress circle by means of Eqs. (9–17) and (9–18). We obtain

$$\left[R_\sigma = R_\epsilon \frac{E}{1 + \nu} \right]$$

$$R_\sigma = (424 \times 10^{-6}) \frac{200 \times 10^9}{1 + 0.30} = 65.2 \text{ MPa}$$

$$\left[(OC)_\sigma = (OC)_\epsilon \frac{E}{1 - \nu} \right]$$

$$(OC)_\sigma = (500 \times 10^{-6}) \frac{200 \times 10^9}{1 - 0.30} = 143 \text{ MPa}$$

Using these values, we plot Mohr's circle of stress as shown in Fig. 9–30. The points are labeled to correspond with those in Fig. 9–28.* Reading from the circle, we see that the principal stresses at D and E are respectively

at D: Max. $\sigma = 143 + 65.2 = 208$ MPa
at E: Min. $\sigma = 143 - 65.2 = 77.8$ MPa

For the element rotated 60° from the X axis, the stress components are given by points F and G, and the values are

at F: $\sigma = 143 - 65.2 \cos 15° = 80.0$ MPa
           $\tau = 65.2 \sin 15° = 16.9$ MPa

at G: $\sigma = 143 + 65.2 \cos 15° = 206$ MPa

Instead of using the transformed circle of stress, we can find the stresses directly from the strains, using Hooke's law for biaxial stress

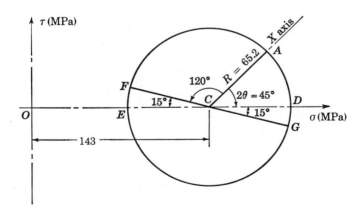

Figure 9–30.   Stress circle.

*A separate figure is unnecessary because the circles could be plotted concentrically, but it would be confusing to do so here.

(see page 47) as expressed by the following equations:

$$\sigma_x = \frac{E(\epsilon_x + \nu\epsilon_y)}{1 - \nu^2}; \quad \sigma_y = \frac{E(\epsilon_y + \nu\epsilon_x)}{1 - \nu^2};$$

$$\tau_{xy} = G\gamma_{xy} = \frac{E}{2(1 + \nu)}\,\gamma_{xy}$$

On substituting the principal strains of 924×10^{-6} and 76×10^{-6} found from the strain circle in Fig. 9–28, the principal stresses are computed to be

$$\text{Max. } \sigma = \frac{(200 \times 10^9)(924 + 0.30 \times 76)(10^{-6})}{1 - (0.30)^2} = 208 \text{ MPa}$$

$$\text{Min. } \sigma = \frac{(200 \times 10^9)(76 + 0.30 \times 924)(10^{-6})}{1 - (0.30)^2} = 77.6 \text{ MPa}$$

These results agree with those found previously; consequently if a Mohr's circle of stress is plotted using these principal stresses, the radius and center will have the values shown in Fig. 9–30.

In a similar manner, the normal and shearing stresses on the element at 60° from the X axis can be determined from the corresponding strain components of $\epsilon_a = 90 \times 10^{-6}$, $\epsilon_b = 910 \times 10^{-6}$, and $\gamma_{ab} = 220 \times 10^{-6}$. Applying Hooke's law, we obtain

$$\sigma_a = \frac{(200 \times 10^9)(90 + 0.30 \times 910)(10^{-6})}{1 - (0.30)^2} = 79.8 \text{ MPa}$$

$$\tau_{ab} = \frac{(200 \times 10^9)(220 \times 10^{-6})}{2(1 + 0.30)} = 16.9 \text{ MPa}$$

A comparison of these two methods of computing stress components should convince you of the advantages of transforming the strain circle to the stress circle.

PROBLEMS

965. As shown in Fig. 9–27, prove that Eqs. (9–17) and (9–18) will transform a strain circle into a stress circle.

966. Starting with an element subjected only to principal stresses, show that the angular deviation β of a line element such as OA in Fig. 9–26 is equal to one-half the shearing strain γ_{ab}.

967. A state of strain is defined by $\epsilon_x = -400 \times 10^{-6}$, $\epsilon_y = 200 \times 10^{-6}$, and $\gamma_{xy} = 800 \times 10^{-6}$. If $E = 200$ GPa and $\nu = 0.30$, determine the principal stresses and stress components on the face whose normal is at $+40°$ from the X axis.

$$\textit{Ans.} \quad \text{Max. } \sigma = 48.3 \text{ MPa; min. } \sigma = -106 \text{ MPa;}$$
$$\sigma = -97.2 \text{ MPa; } \tau = -34.8 \text{ MPa}$$

968. A state of strain is defined by $\epsilon_x = 600 \times 10^{-6}$, $\epsilon_y = -400 \times 10^{-6}$, and $\gamma_{xy} = -600 \times 10^{-6}$. If $E = 200 \times 10^9$ N/m² and $\nu = 0.30$, determine the principal stresses and maximum shearing stress.

969. The strain components at a given point are $\epsilon_x = -800 \times 10^{-6}$, $\epsilon_y = 200 \times 10^{-6}$, and $\gamma_{xy} = -800 \times 10^{-6}$. If $E = 200$ GN/m² and $\nu = 0.30$, find the stress components on the face whose normal is at $+20°$ from the X axis. $\textit{Ans.} \quad \sigma = -105 \text{ MPa; } \tau = -96.6 \text{ MPa}$

9–10 THE STRAIN ROSETTE

The stress in a bar subjected to uniaxial stress can be determined experimentally by attaching a strain gage oriented in the direction of the stress. The stress is then computed, in terms of the strain, from $\sigma = E\epsilon$. The strain is generally small (under 1 part in 1000); hence sensitive instruments are required for measuring it. Originally, strain gages were mechanical or optical, but these have now been almost completely replaced by electrical gages. This type of gage contains a wire or foil element whose electrical resistance varies with its deformation. The gage is cemented to the test specimen, the strain in the specimen being measured as a function of the change in the electrical resistance of the gage element. The wire-type of strain gage has been brought to a high state of perfection by the Baldwin Southwark Division of the Baldwin Locomotive Works. These gages are marketed under the well-known SR-4 trademark of the Baldwin-Lima-Hamilton Company.

As was said above, a single strain gage oriented in the direction of a uniaxial stress is sufficient for computing the stress. For biaxial stress, we might suppose that two strain gages would be sufficient; this would be true if the directions of the principal stresses were known, but this is not usually the case. To determine the direction of the principal stresses in addition to their magnitudes, three values of strain are required. In the preceding article we showed how the values ϵ_x, ϵ_y, and γ_{xy} can be used for this purpose. Unfortunately, there is no equipment that conveniently gives a direct measurement of the shearing strain γ_{xy}, so other methods are needed.

We will now show that a state of strain is uniquely determined by the measurement of three linear strains ϵ_a, ϵ_b, and ϵ_c acting in three arbitrary directions θ_a, θ_b, and θ_c at the same point as in Fig. 9–31. By substituting these strains in Eq. (9–15), we obtain the following set of

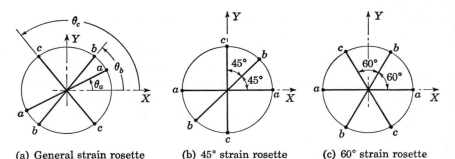

(a) General strain rosette   (b) 45° strain rosette   (c) 60° strain rosette

Figure 9–31.   Strain rosettes.

simultaneous equations:

$$
\begin{aligned}
\epsilon_a &= \frac{\epsilon_x + \epsilon_y}{2} + \frac{\epsilon_x - \epsilon_y}{2} \cos 2\theta_a - \frac{\gamma_{xy}}{2} \sin 2\theta_a \\
\epsilon_b &= \frac{\epsilon_x + \epsilon_y}{2} + \frac{\epsilon_x - \epsilon_y}{2} \cos 2\theta_b - \frac{\gamma_{xy}}{2} \sin 2\theta_b \\
\epsilon_c &= \frac{\epsilon_x + \epsilon_y}{2} + \frac{\epsilon_x - \epsilon_y}{2} \cos 2\theta_c - \frac{\gamma_{xy}}{2} \sin 2\theta_c
\end{aligned}
\quad (a)
$$

The solution of these equations determines the required strain components ϵ_x, ϵ_y, and γ_{xy}.

As a matter of practical convenience, the linear strains are obtained by using either of two combinations of three resistance strain gages: (1) three gages set with their axes at 45° with each other, or (2) three gages whose axes are at 60° with each other as in Fig. 9–31b and c. These combinations are known as *strain rosettes*. The three gages are electrically insulated from each other and are used to determine the strain at the surface of the structure to which they are attached. We shall now construct a Mohr's circle of strain for each of these rosettes.

The 45° or rectangular strain rosette

By substituting $\theta_a = 0°$, $\theta_b = 45°$, and $\theta_c = 90°$ in Eq. (a) and solving, we obtain

$$
\epsilon_x = \epsilon_a, \qquad \epsilon_y = \epsilon_c, \qquad \frac{1}{2}\gamma_{xy} = \frac{\epsilon_a + \epsilon_c}{2} - \epsilon_b \tag{9-19}
$$

thus defining a state of strain from which the strain circle and the stress circle may be constructed as explained in the preceding article.

These results can also be obtained by constructing a Mohr's circle directly from ϵ_a, ϵ_b, and ϵ_c as follows: In Fig. 9–32, the three strains ϵ_a, ϵ_b, and ϵ_c, 45° apart, will be represented by three radii CA, CB, and CD,

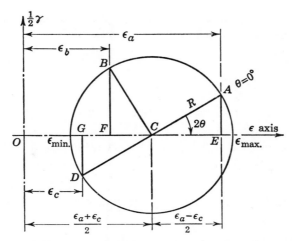

Figure 9–32. Mohr's circle for 45° strain rosette.

90° apart, as shown in the figure. The center C is midway between G and E, so one side (CE) of triangle ACE is known. To construct the circle, the other side AE must be computed. Evidently, triangles CBF and CAF are congruent, so that $AE = CF$. From the geometry of the circle, $CF = OC - OF$; hence

$$AE = \frac{\epsilon_a + \epsilon_c}{2} - \epsilon_b \qquad (b)$$

Also

$$CE = \frac{\epsilon_a - \epsilon_c}{2} \qquad (c)$$

Therefore the radius $R = CA$ is determined from

$$R = \sqrt{(CE)^2 + (AE)^2} \qquad (d)$$

The student should correlate these results with a strain circle constructed from the strain components given by Eq. (9–19).

The 60° or equiangular rosette

In the 60° rosette, the reference angles are $\theta_a = 0°$, $\theta_b = 60°$, and $\theta_c = 120°$. On substituting these values in Eq. (a) and solving, we obtain

$$\left.\begin{aligned}
\epsilon_x &= \epsilon_a \\
\epsilon_y &= \frac{1}{3}(2\epsilon_b + 2\epsilon_c - \epsilon_a) \\
\frac{1}{2}\gamma_{xy} &= \frac{1}{\sqrt{3}}(\epsilon_c - \epsilon_b)
\end{aligned}\right\} \qquad \textbf{(9–20)}$$

From these results, the strain circle and the stress circle can be constructed as described in the preceding article.

PROBLEMS

970. For the 60° strain rosette, show that the expressions in Eq.
(9–20) are correct.

971. Show that in a 60° strain rosette, the principal strains are

$$\epsilon_{\substack{max.\\min.}} = \frac{\epsilon_a + \epsilon_b + \epsilon_c}{3} \pm \frac{2}{3}\sqrt{\epsilon_a(\epsilon_a - \epsilon_b) + \epsilon_b(\epsilon_b - \epsilon_c) + \epsilon_c(\epsilon_c - \epsilon_a)}$$

and the direction of the maximum principal strain is defined by

$$\tan 2\theta = \frac{\sqrt{3}\,(\epsilon_b - \epsilon_c)}{2\epsilon_a - \epsilon_b - \epsilon_c}$$

in which a positive value of θ is measured in a counterclockwise
direction from the direction of ϵ_a.

972. Show that in a 45° strain rosette, the principal strains are

$$\epsilon_{\substack{max.\\min.}} = \frac{\epsilon_a + \epsilon_c}{2} + \frac{1}{\sqrt{2}}\sqrt{(\epsilon_a - \epsilon_b)^2 + (\epsilon_b - \epsilon_c)^2}$$

and the direction of the maximum principal strain is defined by

$$\tan 2\theta = \frac{\epsilon_a + \epsilon_c - 2\epsilon_b}{\epsilon_a - \epsilon_c}$$

973. The three readings on a 45° strain rosette are $\epsilon_a = 400 \times 10^{-6}$, $\epsilon_b = -200 \times 10^{-6}$, and $\epsilon_c = -100 \times 10^{-6}$. If $E = 200$ GPa and $\nu = 0.30$, determine the principal stresses and their directions.
                     Ans.   Max. $\sigma = 109$ MPa at $\theta = -27.2°$

974. Repeat Problem 973 assuming that the strain readings are
$\epsilon_a = 300 \times 10^{-6}$, $\epsilon_b = 600 \times 10^{-6}$, and $\epsilon_c = 100 \times 10^{-6}$.

975. The strains measured on a 60° strain rosette are $\epsilon_a = 300 \times 10^{-6}$, $\epsilon_b = -400 \times 10^{-6}$, and $\epsilon_c = 100 \times 10^{-6}$. If $E = 200$ GN/m² and $\nu = 0.3$, compute the principal stresses and their directions.
                     Ans.   Max. $\sigma = 64.2$ MN/m² at $\theta = -22.0°$

976. A 60° strain rosette attached to the aluminum skin of an
airplane fuselage measures the following strains: $\epsilon_a = 200 \times 10^{-6}$, $\epsilon_b = 200 \times 10^{-6}$, and $\epsilon_c = 400 \times 10^{-6}$. If $E = 70$ GPa and $\nu = \frac{1}{3}$, compute
the principal stresses and their directions.

977. Repeat Problem 976 assuming that the strain readings are
$\epsilon_a = -100 \times 10^{-6}$, $\epsilon_b = 200 \times 10^{-6}$, and $\epsilon_c = -400 \times 10^{-6}$.
                     Ans.   Max. $\sigma = 7.7$ MPa at $\theta = 45°$

9–11 RELATION BETWEEN MODULUS OF RIGIDITY AND MODULUS OF ELASTICITY

In Art. 2–4 we said that $G = E/2(1 + v)$; we are now ready to prove this relation. The state of stress shown in Fig. 9–33a consists of a tensile stress σ_x and a compressive stress σ_y of the same magnitude. Mohr's circle for this (Fig. 9–33b) indicates that an element rotated 45° counterclockwise as in Fig. 9–33c in subjected to pure shear in which τ is numerically equal to σ_x and σ_y. These shearing stresses deform the element *abcd* to the dashed outline *a'b'c'd'* in Fig. 9–33a. The right angle at *a* has decreased to $90° - \gamma$, where γ is the shearing strain. Simultaneously, the right angle at *b* has increased to $90° + \gamma$.

Consider now the deformation of the right triangle *aob* that has equal legs. Since $\sigma_x = -\sigma_y = \tau$, we obtain, with Eqs. (2–9) and (2–10), (see page 47), the normal strains

$$\epsilon_x = \frac{\tau(1 + v)}{E}, \qquad \epsilon_y = -\frac{\tau(1 + v)}{E}$$

Hence the deformed lengths of sides *oa* and *ob* respectively are

$$\overline{oa'} = \overline{oa}\left[1 + \frac{\tau(1 + v)}{E}\right], \qquad \overline{ob'} = \overline{ob}\left[1 - \frac{\tau(1 + v)}{E}\right]$$

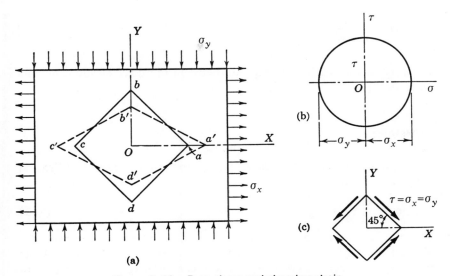

Figure 9–33. Pure shear and shearing strain.

Therefore, from the right triangle $oa'b'$, we have

$$\tan oa'b' = \tan\left(45° - \frac{\gamma}{2}\right) = \frac{\overline{ob'}}{\overline{oa'}} = \frac{1 - \dfrac{\tau(1 + \nu)}{E}}{1 + \dfrac{\tau(1 + \nu)}{E}} \tag{a}$$

From trigonometry, the expanded form of the tangent is

$$\tan\left(45° - \frac{\gamma}{2}\right) = \frac{\tan 45° - \tan\dfrac{\gamma}{2}}{1 + \tan 45° \, \tan\dfrac{\gamma}{2}} = \frac{1 - \dfrac{\gamma}{2}}{1 + \dfrac{\gamma}{2}} \tag{b}$$

since for small angles γ like those that occur with shearing strain, $\tan(\gamma/2)$ is practically equivalent to $\gamma/2$ expressed in radians. Equating this expanded value of $\tan[45° - (\gamma/2)]$ to the right side of Eq. (a) gives

$$\frac{1 - \dfrac{\gamma}{2}}{1 + \dfrac{\gamma}{2}} = \frac{1 - \dfrac{\tau(1 + \nu)}{E}}{1 + \dfrac{\tau(1 + \nu)}{E}}$$

which, with a little algebra, reduces to

$$\gamma = \frac{2\tau(1 + \nu)}{E} \quad \text{or} \quad \frac{\tau}{\gamma} = \frac{E}{2(1 + \nu)}$$

Replacing τ/γ by G, as specified in Hooke's law for shear (Eq. 2–6), we obtain finally

$$G = \frac{E}{2(1 + \nu)} \tag{2–13}$$

which expresses the desired relation between the three elastic constants G, E, and ν.

SUMMARY

The normal stresses caused by a combination of axial and flexural loads are determined from

$$\sigma = \begin{matrix} \oplus \\ \ominus \end{matrix} \frac{P}{A} \pm \frac{My}{I} \tag{9–1}$$

The positive sign refers to tension and the negative sign to compression. The circled signs indicate that the axial stress is uniform and of the same type all over the cross section whereas the magnitude and type of the flexural stress vary with position.

The kern of a section (Art. 9–3) is the part of a cross section through which the resultant compressive force must pass if no tensile stress is to be developed over the section.

For bodies subjected to other than axial and flexural combinations of loading, the elements of the body are subject to both shearing and normal stresses. The stresses on such elements (in fact, on any element) vary with the angular position of the element and are expressed by the following equations which were derived in Art. 9–6.

$$\sigma = \frac{\sigma_x + \sigma_y}{2} + \frac{\sigma_x - \sigma_y}{2} \cos 2\theta - \tau_{xy} \sin 2\theta \qquad (9\text{--}5)$$

$$\tau = \frac{\sigma_x - \sigma_y}{2} \sin 2\theta + \tau_{xy} \cos 2\theta \qquad (9\text{--}6)$$

However, the rules for Mohr's circle (page 380) make it unnecessary to remember these equations and the more important conditions which determine the maximum resultant normal and shearing stresses. A drawing of the circle provides all the information needed to compute the variations in stress at any element. Further applications of Mohr's circle in practical designing are given in Art. 9–8.

The use of Mohr's circle of strain is described in Art. 9–9. The procedure is similar to that with the circle of stress, except that *half values* of shearing strain are plotted as ordinates. A strain circle is readily transformed into a stress circle by means of

$$R_\sigma = R_\epsilon \frac{E}{1 + \nu} \qquad (9\text{--}17)$$

$$(OC)_\sigma = (OC)_\epsilon \frac{E}{1 - \nu} \qquad (9\text{--}18)$$

whence the principal stresses may be easily determined. For experimental determination of stress, the strain rosette (Art. 9–10) is especially useful. The normal strains in any three predetermined directions establish a state of strain from which the strain circle is constructed and converted into the stress circle.

10

Reinforced
Beams

10-1 INTRODUCTION

It was once common to strengthen timber beams by bolting strips of
steel to them. With increasingly lower prices for steel, this practice has
ceased except where timber is plentiful and the cost of transporting steel
to the construction site is high. The most common type of reinforced
beam used today is the concrete beam reinforced with steel rods.

The theory of flexure does not apply to composite beams because
it was based on the assumption that the beam was homogeneous and
that plane transverse sections remained plane, whence the strains varied
directly with their distance from the neutral axis. In investigating the
bending of composite beams, only one assumption is retained: that
plane sections remain plane, that is, the strains vary directly with their
distance from the neutral axis.

The most common method of dealing with a nonhomogeneous
beam is to transform it into an equivalent homogeneous beam to which
the flexure formula may be applied. The basic principle involved here is
that the strains and load capacities must remain unchanged. We shall
discuss first the general case of beams composed of different materials
and then reinforced concrete beams, applying to the latter a procedure

more fundamental and more commonly used than a modification of the flexure formula.

10–2   BEAMS OF DIFFERENT MATERIALS

The timber beam in Fig. 10–1a is reinforced with a steel strip, the steel being securely fastened to the timber so that no slip occurs between them as the beam is bent. From page 153 we see that the ordinary theory of flexure is restricted to beams of homogeneous material and hence does not apply to the beam under consideration. However, by suitable modifications we can obtain an equivalent section in terms of one material to which the theory can be applied.

To obtain an equivalent section, consider a longitudinal steel fiber of the beam at A. Since the steel and wood are assumed to be firmly bolted together, the strains of the steel and wood fibers at A must be equal, that is, $\epsilon_s = \epsilon_w$. Expressing this relation in terms of the stresses and moduli of elasticity, we have

$$\frac{\sigma_s}{E_s} = \frac{\sigma_w}{E_w} \tag{a}$$

Furthermore, in order to be equivalent, the loads carried by the steel fiber and the equivalent wood fiber must be equal, so

$$P_s = P_w$$

or, in terms of the areas,

$$A_s \sigma_s = A_w \sigma_w \tag{b}$$

Combining Eqs. (*a*) and (*b*), we obtain

$$A_s \left(\frac{E_s}{E_w} \right) \sigma_w = A_w \sigma_w$$

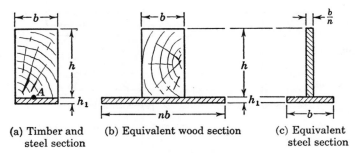

(a) Timber and   (b) Equivalent wood section   (c) Equivalent
   steel section   steel section

Figure 10–1.   Equivalent sections.

from which, by canceling out σ_w and denoting the ratio of the moduli of elasticity E_s / E_w by n, we finally have

$$A_w = nA_s \tag{10-1}$$

This indicates that the area of the equivalent wood is n times the area of the steel. The location of the equivalent area is governed by the condition that the equivalent wood fibers must be at the same distance from the neutral axis as the steel fibers they replace in order to satisfy the criterion of equal deformations in Eq. (a). In other words, the equivalent wood area is n times as wide as the steel it replaces; the equivalent wood section is shown in Fig. 10-1b. If desired, an equivalent steel section can be set up by replacing the original wood by steel $1/n$ as wide, as shown in Fig. 10-1c.

The flexure formula can now be applied directly to either the equivalent wood section or the equivalent steel section. With the equivalent wood section, the actual steel stress is n times the stress in the equivalent wood; with the equivalent steel section, the actual wood stress is $1/n$ times the stress in the equivalent steel.

A similar procedure is used with beams composed of other combinations of materials. For example, consider the experimental section shown in Fig. 10-2a consisting of an aluminum core to which plates of steel and bronze are securely attached. By using the ratio of the moduli of elasticity of the steel and bronze to that of the aluminum, the composite section is transformed into an equivalent section of aluminum (Fig. 10-2b) to which the flexure formula may be applied. In all such cases the neutral axis passes through the centroid of the equivalent section, and the moment of inertia used is the value computed about the centroidal axis of the transformed section.

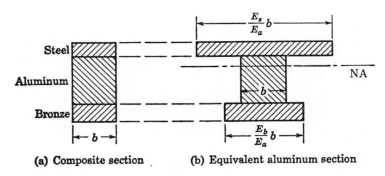

(a) Composite section   (b) Equivalent aluminum section

Figure 10-2.

ILLUSTRATIVE PROBLEM

1001.   A timber beam 150 mm by 300 mm is reinforced, on the bottom only, with a steel strip 75 mm wide by 10 mm thick. Determine the maximum resisting moment if the allowable stresses are $\sigma_s \leqslant 120$ MPa and $\sigma_w \leqslant 8$ MPa. Assume $E_s/E_w = n = 20$.

Solution: Although only rarely is steel used to reinforce one side of a timber beam, this problem illustrates many of the concepts encountered later in reinforced concrete beams. The first of these involves the location of the neutral axis. Since the neutral axis coincides with the centroidal axis of the equivalent section shown in Fig. 10–3, the moments of area about an axis through the base gives

$$[A\bar{y} = \Sigma ay]$$
$$[(45 \times 10^3) + (15 \times 10^3)]\bar{y} = (45 \times 10^3)(160) + (15 \times 10^3)(5)$$
$$\bar{y} = 121 \text{ mm}$$

First finding the moment of inertia about an axis through the top of the flange and then using the transfer formula, we compute the moment of inertia about the neutral axis:

$$\left[I = \Sigma \frac{bh^3}{3}\right]$$
$$I = \frac{150(300)^3}{3} + \frac{1500(10)^3}{3} = 1350 \times 10^6 \text{ mm}^4$$
$$[\bar{I} = I - Ad^2]$$
$$I_{NA} = (1350 \times 10^6) - (60 \times 10^3)(111)^2 = 611 \times 10^6 \text{ mm}^4$$

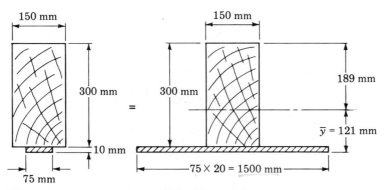

(a) Original section   (b) Equivalent wood section

Figure 10–3.

The resisting moment in terms of the maximum wood stress is

$$\left[M = \frac{\sigma I}{y} \right] \qquad M_w = \frac{(8 \times 10^6)(611 \times 10^{-6})}{189 \times 10^{-3}} = 25.9 \text{ kN·m}$$

In the wood equivalent of the steel, the maximum stress is

$$\sigma_w = \frac{\sigma_s}{n} = \frac{120}{20} = 6 \text{ MPa}$$

whence the resisting moment that will not exceed the permissible steel stress is

$$\left[M = \frac{\sigma I}{y} \right] \qquad M_s = \frac{(6 \times 10^6)(611 \times 10^{-6})}{121 \times 10^{-3}} = 30.3 \text{ kN·m}$$

The smaller resisting moment (i.e., $M_w = 25.9$ kN·m) is the safe resisting moment. In this case there is an excess of steel; hence the beam may be said to be over-reinforced. (The reader will find it instructive to solve this problem using an equivalent steel section.)

PROBLEMS

1002. A timber beam is reinforced with steel plates rigidly attached at the top and bottom as shown in Fig. P–1002. By what amount is the moment increased by the reinforcement if $n = 15$ and the allowable stresses in the wood and steel are 8 MPa and 120 MPa, respectively?   *Ans.*   52.2 kN·m

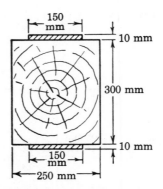

Figures P–1002, P–1003, P–1004.

1003. A simply supported beam 4 m long has the cross section shown in Fig. P–1002. It carries a uniformly distributed load of 20 kN/m over the middle half of the span. If $n = 15$, compute the maximum stresses in the wood and the steel.

1004. Repeat Problem 1002 assuming that the reinforcement consists of aluminum plates for which the allowable stress is 80 MN/m^2. Use $n = 5$.

1005. A timber beam 150 mm by 250 mm is reinforced at the bottom only by a steel plate, as shown in Fig. P–1005. Determine the concentrated load that can be applied at the center of a simply supported span 6 m long if $n = 20$, $\sigma_s \leqslant 120$ MPa, and $\sigma_w \leqslant 8$ MPa. Show that the neutral axis is 170.2 mm below the top and that $I_{NA} = 416 \times 10^6$ mm^4. *Ans.* $P = 13.1$ kN

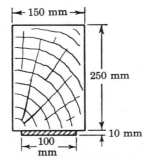

Figures P–1005, P–1006, P–1007.

1006. Determine the width b of the 10-mm steel plate fastened to the bottom of the beam in Problem 1005 that will simultaneously stress the wood and the steel to their permissible limits of 8 MN/m^2 and 120 MN/m^2, respectively.

Ans. $b = 42.5$ mm

1007. A uniformly distributed load of 4 kN/m (including the weight of the beam) is simply supported on a 6-m span. The cross section of the beam is as described in Problem 1005. If $n = 20$, determine the maximum stresses produced in the wood and the steel.

1008. A timber beam 150 mm wide by 250 mm deep is to be reinforced at the top and bottom by steel plates 10 mm thick. How wide should the steel plates be if the beam is to resist a moment of 50 kN·m? Assume that $n = 15$ and the allowable stresses in the wood and steel are 8 MPa and 110 MPa, respectively. *Ans.* $b = 143$ mm

1009. A timber beam 150 mm wide by 200 mm deep is to be reinforced at the top and bottom by aluminum plates 6 mm thick. Determine the width of the aluminum plates if the beam is to resist a moment of 16 kN·m. Assume $n = 5$ and take the allowable stresses as 8 MN/m^2 and 70 MN/m^2 in the wood and aluminum, respectively.

1010.   A pair of C250 × 30 steel channels are securely bolted to a wood beam 200 mm by 254 mm, as shown in Fig. P–1010. (From Table B–2 in Appendix B, the depth of the channel is also 254 mm.) If bending occurs about the axis 1–1, determine the safe resisting moment if the allowable stresses are $\sigma_s = 120$ MPa and $\sigma_w = 8$ MPa. Assume $n = 20$.   *Ans.* $M = 74.7$ kN·m

1011.   In Problem 1010, determine the safe resisting moment if bending occurs about axis 2–2.

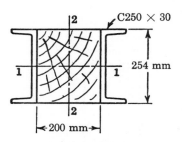

Figures P–1010 and P–1011.

1012.   An aluminum beam having the properties of a W200 × 46 section is reinforced by bolting 6-mm steel plates 203 mm wide to the flanges, as shown in Fig. P–1012. The allowable stresses in the steel and aluminum are 140 MPa and 100 MPa, respectively, and the ratio $E_s/E_a = 3$. Determine (a) the percentage change in the strength of the original unreinforced aluminum section and (b) the percentage change in the flexural rigidity EI.   *Ans.*   (a) +21.4%; (b) +175%

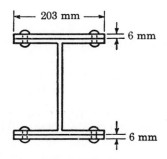

Figure P–1012.

1013.   A solid steel beam 50 mm in diameter is protected against corrosion by a shell of aluminum 6 mm thick firmly bonded to it. Compute the maximum moment the composite section can resist if $\sigma_s \leqslant 120$ MPa and $\sigma_a \leqslant 100$ MPa. Assume $E_s/E_a = 3$.

1014.   A rectangular section 150 mm wide by 250 mm deep resists a bending moment of 140 kN·m. The material of the beam is nonisotropic with the tensile modulus of elasticity twice the compressive modulus. Determine the maximum tensile and compressive stresses in the section.   *Ans.* $\sigma_t = 108$ MPa; $\sigma_c = 76.5$ MPa

1015.   Solve Problem 1014 if the compressive modulus of elasticity is 1.5 times the tensile value.

1016.   An experimental beam is composed of the three materials shown in Fig. P–1016. The materials are firmly fastened together so that there is no relative movement between them. Determine the safe resisting moment if $\sigma_s \leqslant 120$ MN/m², $\sigma_a \leqslant 80$ MN/m², $\sigma_w < 10$ MN/m², and $E_s = 200$ GN/m², $E_a = 70$ GN/m², $E_w = 10$ GN/m².
Ans. $M = 33.8$ kN·m

1017.   In a section similar to that in Fig. 10–2a, on page 418, the width b is 140 mm for each material. The vertical dimensions are 20 mm for steel, 150 mm for aluminum, and 50 mm for bronze. Assuming that the materials are firmly bonded together, determine the maximum stress in each material when the section is resisting a bending moment of 70 kN·m if $E_s = 200$ GPa, $E_a = 70$ GPa, and $E_b = 80$ GPa.

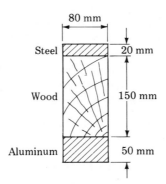

Figure P–1016.

10–3   SHEARING STRESS AND DEFLECTION IN COMPOSITE BEAMS

The formula for horizontal shearing stess [Eq. (5–4), page 189] developed for homogeneous beams applies equally well to the equivalent section of a composite beam because its derivation was based on the difference in normal forces between two adjacent sections. Since the forces on the original composite section and on the equivalent section are the same (see page 417), Eq. (5–4) is valid for either section.

Deflections in composite beams can also be computed, as in homogeneous beams, by using the flexural rigidity EI of the equivalent section. This is true because the deflection is the result of the changes in length of the fibers in the beam, and one of the basic principles of composite beams is that the strains of fibers are identical at corresponding points in the original and equivalent beams.

PROBLEMS

1018. Compute the allowable total vertical shear for a beam having the same cross section as in Problem 1005, if $n = 20$ and the maximum shear stress is 800 kN/m². *Ans.* $V = 23.0$ kN

1019. In the beam section in Problem 1010, assume that the channels are bolted to the wood by two rows of 20-mm bolts spaced 300 mm apart and located 75 mm above and below axis 1–1. Assuming $n = 20$, compute the shearing stress in the bolts caused by a central load of 80 kN applied to a simply supported span 3 m long, if bending takes place (a) about axis 1–1 and (b) about axis 2–2.
Ans. (b) $\tau = 74.8$ MPa

1020. The beam in Problem 1002 carries a uniformly distributed load of 30 kN/m on a simply supported span 5 m long. If $E_s = 200$ GN/m² and $E_w = 10$ GN/m², compute the midspan deflection.
Ans. $\delta = 12.2$ mm

1021. In Problem 1016, determine the shear flow developed between the steel and wood and between the wood and aluminum. Express the results as a function of the vertical shear V measured in newtons. *Ans.* $5.09V$ N/m; $5.12V$ N/m

10–4   REINFORCED CONCRETE BEAMS

Concrete is an excellent building material because it is cheap and fireproof and does not rust or rot. It has about the same strength in compression as soft wood, but its tensile strength is practically zero. For this reason, the tensile side of concrete beams is reinforced with steel bars. Ideally, these steel bars should follow the tensile stress trajectories, but practically they are placed in one layer on the tensile side. Fortunately there is a natural adhesion or bond between concrete and steel; hence no slipping occurs between them during bending,* and the

*Sufficient bond is developed in long beams to permit the steel bars to be laid straight, but in short beams the ends are usually bent over to anchor the steel more securely in the concrete.

principles developed in the preceding article can be used here. Fortunately also, both concrete and steel have about the same coefficient of temperature expansion.

It is usually assumed that the concrete carries no tensile stress, the tensile side of the concrete beam serving merely to position the steel which carries the entire tensile load. The steel is assumed to be uniformly stressed (since it is all at nearly the same distance from the neutral axis), so the line of action of the tensile force acts through the center of the reinforcing steel. The compressive stress in the concrete is assumed to vary linearly from the neutral axis, which places the resultant compressive force at the centroid of the compressive stress triangle (see Fig. 10–4a). The value of E for concrete is usually assumed to be between $\frac{1}{10}$ and $\frac{1}{6}$ that for steel, depending upon the quality of the concrete.

The portion of a reinforced concrete beam in Fig. 10–4a has the cross section shown in Fig. 10–4b. The equivalent section in terms of concrete, shown in Fig. 10–4c, is obtained by using Eq. (10–1) to transform the steel area A_s into the equivalent concrete area nA_s. As before, n is the ratio of the moduli of elasticity, i.e., $n = E_s/E_c$. The shaded portions of Fig. 10–4c indicate the areas that are effective in resisting bending. The distance from the top of the beam to the center of the reinforcing steel is conventionally denoted by the symbol d, and the location of the neutral axis is specified as being a fractional part k of this distance.

If the quantities b, d, A_s, and n are known, the neutral axis (the centroidal axis of the shaded areas in Fig. 10–4c) is located by applying the principle that the moment of area above the neutral axis equals the

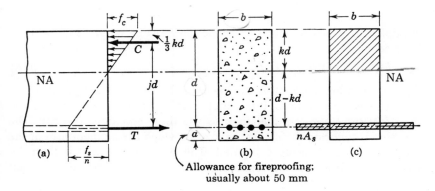

Figure 10–4.   Reinforced concrete beam and equivalent section.

moment of area below this axis:

$$(b\ kd)\left(\frac{kd}{2}\right) = nA_s(d - kd) \tag{10-2}$$

This is a quadratic equation in terms of the distance kd. Sometimes the equation is solved directly for k, but it is more useful to determine kd.

The resultant compressive force C in the concrete acts at the centroid of the stress triangle in Fig. 10-4a, and is therefore at a distance $\frac{1}{3}kd$ from the top of the beam.* The resisting couple, composed of the equal compressive and tensile forces C and T, therefore has a moment arm jd equal to

$$jd = d - \tfrac{1}{3}(kd) \tag{10-3}$$

The notation for stresses in reinforced concrete differs from that used elsewhere in this book, it being conventional to denote the stress in the concrete by f_c and the stress in the steel by f_s. The neutral axis having been located, the moment of inertia of the equivalent section may be computed and the flexure formula applied as in Art. 10-3. However, it is more direct to compute the resisting moment from the couple composed of the compressive force C and the tensile force T; this value is $C(jd)$ or $T(jd)$.

According to this concept, the average compressive stress in the concrete is $\frac{1}{2}f_c$, where f_c is the maximum compressive stress. The compressive force C in the concrete is the product of the average compressive stress multiplied by $b\ kd$ (the area under compression):

$$C = \tfrac{1}{2}f_c(b\ kd) \tag{10-4}$$

The resisting moment based on the maximum compressive stress is therefore

$$M_c = C(jd) = \tfrac{1}{2}f_c(b\ kd)(jd) \tag{10-5}$$

The tensile force T in the steel is the product of the steel area A_s multiplied by the steel stress f_s; hence the resisting moment in terms of the steel is

$$M_s = T(jd) = f_sA_s(jd) \tag{10-6}$$

The safe bending moment is the lower of the two values M_c and M_s.

In solving problems, it is better to follow numerically the steps in the derivations above rather than to use these equations directly. The first step is to locate the neutral axis, after which the moment arm of the resisting couple is easily computed. The resisting moment is then determined by means of the concept of force (C or T) times the moment arm jd.

*This statement and the procedure which follows should be compared with the discussion of a homogeneous rectangular section on pages 158 and 160.

ILLUSTRATIVE PROBLEMS

1022.   In a reinforced concrete beam, $b = 300$ mm, $d = 500$ mm, $A_s = 1500$ mm², $n = 8$. Determine the maximum stresses in the concrete and steel produced by a bending moment of 70 kN·m.

Solution: The equivalent section of the beam is shown in Fig. 10–5. Since the total moment of area about the neutral axis is zero,

$$[\Sigma ay = 0] \qquad 300kd\left(\frac{kd}{2}\right) = (12 \times 10^3)(500 - kd)$$

which reduces to

$$(kd)^2 + 80kd - (40 \times 10^3) = 0$$

from which

$$kd = 164 \text{ mm} = 0.164 \text{ m}$$

The moment arm of the resisting moment is

$$\left[jd = d - \tfrac{1}{3}kd \right] \qquad jd = 500 - \tfrac{1}{3}(164) = 445 \text{ mm} = 0.445 \text{ m}$$

In terms of the concrete, the resisting moment is

$$\left[M = \tfrac{1}{2}f_c(b\ kd)(jd) \right] \qquad 70 \times 10^3 = \tfrac{1}{2}f_c(0.300)(0.164)(0.445)$$

from which the maximum compressive stress is

$$f_c = 6.39 \text{ MPa} \qquad Ans.$$

In terms of the steel, the resisting moment is

$$\left[M = f_s A_s jd \right] \qquad 70 \times 10^3 = f_s(1500 \times 10^{-6})(0.445)$$

from which the steel stress is

$$f_s = 105 \text{ MPa} \qquad Ans.$$

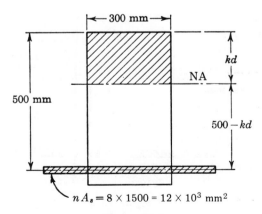

Figure 10–5.

1023.   In a reinforced concrete beam, $b = 250$ mm, $d = 400$ mm, $A_s = 1000$ mm^2, and $n = 8$. If the allowable stresses are $f_c \leqslant 12$ MPa and $f_s \leqslant 140$ MPa, determine the maximum bending moment that may be applied. Is the beam over- or under-reinforced?

Solution: The equivalent section of this beam is shown in Fig. 10–6. Proceeding as in Illustrative Problem 1022, we begin by computing the factors kd and jd:

$$[\Sigma ay = 0] \qquad 250kd\left(\frac{kd}{2}\right) = (8 \times 10^3)(400 - kd)$$

whence

$$kd = 131 \text{ mm} = 0.131 \text{ m}$$
$$\left[jd = d - \tfrac{1}{3}kd\right] \qquad jd = 400 - \tfrac{1}{3}(131) = 356 \text{ mm} = 0.356 \text{ m}$$

To stress the concrete to its maximum limit will require a bending moment

$$\left[M_c = \tfrac{1}{2}f_c(b\,kd)(jd)\right]$$
$$M_c = \tfrac{1}{2}(12 \times 10^6)(0.250)(0.131)(0.356)$$
$$= 70.0 \text{ kN·m}$$

To stress the steel to its limit, the required bending moment is

$$[M_s = f_s A_s jd] \qquad M_s = (140 \times 10^6)(1000 \times 10^{-6})(0.356)$$
$$= 49.8 \text{ kN·m}$$

The safe bending moment is therefore 49.8 kN·m. Since the steel governs, we conclude there is not enough steel; hence the beam is under-reinforced.

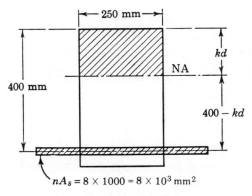

250 mm

kd

NA

400 mm

$400 - kd$

$nA_s = 8 \times 1000 = 8 \times 10^3 \text{ mm}^2$

Figure 10–6.

PROBLEMS

1024. In a reinforced concrete beam, $b = 200$ mm, $d = 400$ mm, $A_s = 1400$ mm². Find the values of kd and jd if (a) $n = 6$ and (b) $n = 10$.   *Ans.* (b) $kd = 177$ mm; $jd = 341$ mm

1025. In a reinforced concrete beam, $b = 250$ mm, $d = 450$ mm, and $n = 10$. The actual maximum stresses developed are $f_c = 6$ MPa and $f_s = 120$ MPa. Determine the applied bending moment and the steel area required.   *Ans.* $M = 45.0$ kN·m; $A_s = 938$ mm²

1026. Repeat Problem 1025 if $d = 540$ mm, all other data remaining unchanged.

1027. Determine the maximum stresses produced in the concrete and steel of a reinforced beam by a bending moment of 70 kN·m if $b = 300$ mm, $d = 500$ mm, $A_s = 1200$ mm², and $n = 8$.
                     Ans. $f_c = 6.91$ MN/m²; $f_s = 130$ MN/m²

1028. In a reinforced concrete beam, $b = 500$ mm, $d = 750$ mm, $A_s = 6000$ mm², and $n = 10$. What are the maximum stresses developed in the concrete and the steel by a bending moment of 270 kN·m?

1029. The dimensions of a reinforced concrete beam are $b = 300$ mm, $d = 450$ mm, $A_s = 1400$ mm², and $n = 8$. If the allowable stresses are $f_c ≤ 12$ MN/m² and $f_s ≤ 140$ MN/m², determine the maximum bending moment that may be applied. In what state of reinforcement is the beam?   *Ans.* $M = 78.4$ kN·m; under-reinforced

1030. In a reinforced beam, $b = 250$ mm, $d = 450$ mm, $A_s = 1400$ mm², and $n = 8$. Determine the safe uniformly distributed load that can be carried on a simply supported span 4 m long if $f_c ≤ 12$ MPa and $f_s ≤ 140$ MPa. Assume 50 mm of concrete below the reinforcing steel, and include the weight of the beam. The density of concrete is 2400 kg/m³.

1031. In a reinforced concrete beam, $b = 300$ mm, $d = 600$ mm, and $n = 9$. If a maximum stress of 5 MPa is developed in the concrete when resisting a moment of 80 kN·m, what stress is developed in the steel? What area of reinforcing steel is required?
                     Ans. $f_s = 90$ MPa; $A_s = 1670$ mm²

1032. Solve Problem 1031 if the bending moment is 70 kN·m, all other data remaining unchanged.

1033. Solve Problem 1027 by computing the moment of inertia of the transformed section and then applying the flexure formula according to the procedure in Art. 10–2. The distance of the equivalent concrete area from the neutral axis of the transformed section can be taken as its radius of gyration with respect to this axis.

1034. Solve Problem 1029, using the procedure outlined in Problem 1033.

10–5 DESIGN OF REINFORCED CONCRETE BEAMS

In the preceding article the dimensions of the reinforced beam were specified. This fixed the location of the neutral axis. Inasmuch as the stresses vary directly with their distance from the neutral axis, the applied bending moment may stress the concrete to its permissible limit while leaving the steel understressed—a condition known as over-reinforcement. The opposite condition, under-reinforcement, may occur when the steel reaches its permissible limit first, the concrete remaining understressed. For maximum economy, both materials should reach their limiting stresses simultaneously—a condition known as *balanced-stress reinforcement*.

In designing a concrete beam with balanced-stress reinforcement, therefore, we start with the assumption that the position of the neutral axis is such that the maximum f_c and the maximum stress f_s/n in the equivalent concrete occur simultaneously; this is shown in the stress diagram in Fig. 10–7. From this, by the proportional relations between the triangles ABC and ADE, we obtain

$$\frac{kd}{d} = \frac{f_c}{\dfrac{f_s}{n} + f_c}$$

or

$$k = \frac{f_c}{\dfrac{f_s}{n} + f_c} \qquad (10\text{--}7)$$

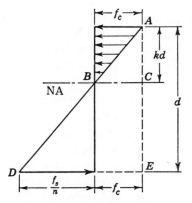

Figure 10–7. Stress distribution.

Having computed k in terms of the allowable stresses, we obtain the value of j by canceling out the term d in Eq. (10–3):

$$j = 1 - \tfrac{1}{3}k \qquad\qquad (10\text{–}8)$$

Once the values of k and j are determined, Eq. (10–5) is used to compute the quantity bd^2. The deeper the beam, the greater will be the moment arm of the resisting couple and the smaller the force. A deep beam therefore requires less concrete and steel than a shallow beam. Because of practical limits to the depth, however, d is usually made about $1.5b$. From this and the now computed value of bd^2, the dimensions b and d are found.

As the final step, the area of the reinforcing steel is computed from Eq. (10–6) or, preferably, from the condition that $C = T = A_s f_s$, in which A_s is now the only unknown. However, the reinforcing rods generally available are not of such size as to equal precisely the steel area required. As a consequence, balanced-stress reinforcement can be only closely approximated.

For most well-designed rectangular beams, the values of k and j are very close to $k = \tfrac{3}{8}$ and $j = \tfrac{7}{8}$. If these values are used, dimensions may be rapidly estimated by substituting them in Eq. (10–5), thereby giving*

$$bd^2 = \frac{6M}{f_c} \qquad\qquad (10\text{–}9)$$

Having thus found bd^2 and assigning values to b and d, we compute the tensile force in the steel and the steel area from

$$T = \frac{M}{\tfrac{7}{8}d} \quad \text{and} \quad A_s = \frac{T}{f_s} \qquad\qquad (10\text{–}10)$$

ILLUSTRATIVE PROBLEM

1035. Design a concrete beam with balanced-stress reinforcement to resist a bending moment of 90 kN·m. The allowable stresses are $f_c = 12$ MPa, $f_s = 140$ MPa, and $n = 8$.

Solution: With balanced-stress reinforcement, the stresses in the concrete and the concrete equivalent of the steel have the values shown in Fig. 10–8. From the proportional relations between the similar triangles ABD and AEF, the value of k is found to be

$$\frac{kd}{d} = \frac{12}{(140/8) + 12}; \qquad k = 0.407$$

---

*The similarity of this result to the flexure formula for rectangular beams, $\sigma = 6M/bh^2$, makes it simple to remember.

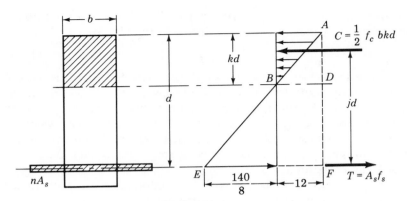

Figure 10–8.

from which the value of j is

$$j = 1 - \tfrac{1}{3}k = 1 - \tfrac{1}{3}(0.407) = 0.864$$

In terms of the concrete, the resisting moment is $C \cdot (jd)$, so

$$\left[M_c = C \cdot (jd) = \left(\tfrac{1}{2}f_c bkd\right)(jd) \right]$$
$$90 \times 10^3 = \tfrac{1}{2}(12 \times 10^6)(bd^2)(0.407)(0.864)$$
$$bd^2 = 0.0427 \text{ m}^3 = 42.7 \times 10^6 \text{ mm}^3 \qquad\qquad (a)$$

Assuming that $d = 1.5b$, we now obtain from Eq. (a), $b = 267$ mm and $d = 400$ mm.

The area of reinforcing steel is now the only unknown. Since the compressive force C in the concrete equals the tensile force T in the steel, we obtain,

$$\left[\tfrac{1}{2}f_c bkd = A_s f_s \right]$$
$$\tfrac{1}{2}(12 \times 10^6)(0.267)(0.407)(0.400) = A_s(140 \times 10^6)$$

whence

$$A_s = 1.86 \times 10^{-3} \text{ m}^2 = 1860 \text{ mm}^2$$

Usually the available stock sizes of reinforcing steel do not produce exactly this area of steel, so the final design only closely approximates balanced-stress reinforcement.

PROBLEMS

1036. A reinforced concrete beam is designed to reach $f_c = 12$ MPa and $f_s = 140$ MPa simultaneously. If $n = 8$ and $d = 450$ mm, compute the moment arm of the resisting couple.

Ans. $jd = 389$ mm

1037.   In a reinforced beam, $d = 600$ mm and $n = 9$. Find the dimensions b and A_s that will resist a bending moment of 80 kN·m with balanced-stress reinforcement, if $f_c = 9 \times 10^6$ N/m^2 and $f_s = 140 \times 10^6$ N/m^2.

1038.   In a reinforced beam, $b = 250$ mm, $d = 450$ mm, and $n = 9$; the allowable stresses are $f_c = 10$ MPa and $f_s = 140$ MPa. Determine A_s for balanced-stress design and the safe resisting moment.

1039.   Design a reinforced concrete beam with balanced-stress reinforcement that will resist a bending moment of 140 kN·m, assuming $d = 1.5b$, $f_c = 12$ MPa, $f_s = 160$ MPa, and $n = 8$.

Ans. $b = 316$ mm; $A_s = 2110$ mm^2

1040.   Solve Problem 1039 if $b = \frac{3}{4}d$.

1041.   A simply supported beam 6 m long is designed to carry a concentrated load of 80 kN at midspan. Compute b and A_s for a depth $d = 600$ mm using balanced-stress reinforcement with $f_c = 8$ MN/m^2, $f_s = 120$ MN/m^2, and $n = 10$. Allow 50 mm of concrete below the steel and include the weight of the beam, assuming the density of concrete is 2400 kg/m^3. (*Hint:* Assume an initial weight per meter for the beam and check this assumption after you have found dimensions.)

Ans. $b = 278$ mm; $A_s = 2220$ mm^2

1042.   A reinforced concrete beam 6 m long and perfectly restrained at the ends is to carry a live load of 20 kN/m in addition to its weight. Assuming $d = 600$ mm, design a beam with balanced-stress reinforcement, using $f_c = 6$ MPa, $f_s = 120$ MPa, and $n = 10$. Allow 50 mm of concrete below the reinforcing steel. The density of concrete is 2400 kg/m^3. (Use the hint in Problem 1041.)

Ans. $b = 219$ mm; $A_s = 1100$ mm^2

1043.   Design a reinforced concrete beam with balanced-stress reinforcement to carry a live load of 80 kN/m over a simple span 4 m long. Use $f_c = 12$ MN/m^2 and $f_s = 140$ MN/m^2, and $n = 8$. Allow 50 mm of concrete below the steel and include the weight of the beam, assuming the density of concrete is 2400 kg/m^3. Assume $b = 200$ mm. (Use the hint in Problem 1041.)

10–6   TEE BEAMS OF REINFORCED CONCRETE

The method used for rectangular reinforced concrete beams becomes quite involved when applied to T beams. Because of the flange of the T, the centroid of the compressive area is no longer $\frac{1}{2}kd$ from the neutral axis, nor is the line of action of the resultant compressive force $\frac{1}{3}kd$ from the top of the beam. As a consequence, it is cumbersome to use the basic procedure described in Art. 10–4, although textbooks on reinforced concrete do develop equations in terms of k and j. It is preferable for the beginner to apply the flexure formula directly to an equivalent section, as indicated in Art. 10–2 and as illustrated in the following problem.

ILLUSTRATIVE PROBLEM

1044. The T beam in Fig. 10–9 is reinforced with 2400 mm² of steel. Assuming $n = 8, f_c \leqslant 12$ MPa, $f_s \leqslant 140$ MPa, determine the maximum safe resisting moment.

Solution: Denoting the distance from the bottom of the flange by y, we compute the moments of area about the neutral axis:

$$[\Sigma ay = 0]$$

$$(750 \times 100)(y + 50) + (400y)\left(\frac{y}{2}\right) - (19.2 \times 10^3)(500 - y) = 0$$

This reduces to

$$y^2 + 471y - (29.25 \times 10^3) = 0$$

from which

$$y = 55.6 \text{ mm}$$

The moment of inertia about the neutral axis is computed by resolving the compressive area into a rectangle 750 mm by 155.6 mm

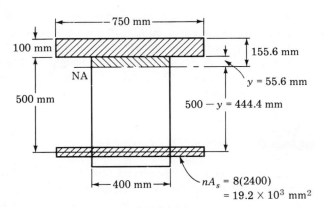

Figure 10–9.

reduced by a rectangle 350 mm by 55.6 mm. Thus we have

$$I = \frac{750(155.6)^3}{3} - \frac{350(55.6)^3}{3} + (19.2 \times 10^3)(444.4)^2$$

$$= 4714 \times 10^6 \text{ mm}^4 = 4714 \times 10^{-6} \text{ m}^4$$

The distance 444.4 mm may be taken as the radius of gyration of the concrete equivalent of the steel.

In terms of the permissible concrete stress, the flexure formula gives

$$\left[M = \frac{\sigma I}{y} \right] \qquad M_c = \frac{(12 \times 10^6)(4714 \times 10^{-6})}{155.6 \times 10^{-3}} = 364 \text{ kN·m}$$

In terms of the concrete equivalent of the steel, the permissible limit is $f_s/n = 140/8 = 17.5$ MPa, and the flexure formula gives

$$\left[M = \frac{\sigma I}{y} \right] \qquad M_s = \frac{(17.5 \times 10^6)(4714 \times 10^{-6})}{444.4 \times 10^{-3}} = 186 \text{ kN·m}$$

This, being smaller than M_c, is the maximum safe resisting moment.

PROBLEMS

1045.   In the reinforced T beam in Fig. P–1045, $b_1 = 500$ mm, $h_1 = 150$ mm, $b = 250$ mm, $h = 500$ mm, $A_s = 3000$ mm², and $n = 10$. Compute the maximum stresses produced in the concrete and the steel by a bending moment of 140 kN·m.

Ans. $f_c = 4.45$ MPa; $f_s = 80.5$ MPa

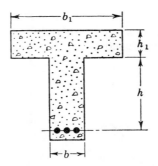

Figures P–1045, P–1046, P–1047.

1046.   The dimensions of the reinforced concrete T beam in Fig. P–1045 are $b_1 = 750$ mm, $h_1 = 100$ mm, $b = 300$ mm, and $h = 450$ mm. If $n = 8$ and $A_s = 3300$ mm², determine the maximum bending moment that may be applied without exceeding $f_c = 12$ MN/m² and $f_s = 140$ MN/m².   *Ans.* $M = 231$ kN·m

1047.   In the reinforced concrete T beam in Fig. P–1045, $b_1 = 900$ mm, $h_1 = 80$ mm, $b = 300$ mm, $h = 520$ mm, and $n = 9$. Find A_s and the maximum resisting moment for a balanced-stress design, using $f_c = 9$ MPa and $f_s = 160$ MPa.

10–7   SHEARING STRESS AND BOND STRESS

In the bending of reinforced concrete beams, the steel is prevented from sliding by the grip of the enveloping concrete. The stress developed by dividing this gripping force by the surface area of the reinforcing bars per linear meter is the *bond stress*. The bond stress is analogous to the shearing stress in a homogeneous beam; it may be computed by applying Eq. (5–4) (page 189) to the equivalent section of concrete shown in Fig. 10–10. Thus we obtain

$$\left[\tau = \frac{V}{Ib} A' \bar{y} \right] \qquad \tau = \frac{V}{Ib'} (nA_s)(d - kd) \tag{a}$$

where b' is the effective width of the steel bars, equivalent to the sum of the perimeters of the steel bars, usually expressed by Σo.

The moment of inertia of the transformed section is found from the flexure formula. Thus, at the concrete equivalent of the steel, the stress is f_s/n, so

$$\left[\sigma = \frac{My}{I} \right] \qquad \frac{f_s}{n} = \frac{M(d - kd)}{I} \tag{b}$$

But $M = T(jd) = A_s f_s jd$; hence Eq. (b) reduces to

$$I = nA_s(d - kd)(jd) \tag{c}$$

Since $nA_s(d - kd) = (bkd)(kd/2)$, as we saw in Eq. 10–2, the moment of inertia can also be expressed by

$$I = \tfrac{1}{2} b(kd)^2 (jd) \tag{d}$$

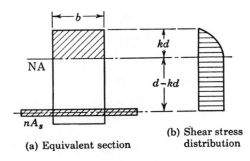

(a) Equivalent section   (b) Shear stress distribution

Figure 10–10.

Substituting in Eq. (a) the value of I from Eq. (c) gives

$$\tau = \frac{V(nA_s)(d - kd)}{\left[nA_s(d - kd)(jd)\right]b'}$$

from which the bond stress is given by

$$\tau = \frac{V}{jd\,b'} = \frac{V}{jd\,\Sigma o} \tag{10-11}$$

The shearing stress at the neutral axis is similarly

$$\left[\tau = \frac{V}{Ib}A'\bar{y}\right] \qquad \text{Max. } \tau = \frac{V}{Ib}(nA_s)(d - kd)$$

By using the value of I defined in Eq. (c), we reduce this to

$$\textbf{Max.}\,\tau = \frac{V}{jd\,b} \tag{10-12}$$

Further examination of Eq. (5-4) shows that the shearing stress in the compressive portion of the concrete varies parabolically as in a homogeneous rectangular section, whereas it remains constant below the neutral axis, as shown in Fig. 10-10b. Comparison of Eqs. (10-11) and (10-12) shows that the bond stress becomes larger than the shearing stress on the neutral surface if Σo is smaller than the width b of the beam. To increase Σo and keep A_s constant, more reinforcing bars of smaller diameter may be used.

PROBLEMS

1048. The reinforced beam in Problem 1028 is subjected to a vertical shear $V = 120$ kN. Calculate the maximum shearing stress and bond stress if the reinforcement consists of six bars 20 mm square.
                                                                   Ans.   373 kPa; 389 kPa

1049. Determine the vertical shear that can be sustained by the beam in Problem 1027 if the reinforcing consists of four bars 10 mm square. Assume that the allowable shearing stress is 350 kN/m² and the allowable bond stress is 550 kN/m².   *Ans.* $V = 39.6$ kN

SUMMARY

Nonhomogeneous beams of two materials, for example, wood and steel, may be transformed into equivalent sections of one material by using the relation

$$A_w = nA_s \tag{10-1}$$

where n is the ratio of the moduli of elasticity of the transformed material to that of the equivalent material. The flexure formula may then be applied directly to the transformed section as shown in Arts. 10–2 and 10–6.

The basic procedure in Art. 10–4 is usually used with reinforced concrete beams. The principles outlined in deriving the following equations should be followed numerically, rather than slavishly applying the equations themselves:

$$(b\ kd)\left(\frac{kd}{2}\right) = nA_s(d - kd) \tag{10–2}$$

$$jd = d - \tfrac{1}{3}(kd) \tag{10–3}$$

The stresses are found from the following equations. If the allowable stresses are specified, the safe resisting moment of the beam is the smaller of the two values:

$$M_c = C(jd) = \tfrac{1}{2}f_c(b\ kd)(jd) \tag{10–5}$$

$$M_s = T(jd) = f_sA_s(jd) \tag{10–6}$$

In designing a concrete beam with balanced-stress reinforcement, the dimensions of the section are unknown, so the above equations cannot be applied. Nevertheless, a stress diagram can be drawn in terms of the specified maximum stresses, as in Fig. 10–7 in Art. 10–5, whence the position of the neutral axis is defined in terms of k as follows:

$$k = \frac{f_c}{\dfrac{f_s}{n} + f_c} \tag{10–7}$$

Equation (10–7) is based on the assumption of balanced-stress reinforcement, i.e., that both concrete and steel reach their allowable stresses simultaneously. This relation between the stresses does not exist when the dimensions of a beam are specified; hence Eq. (10–7) cannot be used as a shortcut substitute for Eq. (10–2) in locating the neutral axis.

The maximum shearing stress in reinforced rectangular concrete beams is given by

$$\text{Max. } \tau = \frac{V}{jd\ b} \tag{10–12}$$

and the bond stress by

$$\tau = \frac{V}{jd\ \Sigma o} \tag{10–11}$$

where Σo is the sum of the perimeters of the reinforcing steel bars.

11

Columns

11-1 INTRODUCTION

A column is a compression member that is so slender compared to its length that under gradually increasing loads it fails by buckling at loads considerably less than those required to cause failure by crushing. In this respect it differs from a short compression member, which, even if eccentrically loaded, undergoes negligible lateral deflection. Although there is no sharp line of demarcation between short compression members and columns, a compression member is generally considered to be a column when its unsupported length is more than 10 times its least lateral dimension.

Columns are usually subdivided into two groups: *long* and *intermediate*; sometimes the short compression block is considered to be a third group. The distinction between the three is determined by their behavior. Long columns fail by buckling or excessive lateral bending; intermediate columns, by a combination of crushing and buckling; short compression blocks, by crushing. We shall now examine these differences in detail.

An ideal column is assumed to be a homogeneous member of constant cross section that is initially straight and is subjected to axial compressive loads. However, actual columns always have small imperfections of material and fabrication, as well as unavoidable accidental

439

eccentricities of load, which produce the effect shown, greatly exaggerated, in Fig. 11–1. The initial crookedness of the column, together with the placement of the load, causes an indeterminate eccentricity e with respect to the centroid of a typical section $m-n$. The loading on this section is similar to that on an eccentrically loaded short strut (Art. 9–3, page 369), and the resultant stress is due to a combination of a direct compressive stress and a flexure stress.

If the eccentricity is small and the member short, the lateral deflection is negligible and the flexural stress is insignificant compared with the direct compressive stress. A long member, however, is quite flexible, because deflection is proportional to the cube of the length; hence a relatively low value of P may cause a large flexural stress accompanied by a negligible direct compressive stress. Thus, at the two extremes, a short column carries principally direct compressive stress, and a long column is subjected primarily to flexural stress. As the length of a column increases, the importance of the direct compressive stress decreases, and that of the flexural stress increases. Unfortunately, in the intermediate column range it has not been possible to determine the rates of change in these stresses or the proportion of each stress that constitutes the resultant stress. It is this indeterminateness that gives rise to the many formulas for intermediate columns; these are discussed in Art. 11–5.

For the present, we have deliberately refrained from establishing any criterion for the difference between long and intermediate columns, except the fact that the long column is subjected principally to flexural stress and the intermediate column to a combination of direct and

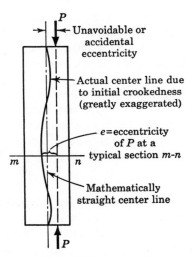

Figure 11–1. Factors contributing to eccentricity of loads in columns.

flexural stress. The distinction between them in terms of actual length can be discussed intelligently only after we have studied the action in a long column.

11-2   CRITICAL LOAD

A long beam is mounted vertically and hinged at the ends so that it is free to bend in any direction. A central horizontal load H is applied to cause bending in its most limber plane, as shown in Fig. 11-2a. Since flexural stress is proportional to deflection, there will be no change in stress if an axial load P is added at each end, as in Fig. 11-2b, H being simultaneously decreased as P increases so that the midspan deflection δ is unaltered. The midspan bending moment is then

$$M = \frac{H}{2}\left(\frac{L}{2}\right) + P\delta$$

which becomes

$$M = (P_{cr})\delta$$

when H has been reduced to zero. Here, as shown in Fig. 11-2c, P_{cr} is the critical load required to maintain the column in its deflected position without any side thrust. Any increase in P beyond this value increases the deflection δ, thereby increasing M, thence δ, etc., until the column buckles or fails. On the other hand, if P is decreased slightly below this critical value, the deflection is decreased, thereby decreasing the bending moment, thence the deflection, etc., and the column

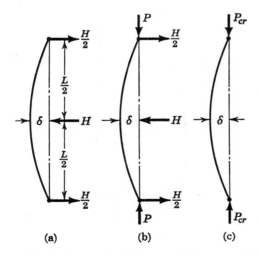

Figure 11-2. Beam and column sustaining equal deflections.

straightens out. A critical load, therefore, can be interpreted as the maximum axial load to which a column can be subjected and still remain straight, although in such an unstable condition that a slight sideways thrust will cause it to bow out, as in Fig. 11–2c. The next article discusses the computation of this critical load.

11–3 LONG COLUMNS BY EULER'S FORMULA

A theoretical analysis of the critical load for long columns was made by the great Swiss mathematician Leonhard Euler in 1757. His analysis is based on the differential equation of the elastic curve $EI(d^2y/dx^2) = M$. As we know now, such an analysis is valid only up to the stress at the proportional limit. In Euler's time, neither the concept of stress nor the limiting stress at the proportional limit had been formulated; hence he did not emphasize the concept of an upper limit to the critical load P. This upper limit is considered in Art. 11–4.

Figure 11–3 shows the center line of a column in equilibrium under the action of its critical load P. The column is assumed to have hinged ends (sometimes called round, pivoted, or pinned) restrained against lateral movement. The maximum deflection δ is so small that there is no appreciable difference between the original length of the column and its projection on a vertical plane. Under these conditions the slope dy/dx is so small that we may apply the approximate differential equation of the elastic curve of a beam, viz.,

$$EI\frac{d^2y}{dx^2} = M = P(-y) = -Py \qquad (a)$$

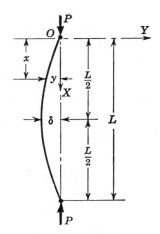

Figure 11–3.

M is negative because in Fig. 11–3 the deflection y is negative. If the column should deflect in the opposite direction so that y is positive, M would still be negative because of the sign convention adopted for bending moment in Art. 4–2.

Equation (a) cannot be integrated directly, as was done in Art. 6–2, because here M is not a function of x. However, we present two methods of solving it. Students who are familiar with dynamics will find Eq. (a) similar to the equation of a simple vibrating body:

$$m\frac{d^2x}{dt^2} = -kx$$

for which the general solution is

$$x = C_1 \sin\left(t\sqrt{\frac{k}{m}}\right) + C_2 \cos\left(t\sqrt{\frac{k}{m}}\right)$$

Hence, by analogy, the solution of Eq. (a) can be written at once as

$$y = C_1 \sin\left(x\sqrt{\frac{P}{EI}}\right) + C_2 \cos\left(x\sqrt{\frac{P}{EI}}\right) \tag{b}$$

Substituting $y = 0$ at $x = 0$ in Eq. (b) gives $C_2 = 0$. If we apply $y = 0$ at $x = L$, we obtain

$$0 = C_1 \sin\left(L\sqrt{\frac{P}{EI}}\right)$$

This is satisfied if $C_1 = 0$ (in which case there is no bending of the column), or by

$$L\sqrt{\frac{P}{EI}} = n\pi \qquad (n = 0, 1, 2, 3, \dots)$$

from which

$$P = n^2\frac{EI\pi^2}{L^2} \tag{c}$$

Students not familiar with dynamics can solve Eq. (a) by rewriting it in the form

$$EI\frac{d}{dx}\left(\frac{dy}{dx}\right) = -Py$$

After multiplying this by 2 dy to obtain perfect differentials, we get, by integration,

$$EI\left(\frac{dy}{dx}\right)^2 = -Py^2 + C_1 \tag{d}$$

Since, according to Fig. 11–3, $y = \delta$ when $dy/dx = 0$, substitution in Eq. (d) gives $C_1 = P\delta^2$, whence Eq. (d) becomes

$$EI\left(\frac{dy}{dx}\right)^2 = P(\delta^2 - y^2)$$

or

$$\frac{dy}{dx} = \sqrt{\frac{P}{EI}}\ \sqrt{\delta^2 - y^2}$$

Separating the variables, we obtain

$$\frac{dy}{\sqrt{\delta^2 - y^2}} = \sqrt{\frac{P}{EI}}\ dx$$

which is integrated to yield

$$\sin^{-1}\frac{y}{\delta} = x\sqrt{\frac{P}{EI}} + C_2$$

To evaluate C_2 we use the relationship $y = 0$ at $x = 0$; hence $C_2 = 0$, so

$$\sin^{-1}\frac{y}{\delta} = x\sqrt{\frac{P}{EI}} \quad\text{or}\quad y = \delta\sin\left(x\sqrt{\frac{P}{EI}}\right) \tag{e}$$

This indicates that the column has the shape of a sine curve. Setting $y = 0$ at $x = L$ in Eq. (e) gives

$$\sin\left(L\sqrt{\frac{P}{EI}}\right) = 0$$

or

$$L\sqrt{\frac{P}{EI}} = n\pi \quad (n = 0, 1, 2, 3, \dots)$$

from which

$$P = n^2\frac{EI\pi^2}{L^2} \tag{f}$$

This agrees with the value found previously in Eq. (c).

The value $n = 0$ is meaningless because then the load P is zero. For the other values of n, the column bends into the shapes shown in Fig. 11–4. Of these, the most important is (a); the others occur with larger loads and are possible only if the column is braced at the middle or third points respectively.* The critical load for a hinged-ended

*Bracing Fig. 11–4b at the midpoint reduces it to the shape in Fig. 11–4a with an equivalent length $\frac{1}{2}L$. Substituting $\frac{1}{2}L$ in place of L in Eq. (11–1) increases the critical load 4 times, which checks Eq. (f) for $n = 2$.

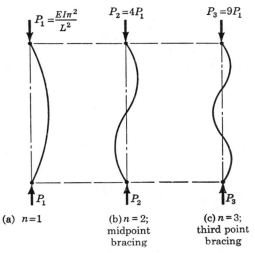

$$P_1 = \frac{EI\pi^2}{L^2} \qquad P_2 = 4P_1 \qquad P_3 = 9P_1$$

(a) $n=1$ (b) $n = 2$;   (c) $n = 3$;
                     midpoint   third point
                     bracing   bracing

Figure 11-4.   Effects of n on loads.

column is therefore

$$P = \frac{EI\pi^2}{L^2} \tag{11-1}$$

The critical load for columns with other end conditions can be expressed in terms of the critical load for a hinged column, which is taken as the fundamental case. Thus, from symmetry, the column with fixed ends in Fig. 11-5a has inflection points at the quarter points of its unsupported length. Since the bending moment is zero at a point of inflection, the free-body diagrams show that the middle half of the fixed-ended column is equivalent to a hinged column having an effective length $L_e = L/2$. If this is substituted in Eq. (11-1), the critical load on a fixed-ended column is

$$P = \frac{EI\pi^2}{L_e^2} = \frac{EI\pi^2}{\left(\dfrac{L}{2}\right)^2} = 4\frac{EI\pi^2}{L^2} \tag{11-2}$$

This is four times the strength of the column if its ends were hinged.

Figure 11-5a also provides a means of determining the load capacity of a column built in at one end and free at the other—the flagpole type of column. The critical loads on it (Fig. 11-5b) and on the fixed-ended column (Fig. 11-5a) are equal, provided the fixed-ended column is four times as long as the flagpole type. In other words, by substituting in Eq. (11-2) an equivalent length L_e that is four times its

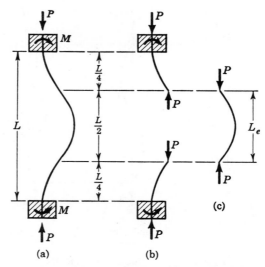

Figure 11–5. Built-in column and free-body diagrams.

actual length, the critical load on a flagpole column is given by

$$P = \frac{4EI\pi^2}{L_e^2} = \frac{4EI\pi^2}{(4L)^2} = \frac{1}{4}\frac{EI\pi^2}{L^2} \tag{11–3}$$

This load is one-quarter the critical load on a hinged column of the same length.

One other type of column is hinged at one end and built in at the other, as in Fig. 11–6. For it, the point of inflection can be shown to be nearly $0.7L$ from the hinged end. Hence substituting an effective length

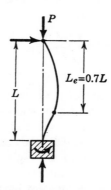

Figure 11–6. Column hinged at one end and built in at the other.

END CONDITION	N = NUMBER OF TIMES STRENGTH OF HINGED COLUMNS	L_e = EFFECTIVE LENGTH
Fixed ends	4	$\frac{1}{2}L$
One end fixed, the other hinged	2	$0.7L$
Both ends hinged	1	L
One end fixed, the other free	$\frac{1}{4}$	$2L$

$L_e = 0.7L$ in Eq. (11–1) gives

$$P = \frac{EI\pi^2}{L_e^2} = \frac{EI\pi^2}{(0.7L)^2} = 2\frac{EI\pi^2}{L^2} \text{ (very nearly)} \qquad (11\text{–}4)$$

The effect of end conditions on the critical load can therefore be expressed in terms of the critical load for the fundamental type of hinged column of the same length. All we need do is apply Eq. (11–1) multiplied by a factor N which varies with end conditions as summarized in the above table, or, preferably, replace L in Eq. (11–1) by the tabulated value of an effective or modified length L_e; that is,

$$P = N\frac{EI\pi^2}{L^2} = \frac{EI\pi^2}{L_e^2}$$

11–4   LIMITATIONS OF EULER'S FORMULA

A column always tends to buckle in its most limber direction. For this reason, and since flexural resistance varies with moment of inertia, the value of I in the column formulas is always the least moment of inertia of the cross section. Any tendency to buckle therefore occurs about the least axis of inertia of the cross section.

Euler's formula also shows that the critical load which causes buckling depends not upon the strength of the material but only upon its dimensions and modulus of elasticity. For this reason, two dimensionally identical slender struts, one of high-strength steel and the other of ordinary structural steel, will buckle under the same critical load because, although their strengths are different, they have the same modulus of elasticity. Good design also requires that a section have as large a moment of inertia as possible. Hence, for a given area, the material should be distributed as far as possible from the centroid and in such a way that the moments of inertia about the principal axes are equal or as nearly equal as possible.

In order for Euler's formula to be applicable, the stress accompanying the bending which occurs during buckling must not exceed the proportional limit. This stress may be found by replacing in Euler's formula the moment of inertia I by its equivalent Ar^2, where A is the cross-sectional area and r the least radius of gyration.* This being done for the fundamental case of a hinged column, Eq. (11–1) becomes

$$\frac{P}{A} = \frac{E\pi^2}{(L/r)^2} \tag{11–5}$$

For other end conditions, substitute in this equation the equivalent length of a hinged column from the table in the preceding article.

Here P/A is the *average stress* in the column when carrying its critical load. This stress is often called the *critical stress*. Its limiting value is the stress at the proportional limit. The ratio L/r is called the *slenderness ratio* of the column. Since an axially loaded column tends to buckle about the axis of least moment of inertia, the least radius of gyration should be used to determine the slenderness ratio.

Conventionally, we define long columns as those for which Euler's formula applies. The limiting slenderness ratio that fixes the lower limit for Euler's formula is easily found by substituting in Eq. (11–5) the known values of the proportional limit and the modulus of elasticity of the specified material. This limiting ratio varies with different materials and even with different grades of the same material.

As an example, for steel that has a proportional limit of 200 MPa and for which $E = 200$ GPa, the limiting slenderness ratio is

$$\left(\frac{L}{r}\right)^2 = \frac{(200 \times 10^9)\pi^2}{200 \times 10^6} \approx 10\ 000 \quad \text{or} \quad \frac{L}{r} \approx 100$$

Below this value, as shown in Fig. 11–7 by the dashed portion of Euler's curve, the Euler unit load exceeds the proportional limit. Hence for $L/r < 100$, Euler's formula is not valid, and the proportional limit is taken as the critical stress. The curve also shows that the critical or allowable stress on a column decreases rapidly as the slenderness ratio increases; hence it is good design to keep the slenderness ratio as small as possible.

Finally, remember that Euler's formulas determine critical loads, not working loads. It is therefore necessary to divide the right side of each formula by a suitable factor of safety—usually 2 to 3, depending on the material—in order to obtain practical allowable values.

*Here, we are using r to denote radius of gyration to conform to AISC notation. Be careful not to confuse this r with the r that is frequently used to denote the radius of a circle.

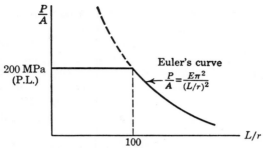

Figure 11-7. Critical or allowable stress is given by the solid line. Dashed portion of Euler's curve is not valid.

ILLUSTRATIVE PROBLEM

1101.   Select the lightest W shape that can be used as a column 7 m long to support an axial load of 450 kN with a factor of safety of 3. Assume (a) both ends hinged and (b) one end fixed and the other hinged. Use σ_{PL} = 200 MPa and E = 200 GPa.

Solution:

Part a.   For steel with a proportional limit of 200 MPa, the specifications for Euler's formula with hinged ends require that $L/r \geqslant$ 100. If $L/r < 100$, the limiting stress is the proportional limit.

The specified working load, when multiplied by the factor of safety, gives a critical Euler load of 1350 kN. Applying Euler's formula and solving for I, we obtain

$$\left[P = \frac{EI\pi^2}{L^2} \right] \quad I = \frac{PL^2}{E\pi^2} = \frac{(1350 \times 10^3)(7)^2}{(200 \times 10^9)(\pi^2)}$$
$$= 33.5 \times 10^{-6} \text{ m}^4 = 33.5 \times 10^6 \text{ mm}^4$$

Also, the slenderness ratio $L/r \geqslant 100$, from which the least r is

$$r \leqslant \frac{L}{100} = \frac{7000}{100} = 70.0 \text{ mm}$$

These criteria establish that the section must have a least $I \geqslant 33.5 \times 10^6$ mm⁴ and a least $r \leqslant 70.0$ mm. This is satisfied by choosing a W250 × 73 section with a least $I = 38.8 \times 10^6$ mm⁴ and a least $r = 64.6$ mm.

If the selection were based on the proportional limit, the section must have a minimum area of 6750 mm² (obtained by dividing the load of 1350 kN by the proportional limit of 200 MPa) and a least r greater than 70.0 mm. These conditions are satisfied by a W310 × 97 section with A = 12 300 mm^2 and least r = 76.9 mm.

The lightest section therefore is the W250 × 73 section.

Part b.   The critical Euler load is 1350 kN, as before. With one end fixed and the other hinged, the effective length of an equivalent hinged column is $0.7L$ = 0.7(7) = 4.9 m. Using this effective length in place of the actual length, we find that the criteria for Euler's formula are

$$I > \frac{PL^2}{E\pi^2} = \frac{(1350 \times 10^3)(4.9)^2}{(200 \times 10^9)\pi^2} = 16.4 \times 10^{-6} \text{ m}^4$$

$$> 16.4 \times 10^6 \text{ mm}^4$$

and

$$r < \frac{L}{100} = \frac{4900}{100} = 49.0 \text{ mm}$$

The lightest section that satisfies these conditions is the W360 × 64 with least I = 18.8 × 10^6 mm^4 and least r = 48.1 mm.

The other set of criteria based on the proportional limit are

$$A > \frac{1350 \times 10^3}{200 \times 10^6} = 6.75 \times 10^{-3} \text{ m}^2 = 6750 \text{ mm}^2 \quad \text{and}$$

$$r > 49.0 \text{ mm}$$

for which the lightest section available is the W250 × 58 with A = 7420 mm^2 and r = 50.4 mm. Comparing the two sets of criteria, we see that the proper section is the W250 × 58.

The unwary reader might be tempted to base his section only on I without checking r and thereby choose a W200 × 52 section with least I = 17.8 × 10^6 mm^4. However, this section has a least r = 51.8 mm and an area of 6660 mm^2, which results in a stress exceeding the proportional limit of 200 MPa. It is therefore not acceptable because it violates the stress–strain proportionality on which Euler's formula is based.

This problem demonstrates the importance of the slenderness ratio in column analysis. In Part a, the selection is governed by elastic stability (i.e., the use of Euler's formula), whereas in Part b the selection is determined by the proportional limit.

PROBLEMS

1102.   A 50-mm by 100-mm timber is used as a column with fixed ends. Determine the minimum length at which Euler's formula can be used if $E = 10$ GPa and the proportional limit is 30 MPa. What central load can be carried with a factor of safety of 2 if the length is 2.5 m?

Ans. $L = 1.66$ m; $P = 32.9$ kN

1103.   An aluminum strut 2 m long has a rectangular section 20 mm by 50 mm. A bolt through each end secures the strut so that it acts as a hinged column about an axis perpendicular to the 50-mm dimension and as a fixed-ended column about an axis perpendicular to the 20-mm dimension. Determine the safe central load, using a factor of safety of 2.5 and $E = 70$ GPa.   *Ans.* $P = 9.24$ kN

1104.   A square aluminum bar is to support a load of 40 kN on a length of 3 m. Assuming pinned ends, determine the length of each side. Use $E = 70$ GPa.   *Ans.*   50.0 mm

1105.   Repeat Problem 1104 assuming that the column is made of wood for which $E = 10$ GPa.

1106.   Two C310 × 45 channels are latticed together so they have equal moments of inertia about the principal axes. Determine the minimum length of a column having this section, assuming pinned ends, $E = 200$ GPa, and a proportional limit of 240 MPa. What safe load will the column carry for a length of 12 m with a factor of safety of 2.5?

Ans. $L = 9.89$ m; $P = 742$ kN

1107.   Repeat Problem 1106 assuming that one end is fixed and the other hinged.

1108.   Select the lightest W shape that will act as a column 8 m long with hinged ends and support an axial load of 270 kN with a factor of safety of 2.5. Assume that the proportional limit is 200 MPa and $E = 200$ GPa.   *Ans.*   W310 × 74

1109.   Select the lightest W shape that will act as a column 12 m long with fixed ends and support an axial load of 700 kN with a factor of safety of 2.0. Assume that the proportional limit is 200 MPa and $E = 200$ GPa.

11-5   INTERMEDIATE COLUMNS. EMPIRICAL FORMULAS

The preceding discussion showed that long columns can be treated by Euler's formula provided that the slenderness ratio is larger than the value at which the average stress reaches the proportional limit. For hinged steel columns, this limit is $L/r \approx 100$ at 200 MPa. Euler's formula is not valid for smaller slenderness ratios. The definition of a short column as one whose length does not exceed 10 times the least lateral dimension sets the upper limit of the slenderness ratio at about 30 for a rectangular section. For practical purposes, the limiting stress

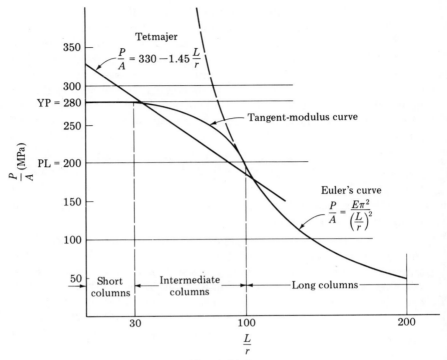

Figure 11–8.

on a short column has been found to be the stress at the yield point; extreme care is required to prevent buckling when stressed to this point. Figure 11–8 shows these conditions for steel having a yield point of 280 MPa and a proportional limit of 200 MPa.

Various methods have been proposed for bridging the gap between the short column range and the long column range. However, none of them has been accepted universally for intermediate columns, partly because of their departure from the stress–strain relationship when the stresses exceed the proportional limit, and partly because of their indeterminate mixture of direct and flexural stresses when loads are reduced by an appropriate factor of safety to bring the stresses below the proportional limit.

Most empirical formulas for intermediate columns have been developed for steel because it is such a common structural material. We shall discuss these first and then indicate their extension to other structural materials.

In one proposed method—that of the tangent-modulus theory—the Euler formula is extended to intermediate columns stressed above the proportional limit by replacing the constant modulus E by a reduced modulus $\bar{E}$, viz.,

$$\frac{P}{A} = \frac{\bar{E}\pi^2}{(L/r)^2} \tag{11-6}$$

The reduced modulus $\bar{E}$, also called the effective or tangent modulus, is obtained by using for $\bar{E}$ the slope of the tangent to the stress–strain diagram at the point corresponding to the average stress in the column. This yields a curve that connects the curves in Fig. 11–8 representing the short and long column formulas. Although this method is empirical because it violates the stress–strain proportionality assumed in the derivation of Euler's formula, actual tests show close agreement with the theoretical curve.[*]

Other methods are frankly empirical. One of the simplest, proposed by T. H. Johnson in 1886, consists of drawing a straight line through the average of the test points obtained by plotting values of P/A (when failure due to buckling appeared imminent) against the corresponding values of L/r. The general equation of this straight-line formula is

$$\frac{P}{A} = \sigma - C\frac{L}{r}$$

where σ is the intercept for $L/r = 0$ and C is the magnitude of the slope.

The results obtained by Tetmajer and Bauschinger with tests on structural steel bars with hinged ends have been widely used. Their results gave for the critical load[†]:

$$\frac{P}{A} = 330 - 1.45\frac{L}{r} \qquad \text{MPa} \tag{11-7}$$

This equation is shown in Fig. 11–8. As mentioned previously, the yield point is the practical limit to P/A; this is recognized in the formula by setting a lower limit to L/r that corresponds to 35 for a yield point of 280 MPa.

Equation (11–7), when divided by a factor of safety of 3, gives an equation for safe working loads that was once widely used in building codes. It is so conservative, however, that it has been largely superseded by others that will be discussed later. We hereafter refer to it in the following form as the straight-line formula:

$$\frac{P}{A} = 110 - 0.483\frac{L}{r} \qquad \text{MPa} \tag{11-8}$$

---

[*]For a discussion of the tangent-modulus method, see F. R. Shanley, *Strength of Materials*, McGraw-Hill, New York, 1957, pp. 582–588.

[†]Numerical values for empirical column formulas given here are approximate conversions to SI units.

This equation is limited to a slenderness range of $30 \leqslant L/r \leqslant 120$ for main members but may be extended to 150 for secondary members used for bracing. Below $L/r = 30$, $P/A = 96.5$ MPa is to be used.

Another widely used empirical column formula is the Rankine–Gordon formula developed about 1860. It assumes that the maximum deflection in a column varies with L^2/c; that is, $\delta_{max.} = \phi L^2/c$, where ϕ is a constant of proportionality depending upon end conditions. Then the maximum stress in a column is

$$\sigma = \frac{P}{A} + \frac{Mc}{I} = \frac{P}{A} + \frac{(P\delta_{max.})c}{Ar^2} = \frac{P}{A}\left[1 + \phi\left(\frac{L}{r}\right)^2\right]$$

whence the formula for average stress is given by

$$\frac{P}{A} = \frac{\sigma}{1 + \phi(L/r)^2}$$

A commonly used form, which we hereafter designate as the Rankine–Gordon formula, is

$$\frac{P}{A} = \frac{124}{1 + \dfrac{1}{18 \times 10^3}\left(\dfrac{L}{r}\right)^2} \qquad \textbf{MPa} \qquad\qquad (11\text{–}9)$$

This formula, which includes a factor of safety, is valid for main members with L/r between 60 and 120 and for secondary members with L/r up to 200. Below $L/r = 60$, a working stress of $P/A = 103$ MPa is specified.

Still another variation of the intermediate column formula is the parabolic type proposed in 1892 by Professor J. B. Johnson (not related to T. H. Johnson of straight-line fame). This formula has the general form

$$\frac{P}{A} = \sigma - C\left(\frac{L}{r}\right)^2$$

in which σ is the stress at the yield point and C is a constant chosen to make the parabola tangent to Euler's curve.

The American Institute of Steel Construction (AISC) defines the limit between intermediate and long columns to be the value of the slenderness ratio C_c given by

$$C_c = \sqrt{\frac{2\pi^2 E}{\sigma_{yp}}}$$

in which E is the modulus of elasticity (200 GPa, for most grades of steel) and σ_{yp} is the yield stress for the particular grade of steel being used. For columns of effective length L_e and minimum radius of gyration r, AISC specifies that for $L_e/r > C_c$, the working stress, σ_w, is given by

$$\sigma_w = \frac{12\pi^2 E}{23(L_e/r)^2} \tag{11-10}$$

(Note that this is Euler's formula with a factor of safety of $23/12 = 1.92$.) For $L_e/r < C_c$, AISC specifies the parabolic formula

$$\sigma_w = \left[1 - \frac{(L_e/r)^2}{2C_c^2}\right]\frac{\sigma_{yp}}{FS} \tag{11-11}$$

where the factor of safety, FS, is given by

$$FS = \frac{5}{3} + \frac{3(L_e/r)}{8C_c} - \frac{(L_e/r)^3}{8C_c^3} \tag{11-12}$$

Observe that the factor of safety is 1.92 at $L_e/r = C_c$, and it becomes smaller for larger values of the slenderness ratio. The variation of σ_w with L_e/r for several grades of steel is shown in Fig. 11-9.

Most column formulas are based on columns with hinged ends. Fixity of the ends increases the load capacity (see Art. 11-3): However, structural columns, which compose the great majority of intermediate columns, practically never have completely rigid ends. Hence it is good practice to assume hinged ends, even though the column is actually riveted or otherwise rigidified at its ends. The effective length may be used for economical column design in the case of partial or complete end restraints.

We should mention one formula that will be developed in Art. 11-7. This formula, called the *secant formula*, assumes a definite eccentricity of load and is theoretically correct but exceedingly cumbersome to use. It is given by

$$\frac{P}{A} = \frac{\sigma_{\max.}}{1 + \frac{ec}{r^2}\sec\left(\frac{L}{2r}\sqrt{\frac{P}{EA}}\right)} \tag{11-13}$$

where $\sigma_{\max.}$ is the maximum stress developed by a load P having a known eccentricity e. The term c is the perpendicular distance from the axis of bending to the extreme fiber, and ec/r^2 is the eccentricity ratio.*

We now consider column formulas for some materials other than steel. The Aluminum Association, Inc., lists column specifications for each of the various types of aluminum alloys. In these specifications, the allowable stress is a constant for short columns, a straight-line relation

---

*It is equivalent to $\delta_{\max.}c/r^2$ in the discussion of the Rankine–Gordon formula on p. 454.

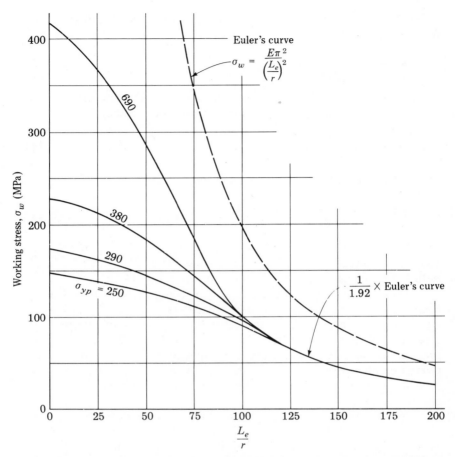

Figure 11-9. Working stress for columns (AISC specifications) for several grades of steel.

approximating the tangent-modulus formula is used for intermediate columns, and Euler's formula is used for long columns. For example, the specifications for 2014–T6 aluminum alloy are[*]

$$\sigma_w = 193 \quad \text{MPa} \qquad\qquad \frac{L}{r} \leqslant 12 \qquad\qquad (11\text{–}14)$$

$$\sigma_w = 212 - 1.59\frac{L}{r} \quad \text{MPa} \qquad 12 < \frac{L}{r} < 55 \qquad (11\text{–}15)$$

$$\sigma_w = \frac{372 \times 10^3}{(L/r)^2} \quad \text{MPa} \qquad \frac{L}{r} \geqslant 55 \qquad (11\text{–}16)$$

[*]The numerical values quoted here are approximate SI conversions of values found in *Specifications for Aluminum Structures*, 3rd ed., Construction Manual Series, Sec. 1, Aluminum Association, New York, April 1976, p. 21.

The column length L in the specifications is defined as the "length of the compression member between points of lateral support, or twice the length of a cantilever column (except where analysis shows that a shorter length can be used)."

For wood columns, the National Lumber Manufacturers Association[*] recommends Euler's formula in the following form:

$$\sigma_w = \frac{\pi^2 E}{2.727(L/r)^2} = \frac{3.619E}{(L/r)^2} \tag{11-17}$$

Of course, load duration and moisture content adjustments must also be made. For rectangular columns with *least lateral dimension d*, $r = \sqrt{d/12}$ and Eq. (11–17) reduces to

$$\sigma_w = \frac{0.3E}{(L/d)^2} \tag{11-18}$$

In addition to the empirical column formulas just presented, a great many other equations are used. However, it is unnecessary to memorize them, for the formula to be used must always be specified. All the equations have one feature in common: they reduce the safe working load as the slenderness ratio increases, although in varying proportions. Depending on the formula specified, the same column may therefore support any one of several safe *legal* loads.

ILLUSTRATIVE PROBLEMS

1110. Using AISC column specifications, determine the safe axial loads on a W360 × 122 section used as a column under the following conditions: (a) hinged ends and a length of 9 m; (b) built-in ends and an unsupported length of 10 m; (c) built-in ends and a length of 10 m braced at the midpoint. Use $\sigma_{yp} = 380$ MPa.

Solution: Table B–2, Appendix B, gives, for a W360 × 122 section, an area $A = 15\ 500$ mm^2 and a least r of 63.0 mm.

Part a. For $\sigma_{yp} = 380$ MPa, the limiting slenderness ratio is

$$\left[C_c = \sqrt{\frac{2\pi^2 E}{\sigma_{yp}}} \right] \qquad C_c = \sqrt{\frac{2\pi^2(200 \times 10^9)}{380 \times 10^6}} = 102$$

Here, the slenderness ratio is $L_e/r = 9000/63.0 = 143$ which is greater

[*]*National Design Specification*, National Lumber Manufacturers Association, Washington, D.C., 1962.

than C_c. Therefore, the working stress is given by

$$\left[\sigma_w = \frac{12\pi^2 E}{23\left(\dfrac{L_e}{r}\right)^2}\right] \qquad \sigma_w = \frac{12\pi^2(200 \times 10^9)}{23(143)^2} = 50.4 \text{ MPa}$$

and the safe axial load is

$$[\,P = \sigma A\,] \qquad P = (50.4 \times 10^6)(15\ 500 \times 10^{-6}) = 781 \text{ kN} \qquad \textit{Ans.}$$

Part b. Using the concept of effective length, we find that a column with built-in or fixed ends is equivalent to a hinged column of half the actual length. Hence, with $L_e = 0.5L = 0.5(10) = 5$ m, the slenderness ratio is $L_e/r = 5000/63.0 = 79.4$, which is less than $C_c = 102$. Therefore, the working stress is determined as follows:

$$\left[\text{FS} = \frac{5}{3} + \frac{3(L_e/r)}{8C_c} - \frac{(L_e/r)^3}{8C_c^3}\right]$$

$$\text{FS} = \frac{5}{3} + \frac{3(79.4)}{8(102)} - \frac{(79.4)^3}{8(102)^3} = 1.90$$

whence

$$\left[\sigma_w = \frac{\left[1 - \dfrac{(L_e/r)^2}{2C_c^2}\right]\sigma_{yp}}{\text{FS}}\right]$$

$$\sigma_w = \frac{\{1 - [(79.4)^2/2(102)^2]\}}{1.90}(380 \times 10^6) = 139 \text{ MPa}$$

Finally, the safe axial load is

$$[\,P = \sigma A\,] \qquad P = (139 \times 10^6)(15\ 500 \times 10^{-6})$$

$$= 2150 \text{ kN} \qquad \textit{Ans.}$$

Part c. Braced at the midpoint, the column is equivalent to one having a length of 5 m, fixed at one end and hinged at the other.

The effective length is $L_e = 0.7L = 0.7(5) = 3.5$ m, whence $L_e/r = 3500/63.0 = 55.6$, which is less than $C_c = 102$. Proceeding as in Part b, we find FS = 1.85 and $\sigma_w = 175$ MPa. Hence, the safe axial load is

$$[P = \sigma A] \qquad P = (175 \times 10^6)(15\,500 \times 10^{-6})$$
$$= 2710 \text{ kN} \qquad Ans.$$

This problem illustrates the increased strength of a column whose ends can be perfectly rigidified. Since this condition is never realized in practice, it is better when determining allowable loads always to assume hinged ends, or to be more realistic in selecting the effective length with fixed ends as about $0.75L$ instead of $0.5L$.

1111. Select the lightest W shape that will support an axial load of 360 kN on an effective length of 4.6 m. Use AISC column specifications with $\sigma_{yp} = 250$ MPa.

Solution: Since both the area A and the least radius of gyration r are unknown and no convenient relation between them can be set up, the selection of the lightest W shape involves a trial-and-error procedure. The steps are (1) assume a working stress; (2) calculate the area required; (3) select the lightest appropriate section based on the area required; and (4) for the section selected, calculate the allowable load based on the column specifications. If the allowable load equals (or is slightly larger than) the applied load, the section selected is the appropriate one. If the allowable load is less than the applied load, a heavier section must be selected and the procedure repeated. Clearly, the number of trials which must be attempted before the correct section is determined depends on how close the initial assumed stress is to the actual stress. One suggestion is to assume an initial working stress of 80% of the stress at $L/r = 0$ determined from the column specifications.

For steel with $\sigma_{yp} = 250$ MPa, the limiting slenderness ratio is

found to be $C_c = \sqrt{\dfrac{2\pi^2 E}{\sigma_{yp}}} = \sqrt{\dfrac{2\pi^2(200 \times 10^9)}{250 \times 10^6}} = 126$

First try. At $L_e/r = 0$, FS = $\frac{5}{3}$, and $\sigma_w = \sigma_{yp}/\text{FS} = 250/\frac{5}{3} = 150$ MPa. Assuming an initial stress of $0.80(150) = 120$ MPa, the required area is

$$A = \frac{P}{\sigma} = \frac{360 \times 10^3}{120 \times 10^6} = 3 \times 10^{-3} \text{ m}^2 = 3000 \text{ mm}^2$$

Therefore, from Table B–2, Appendix B, we select a W200 × 27 with $A = 3390$ mm^2 and least $r = 31.2$ mm. For this section, the slenderness ratio is $L_e/r = 4600/31.2 = 147$ which is greater than $C_c = 126$. Therefore, the working stress for this section is

$$\left[\sigma_w = \frac{12\pi^2 E}{23(L_e/r)^2} \right] \qquad \sigma_w = \frac{12\pi^2(200 \times 10^9)}{23(147)^2} = 47.7 \text{ MPa}$$

Then, the allowable load is $P = \sigma A = (47.7 \times 10^6)(3390 \times 10^{-6}) = 162$ kN. Since this is less than the applied load of 360 kN, the section is inadequate.

Second try.   Next, we select a W200 × 36, which has a larger area and a larger least r. For this section, $A = 4580$ mm^2 and least $r = 40.9$ mm. The slenderness ratio is $L_e/r = 4600/40.9 = 112$, which is less than $C_c = 126$. The working stress for this section is determined as follows:

$$\left[\text{FS} = \frac{5}{3} + \frac{3(L_e/r)}{8C_c} - \frac{(L_e/r)^3}{8C_c^3} \right]$$

$$\text{FS} = \frac{5}{3} + \frac{3(112)}{8(126)} - \frac{(112)^3}{8(126)^3} = 1.91$$

whence

$$\left[\sigma_w = \frac{\left[1 - \dfrac{(L_e/r)^2}{2C_c^2} \right]\sigma_{yp}}{\text{FS}} \right]$$

$$\sigma_w = \frac{\left[1 - \dfrac{(112)^2}{2(126)^2} \right]}{1.91}(250 \times 10^6) = 79.2 \text{ MPa}$$

Then, the allowable load for this section is

$$P = \sigma A = (79.2 \times 10^6)(4580 \times 10^{-6}) = 363 \text{ kN}$$

Since this load is only slightly larger than the applied load of 360 kN, the W200 × 36 is the appropriate section.

The procedure of selecting a section is greatly simplified by using tables giving the allowable axial loads for different sections of various lengths. Such tables are found in a steel handbook such as that published by the AISC. However, this problem illustrates the trial-and-error method which arises frequently in structural design.

PROBLEMS

1112. Determine the slenderness ratio of a 4-m column with built-in ends if its cross-section is (a) circular with a radius of 50 mm and (b) 40 mm square. Use the concept of effective length.

Ans.   (a) 80.0; (b) 173

1113. Using AISC column specifications, determine the maximum length of a W360 × 122 section if it is used as a hinged-end column to support a load of 1200 kN. Use σ_{yp} = 450 MPa.

Ans. L = 7.25 m

1114. Determine the maximum length of a W250 × 167 section used as a hinged-end column to support a load of 1600 kN. Use AISC specifications with σ_{yp} = 380 MPa.

1115. What factor of safety should be used with Euler's column formula so that it will give the same load capacity for steel as is given by the upper limit for main members of (a) the straight-line equation, Eq. (11–8); (b) the Rankine–Gordon equation, Eq. (11–9).

1116. A W360 × 134 section is used as a column with hinged ends. Using AISC specifications, determine the maximum load which can be applied if (a) L = 9 m; (b) L = 15 m. Use σ_{yp} = 290 MPa.

Ans.   (a) 1740 kN; (b) 687 kN

1117. A W200 × 100 section is used as a column 9 m long with built-in ends. (a) Using AISC specifications, compute the safe load that can be applied if the effective length is three-quarters the given length. (b) What is the safe load if the column is also braced at its midpoint? Use σ_{yp} = 380 MPa.   *Ans.*   (a) 837 kN; (b) 2400 kN

1118. Repeat Problem 1117 assuming that the length of the column is 14 m with a W310 × 500 section.

1119. A steel column with an effective length of 10 m is fabricated from two C250 × 45 channels latticed together so that the section has equal moments of inertia about the principal axes. Determine the safe load using AISC specifications. Use σ_{yp} = 380 MPa.

Ans. P = 883 kN

1120. Four 100 × 100 × 10 mm angles are latticed together to form the column section shown in Fig. P–1120. Using AISC specifications with σ_{yp} = 290 MPa, determine the maximum effective length at which a 500-kN load can be safely supported. What should be the spacing between lattice bars if the slenderness ratio of each separate angle is not to exceed three-fourths of that of the fabricated section?

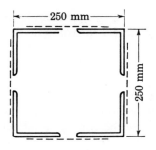

Figure P–1120.

1121. In the bridge truss shown in Fig. P–1121, the end chord *AB* is composed of two C230 × 30 channels latticed together so that the fabricated section has equal moments of inertia about the axes of symmetry. If the safe load *P* on the truss is governed by the strength of member *AB*, determine *P* using AISC specifications with σ_{yp} = 290 MPa.   *Ans.*   *P* = 210 kN

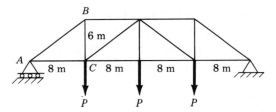

Figure P–1121.

1122. Select the lightest W shape that can be used as a column to support an axial load of 420 kN on an effective length of 4 m. Use AISC specifications with σ_{yp} = 250 MPa.   *Ans.*   W200 × 36

1123. Select the lightest W shape, according to AISC specifications, that can be used as a column to support an axial load of 700 kN on an effective length of 5.5 m. Assume σ_{yp} = 250 MPa.

1124. Repeat Problem 1123 assuming that the axial load is 690 kN and σ_{yp} = 345 MPa.

1125. A hinged-end steel column 10 m long is fabricated from a W200 × 46 beam and two C310 × 45 channels arranged as shown in Fig. P–1125. Determine the safe axial load using AISC specifications with σ_{yp} = 250 MN/m^2.   *Ans.*   *P* = 1440 kN

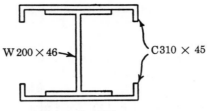

Figure P-1125.

1126. The connecting rod of an engine has a cross section with the following properties: area = 300 mm²; r_x = 3.00 mm; and r_y = 1.40 mm. The wrist pin and crank pin cause the rod to act as a hinged column about the X axis and as a fixed-ended column about the Y axis. Applying the concept of effective length, find the safe load for a length of 250 mm, using the Rankine–Gordon equation, Eq. (11–9).

Ans. P = 25.8 kN

1127. Derive a parabolic formula of the general type $P/A = \sigma - C(L/r)^2$ that will be applicable to aluminum alloy columns with hinged ends. Assume that the parabolic formula will be tangent to an Euler formula with a factor of safety of 2. Use σ = 110 MPa and E = 70 GPa. (*Hint:* For the two formulas, equate their unit loads and also equate their derivatives with respect to the slenderness ratio.)

Ans. $\dfrac{P}{A} = (110 \times 10^6) - 8760\left(\dfrac{L}{r}\right)^2$ for $\dfrac{L}{r} < 79.3$

1128. Four 100 × 100 × 13 mm angles are bolted back to back as shown in Fig. P–1128. Determine the safe load when they are used as a hinged-end column 4 m long. Use AISC specifications with σ_{yp} = 250 MPa.

Figure P-1128.

1129. Determine the safe axial load which can be applied to a 2014–T6 aluminum alloy column if its length is (a) 1 m and (b) 3 m. Assume the geometric properties of the cross section are identical to those of an S310 × 52 steel section.

Ans. (a) 984 kN; (b) 172 kN

1130. Repeat Problem 1129 assuming that the geometric properties of the cross section are identical to those of an S250 × 52 steel section.

1131. Determine the safe axial loads on an oak column 150 mm by 200 mm if the length is (a) 2 m and (b) 4 m. Use $E = 11.5$ GPa.

Ans. (a) 582 kN; (b) 146 kN

1132. Repeat Problem 1131 for a pine column 50 mm by 200 mm for which $E = 11.2$ GPa.

11–6 ECCENTRICALLY LOADED COLUMNS

Columns are usually designed to support axial loads, and the preceding formulas have been presented with this in mind. Under certain conditions, however, columns are subject to loads having a definite eccentricity. This occurs, for example, in the case of a beam connected to the column flange in a building. The secant formula derived in the next article is especially adapted to such cases, but it is so unwieldy that several simplified procedures are currently used.

In the *maximum stress approach*, the eccentrically loaded column is treated as if it were an eccentrically loaded short strut (see Art. 9–3). However, to eliminate the possibility of buckling so that the effect of deflection on the moment arm of eccentric loads may be neglected, the maximum compressive stress is limited to the unit load as computed from a specified column formula. This approach is valid only for moderate slenderness ratios.

Applying this procedure to the column in Fig. 11–10 which supports an axial load P_0 and a load P at an eccentricity e, we find the design criterion to be

$$\sigma > \frac{\Sigma P}{A} + \frac{Mc}{I} = \frac{P_0 + P}{A} + \frac{Pe}{S} \tag{11–19}$$

Here σ is the unit load computed from the specified column formula (*always* use the least radius of gyration to determine the slenderness ratio), I is the moment of inertia with respect to the axis about which the eccentric load causes bending (axis X–X in Fig. 11–10), and S is the section modulus with respect to that axis.

Modern design criteria have refined the maximum stress approach to include the moments, called secondary moments, which are also introduced because the neutral axis is deflected (the so-called P–δ effect). These criteria most often take the form of *interaction equations* which attempt to "weigh in" the relative importance of the axial stress and the bending stress.

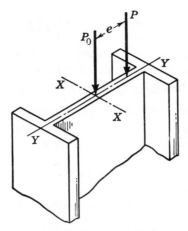

Figure 11–10.   Axial load P_0 and eccentric load P on column.

For example, AISC*recommends that, when the computed axial stress f_a is less than 15% of the actual stress F_a that would be permitted were only axial stress acting, the secondary moments may be neglected and the member must satisfy the following criterion:

$$\frac{f_a}{F_a} + \frac{f_{bx}}{F_{bx}} + \frac{f_{by}}{F_{by}} \leqslant 1.0 \tag{a}$$

When $f_a > 0.15 F_a$, secondary moment effects cannot be neglected. In these cases, AISC requires the following formulas to be satisfied:

$$\frac{f_a}{F_a} + \frac{C_{mx} f_{bx}}{\left(1 - \dfrac{f_a}{F'_{ex}}\right) F_{bx}} + \frac{C_{my} f_{by}}{\left(1 - \dfrac{f_a}{F'_{ey}}\right) F_{by}} \leqslant 1.0 \tag{b}$$

$$\frac{f_a}{0.60 F_y} + \frac{f_{bx}}{F_{bx}} + \frac{f_{by}}{F_{by}} \leqslant 1.0 \tag{c}$$

In Eqs. (a), (b), and (c), the various terms are as follows:

f_a = *computed* axial stress;

F_a = *allowable* axial stress if axial force alone were acting;

f_{bx} = *computed* bending stress about the major axis disregarding the secondary moment;

f_{by} = *computed* bending stress about the minor axis disregarding the secondary moment;

*The notation used here is that used in *Manual for Steel Construction,* American Institute of Steel Construction, New York, 1970.

F_{bx} = *allowable* compressive bending stress about the major axis if moment alone were acting;

F_{by} = *allowable* compressive bending stress about the minor axis if moment alone were acting;

F'_{ex} = Euler buckling stress for buckling about major axis;

F'_{ey} = Euler buckling stress for buckling about minor axis;

C_{mx}, C_{my} = reduction factors to correct for overconservatism in some cases of the amplification factor $[1 - (f_a/F'_e)]$.

For compression members in frames subject to joint translation, or sidesway, C_m may be taken as 0.85.

For compression members in frames braced against sidesway and subject to end moments (but not transverse loads between supports), use $C_m = 0.6 - 0.4(M_1/M_2) \geqslant 0.4$, where M_1/M_2 is the ratio of the smaller to the larger end moment. This ratio is positive when the member is bent in reverse curvature and negative when it is bent in single curvature.

For compression members in frames braced against sidesway in the plane of the loading and subjected to transverse loading between the supports, C_m may be taken as 0.85 for members with restrained ends and as unity for members with no end restraints; C_m may also be determined by rational analysis in this case.

AISC specifications also include formulas for determining the allowable bending stresses F_b as a fraction of the yield stress. The value of F_b depends upon the width/thickness ratio of the section and the bracing intervals.

Interaction formulas similar to the AISC formula have been adopted for other structural materials such as wood and aluminum.

The design of members to carry both axial and bending loads is essentially an iteration procedure. An assumed section is checked for adequacy using the appropriate criteria. This procedure is greatly simplified by the many tables and graphs which are available to assist the designer. Computer programs are also available which will assist the designer in selecting the optimum section that satisfies the interaction formulas.

The problems which follow will illustrate the application of the maximum stress approach. For applications of the interaction equations, the reader is referred to any modern structural design text.[*]

---

[*]See, for example, L. A. Hill, Jr., *Fundamentals of Structural Design: Steel, Concrete, and Timber*, Intext, New York, 1975.

ILLUSTRATIVE PROBLEM

1133. A W360 × 134 section is used as a column with an effective length of 7 m to support one track of a traveling crane in a factory. Determine the maximum permissible reaction P if the column also carries a load of 400 kN from an upper floor, as shown in Fig. 11–11. Use the maximum stress approach (Eq. 11–19) and the AISC column specifications. Assume $\sigma_{yp} = 250$ MPa.

Solution: Table B–2, Appendix B, gives the properties of a W360 × 134 section as $A = 17\ 100$ mm², $S_x = 2330 \times 10^3$ mm³, and least $r = 94.0$ mm. The slenderness ratio is $L_e/r = 7000/94.0 = 74.5$. For $\sigma_{yp} = 250$ MPa, the critical slenderness ratio is

$$C_c = \sqrt{\frac{2\pi^2 E}{\sigma_{yp}}} = \sqrt{\frac{2\pi^2(200 \times 10^9)}{250 \times 10^6}} = 126$$

Since $L_e/r < C_c$, the appropriate AISC formula (Eq. 11–11) determines the working stress as follows

$$\left[FS = \frac{5}{3} + \frac{3(L_e/r)}{8C_c} - \frac{(L_e/r)^3}{8C_c^{\,3}} \right]$$

$$FS = \frac{5}{3} + \frac{3(74.5)}{8(126)} - \frac{(74.5)^3}{8(126)^3} = 1.86$$

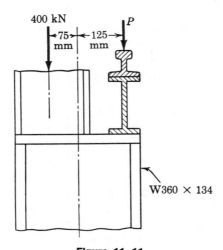

Figure 11–11.

whence

$$\left[\sigma_w = \frac{\left[1 - \frac{(L_e/r)^2}{2C_c^{\,2}} \right]\sigma_{yp}}{FS} \right]$$

$$\sigma_w = \frac{\left[1 - \frac{(74.5)^2}{2(126)^2} \right]}{1.86}(250 \times 10^6) = 111 \text{ MPa}$$

Using the maximum stress approach, we consider the column to act as an eccentrically loaded short compression member limited to this maximum stress of 111 MPa. Applying Eq. (11–19), we obtain

$$\left[\sigma = \frac{\Sigma P}{A} + \frac{M}{S} \right]$$

$$111 \times 10^6 = \frac{(400 \times 10^3 + P)}{(17\ 100 \times 10^{-6})} + \frac{0.125P - 0.075(400 \times 10^3)}{2330 \times 10^{-6}}$$

whence

$$P = 896 \times 10^3 \text{ N} = 896 \text{ kN} \qquad Ans.$$

PROBLEMS

In the following problems, use the maximum stress approach and the AISC column specifications unless otherwise directed.

1134. A W360 × 122 section is used as a column with an effective length of 10 m. Determine the maximum load that can be carried at an eccentricity of 300 mm. Should the load be placed on the X or Y axis? Assume σ_{yp} = 290 MPa. *Ans.* P = 190 kN

1135. Repeat Problem 1134 for a column with an effective length of 4.5 m.

1136. A steel column 50 mm by 75 mm in section has an effective length of 1.5 m. Compute the maximum load that can be carried at an eccentricity of 120 mm from the geometric axis. Assume σ_{yp} = 250 MN/m^2. The column also carries an axial load of 50 kN. *Ans.* P = 25.8 kN

1137. A steel pipe 2.5 m long, built in at its lower end and free at its upper end, supports a sign whose center of gravity is 0.6 m from the axis of the pipe. Applying the concept of effective length, determine the maximum weight of the sign. The outside diameter of the pipe is 140 mm, its area is 2800 mm^2, and its moment of inertia is 6.32 × 10^6 mm^4. Use σ_{yp} = 250 MN/m^2.

1138. A W360 $\times$ 134 section is used as a column whose effective length is 6 m. The column carries an axial load of 260 kN and an eccentric load of 220 kN applied on the minor axis. Determine e, the maximum eccentricity of the load, using $\sigma_{yp} = 250$ MPa.

Ans. $e = 952$ mm

1139. A C310 $\times$ 45 channel is used as a hinged-end column 2.2 m long. How far off center can a load of 50 kN be placed on the X axis? Assume $\sigma_{yp} = 380$ MPa and that the tensile stress is to be limited to 140 MN/m^2. On which side of the Y axis must the load be placed.

Ans. 100 mm

1140. Repeat Problem 1139 using a C310 $\times$ 31 channel.

1141. A W360 $\times$ 134 section is to be used as a column with a length of 9 m. The column supports an axial load of 260 kN and an eccentric load of 360 kN acting on the Y axis. Determine the maximum eccentricity of the 360-kN load using the maximum stress method and the straight-line formula, Eq. (11-8).   *Ans.*   178 mm

1142. Repeat Problem 1141 using a W360 $\times$ 347 section.

11-7 THE SECANT FORMULA

A theoretically correct formula for eccentrically loaded columns can be obtained by extending Euler's analysis in the following manner. Figure 11-12 shows the center line of a column carrying a load P with an eccentricity e on a length L. If this column is extended as indicated by the dashed lines, it becomes equivalent to a hinged Euler column having a length λ. The value of P shown is the critical load for this unknown length λ. Such a column has the shape of a sine curve whose equation with respect to an origin at one end was shown in Eq. (e), Art. 11-3, to be

$$y = \delta \sin\left(x \sqrt{\frac{P}{EI}} \right)$$

Since from Eq. (11-1) $\sqrt{P/EI} = \pi/L$ for the fundamental shape of a hinged column, we obtain

$$y = \delta \sin\left(\frac{\pi x}{L} \right) \tag{a}$$

If the origin is taken at the center, Eq. (a), in terms of the equivalent but unknown length λ, becomes:

$$y = \delta \cos\left(\frac{\pi x}{\lambda} \right) \tag{b}$$

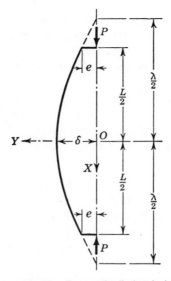

Figure 11-12. Eccentrically loaded column.

Applying the condition that $y = e$ at $x = L/2$ gives

$$e = \delta \cos\left(\frac{\pi L}{2\lambda}\right)$$

from which the value of δ is obtained for substitution in Eq. (*b*), thereby yielding

$$y = e \frac{\cos\left(\frac{\pi x}{\lambda}\right)}{\cos\left(\frac{\pi L}{2\lambda}\right)} \qquad (c)$$

The value of λ is found by applying Euler's formula in Eq. (11-1) with a length λ so that

$$P = \frac{EI\pi^2}{\lambda^2}, \quad \text{or} \quad \lambda = \pi\sqrt{\frac{EI}{P}}, \quad \text{and} \quad \frac{\pi}{\lambda} = \sqrt{\frac{P}{EI}}$$

When this is substituted in Eq. (*c*) we obtain the following equation for the column in Fig. 11-12:

$$y = e \frac{\cos\left(x\sqrt{\frac{P}{EI}}\right)}{\cos\left(\frac{L}{2}\sqrt{\frac{P}{EI}}\right)} \qquad (d)$$

The curvature is found by differentiating Eq. (*d*) twice, whence

$$\frac{d^2y}{dx^2} = -\frac{eP}{EI} \frac{\cos\left(x\sqrt{\dfrac{P}{EI}}\right)}{\cos\left(\dfrac{L}{2}\sqrt{\dfrac{P}{EI}}\right)}$$

Hence, from the differential equation of the elastic curve, the maximum bending moment at $x = 0$ is

$$M = EI\left(\frac{d^2y}{dx^2}\right)_{x=0} = -\frac{eP}{\cos\left(\dfrac{L}{2}\sqrt{\dfrac{P}{EI}}\right)} = -eP\sec\left(\frac{L}{2}\sqrt{\frac{P}{EI}}\right)$$

$$(e)$$

The maximum stress in the eccentrically loaded column is composed of a direct stress and a flexural stress as in a short strut, so

$$\sigma_{\text{max.}} = \frac{P}{A} + \frac{Mc}{I}$$

whence, using $I = Ar^2$ and the value of M from Eq. (*e*), we obtain

$$\sigma_{\text{max.}} = \frac{P}{A}\left[1 + \frac{ec}{r^2}\sec\left(\frac{L}{2r}\sqrt{\frac{P}{EA}}\right)\right] \qquad (11\text{–}20)$$

This equation is known as the *secant formula*. The buckling load P specified in it is converted to a working load P_w by replacing P by fP_w, where f is the factor of safety. Doing this and taking the maximum stress as the yield point we reduce Eq. (11–20) to

$$\sigma_{yp} = \frac{fP_w}{A}\left[1 + \frac{ec}{r^2}\sec\left(\frac{L}{2r}\sqrt{\frac{fP_w}{EA}}\right)\right] \qquad (11\text{–}21)$$

Trial-and-error methods are necessary in using these equations. Their use is facilitated by calculating the values of L/r corresponding to assumed values of P/A for various values of the eccentricity ratio ec/r^2 such as 0.2, 0.4, and so forth, to 1.0. This procedure gives the data in the Table 11–1, from which design curves like those in Fig. 11–13 may be potted.

It is interesting to note that as the slenderness ratio approaches zero, the value of the secant term in Eq. (11–20) approaches unity. Hence in the limit, Eq. (11–20) reduces to

$$\sigma_{\text{max.}} = \frac{P}{A}\left(1 + \frac{ec}{r^2}\right) = \frac{P}{A} + \frac{Mc}{I}$$

which is the equation used for eccentrically loaded short struts.

TABLE 11–1. Design Data for Eq. (11–21) Using
$\sigma_{yp} = 290$ MPa, $f = 2\frac{1}{2}$, and $E = 200$ GPa

$\dfrac{P}{A}$	$\dfrac{L}{r}$				
(MPa)	$\dfrac{ec}{r^2} = 0.2$	0.4	0.6	0.8	1.0
20	193	188	183	178	172
25	171	165	159	153	146
30	155	148	140	133	125
35	142	134	125	116	107
40	131	122	112	102	90.9
45	122	111	100	88.6	74.6
50	113	101	87.9	73.6	56.9
55	106	91.7	76.2	58.4	34.1
60	98.9	82.4	63.7	39.5	—
65	92.1	72.7	49.1	—	—
70	85.3	62.0	28.4	—	—
75	78.1	49.0	—	—	—
80	70.2	30.1	—	—	—
85	60.8	—	—	—	—
90	48.1	—	—	—	—

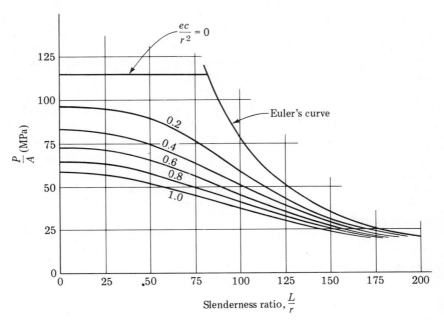

Figure 11–13. Design curves for secant formula with factor of safety $= 2\frac{1}{2}$.

SUMMARY

Long slender columns are solved by Euler's column formula. For columns with hinged ends it is

$$P = \frac{EI\pi^2}{L^2} \qquad\qquad (11\text{–}1)$$

or

$$\frac{P}{A} = \frac{E\pi^2}{(L/r)^2} \qquad\qquad (11\text{–}5)$$

For other end conditions, L in these formulas is replaced by an effective length L_e, values of which are tabulated on page 447.

Euler's formulas are theoretically correct, provided the stress does not exceed the proportional limit. The lower limit of L/r for which they are valid may be obtained by assuming that P/A represents the actual stress in a straight, axially loaded column, and replacing P/A in Eq. (11–5) by the value of the proportional limit.

Columns with a slenderness ratio less than the lower limit for Euler's formulas are known as intermediate columns. No theoretically correct formulas have yet been developed; the closest approach, so far, is the secant formula (Eq. 11–13). However, this formula is too unwieldy; various empirical formulas are used instead. These empirical formulas are specified by the building codes of various communities, and the *legal* specifications of these codes must be adhered to.

Eccentrically loaded columns are analyzed by using either the maximum stress approach or interaction equations. In the maximum stress approach, the columns are treated as eccentrically loaded short struts (see Art. 9–3), except that the value of the working stress is obtained by using a specified column formula. Interaction equations attempt to "weigh in" the relative importance of the axial and bending stresses.

12

Riveted and Welded Connections

12-1 INTRODUCTION

The analysis of riveted and welded connections involves so many indeterminate factors that an exact solution is impossible. Nevertheless, by making certain simplifying assumptions, practical solutions can be readily obtained. The most significant of these assumptions is that when the applied load passes through the centroid of the rivet group, each rivet transmits a load equal to its shear or bearing capacity, depending on which is lower. This assumption, in conjunction with the one that the joint is made from a ductile material, permits us to consider riveted joints as examples of uniform stress distribution.

12-2 TYPES OF RIVETED JOINTS: DEFINITIONS

There are two types of riveted joints: lap joints and butt joints. In a lap joint, the plates to be connected are lapped over one another and fastened together by one or more rows of rivets, as in Fig. 12–1. In a butt joint, the plates are butted together and joined by two cover plates riveted to each of the main plates. (Occasionally only one cover plate is used.) The number of rows of rivets used to fasten the cover plates to each main plate identifies the joint as single-riveted, double-riveted, etc. (Fig. 12–2). Frequently the outer cover plate in a boiler joint is narrower than the inner cover plate, as in Parts (c) and (d) of Fig. 12–2, the outer plate being wide enough to include only the rivet row in which the rivets are most closely spaced. This type of connection is called a *pressure*

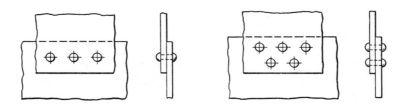

Figure 12–1. Lap joints.

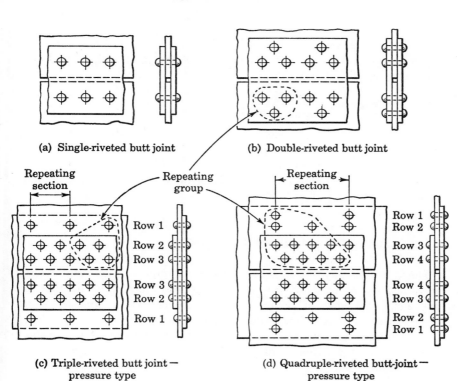

(a) Single-riveted butt joint

(b) Double-riveted butt joint

(c) Triple-riveted butt joint —
pressure type

(d) Quadruple-riveted butt-joint —
pressure type

Figure 12–2. Butt joints.

joint. Caulking along the edge of the outer cover plate to prevent leakage is more effective in this type.

The spacing between the rivets in a given row is called the *pitch*. When the spacing varies in different rows, as in a quadruple-riveted joint, the smallest spacing is known as the *short pitch*, the next as the *intermediate pitch*, and the greatest as the *long pitch*. The spacing between consecutive rows of rivets is called the *back pitch*. When the rivets in consecutive rows are staggered, the distance between the centers of the rivets is the *diagonal pitch*.

In determining the strength of a riveted joint, computations are usually made for a length of joint corresponding to a repeating pattern of rivets. The length of the repeating pattern (more commonly called a *repeating section*) is equal to the long pitch.

Sometimes confusion arises in deciding how many rivets belong in a repeating section. Study of the repeating section in Fig. 12–2c shows that there are five rivets effective in each half of the triple-riveted butt joint: two half rivets in row 1, two whole rivets in row 2, and one whole and two half rivets in row 3. Similarly there are 11 rivets effective in each half of the repeating section of the quadruple-riveted joint in Fig. 12–2d. Inspection of the enclosed repeating rivet groups confirms these statements.

The efficiency of a riveted joint indicates how well the joint has been designed. It compares the strength of the joint with that of the unriveted plate and is defined by

$$\text{efficiency} = \frac{\textbf{strength of joint}}{\textbf{strength of solid plate}} \tag{12–1}$$

The rivet holes in boiler joints are usually drilled or subpunched and reamed out to a diameter 1.5 mm larger than the rivet; however, the rivet is assumed to be driven so tightly that it fills the hole completely. In calculations, therefore, the diameter of the driven rivet is considered equal to that of the rivet hole.

12–3 STRENGTH OF A SIMPLE LAP JOINT

Riveted joints may be considered to be examples of uniform stress governed by the equation $P = A\sigma$. The application of this equation to the elemental types of failure is easily understood by considering a single-riveted lap joint. The failure of a riveted boiler joint is equivalent to any relative movement of the main plates of the joint, because this will destroy its function, which is to maintain a rigid and leakproof connection.

In Fig. 12–3, shear of the rivet evidently permits the main plates to separate. The failure load in shear is given by

$$P_s = A_s \tau = \frac{\pi d^2}{4} \tau \tag{12-2}$$

where d represents the diameter of both the rivet hole and the driven rivet.

Figure 12–4 represents a failure caused by the tearing of the main plate. This failure occurs on a section through the rivet hole, this section evidently having minimum tearing resistance. If p is the width of the plate or the length of a repeating section, the resisting area is the product of the net width of the plate $(p - d)$ and the thickness t. The failure load in tension is

$$P_t = A_t \sigma_t = (p - d) t \sigma_t \tag{12-3}$$

A third type of failure, called a bearing failure, is shown in Fig. 12–5. In this case, relative movement between the main plates may result from a permanent deformation or enlargement of the rivet hole caused by excessive bearing pressure. The rivet itself may also possibly be crushed.

Actually, the intensity with which the rivet bears against the rivet hole is not constant but varies from zero at the edges of the hole to a

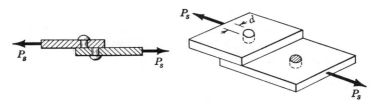

Figure 12–3.   Shear failure.

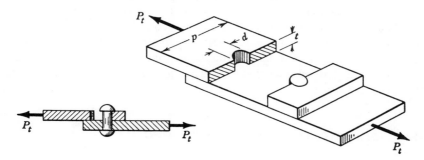

Figure 12–4.   Tear of plate at section through rivet hole.
$$P_t = A_t \sigma_t = (p - d) t \sigma_t$$

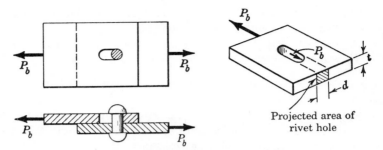

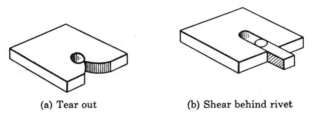

Projected area of
rivet hole

Figure 12-5. Exaggerated bearing deformation of upper plate.
$P_b = A_b\sigma_b = (td)\sigma_b$

(a) Tear out   (b) Shear behind rivet

Figure 12-6. Possible types of failure if rivet hole is too close to edge of plate.

maximum value directly in back of the rivet. It is common practice, however, to use a bearing stress σ_b that is assumed to be uniformly distributed over the projected area of the rivet hole. Then the failure load in bearing is expressed by

$$P_b = A_b\sigma_b = (td)\sigma_b \qquad\qquad (12\text{-}4)$$

Other types of failure are possible but will not occur in a properly designed joint. Among them are tearing of the edge of a plate back of a rivet hole, as shown in Fig. 12-6a, shear failure of the plate behind a rivet hole, as shown in Fig. 12-6b, or a combination of both. Such failures are unlikely to occur if the distance from the edge of the plate to the center of the rivet is $1\frac{3}{4}$ to 2 times the diameter of the rivet. In the problems we shall assume that this distance is great enough to prevent this type of failure.

12-4   STRENGTH OF A COMPLEX BUTT JOINT

The strength of a joint is limited by the capacity of the rivets to transmit load between the plates or by the tearing resistance of the plates themselves, depending on which is smaller. The calculations are divided into two steps: (1) preliminary calculations to determine the load that can be transmitted by one rivet in shear or bearing, and (2) calculations to determine possible methods of failure. The procedure and reasoning are explained in the following illustrative problem.

ILLUSTRATIVE PROBLEM

1201.   A repeating section 180 mm long of a triple-riveted butt joint of the pressure type is illustrated in Fig. 12–7. The rivet hole diameter is $d = 20.5$ mm, the thickness of the main plates is $t = 14$ mm, the thickness of each cover plate is $t' = 10$ mm. The ultimate stresses in shear, bearing, and tension are respectively $\tau = 300$ MPa, $\sigma_b = 650$ MPa, and $\sigma_t = 400$ MPa. Using a factor of safety of 5, determine the strength of a repeating section, the efficiency of the joint, and the maximum internal pressure that can be carried in a 1.5-m-diameter boiler for which this joint is the longitudinal seam.

Solution: The use of ultimate stresses will determine the ultimate load, which is then divided by a suitable factor of safety to determine the safe working load. An equivalent but preferable procedure is to use allowable stresses to determine the safe working load directly. This has the advantage of involving smaller numbers. Thus, dividing the ultimate stresses by the specified factor of safety, we find that the allowable stresses in shear, bearing, and tension respectively are $\tau = 60$ MPa, $\sigma_b = 130$ MPa, and $\sigma_t = 80$ MPa. These values are used in the following computations.

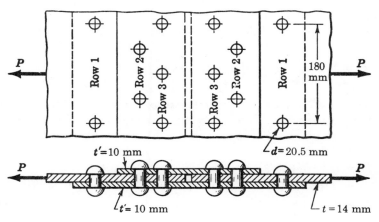

Figure 12–7.

Preliminary Calculations. To single-shear one rivet:

$$P_s = \frac{\pi d^2}{4}\tau = \frac{\pi}{4}(20.5 \times 10^{-3})^2(60 \times 10^6) = 19.8 \text{ kN}$$

To double-shear one rivet:

$$P_s = 2 \times 19.8 = 39.6 \text{ kN}$$

To crush one rivet in the main plate:

$$P_b = (td)\sigma_b = (14 \times 10^{-3})(20.5 \times 10^{-3})(130 \times 10^6) = 37.3 \text{ kN}$$

To crush one rivet in one cover plate:

$$P_b' = (t'd)\sigma_b = (10 \times 10^{-3})(20.5 \times 10^{-3})(130 \times 10^6) = 26.7 \text{ kN}$$

Possible Methods of Failure. Generally, there are only two basic methods of failure. These are determined by (a) capacity of the rivets to transmit load and (b) the tearing resistance of the plates.

(a) *Rivet capacity.* The strength of the single rivet in row 1 in a repeating section is determined by the lowest value of the load that will single-shear the rivet, crush it in the main plate, or crush it in one cover plate. On the basis of the values derived from the preliminary calculations, this value is 19.8 kN.

The strength of each of the two rivets in row 2 depends on the lowest value required to double-shear the rivet, crush it in the main plate, or crush it in both cover plates. The preliminary calculations show this value to be 37.3 kN per rivet, or 74.6 kN for both rivets in row 2.

Each of the two rivets in a repeating section in row 3 transmits the load between the main plate and the cover plates in the same manner as those in row 2, and hence the strength of the two rivets in row 3 is also 74.6 kN.

The total rivet capacity equals the sum of the rivet strengths in all rows:

$$P_r = 19.8 + 74.6 + 74.6 = 169.0 \text{ kN} \qquad (a)$$

(b) *Tearing capacity.* The external load applied to the joint acts directly to tear the main plate at row 1, and a failure would be similar to that shown in Fig. 12–4. The load that will tear the main plate at row 1 is given by

$$P_1 = (p - d)t\sigma_t$$
$$= [(180 \times 10^{-3}) - (20.5 \times 10^{-3})](14 \times 10^{-3})(80 \times 10^6)$$
$$= 178.6 \text{ kN} \qquad (b)$$

The external load applied to the joint does not act directly to tear the main plate at row 2 because part of the load is absorbed or transmitted by the rivet in row 1 from the main plate to the cover plate. Hence, if the main plate is to tear at row 2, the external load must be the sum of the tearing resistance of the main plate at row 2, plus the load transmitted by the rivet in row 1 from the main plate to the cover plate. This statement is illustrated by the free-body diagram in Fig. 12–8 and is further clarified by Fig. 12–9, which shows how the failure may actually occur.

The load transmitted by the rivet in row 1 is the first term on the right-hand side of Eq. (a) and is the value of the rivet strength in row 1. The external load to tear the main plate at row 2 must include this value; it is given by

$$P_2 = (p - 2d)t\sigma_t + \text{rivet strength in row 1}$$
$$= \big[(180 \times 10^{-3}) - 2(20.5 \times 10^{-3})\big](14 \times 10^{-3})(80 \times 10^6)$$
$$+ (19.8 \times 10^3)$$
$$= (155.7 \times 10^3) + (19.8 \times 10^3) = 175.5 \text{ kN} \qquad (c)$$

Similarly, the external load required to tear the main plate at row 3 must include the rivet resistance in rows 1 and 2, or

$$P_3 = \big[(180 \times 10^{-3}) - 2(20.5 \times 10^{-3})\big](14 \times 10^{-3})(80 \times 10^6)$$
$$+ (19.8 \times 10^3) + (74.6 \times 10^3) = 250.1 \text{ kN} \qquad (d)$$

It is obvious now that this computation need not be made because, since the tearing resistance of the main plates at rows 2 and 3 is equal, it gives a larger value than that of Eq. (c). However, this computation

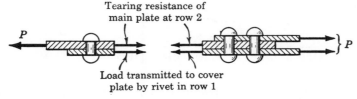

Tearing resistance of
main plate at row 2

Load transmitted to cover
plate by rivet in row 1

Figure 12–8.

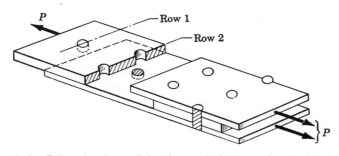

Figure 12–9.   Failure by shear of rivet in row 1 plus tear of main plate in row 2.

illustrates the procedure to be used when a greater number of rivet holes in row 3 than row 2 reduces the tearing resistance of row 3 below that of row 2.

At row 3, the tearing of the *cover plates* is resisted by the tensile strength of the reduced section at that row (see Fig. 12–8). The tensile strength of *one* cover plate is given by

$$P_c = [(180 \times 10^{-3}) - 2(20.5 \times 10^{-3})](10 \times 10^{-3})(80 \times 10^6)$$
$$= 111.2 \text{ kN}$$

In an ordinary butt joint, the tensile capacity of both cover plates is twice this value. In a pressure joint, however, where one cover plate is shorter than the other, the load capacity of the shorter plate must be compared with the rivet load transmitted to it. In this example, the upper cover plate transmits the rivet load of four rivets in single shear, or $4 \times 19.8 = 79.2$ kN, which is less than its tear capacity of 111.2 kN. Hence the load capacity of both cover plates becomes

$$P_c = 79.2 + 111.2 = 190.4 \text{ kN} \tag{e}$$

determined by rivet shear in the upper plate and by tension at row 3 in the lower plate.

The safe load is the lowest value of these several possible methods of failure. Its value is

$$P = 169.0 \text{ kN}$$

determined by the capacity of the rivets to transmit load.

The efficiency is

$$\text{Eff.} = \frac{\text{safe load}}{\text{strength of solid plate}}$$

$$= \frac{169.0 \times 10^3}{(180 \times 10^{-3})(14 \times 10^{-3})(80 \times 10^6)} \times 100$$

$$= 83.8\% \quad Ans.$$

The maximum internal pressure is found by applying the safe load of 169.0 kN in a repeating length of 180 mm:

$$[2P = pDL] \quad 2(169.0 \times 10^3) = p(1.5)(180 \times 10^{-3})$$
$$p = 1.25 \text{ MPa} \quad Ans.$$

Observations Concerning Riveted Joints. We are now ready to appreciate the significance of the following observations. Since the rivets are driven when hot and contract as they cool, there are developed normal forces which press the plates of the joint tightly together. Because of these normal forces, there will be a frictional resistance to any motion of the plates past one another. This frictional resistance must be overcome before there is sufficient deformation of the plates to permit the rivets to bear against the rivet holes. Thus there is an extra margin of strength in the riveted connection. However, because it is difficult to estimate this frictional resistance accurately, it is not considered when computing the strength of a riveted connection.

It is assumed further that each rivet, when driven, expands to fill the rivet hole completely. Only when this is true will all the rivets transmit the load simultaneously. If some of the rivets fill the holes only partially, these rivets will not begin to bear against the plates until there has been sufficient deformation in the remaining rivets and/or plates to take up the slack in the rivet holes.

We have also assumed that the rivets do not bend but remain essentially straight. This is possible only if there are equal elastic deformations of the main and cover plates between adjacent rows of rivets. From the preceding discussion concerning tearing of the main plate at row 2, it is evident that the main plate between rows 1 and 2 carries substantially more load than the cover plate. Hence the physical requirement of equal deformations cannot be true, even if we allow for the thinner thickness of the cover plates. But since the plates are usually specified to be ductile, equal plastic deformations can occur as the stresses approach the yield point.*

Because of these and other reasons, an *exact* analysis of a riveted connection cannot be made. The procedures used here for boiler joints (and those described later for structural joints) give usable values determined by comparatively simple methods.

PROBLEMS

Unless otherwise stated, assume the allowable stresses in the following problems to be $\tau = 60$ MPa, $\sigma_b = 130$ MPa, and $\sigma_t = 80$ MPa.

1202. The longitudinal joint of a boiler having 14-mm plates has a strength of 350 kN in a pitch length of 400 mm. The efficiency of the girth joint is 45% and the allowable tensile stress is 80 MPa. Determine the maximum diameter of the boiler if it is designed to operate at a pressure of 1.4 MPa.   *Ans.* $D = 1.25$ m

1203. A double-riveted lap joint forms the girth seam of a boiler 1.5 m in diameter. Pitch of the rivets is 80 mm; diameter of the rivet holes is 17.5 mm; thickness of the plate is 12 mm. Find the strength of a repeating section, the efficiency, and maximum internal pressure.
                                          Ans. $p = 962$ kPa

*Other examples of inelastic action are discussed in Chapter 14.

1204. The longitudinal seam of a boiler is a triple-riveted lap joint, with the pitch in the outer rows equal to 140 mm and that in the middle row equal to 70 mm. Diameter of the rivet holes is 23.5 mm; thickness of the plate is 12 mm. Determine the strength of a repeating section and the efficiency.

1205. The dimensions of a double-riveted butt joint like that in Fig. 12–2b are: diameter of the rivet holes, 23.5 mm; long pitch, 140 mm; short pitch, 70 mm; thickness of main plate, 14 mm, and of cover plates, 10 mm. Compute the strength of a repeating section and the efficiency.

1206. A double-riveted butt joint is of the pressure type in which the upper cover plate extends over only the inner rows of rivets, whereas the lower cover plate extends over all rows. Its dimensions are: diameter of the rivet holes, 23.5 mm; thickness of main plate, 14 mm; thickness of each cover plate, 10 mm; long pitch, 140 mm; short pitch, 70 mm. Determine the strength of a repeating section and the efficiency of the joint.   *Ans.* $P = 111.6$ kN; 71.2% efficient

1207. If the cover plates in Problem 1206 were each 8 mm thick, determine the method of failure and the efficiency of the joint.

1208. In a double-riveted butt joint of the pressure type in which the upper cover plate extends over only the inner rows of rivets, the thickness of the main plates is 14 mm, that of the shorter upper cover plate is 6 mm, that of the longer lower cover plate is 10 mm. Diameter of rivet holes is 20.5 mm; long pitch is 100 mm; short pitch is 50 mm. Compute the strength of a repeating section.   *Ans.* $P = 75.5$ kN

1209. A triple-riveted butt joint like that in Fig. 12–2c has a long pitch of 200 mm and a short pitch of 100 mm. Diameter of rivet holes is 26.5 mm; thickness of main plate is 16 mm, and of each cover plate, 12 mm. Find the strength of a repeating section and the efficiency.   *Ans.*   Efficiency = 86.4%

1210. A quadruple-riveted joint similar to that in Fig. 12–2d has a long pitch of 350 mm. Diameter of rivet holes is 20.5 mm; thickness of main plate, 10 mm; thickness of each cover plate, 8 mm. Determine the strength of a repeating section and the efficiency.

1211. A quadruple-riveted butt joint like that in Fig. 12–2d has a long pitch of 430 mm. Diameter of rivet holes is 32.5 mm; thickness of main plate is 20 mm, and of each cover plate, 14 mm. Compute the strength of a repeating section, using a factor of safety of 4, based on the ultimate stresses of $\tau = 300$ MPa in single shear and $\tau = 520$ MPa in double shear, $\sigma_b = 660$ MPa, and $\sigma_t = 400$ MPa. If this joint is the longitudinal seam of a boiler carrying an internal pressure of 1.8 MPa, and the girth joint is 50% efficient, what is the maximum allowable boiler diameter?   *Ans.* $D = 2.03$ m

12–5 STRESSES IN RIVETED JOINTS

Sometimes it is necessary to investigate a joint and determine the stresses caused by a given loading. The usual assumption is that each rivet carries a load proportional to its resisting shear area. With this assumption, the tensile load acting across any interior row of rivets is found by subtracting the shear load transmitted by preceding rivet rows from the applied load.

ILLUSTRATIVE PROBLEM

1212. A load of 144 kN acts on the repeating section of the triple-riveted butt joint in Fig. 12–10. Length of section is 200 mm; diameter of rivet holes is 23.5 mm; thickness of main plate is 14 mm, and of each cover plate is 10 mm. Determine the shearing, bearing, and tensile stresses developed in the joint.

Solution: The shear resisting area is that of one rivet in single shear and four rivets in double shear, for a total of nine shear areas. The average load transmitted by one shear area is therefore $\frac{1}{9}(144) = 16.0$ kN. The average shearing stress is

$$\tau = \frac{P_s}{\pi d^2/4} = \frac{16.0 \times 10^3}{\pi(23.5 \times 10^{-3})^2/4} = 36.9 \text{ MPa} \qquad \textit{Ans.}$$

Sketching a free-body diagram of any one of the rivets in rows 2 or 3, where the rivets are in double shear, will disclose that the shear areas cause 32.0 kN to act against the main plate and that 16.0 kN acts against each cover plate. Since the combined thickness of two cover plates is larger than that of the main plate, the maximum average

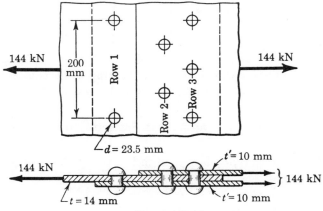

Figure 12–10.

bearing stress occurs in the main plate and is given by

$$\sigma_b = \frac{P_b}{td} = \frac{32.0 \times 10^3}{(14 \times 10^{-3})(23.5 \times 10^{-3})} = 97.3 \text{ MPa} \qquad Ans.$$

The tensile stress in the main plate at row 1 is caused by the entire loading acting across the net section of the plate:

$$\sigma_t = \frac{P_1}{(p - d)t} = \frac{144 \times 10^3}{[(200 \times 10^{-3}) - (23.5 \times 10^{-3})](14 \times 10^{-3})}$$
$$= 58.3 \text{ MPa}$$

The external load, reduced by the amount already transmitted by the single shear area of the rivet in row 1, acts across the net section of the main plate at row 2. The tensile stress at row 2 therefore is

$$\sigma_t = \frac{P_2}{(p - 2d)t} = \frac{(144 - 16) \times 10^3}{[(200 \times 10^{-3}) - 2(23.5 \times 10^{-3})](14 \times 10^{-3})}$$
$$= 59.8 \text{ MPa}$$

The tensile stress at row 3 in the main plate need not be computed because the net load acting there has been still further reduced by the load transmitted by the rivets in row 2, and the net tensile area at row 3 is the same as that at row 2.

At row 3 in the cover plates, the lower cover plate carries the greater load—that transmitted to it by 5 shear areas. Its average tensile stress is

$$\sigma_t = \frac{P}{(p - 2d)t'} = \frac{5(16 \times 10^3)}{[(200 \times 10^{-3}) - 2(23.5 \times 10^{-3})](10 \times 10^{-3})}$$
$$= 52.3 \text{ MPa}$$

The greatest tensile stress in the joint therefore occurs in the main plate at row 2 and is 59.8 MPa.

PROBLEMS

Compute the maximum shearing, bearing, and tensile stresses developed in the following riveted joints under the action of the indicated loads.

1213.   Double-riveted lap joint in Problem 1203: load = 350 kN per meter of length.

1214.   Double-riveted butt joint in Problem 1205: load = 700 kN per meter of length.

Ans. $\tau = 37.7$ MPa; $\sigma_b = 99.3$ MPa; $\sigma_t = 60.1$ MPa

1215.   Triple-riveted lap joint in Problem 1204: load on a repeating section = 90 kN.

1216. Double-riveted butt joint in Problem 1206: load on a repeating section = 90 kN.

Ans. $\tau = 41.5\ \text{MN}/\text{m}^2$; $\sigma_b = 109\ \text{MN}/\text{m}^2$; $\sigma_t = 58.1\ \text{MN}/\text{m}^2$

1217. Triple-riveted butt joint in Problem 1209: load on a repeating section = 200 kN.

1218. Quadruple-riveted joint in Problem 1210: load on a repeating section = 220 kN.

12-6 STRUCTURAL RIVETED JOINTS

Structural riveted joints differ in several ways from those for pressure vessels. The most important differences are as follows: (1) The entire joint is considered because there generally is no repeating pattern of rivets. (2) Cover or splice plates are usually the same length because there is no need to shorten one plate to permit tighter caulking against leakage. (3) Each rivet is assumed to transmit its proportional share of the applied load. (4) The diameter of the rivet hole is taken as 3 mm larger than the diameter of the undriven rivet.

For condition (3) to be true, it is necessary that the applied load pass through the centroid of the rivet group (see page 7). Eccentric loading of riveted connections is discussed in the next article. Condition (4) results from the fact that the parts to be joined are punched separately, the diameter of the punched hole being 1.5 mm larger than the rivet diameter. Another 1.5 mm is added to compensate for the probable damage done to the metal around the hole by punching. For shearing and bearing calculations, however, the diameter of the undriven rivet is used, because the rivet holes in plates that are punched separately are unlikely to match perfectly and consequently the cross section of the rivet will be less than that of the rivet hole.

ILLUSTRATIVE PROBLEM

1219. Using the structural method of assuming all rivets to carry a proportional share of the load, compute the tensile stress in the main plate at row 3 of the quadruple-riveted butt connection shown in Fig. 12–11. The connection transmits a load $P = 360$ kN across a plate width $p = 250$ mm. Also compute the width of the cover or strap plates at row 2 if the tensile stress is not to exceed 100 MPa. Rivet diameter is 19 mm; thickness of main plate is 14 mm, and of each cover plate, 8 mm.

Solution: Figure 12–11 shows the entire connection; there is no repeating pattern of rivets as in a boiler joint. As there are 10 rivets to

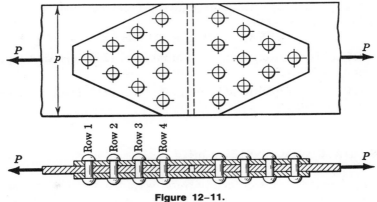

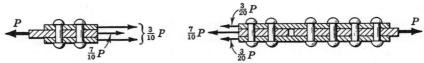

Figure 12–11.

Figure 12–12.   Free-body diagram of section between rows 2 and 3.

carry the load, each rivet may be assumed to carry $\frac{1}{10}$ of the load. Since the single rivet in row 1 transmits $\frac{1}{10}$ of the load to the cover plates, and the two rivets in row 2 transmit $\frac{2}{10}$ of the load, $\frac{7}{10}$ of the load is left to tear the main plate at row 3. This is shown in the free-body diagram of a section between rows 2 and 3 in Fig. 12–12. Thus we obtain for the tensile stress in the main plate at row 3, where 3 rivet holes reduce the net section (note that the rivet hole diameter = rivet diameter + 3 mm),

$$[P = A\sigma = (p - 3d)t\sigma_t]$$
$$\tfrac{7}{10}(360 \times 10^3) = [(250 \times 10^{-3}) - 3(22 \times 10^{-3})](14 \times 10^{-3})\sigma_t$$
$$\sigma_t = 97.8 \text{ MPa} \qquad Ans.$$

Figure 12–12 also shows that the cover plates at row 2 are subject to $\frac{3}{10}$ of the applied load. Hence, the width of these cover plates required to develop the maximum tensile stress permitted is

$$[P = A\sigma = (p - 2d)(2t')\sigma_t]$$
$$\tfrac{3}{10}(360 \times 10^3) = [p - 2(22 \times 10^{-3})][2(8 \times 10^{-3})](100 \times 10^6)$$

whence
$$p = 0.112 \text{ m} = 112 \text{ mm} \qquad Ans.$$

PROBLEMS

1220. Determine the safe load on the butt connection in Fig. 12–11 if the allowable stresses are $\tau = 100 \text{ MN/m}^2$, $\sigma_t = 140 \text{ MN/m}^2$, and $\sigma_b = 280$ MN/m^2. Use 19-mm rivets; plate width $p = 280$ mm; thickness of main plate = 14 mm and of each strap plate = 10 mm.

Ans. $P = 506$ kN

1221.   In the joint in Illustrative Problem 1219, if the rivet at row 1 is omitted, determine the maximum stresses in shear, bearing, and tension in the main plate for a load of 260 kN. Also compute the minimum width of the strap plates in rows 2 and 3 if the tensile stress is limited to 100 MPa.

1222.   Omitting row 4 of the butt connection shown in Fig. 12–11, compute the safe load if the allowable stresses are $\tau = 90$ MPa, $\sigma_t = 120$ MPa, and $\sigma_b = 190$ MPa. Use 25-mm rivets; plate width $p = 230$ mm; thickness of main plate = 14 mm and of each strap plate = 10 mm.

1223.   Two plates are joined by four rivets 25 mm in diameter, as shown in Fig. P–1223. Find the allowable load P if the working stresses are $\tau = 70$ MN/m^2, $\sigma_t = 100$ MN/m^2, and $\sigma_b = 140$ MN/m^2.

Ans. $P = 122$ kN

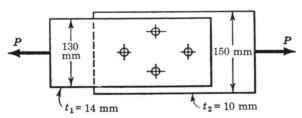

Figures P–1223 and P–1224.

1224.   Repeat Problem 1223 assuming that the allowable stresses are $\tau = 100$ MN/m^2, $\sigma_t = 140$ MN/m^2, and $\sigma_b = 220$ MN/m^2.

1225.   Find the safe load on the lap connection shown in Fig. P–1225 if the rivets are of 19-mm diameter and the plates are 8 mm thick. Use allowable stresses of $\tau = 95$ MPa, $\sigma_t = 140$ MPa, and $\sigma_b = 220$ MPa.

Ans. $P = 242$ kN

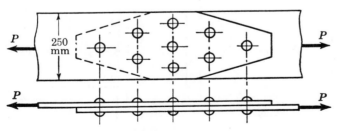

Figures P–1225 and P–1226.

1226.   Repeat Problem 1225 assuming that the rivets are of 22-mm nominal diameter and the plates are 10 mm thick.

1227.   Two plates 250 mm wide and 20 mm thick are to be connected by a riveted butt joint with two strap plates using rivets of 22-mm nominal diameter. The axial tensile load on the joint is 400 kN. If the allowable stresses are $\tau = 70$ MPa, $\sigma_t = 110$ MPa, and $\sigma_b = 130$ MPa, determine: (a) the minimum number of rivets; (b) the minimum rows of rivets and the best distribution of rivets in each row; (c) the minimum thickness of each strap plate, using the distribution in part (b).   *Ans.*   (c) $t = 10.4$ mm

1228.   Solve Problem 1227, using 19-mm nominal diameter rivets and allowable stresses of $\tau = 110$ MPa, $\sigma_t = 140$ MPa, and $\sigma_b = 220$ MPa.

12–7   ECCENTRICALLY LOADED RIVETED CONNECTIONS

Occasionally it is impossible to load a riveted connection so that the load passes through the centroid of the rivet group. Such a condition is called *eccentric loading;* the load is not distributed equally over all the rivets (see Fig. 12–13a). However, by adding a pair of equal, oppositely directed, and collinear forces of magnitude P (shown dashed) at the centroid of the rivet group, the applied eccentric load P is replaced by a central load P and the torsional couple $T = Pe$, as shown in Fig. 12–13b.

The effect of the central load P is resisted equally by the direct load $P_d = P/n$ acting on each of the n rivets, as shown by the free-body diagram of the plate in Fig. 12–14a. The torsional couple T is resisted by torsional loads P_t (Fig. 12–14b), which act perpendicular to the radius ρ from the centroid of the rivet group and vary directly with the distance of the rivets from the centroid. To determine the torsional load on any rivet, we may consider that the connection is equivalent to a flanged coupling consisting of three concentric circles of rivets, and use

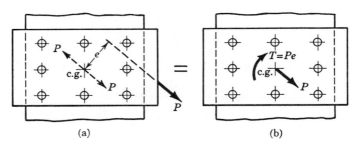

(a)   (b)

Figure 12–13.   Eccentrically loaded riveted connection.

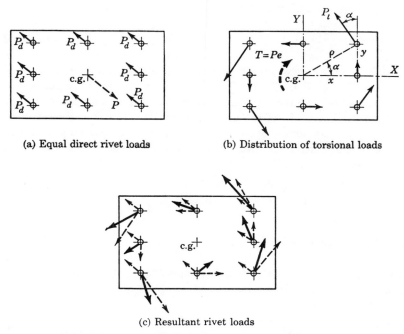

(a) Equal direct rivet loads   (b) Distribution of torsional loads

(c) Resultant rivet loads

Figure 12-14. Analysis of eccentrically loaded connection.

the method outlined in Art. 3–3 (see page 87). Then the resultant load on any rivet is the vector sum of the direct and torsional loads on that rivet, and appears as shown in Fig. 12–14c.

A better method of determining the torsional load is to apply the torsion formula $\tau = T\rho/J$. Here τ represents the average shearing stress on any rivet, ρ is the radial distance from its center to the centroid of the rivet group, and J may be expressed as

$$J = \Sigma A\rho^2 \qquad (a)$$

Since all the rivets have the same area A, and since ρ^2 may be expressed in terms of the x and y coordinates of any rivet so that $\rho^2 = x^2 + y^2$ (see Fig. 12–14b), we may rewrite Eq. (a) as

$$J = A(\Sigma x^2 + \Sigma y^2) \qquad (b)$$

whence the torsion formula becomes

$$\tau = \frac{T\rho}{A(\Sigma x^2 + \Sigma y^2)} \qquad (c)$$

Transposing A to the left side of this equation determines the torsional load P_t on any rivet from $P_t = A\tau$, so we finally obtain

$$P_t = \frac{T\rho}{\Sigma x^2 + \Sigma y^2} \tag{12–5}$$

The resultant load on a typical rivet is obtained as the vector sum of P_d and P_t. See Fig. 12–14c. This vector addition is easily performed analytically by resolving P_d and P_t into x and y components. The components P_{d_x} and P_{d_y} of the direct load are constant for all rivets. The components of the torsional load P_t are obtained by observing from Fig. 12–14b that the angle α between the radius ρ and the X axis equals the angle between P_t and the Y axis; hence

$$P_{t_x} = P_t \sin \alpha = P_t \frac{y}{\rho}$$

and

$$P_{t_y} = P_t \cos \alpha = P_t \frac{x}{\rho}$$

since $\sin \alpha = y/\rho$ and $\cos \alpha = x/\rho$. Replacing P_t in these relations by its value from Eq. (12–5), we obtain

$$\left. \begin{aligned} P_{t_x} &= \frac{T}{\Sigma x^2 + \Sigma y^2} \cdot y \\ P_{t_y} &= \frac{T}{\Sigma x^2 + \Sigma y^2} \cdot x \end{aligned} \right\} \tag{12–6}$$

where x and y are the coordinates of a rivet measured from the centroid of the rivet group.

The maximum load on any rivet occurs when P_{d_x} and maximum P_{t_x} as well as P_{d_y} and maximum P_{t_y} are additive, as at the upper right corner; whence the resultant rivet load is found from

$$P_r = \sqrt{\left(P_{d_x} + P_{t_x}\right)^2 + \left(P_{d_y} + P_{t_y}\right)^2} \tag{12–7}$$

The use of these equations is illustrated in the following illustrative problem.

ILLUSTRATIVE PROBLEM

1229.   On the connection of 12 rivets shown in Fig. 12–15, the load $P = 200$ kN passes through the center of rivet C and has a slope of $4/3$. Determine the resultant load on the most heavily loaded rivet.

Solution: The effect of the applied load is equivalent to an equal central load acting through the centroid of the rivet group plus a

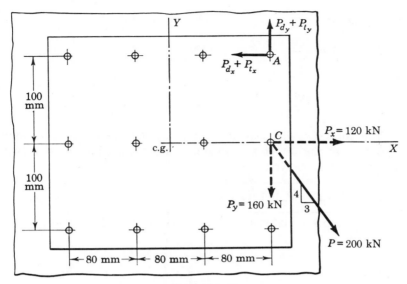

Figure 12–15.

torsional couple equal to the moment of P about the centroid of the rivet group. Replacing P by its components $P_x = 120$ kN and $P_y = 160$ kN, and noting that the moment of P is equal to the moment sum of its components, we find that the torsional couple is

$$T = (160 \times 10^3)(120 \times 10^{-3}) = 19.2 \text{ kN} \cdot \text{m}$$

Before applying Eq. (12–6) we compute the value of Σx^2 and Σy^2. There are six rivets whose x coordinate is 40 mm and six rivets whose x coordinate is 120 mm. Also there are eight rivets whose y coordinate is 100 mm. Therefore,

$$\Sigma x^2 + \Sigma y^2 = \left[6(40)^2 + 6(120)^2 \right] + 8(100)^2$$
$$= 0.176 \times 10^6 \text{ mm}^2 = 0.176 \text{ m}^2$$

Applying Eq. (12–6) gives the maximum components of the torsional load as

$$\left[P_{t_x} = \frac{T}{\Sigma x^2 + \Sigma y^2} \cdot y \right] \qquad P_{t_x} = \frac{19.2 \times 10^3}{0.176}(100 \times 10^{-3})$$
$$= 10.9 \text{ kN}$$

$$\left[P_{t_y} = \frac{T}{\Sigma x^2 + \Sigma y^2} \cdot x \right] \qquad P_{t_y} = \frac{19.2 \times 10^3}{0.176}(120 \times 10^{-3})$$
$$= 13.1 \text{ kN}$$

The x and y components of the direct load on any rivet are found by dividing the x and y components of the applied load by the number of rivets. Thus

$$P_{d_x} = \frac{P_x}{n} = \frac{120 \times 10^3}{12} = 10.0 \text{ kN}$$

and

$$P_{d_y} = \frac{P_y}{n} = \frac{160 \times 10^3}{12} = 13.3 \text{ kN}$$

The most heavily loaded rivet is at A, where the maximum components of the direct and torsional loads are additive as shown. Applying Eq. (12–7), we have

$$P_r = \sqrt{\left(P_{d_x} + P_{t_x}\right)^2 + \left(P_{d_y} + P_{t_y}\right)^2}$$
$$= \sqrt{(10.0 + 10.9)^2 + (13.3 + 13.1)^2} = 33.7 \text{ kN} \qquad Ans.$$

PROBLEMS

1230. Compute the resultant load on the least loaded rivet in Illustrative Problem 1229.

1231. A gusset plate is riveted to a larger plate by four 22-mm rivets arranged and loaded as shown in Fig. P–1231. Determine the maximum and minimum shear stress developed in the rivets.

          Ans.   Max. $\tau = 47.4$ MPa; min. $\tau = 29.4$ MPa

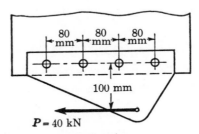

Figure P–1231.

1232. In the gusset plate connection shown in Fig. P–1232, each rivet has a cross-sectional area of 300 mm². The allowable load P was designed for a shearing stress of 70 MPa. What will be the maximum shearing stress in the rivets if the rivet at A is improperly driven so that it cannot carry any load?

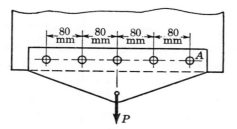

Figure P–1232.

1233. If the maximum load permitted on any single rivet in the connection shown in Fig. P–1233 is 15 kN, compute the safe value of P.

Ans. $P = 37.9$ kN

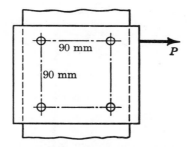

Figures P–1233 and P–1234.

1234. Repeat Problem 1233 assuming that the upper left rivet is improperly driven so that it does not carry any load.

1235. Rivets 22 mm in diameter are used in the connection shown in Fig. P–1235. If $P = 90$ kN, what thickness of plate is required so as not to exceed a bearing stress of 140 MPa? *Ans.* $t = 9.90$ mm

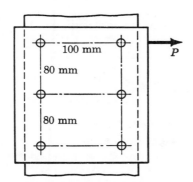

Figures P–1235, P–1236, and P–1237.

1236. For the riveted connection in Problem 1235, compute the load *P* which will cause a maximum rivet load of 20 kN if the lower right rivet is removed. *Ans.* *P* = 55.6 kN

1237. Resolve Problem 1235 assuming that the load *P* is replaced by a 90-kN load which passes through the center of the upper right rivet with a slope of 3 vertically upward to 4 horizontally to the right.

1238. In the gusset plate connection shown in Fig. P–1238, if *P* = 60 kN, determine the shearing stress in the most heavily loaded of the four 22-mm rivets. *Ans.* τ = 70.0 MN/m²

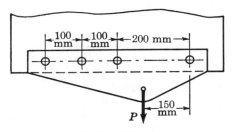

Figure P–1238.

1239. For the connection shown in Fig. P–1239, determine the shearing stress in the most heavily loaded of the three 22-mm rivets.
Ans. τ = 159 MPa

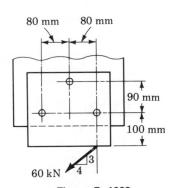

Figure P–1239.

1240. For the riveted connection in Fig. P–1240, determine the allowable load *P* if the shearing stress in the 25-mm rivets is limited to 140 MN/m².

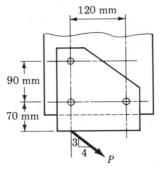

Figure P-1240.

12-8   WELDED CONNECTIONS

The reliability of welded connections has increased to the point where they are used extensively to supplement or replace riveted or bolted connections in structural and machine design. It is frequently more economical to fabricate a member by welding simple component parts together than to use a complicated casting.

Welding is a method of joining metals by fusion. With heat from either an electric arc or an oxyacetylene torch, the metal at the joint is melted and fuses with additional metal from a welding rod. When cool, the weld material and the base metal form a continuous and almost homogeneous joint. To protect the weld from excessive oxidation, a heavily coated welding rod is used which releases an inert gas that envelopes the arc stream; this technique is called the *shielded arc process*.*

The two principal types of welds are butt welds and fillet welds (see Fig. 12–16). The strength of a butt weld is equal to the allowable stress multiplied by the product of the length of the weld times the thickness of the thinner plate of the joint. The allowable stress is taken to be the same as that of the base metal.

The strength of side or transverse fillet welds is assumed to be determined by the shearing resistance of the throat of the weld regardless of the direction of the applied load. In the 45° fillet weld in Fig. 12–17, with leg equal to t mm, the shearing area through the throat is the length L mm of the weld times the throat depth, or $A = Lt \sin 45° = L(0.707t)$ mm^2. The allowable stresses for fillet welds specified by the

---

*A complete description of the welding process may be found in most books on structural design. See, for example, B. Bressler, T. Y. Lin, and J. B. Scalzi, *Design of Steel Structures*, 2nd ed., Wiley, New York, 1968.

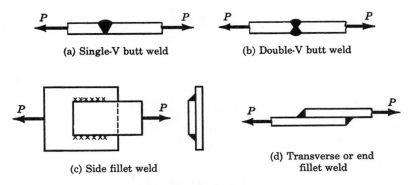

Figure 12–16.   Types of welds.

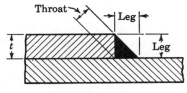

Figure 12–17.

AISC (based upon recommendations of the American Welding Society) depend upon the electrode used in the welding process and upon the grade of steel being welded. For example, if E–70 electrodes are used to weld A36 steel (one of the more common grades of structural steel used today), the allowable shearing stress is 145 MPa.* For this case, the strength of the 45° fillet weld, in newtons, is

$$P = \tau A = (145 \times 10^6)(0.707tL \times 10^{-6}) = 103tL$$

Usually, however, the strength of a fillet weld is expressed in terms of the allowable force q per millimeter of weld length, given by

$$q = \frac{P}{L} = 103t \qquad \text{N/mm}$$

where, to repeat, t is expressed in millimeters.

As a rule, special precautions are necessary to ensure that the length of the leg of a fillet weld along an edge is actually equal to the thickness of the edge. One reason is that edges of rolled shapes are rounded and the length of the leg would be less than the nominal thickness of the shape. Another reason is that, during welding, the corner of the edge may melt into the weld, which would reduce the

---

*Numerical values for stresses and lengths cited in this section are approximate SI conversions of the specifications in *Manual of Steel Construction*, American Institute of Steel Construction, New York, 1973.

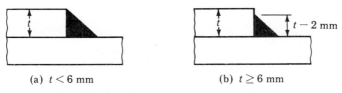

(a) $t < 6$ mm   (b) $t \geq 6$ mm

Figure 12–18. Maximum size of fillets.

length of the leg. Therefore, AISC specifications require that the maximum size of a fillet weld should be 2 mm less than the material thickness along edges 6 mm or more thick. For edges less than 6 mm thick, the maximum size of the weld may equal the edge thickness. These specifications are illustrated in Fig. 12–18. Weld sizes may exceed these specifications if the designer stipulates that the weld is to be built out to obtain full throat thickness.

ILLUSTRATIVE PROBLEM

1241. A $100 \times 100 \times 10$ mm angle is to be welded to a gusset plate. The angle carries a load of 190 kN applied along its centroidal axis. (a) Determine the lengths of side fillet welds required at the heel and toe of the angle. The weld at the heel is to be the same size as the maximum permissible weld at the toe. (b) Determine the lengths of the side fillet welds if a transverse fillet weld is added at the end of the angle. Assume that the allowable shearing stress through the throat of each weld is 145 MPa.

Solution:

Part a. Figure 12–19 shows the forces that keep the angle in equilibrium: P_1 and P_2 are the resisting forces exerted by the welds at the heel and toe, respectively. They are assumed to act along the edges of the angle. Taking moments about a center on the line of action of P_2,

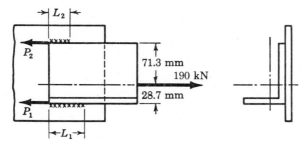

Figure 12–19.

we obtain

$$[\Sigma M_{P_2} = 0] \qquad 100P_1 = 190(71.3) \qquad P_1 = 135.5 \text{ kN}$$

With respect to a moment center on the line of action of P_1, we have

$$[\Sigma M_{P_1} = 0] \qquad 100P_2 = 190(28.7) \qquad P_2 = 54.5 \text{ kN}$$

Since the thickness of the angle is 10 mm (which is greater than 6 mm), the maximum size of the fillet weld at the toe of the angle is $10 - 2 = 8$ mm. The strength per millimeter of this size weld with $\tau = 145$ MPa is $q = 103t = 103(8) = 824$ N/mm. Hence the required lengths of weld are

$$\left[L = \frac{P}{q} \right] \qquad L_1 = \frac{135.5 \times 10^3}{824} = 164 \text{ mm} \qquad Ans.$$

and

$$L_2 = \frac{54.5 \times 10^3}{824} = 66.1 \text{ mm} \qquad Ans.$$

These values should be increased by a small amount to provide for starting and stopping of the weld.

Part b. If a transverse fillet weld is added along the edge at the end of the angle, its maximum size is again 8 mm. This weld should be symmetrically applied about the action line of the applied load in order to avoid eccentricity of loading. As shown in Fig. 12–20, this limits the transverse weld to twice the dimension of 28.7 mm, and this length will sustain a force of $qL = (824)(2 \times 28.7) = 47.3$ kN acting at its midpoint.*

Taking moments first about a center on the line of action of P_2 and then about a center on the line of action of P_1, we obtain

$$[\Sigma M_{P_2} = 0] \qquad 100P_1 = (190 - 47.3)(71.3) \qquad P_1 = 101.7 \text{ kN}$$
$$[\Sigma M_{P_1} = 0] \qquad 100P_2 = (190 - 47.3)(28.7) \qquad P_2 = 41.0 \text{ kN}$$

Thus, the lengths of weld required at the heel and toe are respectively:

$$\left[L = \frac{P}{q} \right] \qquad L_1 = \frac{101.7 \times 10^3}{824} = 123 \text{ mm} \qquad Ans.$$

$$L_2 = \frac{41.0 \times 10^3}{824} = 49.8 \text{ mm} \qquad Ans.$$

A larger size of weld may be used at the heel if it is necessary to reduce the length L_1.

*It is not uncommon, however, to let the transverse fillet weld run completely across the leg of the angle and ignore the resulting eccentricity of loading. Eccentrically loaded welded connections are discussed in the next article.

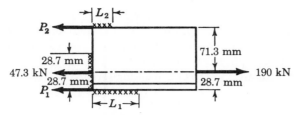

Figure 12–20.

PROBLEMS

1242. A plate 150 mm wide and 14 mm thick is lapped over and welded to a gusset plate. Determine the minimum length of an 8-mm side fillet weld that will be necessary if the plate is subjected to an axial stress of 140 MPa. Use an allowable shearing stress through the throat of the weld of 145 MPa.   *Ans.*   179 mm on each side

1243. Solve Problem 1242 using the maximum size of welds permitted.

1244. A $150 \times 100 \times 13$ mm angle is to be welded to a gusset plate with the 150-mm leg against the plate. If the angle carries a centroidal load of 400 kN, what lengths of 8-mm side fillet welds will be required along the toe and heel of the angle. Assume the allowable shearing stress through the throats of the welds is 145 MPa.
                                          Ans.   161 mm and 324 mm

1245. Solve Problem 1244 using a 12-mm weld at the heel of the angle and the maximum permissable size at the toe.

1246. A 16-mm plate is lapped over and secured, as shown in Fig. P–1246, by transverse fillet welds on the inside and outside to form a penstock 1.5 m in diameter. Determine the safe internal pressure, assuming allowable stresses of $\sigma_t = 160$ MN/m² for the plate and $\tau = 120$ MN/m² through the throats of the welds. Use the maximum size of welds permitted.   *Ans.* $p = 3.17$ MN/m²

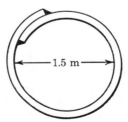

Figure P–1246.

1247. A tank is fabricated by welding two caps, as shown in Fig. P-1247, to the ends of a cylinder 1.2 m in diameter. If the caps and cylinder are 10 mm thick, determine the safe internal pressure that will not exceed a shearing stress of 110 MPa in the throat of the maximum size of fillet weld around the entire circumference.

Ans. $p = 2.07$ MPa

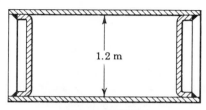

Figure P-1247.

12-9 ECCENTRICALLY LOADED WELDED CONNECTIONS

In the preceding article the methods of static equilibrium were used to determine the force to be resisted by a weld. This analysis was based on the concept that the weld is uniformly loaded along its length. This assumption is reasonable if all welds are of the same size and if the applied load passes through the centroid of the weld lines. If the resultant load P does *not* pass through this centroid, the welds are not uniformly loaded per millimeter of length. This causes a variable elastic deformation to exist in the welds. The following simplified analysis shows how to determine the maximum intensity of loading per millimeter of weld on which the size of the weld is based.

Proceeding as in the case of eccentrically loaded riveted connections, we add a pair of equal, oppositely directed, and collinear forces P (shown dashed in Fig. 12-21a) at the centroid C, of the weld lines, thereby reducing the eccentric load P to a central load P and a torsional couple $T = Pe$. In Fig. 12-21b, the central load P is resisted by the *direct* force q_d per millimeter of weld. This direct force is uniformly distributed over all the welds and is given by

$$q_d = \frac{P}{\Sigma L} \tag{12-8}$$

where ΣL is the total length of all the welds.

In Fig. 12-21c, the torsional couple is resisted by a variable *torsional* force q_t per millimeter of weld. Assuming elastic action of the welds, but that the plate is rigid and twists about the centroid C, we may find the intensity of this torsional force by applying the torsion formula with a modified value of J.

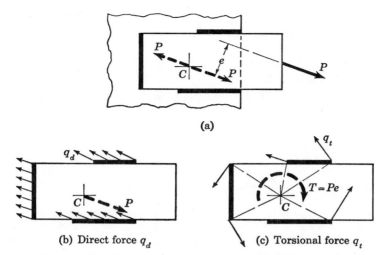

(a)

(b) Direct force q_d (c) Torsional force q_t

Figure 12–21.   Analysis of eccentrically loaded welded connections. Part (a) is the vector sum of parts (b) and (c).

To determine J, consider the arrangement of straight welds shown in Fig. 12–22. For any weld of length L, the centroidal value $\bar{J}$ is the sum of the rectangular moments of inertia with respect to axes through its center directed along and perpendicular to its length. These values are respectively zero and $\frac{1}{12}L^3$. Applying the transfer formula, we obtain, with respect to the centroid C of the weld group,

$$\left[J = \bar{J} + Ld^2\right]$$

$$J = \tfrac{1}{12}L^3 + L\bar{\rho}^{\,2} = \tfrac{1}{12}L^3 + L(\bar{x}^2 + \bar{y}^2)$$

Repeating this computation for every weld in the connection and adding the results, we find that the modified J of the torsion formula becomes

$$J = \Sigma L\left(\tfrac{1}{12}L^2 + \bar{x}^2 + \bar{y}^{\,2}\right)$$

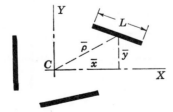

Figure 12–22.   Evaluation of J.

Applying the torsion formula gives the torsional force q_t acting perpendicular to the radial location ρ of any point on a weld as

$$q_t = \frac{T\rho}{\Sigma L\left(\frac{1}{12}L^2 + \bar{x}^2 + \bar{y}^2\right)} \qquad (12\text{–}9)$$

More useful are the following expressions for the components of q_t, obtained as in the analysis of riveted connections:

$$\left.\begin{aligned} q_{t_x} &= \frac{Ty}{\Sigma L\left(\frac{1}{12}L^2 + \bar{x}^2 + \bar{y}^2\right)} \\[2mm] q_{t_y} &= \frac{Tx}{\Sigma L\left(\frac{1}{12}L^2 + \bar{x}^2 + \bar{y}^2\right)} \end{aligned}\right\} \qquad (12\text{–}10)$$

in which x and y are the coordinates of any selected point of any weld.

The maximum intensity of the weld force occurs at the point where q_{d_x} and maximum q_{t_x}, as well as q_{d_y} and maximum q_{t_y}, are additive. Combining these values vectorially gives

$$q = \sqrt{\left(q_{d_x} + q_{t_x}\right)^2 + \left(q_{d_y} + q_{t_y}\right)^2} \qquad (12\text{–}11)$$

Frequently this value of q is used to determine the size of all welds, but occasionally the size of each individual weld is based on the value of the highest stressed point in that weld.

ILLUSTRATIVE PROBLEM

1248. A plate is attached to the frame of a machine by two side fillet welds as shown in Fig. 12–23a. Determine the size of the welds to resist a vertical load of 40 kN. Use an allowable shearing stress of 145 MPa through the throats of the welds.

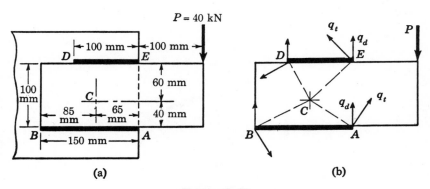

Figure 12–23.

Solution: The centroid of the weld lines, with respect to an origin at A, is found to be

$$[L\bar{x} = \Sigma lx] \quad (150 + 100)\bar{x} = 150(75) + 100(50) \quad \bar{x} = 65 \text{ mm}$$

$$[L\bar{y} = \Sigma ly] \quad\quad\quad\quad 250\bar{y} = 100(100) \quad\quad\quad\quad \bar{y} = 40 \text{ mm}$$

Using these values, we locate the centroid C as shown in Fig. 12–23a. The moment of P about this centroid determines the torsional couple to be

$$T = Pe = 40(100 + 65) = 6600 \text{ kN} \cdot \text{mm}$$

The modified J for the weld group is the sum of the J values for each weld. Remembering that $\bar{x}$ and $\bar{y}$ are the coordinates of the center of each weld relative to the common centroid C, we obtain

$$\left[J = L\left(\tfrac{1}{12}L^2 + \bar{x}^2 + \bar{y}^2\right) \right]$$

$$J_{AB} = 150\left[\frac{(150)^2}{12} + (10)^2 + (40)^2 \right] = 0.536 \times 10^6 \text{ mm}^3$$

$$J_{DE} = 100\left[\frac{(100)^2}{12} + (15)^2 + (60)^2 \right] = 0.466 \times 10^6 \text{ mm}^3$$

Their sum determines the modified J to be

$$J = \Sigma J = (0.536 + 0.466) \times 10^6 = 1.00 \times 10^6 \text{ mm}^3$$

The components of the direct load are

$$q_{d_y} = \frac{P}{\Sigma L} = \frac{40 \times 10^3}{250} = 160 \text{ N/mm} \uparrow \quad \text{and} \quad q_{d_x} = 0$$

These values are to be combined with the components of the torsional forces at A and E. These are the highest stressed points in the welds AB and DE as revealed by an inspection of Fig. 12–23b.

Applying Eq. (12–10), we obtain

$$\left[q_{t_x} = \frac{Ty}{J} \right] \quad\quad \text{At } E: \quad q_{t_x} = \frac{(6600 \times 10^3)(60)}{1.00 \times 10^6}$$

$$= 396 \text{ N/mm} \leftarrow$$

$$\text{At } A: \quad q_{t_x} = \frac{(6600 \times 10^3)(40)}{1.00 \times 10^6}$$

$$= 264 \text{ N/mm} \rightarrow$$

$$\left[q_{t_y} = \frac{Tx}{J} \right] \quad \text{At } E \text{ and } A: \quad q_{t_y} = \frac{(6600 \times 10^3)(65)}{1.00 \times 10^6}$$

$$= 429 \text{ N/mm} \uparrow$$

Combining the direct and the torsional force components shows that the highest stressed values in the welds are

$$\left[q = \sqrt{(\Sigma q_x)^2 + (\Sigma q_y)^2} \right]$$

$$q_E = \sqrt{(396)^2 + (160 + 429)^2} = 710 \text{ N/mm}$$

$$q_A = \sqrt{(264)^2 + (160 + 429)^2} = 645 \text{ N/mm}$$

We now apply the AISC specification that the allowable force per millimeter of weld for $\tau = 145$ MPa is $103t$ (regardless of the direction of the force), where t is the length of the leg of the weld in millimeters. Hence the size of the welds, based on the highest stressed point, is

$$\left[q = 103t \right] \qquad q_E = 710 = 103t \qquad t = 6.89 \text{ mm}$$

which requires the use of 7-mm welds.

If desired, a slightly smaller weld may be used in AB, based on $q_A = 645$ N/mm.

PROBLEMS

1249. A bracket is welded to the frame of a machine as shown in Fig. P–1249. Determine the size of fillet weld to be specified to the nearest millimeter. Use $\tau = 145$ MPa through the throats of the welds.

Ans. $t = 5$ mm

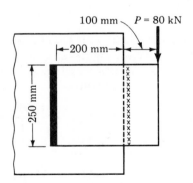

Figure P–1249.

1250. A plate is lapped over and welded to a gusset plate as shown in Fig. P–1250. Determine the size of fillet welds to be specified to the nearest millimeter. What maximum value of P can be applied vertically as shown if 8-mm welds are used? Use $\tau = 145$ MPa through the throats of the welds.

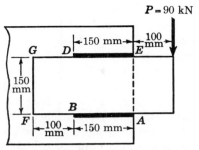

Figures P–1250, P–1251, and P–1252.

1251. Referring to Fig. P–1250, find the maximum force per millimeter of weld if a transverse fillet weld is added along the entire 150-mm width of the plate at A.   *Ans.* $q = 531$ N/mm

1252. In Fig. P–1250, transverse fillet welds are added along the entire 150-mm width of the plate at both A and F. Determine the maximum force developed per millimeter of weld.

1253. An angle is welded to a plate to resist a load P acting through its centroid. The specified lengths of 8-mm welds are as shown at (a) in Fig. P–1253, but a welder applies them as shown in (b). Using the design load determined from (a), find the maximum load per millimeter of weld in (b), assuming elastic action to occur only in the welds. Use $\tau = 145$ MPa through the throats of the welds.

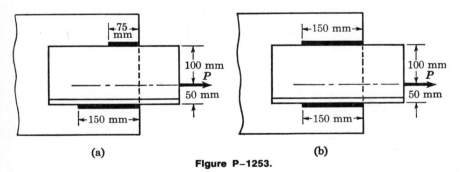

Figure P–1253.

1254. Solve Illustrative Problem 1248 assuming that a transverse weld is added along the entire 100-mm width of the plate at A.
                Ans.   Max. $q = 391$ N/mm; use 4-mm welds

SUMMARY

There are two classes of riveted joints: those used for boilers and those used in structures. In the former, the diameter of the rivet hole determines the diameter used for shearing and bearing calculations; in the latter, the actual rivet diameter is used for shearing and bearing, the rivet hole being assumed to be 3 mm larger.

Further differences between these two types of riveted joints are given by the methods used to compute their strength. For boiler joints, it is preferable to begin with the strength of a single rivet in shear and bearing, and then, depending upon the conditions in each row of rivets, compute the governing load per rivet in each row. The sum of these governing loads is the total rivet strength. The tensile strength of the main plate in any row is assumed to be increased by the strength of the rivets in the rows between that row and the externally applied load. In structural joints, all the rivets are assumed to carry their proportional share of the load. In either type of joint, the effect of frictional resistance is neglected.

Occasionally it is impossible to have the load on a riveted connection pass through the centroid of the rivet group. If the eccentric load is replaced by a central load P and a torsional couple $T = Pe$, the resultant load on any rivet is the vector sum of the direct rivet load $P_d = P/n$ (n being the number of rivets) and a torsional rivet load P_t expressed by

$$P_t = \frac{T\rho}{\Sigma x^2 + \Sigma y^2} \tag{12-5}$$

However, the resultant rivet load P_r is more conveniently determined by combining the components of the direct rivet load and the torsional load. The components P_{d_x} and P_{d_y} are constant for all rivets, and the components of the torsional load are given by

$$\left. \begin{aligned} P_{t_x} &= \frac{T}{\Sigma x^2 + \Sigma y^2} \cdot y \\[2mm] P_{t_y} &= \frac{T}{\Sigma x^2 + \Sigma y^2} \cdot x \end{aligned} \right\} \tag{12-6}$$

where x and y are the coordinates of a rivet measured from the centroid of the rivet group.

The maximum rivet load is found by combining the components of the direct load with the maximum additive components of the torsional couple in accordance with

$$P_r = \sqrt{(P_{d_x} + P_{t_x})^2 + (P_{d_y} + P_{t_y})^2} \tag{12-7}$$

The length of a weld is determined by computing the resisting force required to satisfy the conditions of static equilibrium, and then dividing the resisting load by the strength of the weld in newtons per millimeter. AISC specifications for weld strength depend upon the electrode used in the welding process and upon the grade of steel being welded. The maximum size of a fillet weld is 2 mm less than the material thickness along edges 6 mm or more thick. For edges less than 6 mm thick, the maximum size of the weld may equal the edge thickness.

Eccentrically loaded weld groups, in which the applied load does not pass through the centroid of the weld lines, are treated in a similar manner as are eccentrically loaded riveted connections. The eccentric load is replaced by a central load P passing through the centroid of the weld lines and a torsional couple T equal to the moment of P about this centroid. The intensity of loading is then determined as the vector sum of the uniform direct loading $q_d = P/\Sigma L$ (ΣL being the total length of all welds) and a torsional loading q_t expressed by

$$q_t = \frac{T\rho}{\Sigma L\left(\frac{1}{12}L^2 + \bar{x}^2 + \bar{y}^2\right)} \qquad (12\text{-}9)$$

in which ρ is the radial distance from the centroid of the weld lines to any point on the weld group, and $\bar{x}$ and $\bar{y}$ are the centroidal coordinates of the center of each weld line.

Usually it is more convenient to determine the components of the torsional loading as expressed by

$$\left.\begin{array}{l} q_{t_x} = \dfrac{Ty}{\Sigma L\left(\frac{1}{12}L^2 + \bar{x}^2 + \bar{y}^2\right)} \\[4mm] q_{t_y} = \dfrac{Tx}{\Sigma L\left(\frac{1}{12}L^2 + \bar{x}^2 + \bar{y}^2\right)} \end{array}\right\} \qquad (12\text{-}10)$$

in which x and y are the coordinates of any selected point of any weld.

The maximum intensity of the weld force occurs at the point where q_{d_x} and maximum q_{t_x}, as well as q_{d_y} and maximum q_{t_y}, are additive. Combining these values vectorially gives

$$q = \sqrt{(q_{d_x} + q_{t_x})^2 + (q_{d_y} + q_{t_y})^2} \qquad (12\text{-}11)$$

13

Special Topics

13-1 INTRODUCTION

The preceding chapters have dealt with topics that comprise the usual undergraduate course in strength of materials, but the subject does not end there. We now consider briefly some additional topics* that properly belong in an advanced course in strength of materials; some, such as photoelasticity and finite element methods, constitute a field of study in themselves. Each of the topics is complete in itself so that all or any of them may be studied independently of the others.

13-2 REPEATED LOADING; FATIGUE

Many machine parts are subjected to varying stresses caused by repeated loading and unloading. Parts subjected to such loading

*For extended discussions of these topics, see one or more of the following books: A. P. Bores, O. N. Sidebottom, F. B. Seely, and J. O. Smith, *Advanced Mechanics of Materials*, 3rd ed., Wiley, New York, 1978; S. Timoshenko, *Strength of Materials*, Vols. I and II, Van Nostrand Reinhold, New York; 1955, G. Murphy, *Mechanics of Materials*, McGraw-Hill, New York, 1948.

frequently fail at a stress much smaller than the ultimate strength determined by a static tensile test. Failures of this type are known as *fatigue failures*. In order to properly design members that are subjected to stress reversals, it is necessary to know the stress that can safely be carried an indefinite number of times (or a somewhat higher stress that can be carried safely for a limited number of reversals, as when a machine is used only occasionally and may therefore have a long life).

Testing to determine these values is called *fatigue testing*. The simplest method involves reversed bending. In it, a round specimen S is mounted in bearings, as shown in Fig. 13–1, and subjected to bending couples by the load W. As the motor M rotates the specimen, a fiber that was originally on top passes from compression to tension and back to compression, thereby undergoing a complete reversal of stress for each revolution. A revolution counter, R, registers the number of revolutions until failure occurs. The motor then stops automatically. In testing a given material, about a dozen identical specimens are prepared and rotated in the machine, each at a different load W, until failure occurs or until four to five million cycles have been recorded. A typical result obtained in this way is shown in the $\sigma - N$ diagram in Fig. 13–2, where stress versus number of cycles is plotted with semilogarithmic

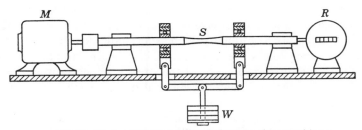

Figure 13–1. Rotating beam fatigue-testing machine.

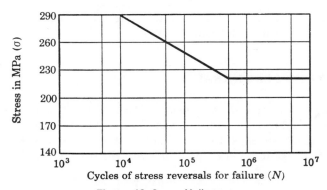

Figure 13–2. $\sigma - N$ diagram.

scales. The point at which this diagram* levels off is called the *endurance limit*, denoted by σ_e. Although no definite relation exists between the endurance limit and the ultimate strength, tests show that the endurance limit is between 40 and 50% of the ultimate strength.

When a ductile steel specimen is subjected to a gradually increasing load, yielding of the specimen is evident considerably before actual failure occurs. But a specimen of the same material that is subjected to stress reversals fails suddenly, without any plastic deformation or any other warning. Thus the fatigue failure of ductile steel is similar to the static failure of a brittle material.

At first it was thought that repeated applications of the load changed the crystalline structure of the material, but we now know that this is not true. Fatigue failure is explained more satisfactorily by the localized stress theory, which is based on the stress concentrations that occur (a) inside a material because of discontinuities in the material itself and (b) at the surface of a material because of abrupt changes in section. These stress concentrations are not serious when a ductile material is subjected to a static load; but when the load is repeatedly applied, they cause minute cracks which spread with each repetition until the member suddenly fractures.

For example, a flat steel bar with a small central hole is loaded by axial tensile forces. As shown in the next article, the stress at the edge of the hole is three times the stress across the full section. However, if this bar is subjected to a gradually increasing static tensile load, it will sustain practically the same ultimate load as a similar bar without a hole. The apparent insignificance of the stress concentration at the hole is due to the fact that although the material there does yield at one-third of the load at which the whole bar yields, this yielding is purely local and the stress at the edge of the hole remains constant during it. With increasing load, the whole bar finally yields; then the effect of a small hole on the strength of the bar is negligible. But if two bars, one solid and the other with a small hole, were subjected to a tensile fatigue test, the failure load of the bar with the hole would actually be about one-third that of the solid bar.

13-3 STRESS CONCENTRATION

As we said in the preceding article, the effect of stress concentration on ductile steel subjected to repeated loading is similar to its effect on a brittle material subjected to static loading. We consider now the effect of abrupt change in section upon the stress distribution. In Fig. 13–3, a

*A log-log plot may be used for this diagram, but a Cartesian plot does not show the endurance limit so clearly.

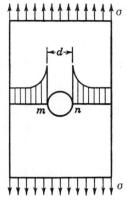

Figure 13-3.

small circular hole in a rectangular plate subjected to a uniform tensile stress σ causes a stress distribution across a section through the hole as shown by the shaded area. This stress distribution is expressed by

$$\sigma' = \frac{\sigma}{2}\left(2 + \frac{d^2}{4r^2} + \frac{3}{16}\frac{d^4}{r^4}\right) \qquad (a)$$

where d is the diameter of the hole and r is the distance from the center of the hole.* From Eq. (a) we find the stresses at points m and n to be 3σ. Because of bending action around the hole, compressive stresses of magnitude σ are created at the top and bottom points of the hole.

A similar stress concentration is caused by the small elliptical hole shown in Fig. 13-4. The maximum stress at the ends of the horizontal

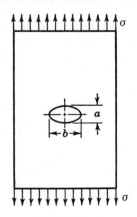

Figure 13-4.

---

*See S. Timoshenko and J. N. Goodier, *Theory of Elasticity*, 2nd ed., McGraw-Hill, New York, 1951, p. 81.

axis of the hole is given by

$$\sigma_{max.} = \sigma\left(1 + 2\frac{b}{a}\right) \tag{b}$$

This stress increases with the ratio b/a; hence a very high stress concentration is produced by a narrow hole or crack perpendicular to the direction of the tensile stress and therefore such cracks tend to spread. This spreading may be stopped by drilling small holes at the ends of the crack, thus replacing a high stress concentration by a relatively smaller one.

Small semicircular grooves in a plate (Fig. 13–5) produce stress concentrations at point m and n that are about three times the average stress σ applied at the ends of the plate.

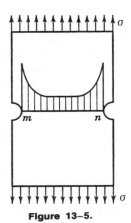

Figure 13–5.

Values* of stress concentration factors for several other cases of abrupt change in section are listed in Table 13–1. If k denotes the factor of stress concentration, the maximum stresses for axial, torsional, and flexural loads are given by

$$\sigma = k\frac{P}{A}, \qquad \tau = k\frac{Tr}{J}, \qquad \sigma_f = k\frac{Mc}{I}$$

Factors of stress concentration for repeated loading are sometimes but not always lower than the theoretical values for static loading given in Table 13–1. In alloy steels and quenched-carbon steels, these theoretical values are recommended for repeated loading; somewhat lower

*These values are taken from a more complete tabulation for these and other cases on pp. 382–406 of R. J. Roark's *Formulas for Stress and Strain*, 4th ed., McGraw-Hill, New York, 1965. For general expressions of these stress concentration factors, see pp. 590–603 of the 5th ed. of the same book by R. J. Roark and W. C. Young (1975).

TABLE 13-1.   Stress Concentration Factors

I. Square shoulder with fillet in rectangular bar

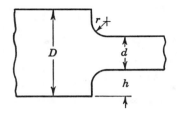

(a) Tension	h/r	r/d					
		0.05	0.10	0.20	0.27	0.50	1.0
	0.5	1.70	1.60	1.53	1.47	1.39	1.21
	1.0	1.93	1.78	1.67	1.59	1.42	1.22
	1.5	—	1.89	1.72	1.65	1.43	1.23
	2.0	—	1.95	1.80	1.70	1.44	1.23

(b) Bending	h/r	r/d					
		0.05	0.10	0.20	0.27	0.50	1.0
	0.5	1.61	1.49	1.39	1.34	1.22	1.07
	1.0	1.91	1.70	1.48	1.38	1.22	1.08
	1.5	2.00	1.73	1.50	1.39	1.23	1.08
	2.0	—	1.74	1.52	1.39	1.23	1.09

II. Square shoulder with fillet in circular shaft

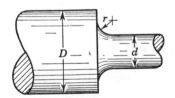

(a) Tension: Approximately same as Case I(a)
(b) Bending: Approximately same as Case I(b)

(c) Torsion	D/d	r/d					
		0.005	0.01	0.02	0.03	0.04	0.10
	2.00	—	3.0	2.25	2.00	1.82	1.44
	1.33	—	2.7	2.16	1.91	1.76	1.40
	1.20	3.00	2.5	2.00	1.75	1.62	1.34
	1.09	2.20	1.88	1.53	1.40	1.30	1.15

(Continued)

TABLE 13–1. (*Continued*)

III. Semicircular notch in circular shaft.

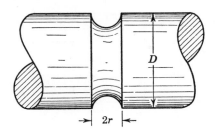

(*a*) Tension	$\dfrac{r}{D - 2r}$	0.05	0.15	0.30	0.40	0.52	0.75
	k	2.57	2.16	1.81	1.65	1.51	1.36

(*b*) Bending	$\dfrac{r}{D - 2r}$	0.05	0.10	0.20	0.30	0.50	0.75
	k	2.20	1.86	1.59	1.45	1.30	1.18

(*c*) Torsion: $k = \dfrac{2D}{D + 2r}$

values are permissible for carbon steels not quenched. The fatigue factor of stress concentration also varies with the size and type of material.*

13–4 THEORIES OF FAILURE

Various theories of failure have been proposed, their purpose being to establish, from the behavior of a material subjected to simple tension or compression tests, the point at which failure will occur under any type of combined loading. By failure we mean either yielding (resulting in excessive permanent deformation) or actual rupture, whichever occurs

---

*A more complete discussion of fatigue will be found in *Prevention of Fatigue of Metals*, Battelle Memorial Institute, published by Wiley. See also Timoshenko's *Strength of Materials*, Vol. II, Arts. 78 to 81, for a discussion of fatigue, and his Chap. 8 for stress concentration. A fairly complete bibliography on the subject of stress concentration is given in Roark and Young's *Formulas for Stress and Strain*, 5th ed., pp. 604–606.

first. Failure resulting from local crippling or elastic instability is not considered here.

The beginning of plastic flow (i.e., yielding) is indicated in a uniaxial tensile test by the deviation from proportionality of stress to strain. Practically, yielding begins at the yield strength at which plastic deformation becomes significant. When several components of stress occur, yielding depends on some combination of these components. Although no theoretical method has been devised which correlates yielding in the uniaxial tensile test with yielding in more complex loadings, several theories of failure, based on predicting the onset of yielding, have been proposed.

The maximum stress theory

The maximum stress theory proposed by Rankine is the oldest as well as the simplest of all the theories. It is based on the assumption that failure occurs when the maximum principal stress on an element reaches a limiting value, the limit being the yield point in a simple tension test (or ultimate strength, if the material is brittle). The theory disregards the effect of possible other principal stresses and of the shearing stresses on other planes through the element. For example, Mohr's circles for the pure shear and pure tension in parts (a) and (b) in Fig. 13–6 show that although both are equally strong according to the maximum principal stress theory, part (a) has twice the maximum shearing stress of part (b). This indicates that the maximum tensile or compressive stress alone is not sufficient to define yielding. Nevertheless, this theory does give results that agree well with test results from brittle materials.

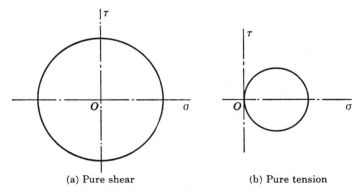

(a) Pure shear   (b) Pure tension

Figure 13–6. Although principal stresses are equal in parts (a) and (b), (a) has twice the shearing stress of (b).

The maximum strain theory

According to the maximum strain theory, which is credited to B. de Saint Venant, a ductile material begins to yield when the maximum principal strain ϵ reaches the strain at which yielding occurs in simple tension or when the minimum principal strain (i.e., the compressive strain) equals the yield point strain in simple compression. However, if we examine Hooke's law for triaxial stress expressed by the following equations:

$$\left.\begin{aligned} \epsilon_x &= \frac{1}{E}\big[\sigma_x - \nu(\sigma_y + \sigma_z)\big] \\ \epsilon_y &= \frac{1}{E}\big[\sigma_y - \nu(\sigma_z + \sigma_x)\big] \\ \epsilon_z &= \frac{1}{E}\big[\sigma_z - \nu(\sigma_x + \sigma_y)\big] \end{aligned}\right\} \tag{2-12}$$

we see that when $\sigma_x = -\sigma_y = -\sigma_z$, the maximum strain is $(1 + 2\nu)\sigma/E$. On the other hand, when $-\sigma_x = -\sigma_y = -\sigma_z$, as in hydrostatic compression, the maximum strain is $(1 - 2\nu)\sigma/E$. Thus, different strains may appear with the same maximum stress.

The maximum shear theory

Sometimes called Guest's theory, the maximum shear theory assumes that yielding begins when the maximum shearing stress equals the maximum shearing stress developed at yielding in simple tension. Since the maximum shearing stress is equal to one-half the difference between the principal stresses, the condition for yielding is

$$\tau_w = \tfrac{1}{2}(\sigma_{\text{max.}} - \sigma_{\text{min.}}) = \tfrac{1}{2}\sigma_{\text{yp}}$$

The Mises yield theory

The Mises yield theory, also known as the maximum shear distortion theory, assumes that yielding can occur in a general three-dimensional state of stress when the root mean square of the differences between the principal stresses is equal to the same value in a tensile test. If $\sigma_1 > \sigma_2 > \sigma_3$ are the principal stresses and σ_{yp} is the yield strength in simple tension, this concept gives

$$\begin{aligned} \tfrac{1}{3}&\big[(\sigma_1 - \sigma_2)^2 + (\sigma_2 - \sigma_3)^2 + (\sigma_3 - \sigma_1)^2\big] \\ &= \tfrac{1}{3}\big[(\sigma_{\text{yp}} - 0)^2 + (0 - 0)^2 + (0 - \sigma_{\text{yp}})^2\big] \\ &= \tfrac{2}{3}\sigma_{\text{yp}}^{\,2} \end{aligned}$$

from which we obtain

$$2\sigma_{yp}^2 = (\sigma_1 - \sigma_2)^2 + (\sigma_2 - \sigma_3)^2 + (\sigma_3 - \sigma_1)^2$$

Summary

Of these several theories of failure, experimental work shows best agreement with the Mises yield theory when applied to ductile materials. For such materials the maximum shear theory also gives good agreement. For rupture in brittle materials, such as cast iron, the maximum stress theory is generally preferred.

13–5 ENERGY METHODS

By equating the external work done by applied loads as they deform an elastic body to the internal strain energy stored in the body, we obtain a method of determining deflections that is based on the principle of conservation of energy. This method will be shown to be extremely versatile. We begin by obtaining expressions for the strain energy U stored in a body under various loadings.

Axial loading

As an axial load is gradually applied to a bar of constant cross section, the load P increases from zero to its applied value as the bar undergoes a deformation δ. (The case of suddenly applied loads is considered in the next article.) The work done, stored as strain energy in the body, is the product of the average force $\frac{1}{2}P$ and the deformation $\delta = PL/AE$. Hence we obtain

$$U = \frac{1}{2}\frac{P^2 L}{AE} \tag{13–1}$$

If the bar varies in cross section, this result may be applied to segments of length dx and integrated over the length of the bar to obtain

$$U = \int_0^L \frac{1}{2}\frac{P^2\,dx}{AE} \tag{13–1a}$$

Torsional loading

For a circular bar of constant cross section, the strain energy stored in the body is equal to the product of the average torque and the angular deformation, that is,

$$U = \frac{1}{2} T\theta = \frac{1}{2} T\left(\frac{TL}{JG}\right) = \frac{1}{2}\frac{T^2L}{JG} \tag{13-2}$$

When the torque varies, this result may be applied to segments of length dx and integrated over the length of the bar to obtain

$$U = \int_0^L \frac{T^2\,dx}{2JG} \tag{13-2a}$$

Flexural loading

In the portion of a beam shown in Fig. 13-7, consider the differential element isolated by two transverse sections a distance dx apart. Treating this element as an axially loaded bar, where $P = \sigma_f\,dA = (My/I)\,dA$, the energy stored in it is

$$\left[U = \frac{P^2L}{2AE}\right] \qquad \frac{P^2\,dx}{2AE} = \frac{M^2y^2}{I^2}(dA)^2 \cdot \frac{dx}{2(dA)E} = \frac{M^2\,dx}{2EI^2}\cdot y^2\,dA$$

Summing this result to include all elements across the depth of the beam gives the energy stored in the differential length dx of the beam as

$$dU = \frac{M^2\,dx}{2EI^2}\int y^2dA = \frac{M^2\,dx}{2EI}$$

whence for the entire length of the beam we finally obtain

$$U = \int \frac{M^2\,dx}{2EI} \tag{13-3}$$

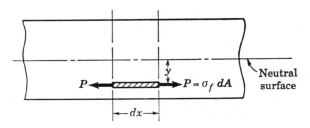

Figure 13-7. Energy stored in a differential element of a beam:

$$P = \sigma_f\,dA = (My/I)\,dA$$

Deflections

Instead of directly equating external work to the foregoing expressions for strain energy, considerable simplification is obtained by applying *Castigliano's theorem* or its close equivalent, the *virtual-work method*.

Castigliano's theorem states that the deflection caused by any external force is equal to the partial derivative of the strain energy with respect to that force. To prove the theorem, consider the elastic body in Fig. 13–8 supported on rollers at A and B and by a hinge at C. As the loads $P, Q, F, \ldots$ are each applied independently to the body, they cause a strain energy U to be stored in the body that is exactly equivalent to the work done on the body by these loads. Note that the reactions do no work, because the roller reactions are perpendicular to the displacements of the rollers and the hinge reaction is fixed and cannot move.

Assume now that one of the loads, Q, is increased by a small amount dQ. The increase in strain energy will be

$$dU = \frac{\partial U}{\partial P} dP + \frac{\partial U}{\partial Q} dQ + \frac{\partial U}{\partial F} dF + \cdots$$

However, since only Q is assumed to get an increment, $dP = dF = \ldots = 0$, and the change in U therefore is

$$dU = \frac{\partial U}{\partial Q} dQ \tag{a}$$

On the assumption that the principle of superposition applies, if dQ had been acting initially on the body and $P, Q, F, \ldots$ had been subsequently applied, the body would contain the same strain energy as before. If δ is the displacement of the point of application of Q in the direction of Q, then the increment dQ rides through the distance δ as the actual loads are applied and produces the external work increment $dU = \delta\, dQ$. On equating this external work to the internal work given by Eq. (a), we obtain

$$dU = \delta\, dQ = \frac{\partial U}{\partial Q} dQ$$

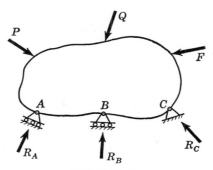

Figure 13–8.

or

$$\delta = \frac{\partial U}{\partial Q} \tag{13-4}$$

This is interpreted as follows: The partial derivative of the strain energy with respect to one of the external loads equals the displacement of the point of application of that load in the direction of that load.

In applying Castigliano's theorem to a bar in which the strain energy is the sum of the following integrals,

$$U = \int \frac{P^2\,dx}{2AE} + \int \frac{T^2\,dx}{2JG} + \int \frac{M^2\,dx}{2EI}$$

the deflection, $\delta = \partial U/\partial Q$, is best evaluated by differentiating inside the integral sign before integrating. This procedure is permissible because Q is not a function of x. With this simplification, the deflection at Q in the direction of Q is given by

$$\delta = \frac{\partial U}{\partial Q} = \int \frac{P\dfrac{\partial P}{\partial Q}\,dx}{AE} + \int \frac{T\dfrac{\partial T}{\partial Q}\,dx}{JG} + \int \frac{M\dfrac{\partial M}{\partial Q}\,dx}{EI} \tag{13-5}$$

If no load is acting at the point where the deflection is desired, a dummy load Q in the direction of the desired deflection may be added at that point. Then, *after* differentiating but *before* integrating, the dummy load is set equal to zero.

In the virtual-work method, a dummy load of 1.0 N is used in place of a dummy load Q. Then the derivatives $\partial P/\partial Q$ and $\partial M/\partial Q$ are designated respectively as u and m. It is convenient to think of u and m as the axial force and bending moment caused by the unit dummy load. The resulting inconsistency in units is eliminated by multiplying δ by 1.0 N thereby

$$1.0\text{ N} \times \delta = \int \frac{Pu\,dx}{AE} + \int \frac{Mm\,dx}{EI} \tag{13-6}$$

This result can be obtained by a different process of reasoning based on equating total external work to total internal work. For brevity we have developed it here as a special case of Castigliano's theorem which enables us to show the equivalence of the methods. The following illustrative problems show how both methods are used and also give a physical interpretation to the integral $\int (Mm/EI)\,dx$ which simplifies the computations.

ILLUSTRATIVE PROBLEMS

1301.   A bar is bent into a circular arc of radius R and held in a horizontal plane as shown in Fig. 13–9a. Find the deflection at A caused by a vertical load P applied there.

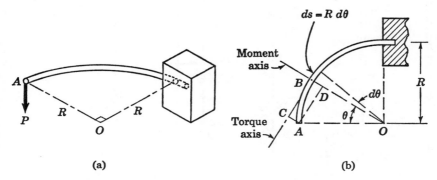

Figure 13–9. Deflection of free end of bar bent into a circular arc in a horizontal plane.

Solution: The vertical load causes both twisting and bending over the element of length $ds = R\, d\theta$. As shown by the top view in Fig. 13–9b, the moment arm about the torque axis BC is $AC = BD = OB - OD = R - R\cos\theta$; and the moment arm about the moment axis BO is $AD = R\sin\theta$. Hence we have

$$T = PR(1 - \cos\theta) \quad \text{and} \quad \frac{\partial T}{\partial P} = R(1 - \cos\theta)$$

$$M = PR\sin\theta \qquad\qquad \text{and} \quad \frac{\partial M}{\partial P} = R\sin\theta$$

Applying Eq. (13–5), we have

$$\delta = \frac{1}{JG}\int T\frac{\partial T}{\partial P}\,ds + \frac{1}{EI}\int M\frac{\partial M}{\partial P}\,ds$$

$$= \frac{1}{JG}\int_0^{\pi/2} PR(1 - \cos\theta)\big[R(1 - \cos\theta)\big]R\,d\theta$$

$$+ \frac{1}{EI}\int_0^{\pi/2} PR\sin\theta(R\sin\theta)R\,d\theta$$

which is evaluated to give

$$\delta = \frac{PR^3}{JG}\left(\frac{3\pi - 8}{4}\right) + \frac{PR^3}{EI}\left(\frac{\pi}{4}\right) \qquad Ans.$$

1302.   A rigid frame, loaded as shown in Fig. 13–10a, is hinged at A and supported on rollers at D. Assuming EI is constant throughout

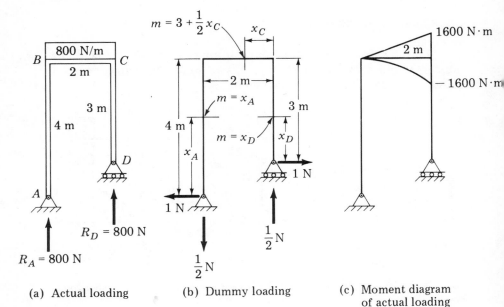

(a) Actual loading   (b) Dummy loading   (c) Moment diagram
                                                              of actual loading

Figure 13–10.

the frame, determine the horizontal deflection at the roller D. Neglect
the effect of axial deformations.

Solution: In this example, we use the virtual work relation*

$$\delta = \int \frac{Mm \, dx}{EI}$$

Because discontinuities in the moment equation occur at the change of
loading points, we rewrite the integral in the following form:

$$\delta_D = \int_A^B \frac{Mm \, dx}{EI} + \int_B^C \frac{Mm \, dx}{EI} + \int_C^D \frac{Mm \, dx}{EI} \qquad (a)$$

The various values of m are found from the free-body diagram of
the frame (Fig. 13–10b) in which a unit dummy load has been applied
at D in the direction of the desired deflection. The conditions of static
equilibrium determine the reactions to this unit load to be as shown.
The values of m shown are now 'determined from the definition of
bending moment, which specifies that we may take moments about an
exploratory section of the forces lying to one side of the section.

*The   coefficient 1.0 N that should precede δ is usually omitted but is
understood to be included so as to insure dimensional homogeneity.

Substituting these values of m in Eq. (a) and noting that EI is constant we obtain

$$EI\delta_D = \int_A^B x_A M \, dx + \int_B^C \left(3 + \frac{1}{2}x_C\right) M \, dx + \int_C^D x_D M \, dx \qquad (b)$$

These integrals have the same physical meaning that we encountered in our earlier study of the area-moment method (page 229), and hence we may express them in the equivalent form

$$EI\delta_D = (\text{area})_{AB} \cdot \bar{x}_A + 3(\text{area})_{BC} + \tfrac{1}{2}(\text{area})_{CB} \cdot \bar{x}_C$$
$$+ (\text{area})_{DC} \cdot \bar{x}_D \qquad (c)$$

in which, for example, $(\text{area})_{AB} \cdot \bar{x}_A$ represents the moment of area about A of the actual moment diagram for the segment AB. Observe that moments of area are taken about the point from which x was measured in finding the various values of m in Fig. 13–10b.

The moment diagram by parts of the actual loading is now drawn as in Fig. 13–10c. Observe that no moment exists in the legs AB and CD; hence in Eq. (c) the first and last terms on the right side are zero. Evaluating Eq. (c), we now obtain

$$EI\delta_D = 3\left[\frac{1600 \times 2}{2} - \frac{1600 \times 2}{3} \right]$$
$$+ \frac{1}{2}\left[\left(\frac{1600 \times 2}{2}\right)\left(\frac{2}{3}\right) - \left(\frac{1600 \times 2}{3}\right)\left(\frac{2}{4}\right) \right]$$

from which

$$EI\delta_D = 1867 \text{ N} \cdot \text{m}^3 \qquad Ans.$$

If this result had been negative, it would merely mean that the deflection at D was opposite to the direction of the unit load at D.

PROBLEMS

1303. Determine the midspan value of $EI\delta$ for a simply supported beam L meters long carrying a uniformly distributed load of w N/m over its right half.

1304. As shown in Fig. P–1304, two aluminum rods AB and BC, hinged to rigid supports, are pinned together at B to carry a vertical load of 20 kN. If each rod has a cross-sectional area of 400 mm^2 and $E = 70$ GPa, compute the horizontal and vertical deflections of point B. Assume $\alpha = 30°$ and $\theta = 30°$.   *Ans.* $\delta_h = 0.412$ mm; $\delta_v = 3.57$ mm

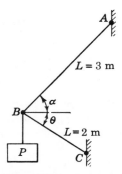

Figures P–1304 and P–1305.

1305. Solve Problem 1304 if rod AB is of steel, with $E = 200$ GPa. Assume $\alpha = 45°$ and $\theta = 30°$; all other data remain unchanged.   *Ans.* $\delta_h = 0.417$ mm; $\delta_v = 1.37$ mm

1306. A circular bar is bent into the shape of a quarter-ring and supported in a vertical plane as shown in Fig. P–1306. Compute the horizontal and vertical displacements of point A.

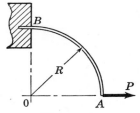

Figure P–1306.

1307. A circular bar is bent into the shape of a half-ring and supported in a vertical plane as shown in Fig. P–1307. Determine the horizontal movement of point C and the vertical movement of point B.

$$Ans. \quad \delta_{C_h} = \frac{PR^3\pi}{2EI} \; ; \; \delta_{B_v} = \frac{PR^3}{2EI}$$

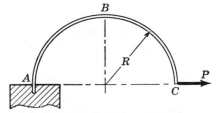

Figures P–1307, P–1308, P–1309.

1308. Repeat Problem 1307 assuming that the load P is applied vertically downward at C.

1309. In Problem 1307, let the load P be applied at C and perpendicular to the plane of ABC. What will be the displacement of C in the direction of the load?   *Ans.* $\dfrac{PR^3}{JG}\left(\dfrac{3}{2}\pi\right) + \dfrac{PR^3}{EI}\left(\dfrac{\pi}{2}\right)$

1310. A vertical load P is applied to the rigid cantilever frame shown in Fig. P–1310. Assuming EI to be constant throughout the frame, determine the horizontal and vertical displacements of points B and C. Neglect axial deformations.

Ans. $\delta_{B_v} = 0; \ \delta_{B_h} = \delta_{C_h} = \dfrac{Pba^2}{2EI}; \ \delta_{C_v} = \dfrac{Pb^2}{EI}\left(a + \dfrac{b}{3}\right)$

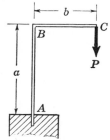

Figures P–1310 and P–1311.

1311. In Problem 1310, let the load P be applied at C and perpendicular to the plane of ABC. Find the displacements of points B and C in the direction of the load.   *Ans.* $\delta_C = \dfrac{Pab^2}{JG} + \dfrac{P}{3EI}(a^3 + b^3)$

1312. The rigid frame shown in Fig. P–1312 is supported by a hinge at A and a roller at D and carries the triangularly distributed load. Assuming constant EI, compute $EI\delta$ at the roller D. Neglect axial deformation.   *Ans.* $EI\delta = 2970 \ \text{N} \cdot \text{m}^3$

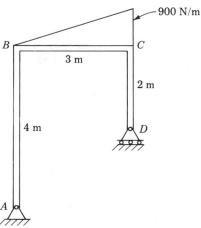

Figure P–1312.

1313. The rigid frame in Fig. P–1313 is subjected to a horizontal load and a vertical load as shown. Assuming constant EI and neglecting axial deformations, determine the value of $EI\delta$ at the roller support D.

Ans. $EI\delta = 58.2 \text{ kN} \cdot \text{m}^3$

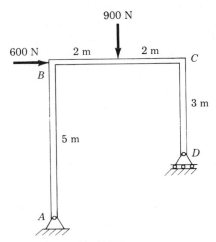

Figure P–1313.

13–6 IMPACT OR DYNAMIC LOADING

The deformations produced in elastic bodies by impact loads cause them to act as springs, although that is not their designed function. If the equivalent spring constant for such members is defined as the load required to cause a unit deformation, the spring constant in each case can be determined from our earlier study of deformations. Actually, however, as we shall see, it is unnecessary to determine the equivalent spring constant. For the present, we may consider the problem of impact as analogous to that of a falling body stopped by a spring (Fig. 13–11). The mass m has zero velocity when first dropped, and also when the spring is deflected through the maximum dynamic deflection δ. Equating the resultant work done on mass m to the zero change in kinetic energy, we therefore obtain*

$$mg(h + \delta) - \tfrac{1}{2}k\delta^2 = 0 \qquad (a)$$

where $mg(h + \delta)$ is the work done by gravity on the body, and $\tfrac{1}{2}k\delta^2$ is the resisting work done by the equivalent spring.

*The mass is assumed to remain in contact with the spring. Also, some energy is dissipated by the impact, so the actual deflection is always less than that given by Eq. (a).

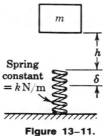

Figure 13-11.

If Eq. (a) is rearranged in the form

$$\delta^2 - 2\frac{mg}{k}\delta - 2\frac{mg}{k}h = 0$$

and mg/k is replaced by δ_{st}, which is the static deformation produced by a gradual application of the weight mg, the following general value of δ is obtained:

$$\delta = \delta_{st} + \sqrt{(\delta_{st})^2 + 2\delta_{st}h} \qquad (b)$$

Two extreme cases are of interest. If h is large compared with δ, we may neglect the work $mg\delta$ in Eq. (a), which then reduces to

$$\delta = \sqrt{2\frac{mg}{k}h} = \sqrt{2\delta_{st}h} \qquad (c)$$

In the other extreme case, $h = 0$ (i.e., the load is suddenly applied) and Eq. (a) reduces to

$$\delta = 2\frac{mg}{k} = 2\delta_{st} \qquad (d)$$

Because of a suddenly applied load, the deflection and consequently the stress which is directly proportional to it are therefore twice as great as that caused by the same load gradually applied.

The ratio of the maximum dynamic deformation δ to the static deformation δ_{st} gives a value which may be called the *impact factor*. This is easily determined by rearranging Eq. (b) in the form

$$\delta = \delta_{st} + \delta_{st}\sqrt{1 + \frac{2h}{\delta_{st}}} = \delta_{st}\left(1 + \sqrt{1 + \frac{2h}{\delta_{st}}}\right)$$

Hence the impact factor is

$$\frac{\delta}{\delta_{st}} = 1 + \sqrt{1 + \frac{2h}{\delta_{st}}} \qquad (13\text{--}7)$$

Multiplying mg by this factor gives an equivalent impact load P, which may be used in the formulas for static loading to compute the maximum stress and deflection. Or, if preferred, the static stress due to a gradual application of mg may be multiplied by the impact factor to give the maximum stress:

$$\sigma_{\text{max.}} = \sigma_{\text{st}}\left(1 + \sqrt{1 + \frac{2h}{\delta_{\text{st}}}}\,\right) \tag{13-8}$$

We now apply these results to various types of impact loading.

Tension

The most usual type of impact loading is shown in Fig. 13–12. A mass m drops freely through a height h before striking a stop on the end of the rod, thereby producing the dynamic deflection δ. Assuming that the stresses remain within the elastic range and that δ is negligible compared with h, we replace δ_{st} in Eq. (c) by its value from the deformation equation $\delta_{\text{st}} = mgL/AE$, whence we obtain

$$\delta = \sqrt{\frac{2L}{AE}\,mgh} \tag{e}$$

The corresponding stress in the rod is

$$\sigma = \frac{\delta}{L}E = \sqrt{\frac{2E}{AL}\,mgh} \tag{f}$$

By replacing mgh by the kinetic energy $\frac{1}{2}mv^2$, this may also be used to determine the shock stress caused by the sudden stopping of a mass m that is moving with a velocity v:

$$\sigma = \sqrt{\frac{2E}{AL}\cdot\frac{mv^2}{2}} \tag{g}$$

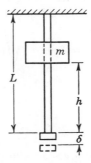

Figure 13–12. Impact loading of a rod.

These equations show that the stress due to impact can be reduced by using a material with a lower value of E or by increasing the area A or the length L of the rod. This is quite different from static tension where the stress is independent of E or L.

The foregoing discussion assumes that the stress remains below the proportional limit. When the stress is above the proportional limit the problem is more complex, because the elongation is no longer proportional to the load. Nevertheless, we can still find a basis for determining actual rupture due to impact. Thus, it being assumed that the shape of a tensile test diagram does not depend upon the speed with which the bar is strained, the area $OABC$ in Fig. 13–13 represents the work done upon the bar to produce an elongation δ; this must be equal to the work $mg(h + \delta)$ done by the falling weight mg. When $mg(h + \delta)$ equals or exceeds the total area $OADE$, the falling weight will rupture the bar.

A bar's resistance to impact also depends upon the ductility of the material. Figure 13–14 shows the tensile test diagram of a high-strength steel of low ductility superimposed upon the diagram for a steel of lower strength but high ductility. The horizontally shaded area A_1 is much larger than the vertically shaded area A_2, showing that the more ductile steel will absorb more energy before rupture than the less ductile steel. For this reason, ductile materials are usually selected for members subject to impact or shock loading.

In connection with the above, the total area of a stress-strain diagram is called the *modulus of toughness*; it represents the energy absorbed per unit volume. Its value is equal approximately to

$$U_r = \frac{\sigma_y + \sigma_u}{2} \cdot \epsilon_u \tag{h}$$

where σ_y and σ_u are respectively the yield point and the ultimate strength, and ϵ_u is the ultimate strain. The partial area of a stress–strain diagram up to the stress σ_e at the elastic limit is called the *modulus of resilience*; it represents the energy that can be absorbed per unit volume

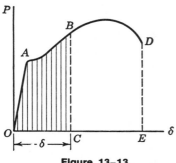

Figure 13–13.

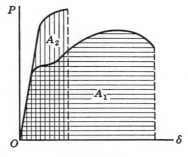

Figure 13–14.

without creating a permanent distortion. If the stress–strain relation is linear, its value is

$$U = \frac{1}{2}\sigma_e\epsilon_e = \frac{1}{2}\sigma_e \cdot \frac{\sigma_e}{E} = \frac{\sigma_e^2}{2E} \qquad (i)$$

Flexure

In Fig. 13–15 a simply supported beam is subjected to the impact of a mass m falling freely through the height h before striking the midpoint of the beam and causing the dynamic deflection δ. If the proportional limit is not exceeded, Eqs. (b), (c), and (d) apply here also. Assuming that h is large compared with δ_{st}, we obtain

$$\delta = \sqrt{2h\delta_{st}}$$

Hence the impact factor for a centrally loaded simple beam where $\delta_{st} = mgL^3/48EI$ (Case 6, Table 6–2, page 270) becomes

$$\frac{\delta}{\delta_{st}} = \sqrt{\frac{2h}{\delta_{st}}} = \sqrt{\frac{96EIh}{mgL^3}}$$

The static stress found from the flexure formula is

$$\sigma_{st} = \frac{Mc}{I} = \frac{mgL}{4} \cdot \frac{c}{I}$$

and the maximum stress is

$$\sigma = \frac{\delta}{\delta_{st}}\sigma_{st} = \sqrt{\frac{6mghEc^2}{LI}} \qquad (j)$$

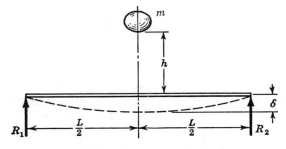

Figure 13–15. Impact loading of a beam.

Limitations

The preceding discussions assumed that the work done by a falling body or the kinetic energy of a moving body can be stored in the resisting member in the form of strain energy. This assumption can never be realized for the following reasons: If the velocity of impact is high, the deceleration of the moving body may be so rapid as to transform the kinetic energy into partly heat and partly local deformations, largely inelastic, of both the moving body and the resisting member. Even if the velocity of impact is low, the resisting member may have great stiffness, causing the same results. Finally, if the mass of the resisting member is large compared to that of the moving body, the *inertia* of the resisting member may also cause the same result.

PROBLEMS

1314.   A 50-kg mass falls through 2 m and is then caught on the end of a wire rope 30 m long having a cross-sectional area of 250 mm². Compute the maximum stress in the rope. Assume $E = 100$ GPa.

Ans. $\sigma = 164$ MPa

1315.   An elevator having a mass of 2 Mg is being lowered at the rate of 2 m/s. The hoisting drum is stopped suddenly when 30 m of cable have been unwrapped. If the cross-sectional area of the cable is 600 mm² and $E = 100$ GN/m², compute the maximum stress in the cable. Neglect the weight of the cable.

1316.   A 6-kg mass falls 0.8 m and strikes the head of the steel bolt shown in Fig. P–1316. Assuming all the energy is absorbed by the bolt, compute the required thickness t of its head if the shearing stress on the cylindrical surface through the head is not to exceed 80 MN/m². Assume $E = 200$ GN/m².   *Ans.* $t = 12.5$ mm

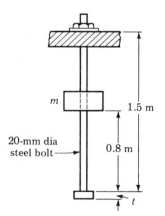

20-mm dia steel bolt

m

1.5 m

0.8 m

t

Figure P–1316.

1317. A simply supported rectangular beam of length L and cross-sectional area A is struck at the center by a mass m falling through a height h. Show that the maximum bending stress is given by $\sigma^2 = 18mghE/AL$.

1318. Compute the impact factor for a simply supported beam 3 m long subject to an impact by a 900-kg mass that drops from a height of 2.5 m onto a point 1.0 m from one end. Assume the beam section is rectangular, 40 mm wide and 90 mm deep, and $E = 200 \times 10^9$ N/m^2. Neglect the mass of the beam.   *Ans.*   25.9

1319. A rectangular beam 60 mm wide and 100 mm deep is used as a cantilever 2 m long. A 40-kg mass falls through a height of 0.2 m before striking the free end. Compute the maximum bending stress and deflection caused by the impact. Neglect the mass of the beam and assume that the 40-kg mass remains in contact with the beam. Assume $E = 200$ GPa.   *Ans.* $\sigma = 162$ MPa; $\delta = 21.5$ mm

1320. A freight car having a mass of 12 Mg is moving at 1.2 m/s when it is stopped by a nest of eight bumper springs. Each spring is 10 turns of 25-mm-diameter steel wire on a mean coil radius of 100 mm. Use the Wahl formula [Eq. (3–10) on page 97] to determine the maximum stress developed in the springs. Assume $G = 80$ GPa.

13–7   SHEARING STRESSES IN THIN-WALLED MEMBERS SUBJECTED TO BENDING; SHEAR FLOW

The formula developed in Art. 5–7 (page 188) for determining the shearing stress induced by flexure can also be used to determine the shearing stress across the flanges of wide flange beams or channels or other sections. The existence of this shearing stress is explained in Fig. 13–16, which shows the free-body diagrams of portions of the flanges cut out by two adjacent sections. The dashed lines indicate a phantom view of the cantilever beam from which these sections were cut. When the external vertical shear acts downward, the upper flange is in tension, the tensile force T_2 being larger than T_1 because the bending moment is greater at section 2 than at section 1. For equilibrium of the upper flange segment, a longitudinal shearing force F_s must act as shown, thereby inducing the lateral shear force H_1. The direction of H_1 determines the directions of the shearing stresses in the upper flange. Similarly, the compressive forces C_2 and C_1 developed in the lower flange segment required a leftward lateral shear H_2, and hence the shearing stresses are here directed leftward. Because of tension in the upper flange and compression in the lower flange, the shearing stresses in these flanges are in opposite directions.

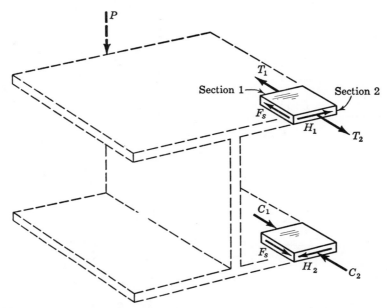

Figure 13-16. Lateral shear forces H_1 and H_2 in flanges of cantilever beam. External vertical shear acts downward.

The magnitude of the longitudinal shearing stress across the flange is given by Eq. (5–4) (page 189) if the flange is assumed to be relatively thin so that, as was done in deriving this equation, the shearing stress may without serious error be considered uniform across the thickness of the flange. Even more convenient is the concept of shear flow developed in Art. 3–5 and again in Art. 5–7 as Eq. (5–4a). Thus at a distance z from the free edge of the flange in either part (a) or (b) of Fig. 13–17, Eq. (5–4a) gives

$$q = \frac{V}{I} Q = \frac{V}{I}(tz)y = \left(\frac{Vht}{2I} \right)z \qquad (a)$$

This shows that the shear flow in the flanges varies linearly with the distance from the free edge.

The variation in shear flow and its direction are shown graphically in Fig. 13–18a and b, in which the external vertical shear V is assumed to act downward.

Similarly, for the split tube in Fig. 13–17c, the shear flow across any radial section defined by ϕ is

$$q = \frac{V}{I} Q = \frac{V}{I} \int_0^\phi r \sin \phi(tr\, d\phi) = \frac{Vtr^2}{I}(1 - \cos \phi) \qquad (b)$$

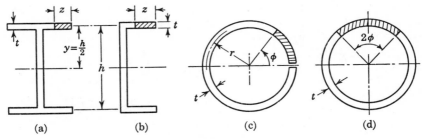

Figure 13–17.

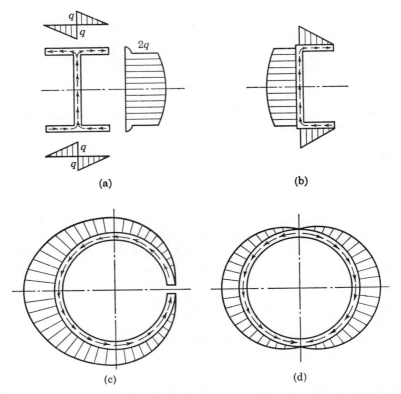

Figure 13-18.   Variation in shear flow. In parts (a) and (b), external vertical shear *V* is down; in parts (c) and (d), external *V* is up.

When the external vertical shear acts upward, Fig. 13–18c shows how the shear flow varies from zero at the split to a maximum value at a section opposite the split. As an exercise, verify the variation in shear flow for the solid tube in Fig. 13–18d.

13–8   SHEAR CENTER

We are now ready to consider the bending of members that have only one axis of symmetry, the loading being such as to cause this axis to become the neutral axis. On any section of a beam subjected to other than pure bending, there exist shearing stresses. These stresses create internal shearing forces whose resultant must be equal, opposite, and collinear to the external shear; otherwise the bending is accompanied by twisting of the beam. Bending without twisting occurs only when the resultant of the shearing forces passes through the shear center (also called the center of twist and sometimes the flexural center). The shear center is defined as the point in the cross section of a beam through which the plane of the transverse bending loads must pass so that the beam will bend without twisting.

We begin by considering a channel section used as a cantilever; the free-body diagram is shown in Fig. 13–19. The resisting forces consist of the resisting vertical shear V_r considered as acting through the web of the channel (as shown for a wide flange beam in Illustrative Problem 566), the resisting couple M_r composed of the tensile and compressive forces T and C (shown for convenience as acting through and normal to the flanges), and the horizontal flange forces H, which

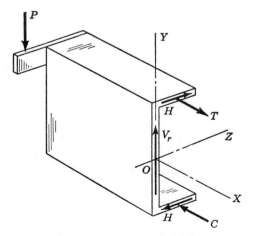

Figure 13–19.   Free-body diagram of channel section.

are the resultants of the shearing stresses in the flanges computed as shown in the preceding article. It may seem surprising that the load P does not act through the longitudinal centroidal plane of the section; but the equations of equilibrium show why P must act as shown. Later we shall compute its position. The six equations of equilibrium, and the reasons they are satisfied, are as follows:

1. $\Sigma X = 0$, satisfied by balance between the equal and oppositely directed tensile and compressive forces T and C.

2. $\Sigma Y = 0$, satisfied by the resisting vertical shear V_r balancing the vertical shear V caused by P.

3. $\Sigma Z = 0$, satisfied by the balance of the equal and oppositely directed flange forces H.

4. $\Sigma M_y = 0$, satisfied because vertical loads cause no moment about the Y axis, and the moments of the horizontal forces T and C about Y cancel each other, as do the flange forces H.

5. $\Sigma M_z = 0$, satisfied because the applied bending moment M is balanced by the resisting moment M_r supplied by T and C.

6. $\Sigma M_x = 0$; this condition must be satisfied to prevent the beam from twisting as it bends. It can be satisfied only if the moment of the applied load balances the moments of the shearing forces developed over the section. Selecting the X axis through the web eliminates the moment of V_r. In the end view of Fig. 13–19, shown in Fig. 13–20, we set a moment summation about O equal to zero and obtain

$$[\Sigma M_O = 0] \qquad Ve = Hh \qquad\qquad (a)$$

The value of the flange force H is the product of the average shear flow in the flange multiplied by the length of the flange. Using Eq. (a) of Art. 13–7, we have

$$H = q_{\text{ave.}} \cdot L = \left(\frac{1}{2} \cdot \frac{Vhtb}{2I} \right) b = \frac{Vhtb^2}{4I}$$

This value of H may now be substituted in Eq. (a), yielding

$$e = \frac{Hh}{V} = \frac{h^2b^2t}{4I} \qquad\qquad (13\text{–}9)$$

Point C on the neutral axis (NA) located a distance e from the center of the web is the shear center for the channel.

In a wide flange beam or an I beam, flexure loads develop a lateral shear in each outstanding flange as in a channel section, but the shearing forces in each flange are equal and oppositely directed, as in Fig. 13–21. (See also Fig. 13–18a for the direction of the shear flow.) Hence, moments of the shearing forces about the centroid of the section balance out, and the plane of the loads must also contain the centroid if

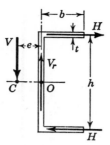

Figure 13–20.

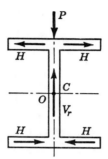

Figure 13–21. In a section having two axes of symmetry, the shear center C coincides with the centroid O.

twisting is to be avoided. We conclude that the shear center coincides with the centroid of the wide flange section. In general, the shear center is located at the intersection of two axes of symmetry, if they exist.

For an unequal flanged H section with relatively thin flanges and web, like that in Fig. 13–22, the bending resistance of the web may be assumed to be negligible and the total vertical shear V may be assumed to be resisted by the internal shears V_1 and V_2 acting along the center lines of the flanges. To prevent twisting, the resultant V_r of the flange shears must be equal, opposite, and collinear with the external shear V. Hence, taking moments about the shear center C, we have

$$V_1 b_1 = V_2 b_2 \qquad (b)$$

Another relation between V_1 and V_2 may be found from the fact that the two flanges bend as though they were separate beams which have identical radii of curvature. Hence, applying $\rho = EI/M$ gives

$$\frac{\rho}{E} = \frac{I_1}{M_1} = \frac{I_2}{M_2} \qquad (c)$$

But the bending moments M_1 and M_2 at any section along the beam are equal respectively to the product of the vertical shears V_1 and V_2 in the

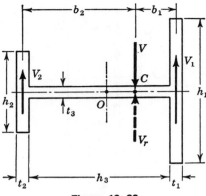

Figure 13–22.

flanges and the distance to the section. Hence Eq. (*c*) becomes

$$\frac{I_1}{V_1} = \frac{I_2}{V_2} \qquad\qquad (d)$$

which when combined with Eq. (*b*) yields

$$\frac{b_1}{b_2} = \frac{I_2}{I_1} \qquad\qquad (13\text{–}10)$$

Therefore, since the ratio b_1/b_2 and the sum of b_1 and b_2 are known, the position of the shear center is easily located. It lies between the centroid of the section and the centroid of the flange that has the larger moment of inertia.

When there is only one flange, as in the T section in Fig. 13–23, if the bending resistance of the web is again assumed to be negligible, the shear center coincides with the centroid of the flange. In general, for any section composed of two narrow rectangles where the shear flow is along the longer center lines of the rectangles, the shear center is at the intersection of these center lines, as shown in Fig. 13–24.

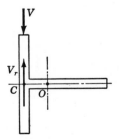

Figure 13–23.

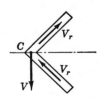

Figure 13–24.

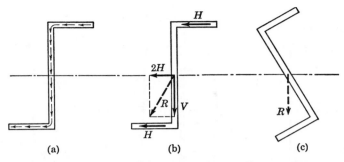

Figure 13–25.   Shear center of a Z section coincides with the centroid.

The procedure for a Z section is the same as for a channel section; the shear flow is shown in Fig. 13–25a, and the resultant shear forces in Fig. 13–25b. The resultant of the two flange forces is $2H$ acting through the centroid of the section. Combining this resultant with the shear force in the web gives the resultant shear force R in the section. Evidently the shear center C coincides with the centroid O. The fact that the plane of loading must coincide with R requires that the section be oriented as shown in Fig. 13–25c if we expect a vertical loading to cause bending in the vertical plane. In the next article we explore more completely the general relation between the plane of loading and the plane of bending.

ILLUSTRATIVE PROBLEM

1321.   If the vertical shearing force acting on the thin-walled channel section in Fig. 13–26 is 2000 N, compute and illustrate the shear flow and determine the shear center.

Solution: The moment of inertia about the neutral axis is computed from

$$\left[I = \Sigma \left(\frac{bh^3}{12} + Ad^2 \right) \right] \qquad I = \frac{(2.5)(250)^3}{12} + 2(2.5)(100)(125)^2$$

$$= 11.07 \times 10^6 \text{ mm}^4$$
$$= 11.07 \times 10^{-6} \text{ m}^4$$

Because of the thin wall, the shear flow may be assumed to act along the center line $ABCDE$ of the section, as shown in Fig. 13–26b. At A, the shear flow is zero; at B, Eq. (5–4a) gives

$$q_B = \frac{V}{I} Q_{AB} = \frac{2000}{11.07 \times 10^{-6}} (2.5 \times 10^{-3})(100 \times 10^{-3})$$
$$\times (125 \times 10^{-3})$$
$$= 5646 \text{ N/m}$$

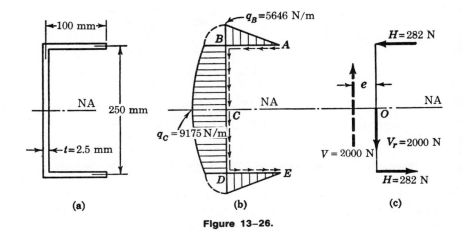

Figure 13–26.

The shear flow at any other point C may be found directly from $q_C = (V/I)Q_{AC}$. However, since Q_{AC} is the moment of area from A to C, which is equivalent to the sum of the moment of area from A to B plus that from B to C, we may write $q_C = (V/I)(Q_{AB} + Q_{BC})$, which then reduces to the more convenient form $q_C = q_B + (V/I)Q_{BC}$. Thus the shear flow at C is

$$q_C = q_B + \frac{V}{I}Q_{BC} = 5646 + \left[\frac{2000}{11.07 \times 10^{-6}}(2.5 \times 10^{-3}) \right.$$
$$\left. \times (125 \times 10^{-3})\left(\frac{125 \times 10^{-3}}{2}\right)\right]$$

$$= 5646 + 3529 = 9175 \text{ N/m}$$

As shown in Fig. 13–26b, the shear flow from A to B varies directly with the distance from A; but from B to C to D it varies along a parabolic arc. The average shear flow in the web is therefore $5646 + \frac{2}{3}(3529) = 8000$ N/m. The shear force in the web is $V_r = q_{ave} \cdot L = 8000(0.250) = 2000$ N, which agrees with the applied vertical shear $V = 2000$ N. The shear force in each flange is $H = q_{ave} \cdot L = (\frac{1}{2} \times 5646)(0.100) = 282$ N.

To avoid twisting of the section, the external shear V must lie a distance e to the left of O (Fig. 13–26c) so that the twisting moments exerted by the internal shear forces will be balanced. Hence,

$$[\Sigma M_O = 0] \qquad 2000\,e = 282(250) \qquad e = 35.3 \text{ mm}$$

The value of e can be computed more easily from Eq. (13–9):

$$e = \frac{h^2b^2t}{4I} = \frac{(250)^2(100)^2(2.5)}{4(11.07 \times 10^6)} = 35.3 \text{ mm}$$

However, the above numerical computations are presented in order to emphasize the principle of shear flow and to indicate its extension to the more complex problems that follow.

PROBLEMS

1322.   Locate the shear center for the section shown in Fig. 13–22 if $t_1 = t_2 = t_3 = 10$ mm, $h_1 = 150$ mm, $h_2 = 100$ mm, $h_3 = 180$ mm.

1323.   Determine the position of the shear center for a section composed of a thin-walled cylinder of thickness t and mean radius r which is split along one longitudinal element, as in Fig. 13–17c.

> *Ans.* $e = 2r$ measured along the axis of symmetry from the center of the cylinder in a direction opposite to the split element.

1324.   Show that the position of the shear center for the semi-circular thin ring in Fig. P–1324 is $e = 4r/\pi$ to the left of O.

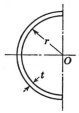

Figure P–1324.

1325.   The thin-walled section shown in Fig. P–1325 consists of a semicircular ring of mean radius r and two straight pieces of length r. Show that the shear center is $e = (tr^4/I)(\pi + 3)$ to the left of O, and hence for $r = 50$ mm and $t = 2.5$ mm, that $e = 86.0$ mm. Need the value of t be specified?

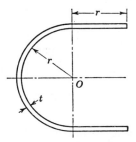

Figure P–1325.

1326. If the vertical shear on the section shown in Fig. P–1326 is 3600 N, construct a shear flow diagram and locate the shear center.

> *Ans.* $q = 3$ and 6 kN/m at the junction of flange and web; $e = 18.8$ mm to the left of the web center

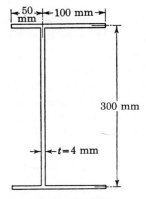

Figure P–1326.

1327. If the vertical shear on the section shown in Fig. P–1327 is 3000 N, construct a shear flow diagram and locate the shear center.

> *Ans.* $q_B = q_E = 3.01$ kN/m; $q_C = q_D = 8.14$ kN/m; $e = 62.8$ mm to the left of web center

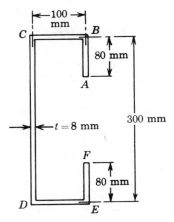

Figure P–1327.

1328. Locate the shear center for the thin-walled section shown in Fig. P–1328.   *Ans.* 27.5 mm to the left of the web

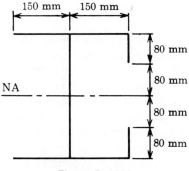

Figure P–1328.

1329. Show that the shear center for the thin-walled split square section shown in Fig. P–1329 is $b/(2\sqrt{2})$ to the left of the corner opposite the split.

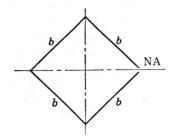

Figure P–1329.

1330. A thin-walled section has the shape shown in Fig. P–1330. The outstanding legs have a slope of 3 vertical to 4 horizontal. Locate the shear center. *Ans.* 16.3 mm to the left of the web

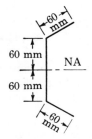

Figure P–1330.

13-9 UNSYMMETRICAL BENDING

The theory of flexure developed in Chapter 5 was restricted to loads lying in a plane that contained an axis of symmetry of the cross section. With this restriction, the neutral axis passes through the centroid of the section and is perpendicular to the plane of loading. The preceding article extended the application of the flexure formula to sections with only one axis of symmetry that were loaded so that this axis became the neutral axis. In either case, bending without twist is possible only if the plane of loading contains the shear center, a requirement that is automatically satisfied when the axis of symmetry coincides with the plane of loading.

There is a further restriction which so far has been adhered to, although not emphasized. The plane of loading must be parallel to or contain a principal axis of inertia of the beam cross section. We first consider the case in which the plane of loading contains an axis of symmetry, such as the Y axis in Fig. 13-27. In deriving the flexure formula (Art. 5-2), we applied the condition of equilibrium that the applied bending moment about the X axis is balanced by the resisting moment exerted by the flexure stresses, i.e., $M_x = \int y(\sigma \, dA)$.

If the bending loads are restricted so that they lie in the longitudinal plane containing the Y axis, the external moment M_y must be zero.* However, the flexure force $\sigma \, dA$ on a typical element of the section has a moment $x(\sigma \, dA)$ about the Y axis. If the Y axis is an axis of symmetry, this moment about Y is neutralized by an identical force (not shown) acting through the point of symmetry. For sections that do not have an axis of symmetry, the resultant moment of the flexure forces about the Y axis is

$$M_y = \int x(\sigma \, dA) = \int x\left(\frac{E}{\rho}y\right)dA = \frac{E}{\rho}\int xy \, dA$$

Therefore M_y will be zero and equilibrium satisfied only if the integral $\int xy \, dA$ is zero. This integral is the product of inertia P_{xy}, which is zero only if X and Y are the principal axes of inertia of the section. We conclude that the flexure formula may be applied only if the bending loads act in a longitudinal plane parallel to or containing one of the principal axes of the section. These planes are called the principal planes of bending.

We are now ready to discuss *unsymmetrical bending*, which is defined as bending caused by loads that are inclined to the principal planes of bending. Examples of unsymmetrical bending are roof purlins

---

*When the section has only one axis of symmetry which becomes the neutral axis, the plane of loading must be offset from, but parallel to, the longitudinal centroidal plane so as to pass through the shear center; but even then M_y is zero.

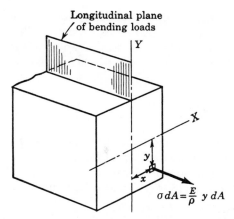

Figure 13-27. Flexure stress causes resisting moment about Y axis as well as about X axis. $M_y = 0$ if Y is an axis of symmetry (or a principal axis if there is no axis of symmetry).

that, because of the inclination of the roof, are subjected to loads whose planes make large angles with the principal axes of inertia of the section; and beams in structures and machines that are subjected to loads which, because of deformation or design, are inclined to the principal planes.

We consider first the case shown in Fig. 13-28, in which a symmetrical section is subjected to loads inclined to the axes of symmetry. Resolving the loading into horizontal and vertical components, we obtain the two loading conditions shown in parts (b) and (c), which can each be solved directly by the flexure formula. In part (b), the X axis is

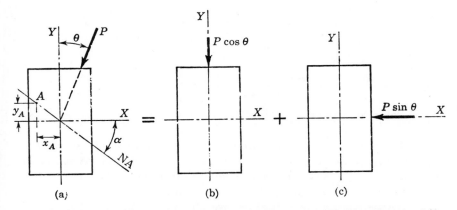

Figure 13-28. Unsymmetrical bending resolved into symmetrical bending about X and Y axes.

the neutral axis, whereas in part (c) the Y axis becomes the neutral axis. Each of these conditions produces flexure stresses that are normal to the cross section; hence the resultant stress at any point is the algebraic sum of the stresses at that point caused by each case considered separately, that is

$$\sigma = \frac{M_x y}{I_x} + \frac{M_y x}{I_y} \tag{13–11}$$

where M_x is the bending moment about the X axis caused by $P \cos \theta$, and M_y is the bending moment about the Y axis due to $P \sin \theta$. In terms of the total bending moment M, it is evident that $M_x = M \cos \theta$ and $M_y = M \sin \theta$, so Eq. (13–11) can also be written as

$$\sigma = \frac{(M \cos \theta) y}{I_x} + \frac{(M \sin \theta) x}{I_y} \tag{13–11a}$$

In applying the algebraic summations of stress indicated by Eq. (13–11) or (13–11a), tabulating the stresses as in Illustrative Problem 1331 below will avoid confusion regarding signs.

To return to Fig. 13–28a, the neutral axis has been shown at an inclination α with the X axis. To determine this inclination, we locate a point A on it by the condition that the neutral axis is the locus of points of zero stress. For the given inclination of the load, these points of zero stress must be in the second and fourth quadrants because only there will the flexural stresses of parts (b) and (c) be of opposite sign. In these quadrants, the coordinates of a point A on the neutral axis will be $-x_A$ and $+y_A$, or $+x_A$ and $-y_A$. Substituting either set in Eq. 13–11a, we obtain

$$\sigma_A = \frac{(M \cos \theta)(-y_A)}{I_x} + \frac{(M \sin \theta)(x_A)}{I_y} = 0$$

from which, by canceling the common term M and rearranging, we get

$$\frac{y_A}{x_A} = \frac{I_x}{I_y} \cdot \frac{\sin \theta}{\cos \theta}$$

which is equivalent to

$$\tan \alpha = \frac{I_x}{I_y} \tan \theta \tag{13–12}$$

From this, we see that unless $I_x = I_y$ or $\tan \theta = 0$ or ∞, the neutral axis is *not* perpendicular to the load. Also observe, as shown in Fig. 13–28a, that the neutral axis is inclined from the X axis in the same angular sense as the plane of loading is from the Y axis.

To determine the deflection, we combine vectorially the deflections caused by the X and Y components of the load. These deflection

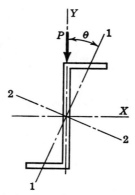

Figure 13-29. Unsymmetrical bending produced in a nonsymmetrical section. Axes 1-1 and 2-2 are principal axes.

components are respectively $\delta_x = P_x L^3 / kEI_y$ and $\delta_y = P_y L^3 / kEI_x$, where k is a factor which depends on how the beam is supported. The inclination of the total deflection with the Y axis is expressed by

$$\frac{\delta_x}{\delta_y} = \frac{P_x/I_y}{P_y/I_x} = \frac{P \sin \theta}{P \cos \theta} \cdot \frac{I_x}{I_y} = \frac{I_x}{I_y} \tan \theta$$

This result is the same as $\tan \alpha$ in Eq. (13-12). Since α was previously defined with respect to the X axis and here is measured from the Y axis, we conclude that the total deflection is perpendicular to the neutral axis.

We now consider the nonsymmetrical section like the vertically loaded Z section in Fig. 13-29. For this section, the principal axes are 1-1 and 2-2; hence the load P is inclined at the angle θ with one of the principal planes of bending. This loading, therefore, also causes unsymmetrical bending. The stresses may be determined, as shown below in Illustrative Problem 1332, by resolving P into components directed along the 1-1 and 2-2 axes and then applying Eq. (13-11). For reference, we direct attention back to Fig. 13-25c, where the resultant load R will cause a vertical deflection if R coincides with the principal axis 1-1. This will cause θ to be zero in Eq. (13-12) and axis 2-2 will become the neutral axis.

ILLUSTRATIVE PROBLEMS

1331. A W250 × 33 section is used as a cantilever beam to support the given loads inclined to the Y axis as shown in Fig. 13-30. Compute the stresses at the corners A, B, C, and D at the wall section. What is the inclination of the neutral axis at the wall?

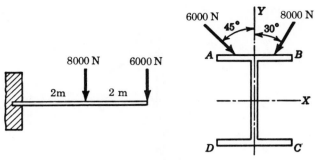

Figure 13–30.

Solution: We begin by resolving the loads into their X and Y components and computing M_x and M_y:

$$M_x = - (6000 \cos 45°)4 - (8000 \cos 30°)2 = -30.8 \text{ kN·m}$$

The negative sign for M_x indicates downward curvature at the wall, and hence the bending causes tension at A and B and compression at C and D.

Taking moments of the X components of the loads about a Y axis at the wall, we obtain

$$M_y = (6000 \sin 45°)4 - (8000 \sin 30°)2 = 8.97 \text{ kN·m}$$

This bending moment causes tension at A and D and compression at B and C.

From Appendix B, Table B-2, we find $S_x = 379 \times 10^3 \text{ mm}^3 = 379 \times 10^{-6} \text{ m}^3$ and $S_y = 64.7 \times 10^3 \text{ mm}^3 = 64.7 \times 10^{-6} \text{ m}^3$. The stresses caused by M_x and M_y are therefore

$$\sigma_x = \frac{M_x}{S_x} = \frac{30.8 \times 10^3}{379 \times 10^{-6}} = 81.3 \text{ MPa}$$

$$\sigma_y = \frac{M_y}{S_y} = \frac{8.97 \times 10^3}{64.7 \times 10^{-6}} = 138.6 \text{ MPa}$$

As indicated by Eq. (13–11), these stresses are combined algebraically. For this purpose, tensile stresses in the accompanying table are denoted by a positive sign and compressive stresses by a negative sign.

Stress in MPa due to:	A	B	C	D
M_x	+81.3	+81.3	−81.3	−81.3
M_y	+138.6	−138.6	−138.6	+138.6
Σ	+219.9	−57.3	−219.9	+57.3

It is not convenient in this example to orient the neutral axis by means of Eq. (13–12) since the plane of loading is not constant along the length of the beam. Instead, we return to Eq. (13–11) and by setting the stress equal to zero for a point on the neutral axis, we obtain

$$\frac{y}{x} = -\frac{I_x}{I_y} \times \frac{M_y}{M_x}$$

whence, replacing y/x by $\tan \alpha$ and substituting the values of I_x and I_y listed in Table B–2 (Appendix B), we have

$$\tan \alpha = -\frac{I_x}{I_y} \times \frac{M_y}{M_x} = -\frac{48.9}{4.73} \times \frac{8.97}{-30.8} = 3.01 \quad \text{or} \quad \alpha = 71.6°$$

In using this method, it is preferable to ignore signs completely and merely substitute numerical quantities. The neutral axis is then oriented so as to be consistent with the signs of the stresses tabulated above. In this instance, the neutral axis is directed up to the right at 71.6° with the X axis.

1332. A cantilever beam 2 m long carries a vertical load $P = 900$ N at the free end, as shown in Fig. 13–31. Compute the maximum stress at the corner A. The properties of the structural Z section are $I_1 = 8.00 \times 10^6$ mm^4, $I_2 = 3.75 \times 10^6$ mm^4, and $I_y = 1.23 \times 10^6$ mm^4.

Solution: We begin by determining the direction and values of the principal moments of inertia. The given properties enable us to plot a Mohr's circle of inertia, as shown in Fig. 13–32, from which we obtain $I_x = 10.53 \times 10^6$ mm^4 and $\theta = 31.4°$.

From Fig. 13–31, the coordinates of point A are

$$x_A = 76.2 \cos 31.4° - 63.5 \sin 31.4° = 32.0 \text{ mm}$$
$$y_A = 76.2 \sin 31.4° + 63.5 \cos 31.4° = 93.9 \text{ mm}$$

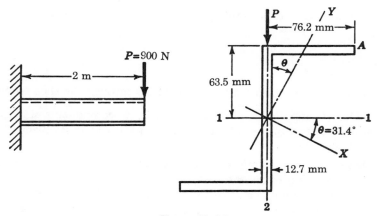

Figure 13–31.

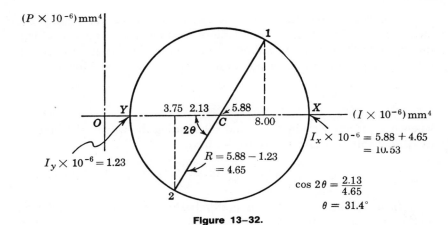

Figure 13–32.

and the components of P are

$$P_x = P \sin \theta = 900 \sin 31.4° = 469 \text{ N}$$
$$P_y = P \cos \theta = 900 \cos 31.4° = 768 \text{ N}$$

Hence the components of the maximum bending moment are

$$M_x = P_y L = 768(2) = 1536 \text{ N·m}$$
$$M_y = P_x L = 469(2) = 938 \text{ N·m}$$

We now apply Eq. (13–11), noting that M_x produces tension and M_y produces compression at A. Substituting the values previously computed, we obtain

$$\sigma_A = \frac{M_x y}{I_x} - \frac{M_y x}{I_y} = \frac{1536(0.0939)}{10.53 \times 10^{-6}} - \frac{938(0.0320)}{1.23 \times 10^{-6}}$$
$$= (13.7 - 24.4) \times 10^6 = -10.7 \text{ MPa} \qquad Ans.$$

The negative sign indicates that the stress at A is compressive.

PROBLEMS

In all these problems the loading passes through the shear center.

1333. Compute the horizontal and vertical components of deflection at the free end of the cantilever beam in Illustrative Problem 1331. Use $E = 200$ GPa.

1334. A beam simply supported at the ends has the cross section and is loaded with a concentrated load P as shown in Fig. P–1334. If the maximum bending stress is not to exceed 120 MPa, determine the maximum safe value of P.   *Ans.* $P = 36.2$ kN

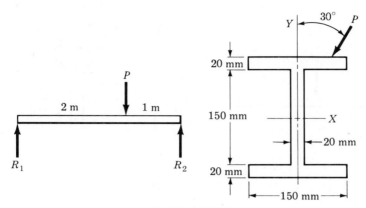

Figure P-1334.

1335. The T section shown in Fig. P–1335 is the cross section of a simply supported beam 5 m long that carries a central concentrated load inclined at 60° to the Y axis. The centroidal X axis is 72.9 mm below the top of the section; $I_x = 34.18 \times 10^6$ mm^4 and $I_y = 5.76 \times 10^6$ mm^4. If $\sigma_c \leqslant 80$ MN/m^2 and $\sigma_t \leqslant 40$ MN/m^2, what is the maximum load that will not overstress the beam? *Ans.* $P = 3.05$ kN

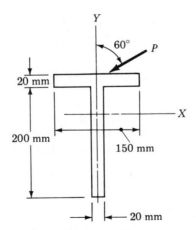

Figure P-1335.

1336. A cantilever beam 3 m long with the same T section as in Problem 1335 carries two concentrated loads applied as shown in Fig. P–1336. Compute the inclination of the neutral axis at the wall, and the maximum compressive and tensile bending stresses.

Ans. $\alpha = 46.6°$; max. $\sigma_c = 61.3$ MPa; max. $\sigma_t = 59.1$ MPa

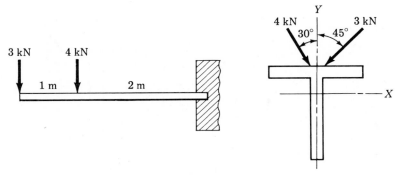

Figure P–1336.

1337. The Z beam in Illustrative Problem 1332 is used as a roof purlin, as indicated in Fig. P–1337. It carries a vertical, uniformly distributed load of 3000 N/m on a simply supported span 4 m long. If the slope of the roof is 1 to 4, compute the stress at corner A if (a) leg AB points up and (b) leg AB points down. Which design is better?

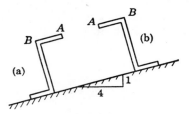

Figure P–1337.

1338. A $150 \times 100 \times 16$ mm angle is used as a cantilever beam 2 m long with the 150-mm leg vertical. It supports a load of 4000 N applied at the free end of the beam. Compute the maximum stress. (*Hint:* Compute the inclination of the neutral axis to determine the point of maximum stress.) *Ans.* 128 MPa

13–10 CURVED BEAMS

Members subjected to bending are not always straight; sometimes, as in the case of crane hooks, they are curved before a bending moment is applied. If the member is sharply curved, the stress distribution is markedly different from that given by the flexure formula $\sigma = Mc/I$, which was derived for beams that are initially straight.

For example, a sharply curved beam is subjected to bending couples, as shown in Fig. 13–33. It is usually assumed that plane radial

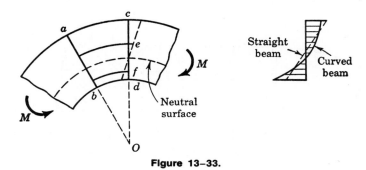

Figure 13–33.

sections remain plane after bending. Although not strictly accurate, this assumption gives results that agree closely with actual strain measurements. In accordance with this assumption, bending causes section cd to rotate, relative to section ab, to the dashed position. Consequently, the two fibers e and f, each equidistant from the neutral surface, will have equal deformations, i.e., $\delta_e = \delta_f$. Applying Hooke's law, $\delta = \sigma L / E$, we have

$$\frac{\sigma_e L_e}{E} = \frac{\sigma_f L_f}{E}$$

From Fig. 13–33 it is evident that the length L_e of fiber e is greater than the length L_f of fiber f, the difference in length depending on how sharply the beam is curved initially. Consequently, σ_e is less than σ_f, and the result is the nonlinear stress distribution shown.

As a consequence of the nonlinear stress distribution, there can be no balance between the tensile and compressive forces over the section if the neutral surface passes through the centroid of the section; the neutral surface must shift from the centroid of the section toward the axis of curvature O. The dashed linear stress distribution shows not only this shift but also the relatively increased stress at the inner fibers and the decreased stress at the outer fibers, in comparison with the stresses computed from the flexure formula.

To determine the shift in position of the neutral axis and to express the stress at any fiber in terms of the applied bending moment, we proceed as follows: In Fig. 13–34, ab and cd represent two adjacent sections of a curved beam. Let $d\theta$ be the angle between these sections before bending and $d\varphi$ be the angle of rotation of cd relative to ab caused by bending. Let y denote the coordinate of a typical element dA with respect to the neutral axis, which is at an as yet undetermined distance e from the centroid of the section; R represents the radius of curvature of the centroidal axis.

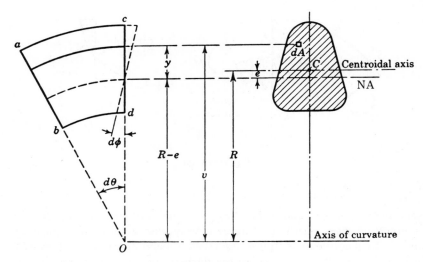

Figure 13–34.

The total elongation of a fiber at a distance y from the neutral axis is $y\,d\varphi$. The original length of this fiber is $(R - e + y)\,d\theta$. Hence, the unit elongation or strain is

$$\epsilon = \frac{\delta}{L} = \frac{y\,d\varphi}{(R - e + y)\,d\theta} \qquad (a)$$

and from Hooke's law, the stress is

$$\sigma = E\epsilon = \frac{E\,d\varphi}{d\theta} \cdot \frac{y}{R - e + y} \qquad (b)$$

If the beam is loaded in pure bending, the conditions of equilibrium require that the sum of the normal forces over a cross section be equal to zero and that the moment of these normal forces balance the applied bending moment. In accordance with the first of these conditions, a force summation over the entire area yields

$$\int \sigma\,dA = \frac{E\,d\varphi}{d\theta} \int \frac{y\,dA}{R - e + y} = 0 \qquad (c)$$

Since $E\,d\varphi/d\theta$ cannot be zero, we obtain

$$\int \frac{y\,dA}{R - e + y} = 0 \qquad (d)$$

in which e is the only unknown. Its value may be found from Eq. (d) by letting v denote the distance from the axis of curvature to the element dA. Then $y = v - (R - e)$, and Eq. (d) is rewritten as

$$\int \frac{y\,dA}{R - e + y} = \int \frac{v - (R - e)}{v}\,dA = \int dA - (R - e)\int \frac{dA}{v} = 0$$

from which we obtain

$$e = R - \frac{A}{\int \frac{dA}{v}}$$
(13-13)

Equating the applied bending moment to the resisting moment gives

$$M = \int y\sigma \, dA = \frac{E \, d\varphi}{d\theta} \int \frac{y^2 \, dA}{R - e + y}$$
(e)

This integral is simplified by adding and subtracting $(R - e)$ to one of the two y's in the numerator so that $y = (R - e + y) - (R - e)$. The integral is then rewritten as

$$\int \frac{y^2 dA}{R - e + y} = \int y \, dA - (R - e) \int \frac{y \, dA}{R - e + y}$$
(f)

The first integral on the right side of Eq. (f) is the moment of the entire cross-sectional area about the neutral axis and equals Ae. The second integral, from Eq. (d), equals zero. Eq. (e) can now be rewritten as

$$\frac{E \, d\varphi}{d\theta} = \frac{M}{Ae}$$

This value of $E \, d\varphi / d\theta$ is then substituted in Eq. (b) to yield finally

$$\sigma = \frac{M}{Ae} \cdot \frac{y}{R - e + y} = \frac{M}{Ae} \cdot \frac{y}{v}$$
(13-14)

Equations (13-13) and (13-14) are theoretically adequate to determine the stresses in curved beams but are limited in usefulness by the difficulty of computing the value of e. This difficulty may be avoided by means of a study made by B. J. Wilson and J. F. Quereau.* These investigators computed the extreme fiber stresses in curved beams of various cross sections with the curved beam theory and with the ordinary flexure formula. From a comparison of these results Wilson and Quereau determined values of a correction factor K by which stresses computed with the flexure formula can be multiplied to give the actual stress in a curved beam. A modified equation for computing the extreme fiber stresses in curved beams is therefore

$$\sigma = K \frac{Mc}{I}$$
(13-15)

Values of K in Eq. (13-15) vary with the ratio R/c, where R is the radius of curvature of the centroidal axis and c is the distance from the

*See "A Simple Method of Determining Stresses in Curved Beams," Circular 16, Engineering Experiment Station, University of Illinois, Urbana, Ill., 1928.

centroidal axis to the inner fiber. As Fig. 13–33 shows, these stress correction factors are greater than unity for the inner fibers and less than unity for the outer fibers. At values of R/c greater than 20, these factors approach unity and the flexure formula may be applied directly to such slightly curved beams. Table 13–2 lists correction factors for various cross sections.

For beams subjected to other than pure bending, as in Fig. 13–35, the system of coplanar forces acting in the plane of curvature is reduced to a single force R acting at the centroid of the section, plus a bending couple M. The moment of this couple is equivalent to the summation of moments about the centroidal axis of the external forces acting to one side of the cross section. The normal stresses produced by this couple are found as in pure bending.

The force R is resolved into two components: a shearing force V in the plane of the cross section, and a normal force N perpendicular to the plane of the cross section. The normal force acting at the centroid of the section produces tensile or compressive stresses, uniformly distributed over the section, of the magnitude $\sigma = N/A$, where A is the

TABLE 13–2. Correction Factors K for Curved Beams for Use In Eq. (13–15)

Cross section	CIRCLE OR ELLIPSE		RECTANGLE		OTHER SECTIONS (AVERAGE VALUES)	
R/c	Inside	Outside	Inside	Outside	Inside	Outside
1.2	3.41	0.54	2.89	0.57	When section is unsym-	
1.4	2.40	0.60	2.13	0.63	metrical, R/c refers to	
1.6	1.96	0.65	1.79	0.67	the inside fiber.	
1.8	1.75	0.68	1.63	0.70	—	—
2.0	1.62	0.71	1.52	0.73	1.63	0.74
3.0	1.33	0.79	1.30	0.81	1.36	0.81
4.0	1.23	0.84	1.20	0.85	1.25	0.86
6.0	1.14	0.89	1.12	0.90	1.16	0.90
8.0	1.10	0.91	1.09	0.92	1.12	0.93
10.0	1.08	0.93	1.07	0.94	1.10	0.94
20.0	1.03	0.97	1.04	0.96	1.05	0.95

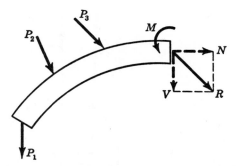

Figure 13-35.   Reduction of applied forces to a single force *R* and a couple *M*.

cross-sectional area. The total normal stress is found by superposing this uniform stress algebraically upon the stresses produced by the bending couple. The shearing stresses produced by the transverse shear V may be calculated as for a straight beam, using $\tau = (V/Ib)Q$.

ILLUSTRATIVE PROBLEM

1339.   The circular link shown in Fig. 13–36 has a rectangular section 100 mm wide by 50 mm thick. Using Eq. (13–15) and values of K from Table 13–2, compute the stresses at A and B and at C and D. Check the stresses at A and B by means of Eqs. (13–13) and (13–14).

Solution: The radius of curvature of the centroidal axis is $R = 100 + 50 = 150$ mm. The ratio $R/c = 150/50 = 3$; hence, from Table 13–2, $K_i = 1.30$ and $K_o = 0.81$. The bending moment at section AB is caused by the moment of P about the centroidal axis of the section. Hence $M = 0.150P = (0.150)(50 \times 10^3) = 7500$ N·m. Applying Eq. (13–15), we obtain

$$\left[\sigma = K\frac{Mc}{I} = K\frac{6M}{bh^2} \right]$$

$$\sigma_A = 1.30\frac{6(7500)}{(0.050)(0.100)^2} = 117 \text{ MPa compression}$$

$$\sigma_B = 0.81\frac{6(7500)}{(0.050)(0.100)^2} = 72.9 \text{ MPa tension}$$

Adding these results (Fig. 13–37a) to the uniform axial stress

$$\sigma_a = -\frac{P}{A} = -\frac{(50 \times 10^3)}{(0.050)(0.100)} = -10.0 \text{ MPa}$$

gives the final values $\sigma_A = -127$ MPa and $\sigma_B = +62.9$ MPa.

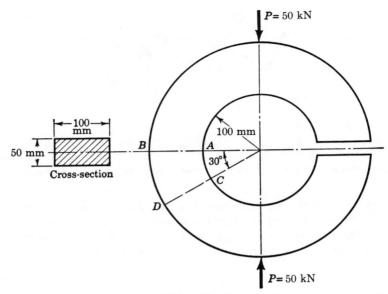

Figure 13–36.

At section CD, the bending moment $M = P(0.150 \cos 30°) = 6495$ N·m. The component of P normal to CD is $N = P \cos 30° = (50 \times 10^3)(0.866) = 43.3$ kN. Hence, the stresses at C and D (Fig. 13–37b) are

$$\left[\sigma = -\frac{N}{A} \pm K\frac{6M}{bh^2} \right]$$

$$\sigma_C = -\frac{43.3 \times 10^3}{(0.050)(0.100)} - 1.30\frac{6(6495)}{(0.050)(0.100)^2}$$

$$= -(8.66 \times 10^6) - (101 \times 10^6) = -110 \text{ MPa}$$

$$\sigma_D = -\frac{43.3 \times 10^3}{(0.050)(0.100)} + 0.81\frac{6(6495)}{(0.050)(0.100)^2}$$

$$= -(8.66 \times 10^6) + (63.1 \times 10^6) = +54.4 \text{ MPa}$$

For a more precise solution, we apply Eq. (13–13) to Fig. 13–38 to obtain

$$e = R - \left(A\Big/\int \frac{dA}{v} \right) = R - \left(bh \Big/ \int_{v_1}^{v_2} \frac{b\,db}{v} \right) = R - \left(h\Big/ \log_e\frac{v_2}{v_1} \right)$$

whence, substituting numerical values,

$$e = 150 - \frac{100}{\log_e(200/100)} = 150 - \frac{100}{0.6931} = 150 - 144.3$$

$$= 5.7 \text{ mm}$$

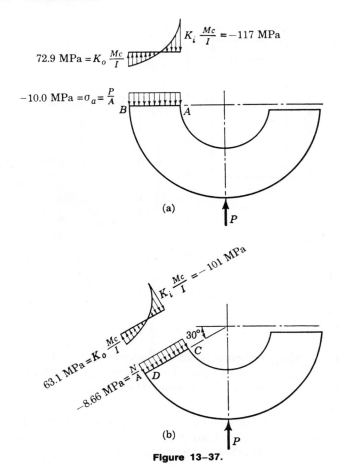

Figure 13-37.

Having determined the value of e, we are now ready to apply Eq.
(13-14), which shows the flexure stresses at A and B to be

$$\left[\sigma = \frac{M}{Ae} \cdot \frac{y}{v} \right]$$

$$\sigma_A = \frac{7500}{(0.050 \times 0.100)(0.0057)} \cdot \frac{(0.050 - 0.0057)}{0.100}$$

$$= 117 \text{ MPa compression}$$

$$\sigma_B = \frac{7500}{(0.050 \times 0.100)(0.0057)} \cdot \frac{(0.050 + 0.0057)}{0.200}$$

$$= 73.3 \text{ MPa tension}$$

To these values we must add the axial stress, -10.0 MPa, giving finally
$\sigma_A = -127$ MPa and $\sigma_B = +63.3$ MPa.

Thus Eq. (13-15) and Table 13-2 give results that agree closely
with Eqs. (13-13) and (13-14), and in addition are simpler to use.

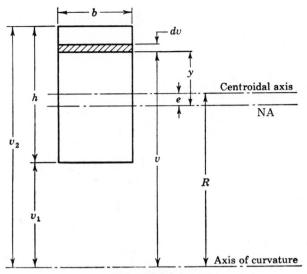

Fig:ire 13–38.

PROBLEMS

1340. A sharply curved beam of rectangular section is 10 mm thick by 50 mm deep. If the radius of curvature $R = 50$ mm, compute the stress in terms of the bending moment M at a point 40 mm from the outer surface.

1341. For the hook of circular section shown in Fig. P–1341, (a) determine the maximum load P that may be supported without exceeding a stress of 120 MN/m². (b) What stress then exists at B?

Ans. (a) $P = 46.1$ kN

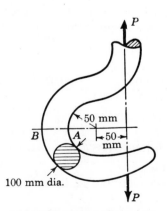

Figures P–1341 and P–1342.

1342.   Repeat Problem 1341 assuming that the hook has a circular section 75 mm in diameter. Obtain the values of K_i and K_o from a graph of K_i and K_o plotted against R/c.

1343.   Determine the diameter of a round steel rod that is to be used as a hook to lift a 10-kN load acting through the center of curvature of the axis of the hook. Assume that $R/c = 4$ and the maximum stress permitted is 110 MPa.   *Ans.* $d = 49.0$ mm

1344.   A crane hook has a cross section that is approximated by the trapezoidal section shown in Fig. P–1344. What is the maximum load P that will not exceed a stress of 80 MPa?

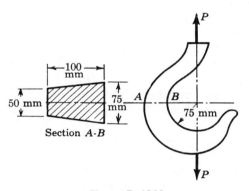

Section A-B

Figure P–1344.

1345.   The cross section of a ring is the T section shown in Fig. P–1345. The inside diameter of the ring is 366 mm. Determine the value of P that will cause a maximum stress of 120 MN/m².

Ans. $P = 80.2$ kN

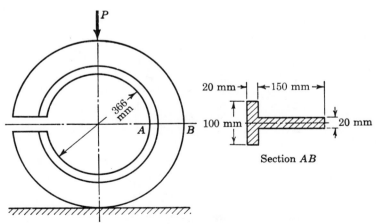

Section AB

Figure P–1345.

13-11   THICK-WALLED CYLINDERS

In the analysis of thin-walled cylinders in Art. 1–6, the forces transmitted across a longitudinal section were determined from a free-body diagram of the section (see Fig. 1–11, page 21). A similar procedure can be used to determine the force transmitted across the longitudinal section of a thick-walled cylinder. Dividing this force by the area over which it acts determines the *average* tangential stress for *either* a thin- or a thick-walled cylinder. The two cases differ in that in a thin-walled cylinder (the wall thickness being equal to or less than one-twentieth of the internal diameter) this average stress is practically equal to the maximum tangential stress, whereas in a thick-walled cylinder it is much smaller than the maximum tangential stress; moreover, the stress distribution is nonlinear.

The problem of determining the tangential stress σ_t and the radial stress σ_r at any point on a thick-walled cylinder in terms of the applied pressures and the dimensions was solved by the French elastician Gabriel Lamé in 1833. The cylinder shown in Fig. 13–39 has radii a and b and is subjected to both a uniformly distributed internal pressure of p_i N/m^2 and an external pressure of p_o N/m^2. This cylinder may be assumed composed of thin shells. Figure 13–40 shows a half-section of a typical shell, the radius of which is r, the thickness dr, and the length unity. The tangential stress in this shell is σ_t; the radial stress on the inner surface is σ_r and that on the outer surface is $\sigma_r + d\sigma_r$, where $d\sigma_r$ is the increment in σ_r due to the variation of pressure across the cylinder wall. The radial stresses are assumed (incorrectly) to be tensile, so a negative result for σ_r will denote compression.

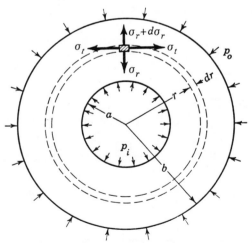

Figure 13–39. Thick-walled cylinder subjected to uniform internal pressure p_i and uniform external pressure p_o.

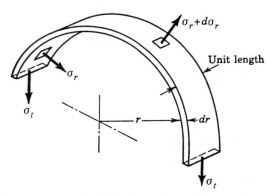

Figure 13–40.   Stresses on half-shell in Fig. 13–39.

This shell may be treated as a thin cylinder; hence, for equilibrium, a vertical summation of forces must equal zero. Thus

$$(\sigma_r + d\sigma_r) \cdot 2(r + dr) - \sigma_r(2r) - 2\sigma_t\, dr = 0$$

The product $dr \cdot d\sigma_r$ being neglected because it is very small compared to the other quantities, this reduces to

$$r\frac{d\sigma_r}{dr} + \sigma_r - \sigma_t = 0 \qquad\qquad (a)$$

Another relation between σ_r and σ_t is obtained from the assumption that plane cross sections remain plane, and hence that the longitudinal strain ϵ_z is constant for all fibers. Applying Hooke's law for triaxial stress (see page 47), we have

$$\epsilon_z = \frac{1}{E}\big[\sigma_z - \nu(\sigma_r + \sigma_t)\big]$$

Since ϵ_z, E, σ_z, and ν are all constant, it follows that $\sigma_r + \sigma_t$ is a constant throughout the cross section. Let this constant be $2A$, so that

$$\sigma_r + \sigma_t = 2A \qquad\qquad (b)$$

An equation involving only σ_r can now be set up by adding Eqs. (a) and (b):

$$r\frac{d\sigma_r}{dr} + 2\sigma_r = 2A$$

or

$$r\frac{d\sigma_r}{dr} = 2(A - \sigma_r)$$

whence, separating the variables, we obtain

$$\frac{d\sigma_r}{A - \sigma_r} = 2\frac{dr}{r}$$

Integration gives

$$-\log_e(A - \sigma_r) = 2 \log_e r + C = \log_e r^2 + C$$

or

$$\log_e(A - \sigma_r)r^2 = -C$$

and

$$(A - \sigma_r)r^2 = e^{-C} = B$$

where B is a more convenient constant than e^{-C}. Solving for σ_r we finally obtain

$$\sigma_r = A - \frac{B}{r^2} \qquad (c)$$

Substituting this value of σ_r in Eq. (b) gives

$$\sigma_t = A + \frac{B}{r^2} \qquad (d)$$

The values of the constants A and B are determined by substituting in Eq. (c) the known values of σ_r at the inner and outer surfaces of the cylinder. These values are

$$\left. \begin{aligned} \sigma_r &= -p_i \quad \text{at} \quad r = a \\ \sigma_r &= -p_o \quad \text{at} \quad r = b \end{aligned} \right\}$$

The minus sign for σ_r indicates a compressive stress.

Applying these values, we obtain

$$\left. \begin{aligned} -p_i &= A - \frac{B}{a^2} \\ -p_o &= A - \frac{B}{b^2} \end{aligned} \right\}$$

which, when solved simultaneously for A and B, produce the following values:

$$A = \frac{a^2 p_i - b^2 p_o}{b^2 - a^2}$$

$$B = \frac{a^2 b^2 (p_i - p_o)}{b^2 - a^2}$$

Substituting these values of A and B in Eqs. (c) and (d) gives the following general expressions for σ_r and σ_t at any point:

$$\left. \begin{aligned} \sigma_r &= \frac{a^2 p_i - b^2 p_o}{b^2 - a^2} - \frac{a^2 b^2 (p_i - p_o)}{(b^2 - a^2)r^2} \\ \sigma_t &= \frac{a^2 p_i - b^2 p_o}{b^2 - a^2} + \frac{a^2 b^2 (p_i - p_o)}{(b^2 - a^2)r^2} \end{aligned} \right\} \qquad \textbf{(13–16)}$$

SPECIAL CASES: MAXIMUM STRESSES

Case I: Internal pressure only

If the internal pressure is p_i and the external pressure is zero ($p_o = 0$), Eq. (13–16) reduces to

$$\left.\begin{aligned}
\sigma_r &= \frac{a^2 p_i}{b^2 - a^2}\left(1 - \frac{b^2}{r^2}\right) \\
\sigma_t &= \frac{a^2 p_i}{b^2 - a^2}\left(1 + \frac{b^2}{r^2}\right)
\end{aligned}\right\} \tag{13–17}$$

Note that σ_r is always a compressive stress, and that σ_t is always a tensile stress. Obviously σ_t is always larger than σ_r and is maximum at the inside surface of the cylinder, where

$$(\sigma_t)_{\text{max.}} = \left(\frac{b^2 + a^2}{b^2 - a^2}\right)p_i \tag{13–18}$$

By representing the ratio b/a by K, Eq. (13–18) may be written:

$$(\sigma_t)_{\text{max.}} = \frac{K^2 + 1}{K^2 - 1}p_i$$

The *average* tangential stress may be found by the method used for thin cylinders in Art. 1–6. Its value is

$$(\sigma_t)_{\text{ave.}} = \frac{ap_i}{b - a} = \frac{p_i}{K - 1}$$

Hence the ratio of the maximum to the average tangential stress is

$$\frac{(\sigma_t)_{\text{max.}}}{(\sigma_t)_{\text{ave.}}} = \frac{K^2 + 1}{K + 1} \tag{13–19}$$

Thus, for a wall thickness equal to one-twentieth of the internal diameter, or $K = b/a = 1.1$, the maximum σ_t is only about 5% larger than the average σ_t. Hence we may assume, without appreciable error, that where the wall thickness is one-twentieth of the internal diameter or less, the tangential stresses are uniformly distributed, which justifies the procedure used in Art. 1–6.

As a Mohr's circle of stress shows, the shearing stress is one-half the difference of the principal stresses, and since the failure of ductile materials like steel (of which most cylinders are made) is assumed to be governed by the maximum shear stress theory (page 518), the value of the maximum shearing stress is important in design. It is maximum at the inner surface of the cylinder where σ_t and σ_r are both maximum and is given by

$$\tau_{\text{max.}} = \frac{(\sigma_t)_{\text{max.}} - (\sigma_r)_{\text{max.}}}{2} = \frac{b^2}{b^2 - a^2}p_i \tag{13–20}$$

Case II: External pressure only

If the external pressure is p_o and the internal pressure $p_i = 0$, Eq. (13–16) reduces to

$$
\left.
\begin{aligned}
\sigma_r &= -\frac{p_o b^2}{b^2 - a^2}\left(1 - \frac{a^2}{r^2}\right) \\[2mm]
\sigma_t &= -\frac{p_o b^2}{b^2 - a^2}\left(1 + \frac{a^2}{r^2}\right)
\end{aligned}
\right\}
\tag{13–21}
$$

In this case, both σ_r and σ_t are always compressive, and σ_t is always larger than σ_r. The maximum compressive stress (σ_t) occurs at the inner surface of the cylinder (at this position σ_r is zero) and is given by

$$
(\sigma_t)_{\text{max.}} = -\frac{2b^2 p_o}{b^2 - a^2}
\tag{13–22}
$$

$(\sigma_t)_{\text{max.}}$ approaches the value $-2p_o$ as b becomes very large compared with a, as in a cylinder with a small central hole.

PROBLEMS

1346. The cylinder for a hydraulic press has an inside diameter of 300 mm. Determine the wall thickness required if the cylinder is to withstand an internal pressure of 40 MPa without exceeding a shearing stress of 80 MPa.   *Ans.* $t = 62.1$ mm

1347. Plot a curve showing the percentage increase in maximum σ_t over average σ_t for ratios of thickness to inside radius of thick-walled cylinders varying from 0 to 3.

1348. A steel hoop shrunk onto a hollow steel tube exerts a contact pressure of 20 MN/m^2 on the tube. An internal pressure of 70 MN/m^2 is then applied to the tube. The inner and outer radii of the tube are 40 mm and 60 mm, and 60 mm and 100 mm for the hoop. Determine the maximum tangential stress in the tube (a) before and (b) after the internal pressure is applied. (c) Find the maximum stress in the hoop due only to the original contact pressure.
      Ans. (a) -72.0 MN/m^2; (b) 24.7 MN/m^2; (c) 42.5 MN/m^2

1349. A thick-walled cylinder is built up by shrinking an aluminum tube 20 mm thick upon a hollow aluminum cylinder having an outside diameter of 150 mm and an inside diameter of 100 mm, thereby causing a contact pressure of 20 MPa. What is the largest internal pressure that can be applied to the assembly without exceeding a tangential stress of 100 MPa at the inner surface? *Ans.* $p = 97.4$ MPa

1350.   The inner and outer radii of a hollow steel shaft are 50 mm and 100 mm, respectively. The hub of a steel gear wheel that is shrunk onto the hollow shaft has an outer radius of 150 mm. The maximum allowable tangential stress developed by shrinking the gear wheel onto the hollow shaft is 200 MPa. The length of hub parallel to the shaft axis is 200 mm, and the coefficient of static friction between the hub and shaft is 0.40. Determine the maximum torque that may be transmitted by the gear wheel without slipping on the shaft.

Ans. $T = 377 \text{ kN} \cdot \text{m}$

14

Inelastic
Action

14-1 INTRODUCTION

The preceding chapters were devoted to analyses in which the stresses
were all in the elastic range; i.e., stress was proportional to strain. The
maximum permissible stress was the proportional limit, even though the
ultimate stress of the material was higher. We now consider a class of
problems in which loads producing inelastic strains may safely be
applied, even though doing so creates permanent deformations in the
material when the loads are removed. We revise our concept of safety to
one involving loads that produce deformations which may be large
compared with elastic deformations, but not so large that they produce
collapse of the structure. In other words, we now consider loads that
may be designated as *limit* loads. A limit load is defined as the
maximum load that can be applied to a structure before collapse begins.
With a ductile material, this condition will not occur until the yield
point has been reached throughout the most highly stressed section in
statically determinate structures. The application of limit loads to indeterminate structures is called *limit analysis* and is considered in Art.
14–5. It must be emphasized that none of the concepts discussed in this
chapter is applicable to brittle materials; some ductility must always be
present.

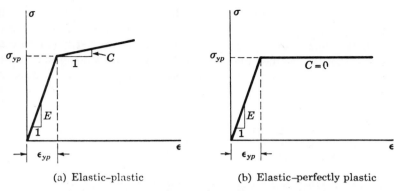

(a) Elastic-plastic   (b) Elastic–perfectly plastic

Figure 14–1.   Idealized stress–strain diagrams.

The stress–strain relation of ductile materials may be approximated by the idealized diagram shown in Fig. 14–1a. The elastic portion of the diagram is a straight line whose slope is E, the modulus of elasticity of the material. The plastic portion is also a straight line beginning at the yield stress, σ_{yp}, and having a slope C. Slope C is smaller than slope E; therefore the increment of stress required to produce a specified increment of strain is less within the plastic region than it is within the elastic region. Such a material is said to *strain-harden*; it does not permit an increase in strain without an increase in stress. A material for which C is zero is called *elastic–perfectly plastic*; for such a material, indefinite plastic flow can occur with no increase in stress beyond the yield point. It has the idealized stress–strain diagram shown in Fig. 14–1b. In our subsequent discussion, the material will be assumed to be of this elastic–perfectly plastic type.

14–2   LIMIT TORQUE

In considering the torsion of circular bars stressed into the plastic range, the only change in the conditions presented in Art. 3–1 is that now the strains may exceed the shear yield strain, γ_{yp}. A transverse section still remains rigid and does not warp; consequently the shearing strain γ remains proportional to its radial distance from the center of the bar.

We now discuss what happens to a circular bar made of an elastic–perfectly plastic material that is twisted progressively through the elastic into the fully plastic range. Until the shear yield point τ_{yp} is reached, the bar is elastic and has the stress distribution shown in Fig. 14–2a. At the beginning of yielding, the torque is given by

$$T_{yp} = \frac{\pi r^3}{2}\tau_{yp} \tag{a}$$

If we twist the shaft beyond this point, the shearing strains continue to increase but the yield stress remains constant, as shown in Fig. 14–1b. Thus at some intermediate radius r_i in the partly plastic case shown in Fig. 14–2b, the outer portion will be subjected to the constant yield stress τ_{yp}; the inner core remains elastic. The torque carried by the elastic core is

$$T_i = \frac{J_i}{r_i}\tau_{\mathrm{yp}} = \frac{\pi r_i^{\,3}}{2}\tau_{\mathrm{yp}}$$

For the plastic portion, it is

$$T_o = \int_{r_i}^{r}\rho(\tau_{\mathrm{yp}}\,dA) = \tau_{\mathrm{yp}}\int_{r_i}^{r}\rho(2\pi\rho\,d\rho) = \frac{2\pi}{3}\left(r^3 - r_i^{\,3}\right)\tau_{\mathrm{yp}}$$

The total torque is their sum

$$T = \frac{\pi r_i^{\,3}}{2}\tau_{\mathrm{yp}} + \frac{2\pi}{3}\left(r^3 - r_i^{\,3}\right)\tau_{\mathrm{yp}}$$

which reduces to

$$T = \frac{\pi r^3}{6}\left(4 - \frac{r_i^{\,3}}{r^3}\right)\tau_{\mathrm{yp}} \tag{14–1}$$

The fully plastic case shown in Fig. 14–2c cannot be reached because it requires an infinite angle of twist. The torque required to produce it is called the limit torque and is denoted by T_L. Its value can be found by setting $r_i = 0$ in Eq. (14–1). This gives

$$T_L = \tfrac{2}{3}\pi r^3\tau_{\mathrm{yp}} \tag{b}$$

which is one-third more than the maximum elastic torque. For future reference, we express this relation as

$$T_L = \tfrac{4}{3}T_{\mathrm{yp}} \tag{14–2}$$

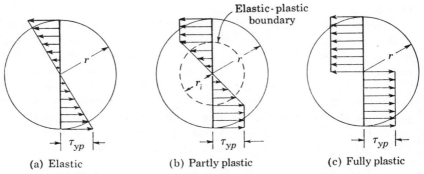

(a) Elastic   (b) Partly plastic   (c) Fully plastic

Figure 14–2. Shear stress distribution as torque is increased.

PROBLEMS

1401. A solid circular shaft 80 mm in diameter is subjected to a torque T. If the yield stress $\tau_{yp} = 140 \text{ MN/m}^2$, determine the maximum elastic torque and the limit torque. If $T = 16 \text{ kN·m}$, to what radius does the elastic action extend?   *Ans.* $r_i = 33.5$ mm

1402. Determine the ratio of the limit torque to the yield torque in a hollow circular shaft whose outer radius is twice the inner radius.
Ans. 1.24

1403. At what fraction of the maximum elastic torque will the elastic region extend to three-quarters of the outer radius in a solid circular shaft?   *Ans.* 1.19

14–3 LIMIT MOMENT

In considering inelastic action in flexure, the assumptions made in Art. 5–1 still apply, except that stresses need not be proportional to strain. Plane transverse sections are still assumed to remain plane; hence strains are proportional to their distance from the neutral axis. However, if the beam is made of an elastic–perfectly plastic material, the stresses remain constant at the yield stress σ_{yp} wherever the strain exceeds the yield strain ϵ_{yp}.

The shaded area in Fig. 14–3a denotes that part of the beam that has been strained into the plastic range as the load P on the rectangular

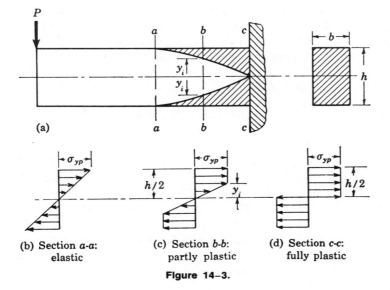

(a)

(b) Section *a-a*:
elastic

(c) Section *b-b*:
partly plastic

(d) Section *c-c*:
fully plastic

Figure 14–3.

cantilever beam is increased. At section a–a, the stresses on the outside fibers have just reached the yield stress, but the stress distribution is still elastic, as shown in part (b). Applying the flexure formula, we find that the resisting moment at this section is

$$M_{yp} = \sigma_{yp} \frac{bh^2}{6} \qquad (a)$$

At section b–b, the section is elastic over the depth $2y_i$, but plastic outside this depth, as shown by the stress distribution in part (c). The stress is constant at σ_{yp} over the plastic portion and varies over the elastic region. The resisting moment of the elastic region as determined by the flexure formula is

$$M = \frac{\sigma_{yp}}{y_i} I_i$$

where I_i is the moment of inertia of the elastic region about the neutral axis. For the plastic region, which here is symmetrical about the neutral axis, the resisting moment is

$$M = 2 \int_{y_i}^{h/2} y(\sigma_{yp}\, dA) = 2\sigma_{yp} \int_{y_i}^{h/2} y\, dA = 2\sigma_{yp} Q$$

where Q is the moment of area of *one* of the plastic regions about the neutral axis. The total resisting moment over a partly plastic symmetrical section therefore is

$$M = \frac{\sigma_{yp} I_i}{y_i} + 2\sigma_{yp} Q \qquad (14\text{–}3)$$

At section c–c, the beam is fully plastic; the stress distribution is constant at the value σ_{yp} over the tensile and compressive portions of the section as shown in part (d). The resisting moment that causes this stress distribution is called the *limit moment* and is given by

$$M_L = 2\sigma_{yp} Q = 2\sigma_{yp} \left(\frac{bh}{2} \cdot \frac{h}{4} \right) = \sigma_{yp} \frac{bh^2}{4} \qquad (b)$$

Comparing Eqs. (b) and (a), we find that

$$M_L = \tfrac{3}{2} M_{yp} \qquad (14\text{–}4)$$

The ratio M_L / M_{yp}, here equal to $\tfrac{3}{2}$, varies with the shape of the cross section. Some values of this ratio for various shapes are listed in Table 14–1. These ratios indicate that the limit moment for rectangular and circular sections is 50 to 70% over the yield moment, whereas structural sections will be only about 10% stronger if inelastic action is permitted.

For beams that are *unsymmetrical* in section, such as the T beam shown in Fig. 14–4, the neutral axis changes its location as the section enters the plastic range. In the fully plastic case where the yield stress is

**TABLE 14–1.   Ratio of Limit Moment
                to Yield Moment**

CROSS SECTION	M_L/M_{yp}
Solid rectangle	1.5
Solid circle	1.7
Thin-walled circular tube	1.27
Typical wide-flange beam	1.1

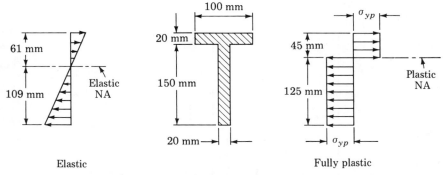

Figure 14–4. Shift in neutral axis in unsymmetrical section.

constant over the section, the equilibrium condition that the total axial force on the section be zero requires that the areas subjected to tension and compression be equal, i.e.,

$$[\,T = C\,] \qquad \sigma_{yp}A_t = \sigma_{yp}A_c \quad \text{or} \quad A_t = A_c \qquad \textbf{(14–5)}$$

Apply this condition to Fig. 14–4 and show that, for the given dimensions, the neutral axis for the fully plastic case is 16 mm above the neutral axis for the elastic case.

PROBLEMS

1404.   Verify the ratios M_L/M_{yp} specified in Table 14–1 for the solid circle and thin-walled tube.

1405.   Compute the ratio of the limit moment to the yield moment for a W200 × 100 beam.   *Ans.*   1.15

1406.   Repeat Problem 1405 for a W360 × 91 beam.

1407.   Repeat Problem 1405 for the T section shown in Fig. 14–4.
                                                                    Ans.   1.77

1408.   The centroidal axis of the section shown in Fig. P–1408 is 202 mm above the bottom, and the moment of inertia about this centroidal axis is 260×10^6 mm⁴. Determine the ratio of the limit moment to the yield moment for a beam having this section. *Ans.*   1.38

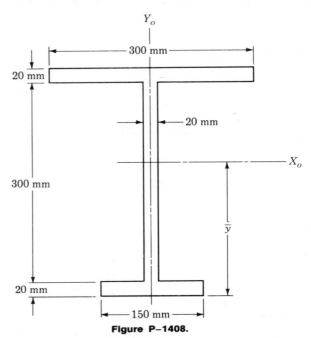

Figure P–1408.

1409. A rectangular beam 50 mm wide and 160 mm deep is made of an elastic–perfectly plastic material for which $\sigma_{yp} = 300$ MPa. Compute the bending moment that will cause one-half of the section to be in the plastic range.   *Ans.* $M = 88.0$ kN·m

1410. In Problem 1409, determine the bending moment that will cause the middle three-fourths of the section to be in the elastic range.

1411. If $\sigma_{yp} = 270$ MPa, compute the limit moment for the section shown in Fig. P–1411.   *Ans.* $M_L = 177$ kN·m

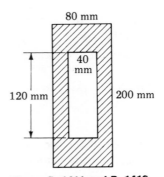

Figure P–1411 and P–1412.

1412. If $\sigma_{\text{yp}} = 270$ MN/m^2, compute the bending moment that will cause the elastic region to extend 40 mm from the neutral axis of the section shown in Fig. P–1411.   *Ans.* $M = 171$ kN·m

1413. A rectangular section is loaded with a bending moment M such that $M_{\text{yp}} < M < M_L$. If k denotes the fractional part of the depth that remains elastic, determine k in terms of M and M_L.

$$Ans.\quad k^2 = 3[1 - (M/M_L)]$$

14–4 RESIDUAL STRESSES

Experiments indicate that a ductile material loaded into the plastic range (curve OAB in Fig. 14–5a) unloads elastically following a path BC that is essentially parallel to the initial elastic path OA. Upon reloading, a slight hysteresis loop is formed; but the material now remains elastic up to the previously strained point B, after which it again becomes plastic (curve CBD). For an idealized elastic–perfectly plastic material to which our treatment is limited, this loading, unloading, and reloading cycle appears as shown in Fig. 14–5b.

The principal effect of unloading a material strained into the plastic range is to create a permanent set. If this permanent set is restricted in any manner, there is created a system of self-balancing internal stresses known as residual stresses. The magnitude and distribution of these residual stresses may be determined by combining the stress pattern (partly or fully plastic) caused by the given loading with the stress pattern created by a load equal to but of opposite sense to the original load. The effect of applying a load equal but opposite to the given load is equivalent to unloading the structure. This unloading creates a stress pattern that is assumed to be fully elastic, as indicated in Fig. 14–5. However, this method of superposition cannot be used if the residual stresses thereby obtained exceed the yield stress.

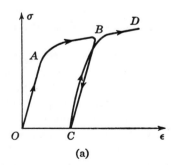
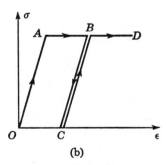

(a)   (b)

Figure 14–5. Unloading and reloading of (a) actual ductile material and (b) elastic–perfectly plastic material. In part (b) the unloading and reloading lines actually coincide but are shown slightly separated for better comparison with part (a).

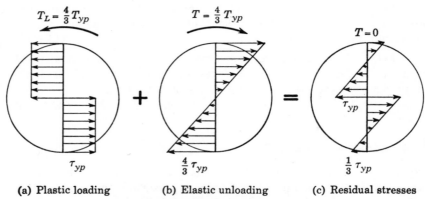

| (a) Plastic loading | (b) Elastic unloading | (c) Residual stresses |

Figure 14–6.   Residual stresses in torsion.

As a first example of residual stress, we consider a circular bar strained into the fully plastic state by the limit torque. As we saw in Art. 14–2, the limit torque is $\frac{4}{3}$ the yield torque, and the stress distribution is as shown in Fig. 14–6a. To unload the bar, we now apply an opposite torque as shown in Fig. 14–6b. Recalling that the unloading is assumed to be elastic, we obtain the stress distribution shown. Superimposing the loadings and stress patterns of parts (a) and (b), we obtain the unloaded bar with the residual stress distribution shown in part (c).

An interesting phenomenon of residual stresses is that the bar now behaves elastically if the original limit torque is now reapplied, as shown in Fig. 14–7. Combining parts (a) and (b) of Fig. 14–7, we obtain the original plastic state shown in part (c). On the other hand, with reversed reloading, the residual stresses are unfavorable and no more than two-thirds of the yield torque can be reapplied in the opposite sense before the bar reaches its plastic limit; this is shown in Fig. 14–8. Evidently, further plastic yielding occurs if the sum of the original and reversed loading exceeds twice the value of the maximum yield torque. Under these conditions, the additional yield that occurs after each cycle soon results in rupture, as we know from bending a strip of metal back and forth plastically for a few times.*

As a second example of residual stress, we consider a rectangular section of a beam that is strained into the fully plastic state by the limit moment. For the rectangular section, the limit moment is $\frac{3}{2}$ the yield moment and produces the stress distribution shown in Fig. 14–9a. Releasing the load is equivalent to adding the equal but reversed loading of part (b) to part (a); this results in an unloaded bar having the residual stress distribution shown in part (c). Remember that the un-loading is assumed to be elastic, which produces the linear stress

*For further detail, see J. A. Van den Broek, *Theory of Limit Design*, Wiley, New York, 1948, especially pp. 23–25.

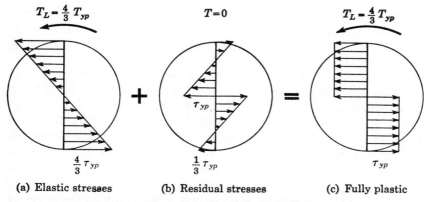

(a) Elastic stresses   (b) Residual stresses   (c) Fully plastic

Figure 14–7.   Original torque reapplied to part (c) of Fig. 14–6.

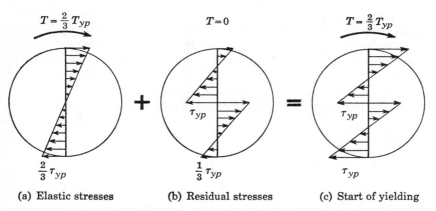

(a) Elastic stresses   (b) Residual stresses   (c) Start of yielding

Figure 14–8.   Effect of reversed reloading on part (c) in Fig. 14–6.

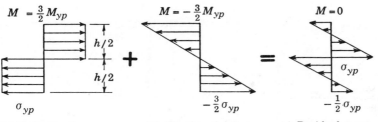

(a) Fully plastic   (b) Elastic unloading   (c) Residual stresses

Figure 14–9.   Residual stresses in flexure.

distribution shown in part (b). Observe that although residual stresses are self-balancing, if some of the material is removed, an unbalance is created. This explains why members which are cold-formed distort after machining.

As we saw in the case of torsion, a beam that has been unloaded from the fully plastic state may be reloaded in the same sense, the beam now remaining elastic until the limit moment is reached. For reversed reloading, the beam also remains elastic, the only restriction being that the sum of the original and reversed loads must not exceed twice the maximum elastic moment M_{yp} if further yielding is not to occur. Since the limit moment is $\frac{3}{2} M_{yp}$, this restriction makes the maximum reversed moment $\frac{1}{2} M_{yp}$ for rectangular sections.

As a final example of residual stress, we consider the effect of bending a straight rectangular bar about a circular die as shown in Fig. 14–10a. When the bar is released, it springs back through the angle θ_s as shown in Fig. 14–10b. This springback angle is of great importance in metal-forming operations. Its value and the relation between the forming radius of curvature R_o and the final radius of curvature R_f can be found by combining the plastic strain caused by loading with the elastic strain caused by unloading. This procedure duplicates that used to determine residual stresses.

As we saw in Art. 5–2, the strain in bending is $\epsilon = y/\rho$; hence at an outside element of the bar, the initial plastic strain in part (a) is

$$\epsilon_o = \frac{h/2}{R_o}$$

and the final residual strain in part (b) is

$$\epsilon_f = \frac{h/2}{R_f}$$

Unloading the bar is equivalent to applying a moment opposite to the limit moment to the deformed bar in part (a). As we saw in Fig. 14–9b, the maximum stress in this unloading is $\frac{3}{2}\sigma_{yp}$ and hence the correspond-

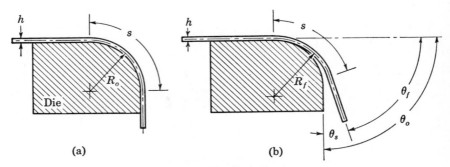

(a)   (b)

Figure 14–10.   Springback.

ing elastic strain is

$$\epsilon_e = \frac{\sigma}{E} = \frac{\frac{3}{2}\sigma_{yp}}{E}$$

Superimposing these strains gives the residual strain as

$$\epsilon_f = \epsilon_o - \epsilon_e$$

or

$$\frac{h/2}{R_f} = \frac{h/2}{R_o} - \frac{\frac{3}{2}\sigma_{yp}}{E}$$

which reduces to

$$\frac{1}{R_f} = \frac{1}{R_o} - \frac{3\sigma_{yp}}{Eh} \tag{14-6}$$

The angular change θ_f caused by the final radius of curvature R_f is found by applying $ds = R_f \, d\theta$, rewritten as

$$d\theta = \frac{1}{R_f} \, ds$$

and then integrated to give

$$\theta_f = \int_0^s \frac{1}{R_f} \, ds$$

Since $1/R_f$ is constant and the length s of the bend as shown in part (a) is $s = R_o\theta_o$, we obtain

$$\theta_f = \frac{R_o}{R_f}\theta_o$$

The springback angle may now be evaluated as

$$\theta_s = \theta_o - \theta_f = \theta_o\left(1 - \frac{R_o}{R_f}\right)$$

from which, by using Eq. (14-6), we finally obtain

$$\boldsymbol{\theta_s = \theta_o R_o\left(\frac{3\sigma_{yp}}{Eh}\right)} \tag{14-7}$$

This result indicates that the relative amount of springback may be reduced by using a smaller forming radius, or thicker bars, or material having a low yield strain σ_{yp}/E. It also indicates the amount by which the forming angle θ_o must be modified to produce a final bend of a specified amount.

In circular bars twisted into the plastic state, springback also occurs after the torque is removed. In this case, the elastic springback is equal to the angle of twist caused by elastic unloading.

ILLUSTRATIVE PROBLEM

1414. The outer bars in Fig. 14–11 are of 2024–T4 aluminum alloy for which $\sigma_{yp} = 330$ MPa; the center bar is of steel for which $\sigma_{yp} = 290$ MPa. The cross-sectional area of each aluminum bar is 600 mm²; for the steel bar, it is 900 mm². If P is the limit load, determine the residual stress after P is removed. Assume that the bars are securely attached to the rigid end plates and that $E_a = 70$ GPa and $E_s = 200$ GPa.

Solution: The limit load is the load at which yielding begins in both materials. Its value is

$$[P = \Sigma A \sigma_{yp}] \qquad P_L = 2(600 \times 10^{-6})(330 \times 10^6)$$
$$+ (900 \times 10^{-6})(290 \times 10^6)$$
$$= [2(198) + 261] \times 10^3 = 657 \text{ kN}$$

Applying an equal opposite load of this amount is equivalent to removing the load. Assuming this action to be elastic, we apply the procedure discussed in Art. 2–5 for statically indeterminate members. One relation between the bar loads is found from the free-body diagram, viz.,

$$[\Sigma F = 0] \qquad 2P_a + P_s = 657 \text{ kN} \qquad (a)$$

Another relation is found from the fact that the bars undergo equal deformations:

$$\left[\left(\frac{PL}{AE} \right)_a = \left(\frac{PL}{AE} \right)_s \right]$$
$$\frac{P_a(250)}{(600)(70)} = \frac{P_s(350)}{(900)(200)} \quad \text{or} \quad P_s = 3.06 P_a \qquad (b)$$

Solving Eqs. (a) and (b), we find the unloading forces to be

$$P_a = 130 \text{ kN} \quad \text{and} \quad P_s = 398 \text{ kN}$$

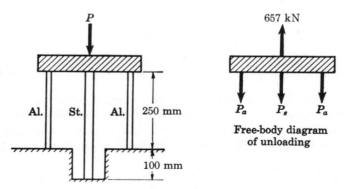

Figure 14–11.

An algebraic summation of these results and the original loads determines the residual forces to be

$$P_a = 130 - 198 = -68 \text{ kN}$$
$$P_s = 398 - 261 = +137 \text{ kN}$$

and the residual stresses are

$$\left[\sigma = \frac{P}{A} \right] \qquad \sigma_a = -\frac{68 \times 10^3}{600 \times 10^{-6}} = -113 \text{ MPa} \quad Ans.$$

$$\sigma_s = \frac{137 \times 10^3}{900 \times 10^{-6}} = +152 \text{ MPa} \qquad Ans.$$

Observe that after this prestressing, the system remains fully elastic as any load value up to the original limit load is reapplied.

PROBLEMS

1415. In Illustrative Problem 1414, let the area of the steel bar be changed to 1200 mm². If a load $P = 600$ kN is applied and then removed, determine the residual force in each bar.

$$Ans. \quad P_s = 54.7 \text{ kN}$$

1416. The bar shown in Fig. P–1416 is firmly attached to rigid supports. The yield strengths for steel and aluminum are, respectively, 290 and 330 MPa. Determine the residual stresses if the limit load is applied at P and then removed.

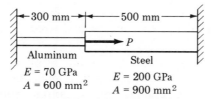

Figure P–1416 and P–1417.

1417. Solve Problem 1416 if a load $P = 350$ kN is applied and then removed. $Ans.$ $\sigma_a = -15.0 \text{ MN/m}^2$; $\sigma_s = -10.0 \text{ MN/m}^2$

1418. A sandwich beam is made by bonding 4-mm strips of 2024–T3 aluminum alloy between layers of foam plastic to form the section shown in Fig. P–1418. The foam plastic acts only to separate the aluminum strips; its effect on bending resistance is negligible. A positive bending moment of 16 kN·m is applied and then removed. Determine the residual stresses if $\sigma_{yp} = 300$ MPa.

$Ans.$ ∓20 MPa on outer strips; ±40 MPa on inner strips

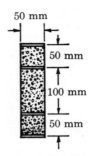

50 mm

50 mm

100 mm

50 mm

Figure P–1418.

1419. The outer diameter of a hollow circular shaft is twice the inner diameter. Determine the residual stress pattern after the limit torque has been applied and removed.

1420. The torque applied to a solid circular bar of radius r causes the elastic region to extend to one-half the radius. Determine the residual stress pattern after the torque is removed. What will be the residual angle of twist? *Ans.* $\sigma_r = \dfrac{17}{24} \dfrac{L\tau_{yp}}{Gr}$

1421. A 150-mm length at each end of a straight shaft 1 m long and 10 mm in diameter is bent at right angles to the shaft. Determine the angle through which one of the bent ends must be twisted relative to the other so that they will be exactly 90° apart after the twist load is removed. Assume that $\tau_{yp} = 140 \text{ MN/m}^2$ and $G = 80 \text{ GN/m}^2$.
Ans. 108.7°

1422. A rectangular bar 50 mm wide by 90 mm deep is subjected to a bending moment that makes two-thirds of the bar plastic. If $\sigma_{yp} = 260 \text{ MN/m}^2$, determine the residual stress pattern after the moment is removed.

1423. A rectangular bar 30 mm wide by 60 mm deep is loaded by a bending moment of 6 kN·m, which is then removed. If $\sigma_{yp} = 280$ MN/m^2, determine the residual stress at 20 mm from the neutral axis. (*Hint:* Refer to Problem 1413.) *Ans.* $\sigma_r = 15.1 \text{ MN/m}^2$

1424. If a beam of unsymmetrical section is loaded into the fully plastic state, show that any release of the load will cause a residual stress in excess of the yield stress which, being impossible, means that the theory of elastic unloading cannot be applied to beams of unsymmetrical section.

1425. A beam with the cross-section shown in Fig. P–1425 is loaded with a bending moment that causes the elastic region to extend for 60 mm from the neutral axis. If $\sigma_{yp} = 270$ MPa, determine the residual stress pattern after this moment is removed.

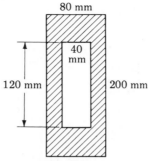

Figure P-1425.

1426. A sheet of steel 10 mm thick is bent over a 90° arc of a circular die 100 mm in radius. If $\sigma_{yp} = 270$ MPa and $E = 200$ GPa, determine the residual radius of curvature.   *Ans.* $R = 110$ mm

1427. Determine the angle of contact with the circular die so that the sheet described in Problem 1426 will have a permanent bend angle of 90°.

1428. A circular die with a radius of 250 mm is used to bend a 2024–T4 aluminum alloy plate 10 mm thick. Determine the angle of contact so that the plate will have a permanent bend angle of 180°. Assume that $\sigma_{yp} = 330$ MN/m^2 and $E = 70$ GN/m^2. *Ans.* $\theta_o = 282°$

14-5 LIMIT ANALYSIS

We now consider the application of limit load, limit torque, and limit moment to the analysis of statically indeterminate structures. This procedure, known as *limit analysis*,* is the method of determining the loading that causes actual collapse of the structure to impend or results in excessively large deformations. It is applicable only to ductile materials, which in this simplified presentation are assumed to be elastic–perfectly plastic (see Fig. 14–1b). The method is surprisingly simple, since it consists of only two steps. The first step is a geometric study of the structure to determine what part or parts of it must become fully plastic to permit the structure as a whole to undergo large deformations. The second step is an equilibrium analysis to determine the external loading that creates such localized fully plastic parts.

*See J. A. Van den Broek, *Theory of Limit Design*, Wiley, New York, 1948, for a concise justification of the principles and an extended application to redundant beams and other structures.

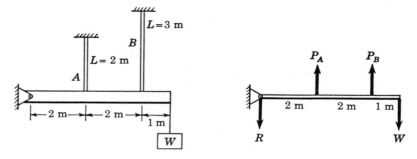

Figure 14–12.

One example of limit analysis applied to axially loaded members has already been presented in Illustrative Problem 1414. As another example, consider the rigid beam in Fig. 14–12 which is supported by two steel rods of different lengths. In the elastic solution, the deformations of the rods are proportional to the distances of the rods from the hinge. This condition results in one relation between P_A and P_B, after which the equation of static equilibrium $\Sigma M_R = 0$ can be applied to determine the maximum load that will not overstress either rod.

In limit analysis, however, the capacity of each rod is determined by the load at which yielding begins, i.e., $P = A\sigma_{yp}$. Thus if we assume the areas of the steel rods A and B to be 300 mm² and 400 mm², respectively, and the yield stresses to be 330 MPa and 290 MPa, respectively, a moment summation about the hinge determines the maximum value of W to be

$$[\Sigma M_R = 0]$$
$$5W = 2(300 \times 10^{-6})(330 \times 10^6) + 4(400 \times 10^{-6})(290 \times 10^6)$$
$$W = 132 \text{ kN}$$

Observe that yield deformations will not become excessive until this value of W is reached. Usually this limit load is divided by a suitable factor of safety to guard against excessive deformation. For comparison, if both steel rods must remain in the elastic range, the maximum load will be reduced to 119 kN.

As an example of limit analysis in torsion, consider the compound shaft attached to rigid supports shown in Fig. 14–13. The problem is to determine the maximum torque T that can be applied at the junction of the aluminum and steel segments before uncontrolled rotation begins.

Since we are assuming the materials to be elastic–perfectly plastic, remember that yielding is assumed to progress unimpeded after the yield stress is reached. While either segment remains in the elastic range, angular deformations will remain small. Hence excessive yielding will

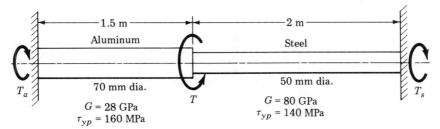

Figure 14–13.

not occur until *both* segments have reached their limit torques. The limit torque is

$$T_L = \frac{4}{3} T_{yp} = \frac{4}{3}\left(\frac{\pi r^3}{2}\right)\tau_{yp}$$

Hence a moment summation about the axis of the shaft gives

$$[\Sigma T = 0]$$

$$T = \frac{4}{3}\left(\frac{\pi}{2}\right)(0.035)^3(160 \times 10^6) + \frac{4}{3}\left(\frac{\pi}{2}\right)(0.025)^3(140 \times 10^6)$$

$$= 18.9 \text{ kN} \cdot \text{m} \qquad Ans.$$

For comparison, if both materials must remain elastic, the maximum torque will be reduced to 9.60 kN·m.

Finally we discuss the limit analysis of beams. If we reconsider the cantilever beam in Fig. 14–3 (page 573), the most highly stressed section at the wall changes successively from fully elastic to partly plastic and finally to fully plastic as the load is increased. Other sections, such as *b–b*, become partly plastic; to the left of section *a–a* the beam remains fully elastic. Only the wall section approaches the fully plastic state because, although ductile yielding has begun between sections *a–a* and *c–c*, the end of the beam will not deflect uncontrollably as long as the wall section can absorb an increase in bending moment. Once a section becomes fully plastic, all its fibers yield without further increase of stress, thereby permitting the parts of the beam on either side of this section to rotate relative to each other.* For this reason, a fully plastic section is called a *plastic hinge*, and the bending moment which creates it is assumed to be the limit moment M_L.

Collapse of a statically determinate beam is considered to be synonymous with the formation of a plastic hinge. The outline of a beam when uncontrolled deformation can occur is called a *collapse mechanism*. Several examples are shown in Fig. 14–14. In each case the dashed outlines represent the collapse mechanism.

*We exclude strain-hardening by assuming the material to be elastic–perfectly plastic.

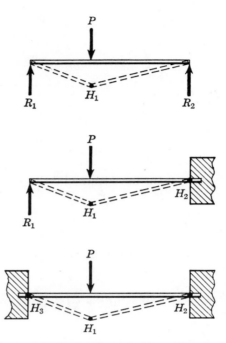

Figure 14–14.   Plastic hinges *H* form at sections of maximum moment.

In general, a plastic hinge will form at a section of zero shear, that is, where the bending moment is a maximum. The location of plastic hinges is therefore obvious for beams subjected to concentrated loads and reactions. For indeterminate beams carrying distributed loads, the location of plastic hinges is more complex. Sometimes more than one collapse mechanism is possible, in which case we must compute the limit load for each possibility and then use the smallest limit load. These concepts are discussed in the following illustrative problems.

ILLUSTRATIVE PROBLEMS

1429.   A beam perfectly restrained at the ends carries a uniformly distributed load of w N/m over its length, as shown in Fig. 14–15. Determine the limit load and compare it with the maximum elastic load.

Solution: From symmetry, it is evident that sections of zero shear and of maximum moment occur at midspan and at the ends. The collapse mechanism shown by dashed lines will occur when plastic hinges form at A, B, and C as the moments at these positions each approach the limit moment M_L. The redundant supports V_A and M_A can now be

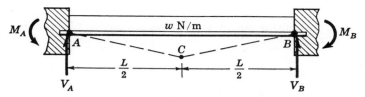

Figure 14–15.

found by applying the equations of static equilibrium to this collapse mechanism. However, an equivalent and preferable procedure is to apply the definition of bending moment.

Since the values of M_A, M_B, and M_C all become M_L, but the sign convention of bending moment makes M_A and M_B negative, we obtain*

$$[M_C = (\Sigma M)_L] \qquad M_L = V_A \frac{L}{2} - M_L - \left(\frac{wL}{2}\right)\left(\frac{L}{4}\right) \qquad (a)$$

$$[M_B = (\Sigma M)_L] \qquad -M_L = V_A L - M_L - (wL)\left(\frac{L}{2}\right) \qquad (b)$$

from which

$$V_A = \frac{wL}{2} \quad \text{and} \quad M_L = \frac{wL^2}{16} \qquad (c)$$

The elastic solution for this case, previously solved in Chapter 7 and listed in Table 7–2, gives $M_A = -wL^2/12$. Ignoring the minus sign (which merely means that the end moment was originally assumed to be positive) and letting M_A be the maximum elastic moment M_{yp} and w_e be the elastic load, we obtain

$$M_{yp} = \frac{w_e L^2}{12} \qquad (d)$$

On dividing (c) by (d), the ratio between the limit load and the elastic load is

$$\frac{w}{w_e} = \frac{4}{3}\frac{M_L}{M_{yp}} \qquad Ans.$$

The ratio M_L/M_{yp} (listed in Table 14–1) is appreciable for rectangular or circular sections; but for structural sections it is so close to unity that M_L is usually taken as equal to M_{yp}.[†]

1430. A propped beam carries a distributed load of w N/m over its length, as shown in Fig. 14–16a. Determine the relation between the limit load and the limit moment.

---

*It is unfortunate that there are so many L's in Eqs. (a) and (b). Because of the plan of using the first letter of a word as a symbol, L here has three meanings: it stands for *left* in $(\Sigma M)_L$, for *limit* in M_L, and for *length* of beam. We hope this explanation will eliminate confusion.

†See Van den Broek, *Theory of Limit Design*, p. 39.

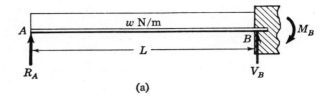

(a)

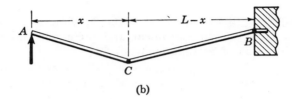

(b)

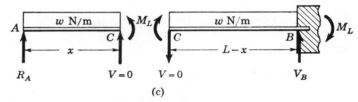

(c)

Figure 14–16.

Solution: The collapse mechanism is shown in Fig. 14–16b. The location of the plastic hinge at C is unknown, but it may be found from the condition that when the moment at C is a maximum, the vertical shear is zero. This condition was developed from statics (see Art. 4–4) and is valid whether or not the stresses or strains are in the elastic range. Drawing the free-body diagrams of the beam segments as in Fig. 14–16c, where the moments at C and B are the limit moments, we have from the segment AC

$$\left[M_A = (\Sigma M)_R \right] \qquad 0 = M_L - \frac{wx^2}{2} \tag{a}$$

and from the segment CB

$$\left[M_B = (\Sigma M)_L \right] \qquad -M_L = M_L - \frac{w}{2}(L - x)^2 \tag{b}$$

Solving these equations, we obtain

$$x = 0.414L \quad \text{and} \quad M_L = 0.0858wL^2 \quad \textit{Ans.}$$

1431.   Two cantilever beams separated by a roller jointly support a uniformly distributed load as shown in Fig. 14–17. Determine the limit load.

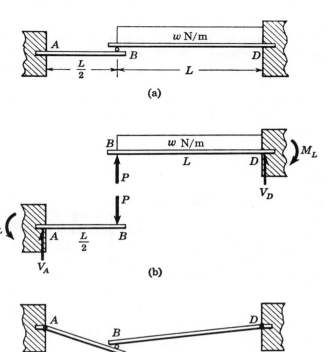

Figure 14–17.

Solution: In this variation of the preceding example, the prop support is replaced by a cantilever support. This introduces the possibility of a plastic hinge forming at A. If the hinge does not form, we have the situation discussed previously in which $M_L = 0.0858wL^2$. If the hinge forms, collapse is possible by rotation about plastic hinges at A and D as shown in part (c).

To determine the limit load for this possibility, we draw the free-body diagrams as shown in part (b), where the wall moments become the limit moments M_L. Expressing these moments in terms of the common contact force P, we have

$$\left[M_A = (\Sigma M)_R \right] \qquad -M_L = -P\frac{L}{2}$$

$$\left[M_D = (\Sigma M)_L \right] \qquad -M_L = PL - \frac{wL^2}{2}$$

From this, by eliminating P we obtain

$$M_L = \frac{wL^2}{6} = 0.167wL^2$$

Since this gives a smaller limit load than that found from $M_L = 0.0858wL^2$, it determines the desired limit load. Note that two collapse mechanisms were possible in this problem—one occurring as described above, and the other occurring from collapse of BD while AB acts as a prop support. The actual collapse mechanism is determined by the smaller limit load. It can be shown that the minimum length of AB at which collapse must occur as shown in Fig. 14–17c is $0.207L$. (See Problem 1435.)

PROBLEMS

1432.   A bracket is fastened to a rigid wall by three identical bolts as shown in Fig. P–1432. The cross-sectional area of each bolt is 150 mm². If $\sigma_{yp} = 300$ MPa, compute the maximum moment M that can be applied to the bracket. Assume the bracket to be rigid so that the deformations of the bolts is caused by rotation of the bracket about O. Determine the ratio of the limit moment to the yield moment.

Ans. $M = 13.5$ kN·m; 1.29

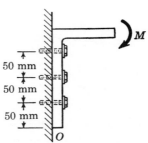

Figure P–1432.

1433.   Three rods, each 300 mm² in area, jointly support a load W as shown in Fig. P–1433. Assume there is no slack or stress in the rods before the load is applied. Determine the ratio of the limit load to the maximum elastic load. For bronze, assume that $\sigma_{yp} = 140$ MN/m² and $E = 83$ GN/m². For steel, assume that $\sigma_{yp} = 240$ MN/m² and $E = 200$ GN/m².

Ans.   1.31

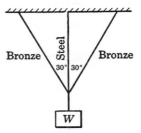

Figure P–1433.

1434. Determine the maximum torque that can be applied at 1.0 m from the right end of the shaft in Fig. 14–13 on page 587 without causing excessive rotation.

1435. Prove the statement made at the end of Illustrative Problem 1431 that the minimum length of AB at which collapse must occur as shown in Fig. 14–17c is $0.207L$.

1436. Determine the limit load for the propped beam loaded as shown in Fig. P–1436. $Ans.$ $P_L = \dfrac{M_L}{ab}(L + a)$

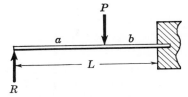

Figure P–1436.

1437. Two cantilever beams are separated by a roller as shown in Fig. P–1437. If both beams have the same limit moment M_L, determine the load P at which collapse impends. $Ans.$ $P = 2M_L/a$

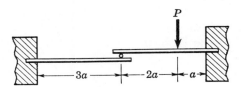

Figure P–1437.

1438. Two cantilever beams separated by a roller support a load P as shown in Fig. P–1438. Each beam is rectangular and has the cross section shown. Determine the limit load that can be applied if $\sigma_{yp} = 300$ MN/m². What is the collapse mechanism if the beam cross sections are interchanged? What is the situation if both beams have the same cross section? $Ans.$ $P = 31.6$ kN

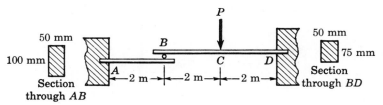

Figure P–1438.

1439. A load P is supported by a cantilever resting on a simple beam as shown in Fig. P–1439. If the limit moment of the simple beam is three-quarters that of the cantilever beam, determine the load P at which collapse impends.   *Ans.* $P = 3M_L/a$

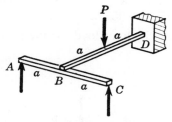

Figure P–1439.

1440. Two steel beams are mounted at right angles and in contact with each other at their midpoints. The upper beam is a W200 × 27 section simply supported on a 3-m span; the lower beam is a W250 × 33 section simply supported on a 4-m span. At their cross-over point they jointly support a load P. If $\sigma_{yp} = 290$ MPa, determine the load at which collapse impends. Assume $M_L = M_{yp}$.
Ans. $P = 206$ kN

1441. A restrained beam is loaded as shown in Fig. P–1441. If K denotes the ratio of the limit moment to the yield moment, determine the ratio of the limit load to the maximum load at the beginning of yielding. (*Hint*: Refer to Problem 713 on page 293.)   *Ans.* $\frac{4}{3}K$

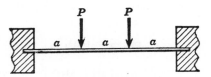

Figure P–1441.

1442. Repeat Problem 1441, using the restrained beam loading of case 6 of Table 7–2 on page 307.   *Ans.* $\frac{5}{4}K$

1443. Determine the load P in terms of the limit moment at which collapse will impend for the restrained beam loaded as shown in Fig. P–1443.   *Ans.* $P = 2M_L/3$

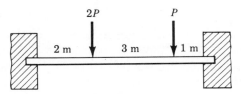

Figure P–1443.

1444. A continuous beam is simply supported over two spans each of length L. It carries a uniformly distributed load of w N/m over its entire length, as shown in Fig. P–1444. Determine w in terms of the limit moment.

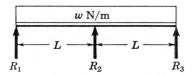

Figure P–1444 and P–1445.

1445. Repeat Problem 1444 assuming that both ends of the continuous beam are perfectly restrained.

1446. Determine the load P in terms of the limit moment for the continuous beam loaded as shown in Fig. P–1446.

Ans. $P = 2.5 M_L / a$

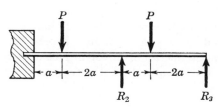

Figure P–1446.

SUMMARY

Inelastic action is applicable only to ductile materials. In this introductory presentation, the material is restricted to the elastic–perfectly plastic type so that strain-hardening effects are not considered.

For *solid circular shafts* twisted into the plastic range, the torque is

$$T = \frac{\pi r^3}{6}\left(4 - \frac{r_i^3}{r^3}\right)\tau_{yp} \tag{14–1}$$

where r_i defines the elastic–plastic boundary. For the fully plastic case, the limit torque is

$$T_L = \tfrac{4}{3} T_{yp} \tag{14–2}$$

For *symmetrical* beams bent into the plastic range, the bending moment is

$$M = \frac{\sigma_{yp} I_i}{y_i} + 2\sigma_{yp} Q \tag{14–3}$$

where y_i defines the elastic–plastic boundary, I_i is the moment of inertia of the elastic core, and Q is the moment of area about the neutral axis of *one* of the plastic regions.

In a *rectangular* section, the limit moment for the fully plastic case is

$$M_L = \tfrac{3}{2}M_{yp} \tag{14–4}$$

For other sections, the ratio of the limit moment to the yield moment changes as listed in Table 14–1. For structural sections, limit moment is essentially equal to yield moment.

For *unsymmetrical* sections, the neutral axis changes its location as the section enters the plastic range. For the fully plastic case, the position of the neutral axis is determined by the condition that the tensile and compressive areas of the section must be equal.

Residual stresses remain in a structure after it is released from being loaded into the plastic range. They are computed by combining the stress pattern caused by the actual loading with a stress pattern, assumed to be fully elastic, that is caused by an equal opposite load. The net effect of these two loadings is equivalent to unloading the structure. Of special importance is the fact that this process will not apply if the residual stresses thus obtained exceed the yield stress.

In metal-forming operations involving rectangular sections, elastic springback causes a residual radius of curvature expressed by

$$\frac{1}{R_f} = \frac{1}{R_o} - \frac{3\sigma_{yp}}{Eh} \tag{14–6}$$

and a springback angle given by

$$\theta_s = \theta_o R_o\left(\frac{3\sigma_{yp}}{Eh}\right) \tag{14–7}$$

in which h is the depth of the section and R_o and R_f are the radii to the neutral axis.

Limit analysis is the process by which we determine the loading that causes actual collapse of the structure to impend or results in excessively large deformations. Two steps are involved. The first determines what part or parts of a structure must become fully plastic in order to permit the structure as a whole to undergo large deformations. The second applies the conditions of static equilibrium to determine the external loading that causes the localized fully plastic sections to occur.

Appendix A

Moments of Inertia

A-1 DEFINITION OF MOMENT OF INERTIA

Many engineering formulas, such as those relating to strength of beams, columns, and deflection of beams, involve the use of a mathematical expression of the form $\int \rho^2 dA$, where ρ is the perpendicular distance from dA to the axis of inertia. This integral appears so frequently that it has been named *moment of inertia*.* Moment of inertia applied to areas has no real meaning when examined by itself; it is merely a mathematical expression usually denoted by the symbol I. However, when used in

*The term *moment of inertia* is derived as follows: force is related to the mass (i.e., inertia) of a body and its acceleration by the equation $F = Ma$. The equation relating applied forces to the angular acceleration α of rotating bodies is $F \cdot d = [\int \rho^2 dM] \cdot \alpha$. If the first equation is stated as force equals inertia times acceleration, then by analogy the second equation may be stated as moment of force equals moment of inertia times acceleration. By comparison of the above statements, the expression $\int \rho^2 dM$ is termed moment of inetia. Similarly, for areas, the expression $\int \rho^2 dA$ is known as the moment of inertia of the area.

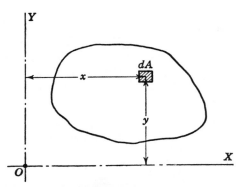

Figure A–1.

combination with other terms, as in the flexure formula for beam stresses, $\sigma = Mc/I$, it begins to have significance.

The mathematical definition of moment of inertia, $I = \int \rho^2 \, dA$, indicates that an area is divided into small parts such as dA, and each area is multiplied by the square of its moment arm about the reference axis. Thus, as shown in Fig. A–1, if the coordinates of the center of the differential area dA are (x, y), the moment of inertia about the X axis is the summation of the product of each area dA by the square of its moment arm y. This gives

$$I_x = \int y^2 \, dA \tag{A–1}$$

Similarly, the moment of inertia about the Y axis is given by

$$I_y = \int x^2 \, dA \tag{A–2}$$

The moment of inertia (of area) is sometimes called the *second moment of area* because each differential area multiplied by its moment arm gives the moment of area; when multiplied a second time by its moment arm it gives the moment of inertia. The term *second moment of area* is preferable to the expression *moment of inertia*; the latter is confusing when applied to an area having no inertia. The term moment of inertia, however, is long established and is not likely to be superseded by the other.

Units and signs

Examination of the integral $\int \rho^2 dA$ shows it to be a fourth-dimensional term because it is composed of a distance squared multiplied by an area. Thus if L is the unit of distance, the unit of I is L^4. A convenient unit of L is millimeters; this gives quartic millimeters (mm^4) as the dimensional unit of I.

   The sign of I is obviously independent of the sign of the moment arm ρ (since if ρ is minus, squaring it makes it plus); it depends entirely on the sign of the area. We shall define a positive area as one which adds to the area of a figure, and a negative area as one which reduces the area of the figure. For a net area, the moment of inertia must always be positive.

A–2   POLAR MOMENT OF INERTIA

The moment of inertia for an area relative to a line or axis perpendicular to the plane of the area is called the polar moment of inertia and is denoted by the symbol J. In Fig. A–2 the moment of inertia of an area in the XY plane with respect to the Z axis is

$$\left[I = \int \rho^2 \, dA \right] \qquad J_z = \int r^2 \, dA = \int (x^2 + y^2) \, dA$$

$$= \int x^2 \, dA + \int y^2 \, dA$$

Whence from Eqs. (A–1) and (A–2) we finally obtain

$$J_z = I_x + I_y \tag{A–3}$$

   Expressed in words, this equation states that the polar moment of inertia for an area with respect to an axis perpendicular to its plane is equal to the sum of the moments of inertia about any two mutually perpendicular axes in its plane which intersect on the polar axis.

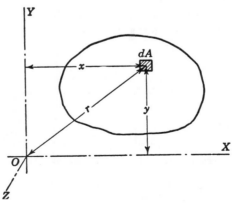

Figure A–2.

A-3 RADIUS OF GYRATION

The term radius of gyration is used to describe another mathematical expression and appears most frequently in column formulas. Radius of gyration is usually denoted either by the symbol k or by the symbol r and is defined by the relation

$$k = \sqrt{\frac{I}{A}} \quad \text{or} \quad I = Ak^2 \tag{A-4}$$

where I is the moment of inertia and A the cross-sectional area.*

The following is a geometric interpretation of this relation. Assume the area of Fig. A-1 to be squeezed into a long narrow strip as shown in Fig. A-3. Each differential element of area dA will then be the same distance k from the axis of inertia. The moment of inertia is given by

$$I = \int \rho^2 \, dA = k^2 \int dA = Ak^2$$

because each differential element has the same moment arm. The strip may be placed on either side of the reference axis, since if k is minus, squaring it will automatically make it plus. Or part of the strip may be at a distance k from one side of the reference axis, and the remainder of the strip at an equal distance k from the other side of the axis.

In view of this discussion, the radius of gyration is frequently considered to be the uniform distance from the reference axis at which the entire area may be *assumed* to be distributed. For an area whose dimensions perpendicular to a reference axis are negligibly small compared with its distance from that axis, the radius of gyration is practically equivalent to the centroidal location of the area.

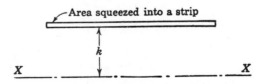

Figure A-3. Concept of radius of gyration.

*In this appendix, radius of gyration will be denoted by k. In structural applications, such as in column theory, it is denoted by r.

A–4   TRANSFER FORMULA FOR MOMENT OF INERTIA

It is often necessary to transfer the moment of inertia from one axis to another parallel axis. The transfer formula affords a method of doing this without further integration. For example, in Fig. A–4, the moment of inertia with respect to a centroidal X axis (X_o) is given by the expression $\bar{I}_x = \int y^2\, dA$. The moment of inertia for the same area with respect to a *parallel* axis (X) located a distance d from the centroidal axis is given by the equation

$$\left[I = \int \rho^2\, dA \right] \qquad I_x = \int (y + d)^2\, dA$$

$$= \int y^2\, dA + 2d \int y\, dA + d^2 \int dA \qquad (a)$$

The d is written outside the integral sign because it is a constant that represents the distance separating the axes. The second of the right-hand terms in Eq. (a) becomes zero because $\int y\, dA = A \cdot \bar{y}$, where $\bar{y}$ represents the distance from the reference axis X_o to the centroid. In this instance $\bar{y}$ has the value of zero because X_o passes through the centroid. We obtain finally,

$$I_x = \bar{I}_x + Ad^2 \qquad\qquad (A–5)$$

Put into words, this equation states that for any area the moment of inertia with respect to any axis in the plane of the area is equal to the moment of inertia with respect to a *parallel centroidal axis* plus a transfer term composed of the product of the area multiplied by the square of the distance between the axes. Evidently the least moment of inertia for any given direction of an axis is the centroidal moment of inertia. Note carefully that the centroidal axis involved in the transfer formula is always the centroidal axis of the area used in the transfer term Ad^2.

A similar relation exists between the radii of gyration with respect to parallel axes, one of which is a centroidal axis. Replacing I_x by Ak_x^2

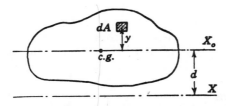

Figure A–4.   Moments of inertia between parallel axes.

and $\bar{I}_x$ by $A\bar{k}_x{}^2$ in Eq. (A–5), we obtain

$$Ak_x{}^2 = A\bar{k}_x{}^2 + Ad^2$$

whence

$$k_x{}^2 = \bar{k}_x{}^2 + d^2 \tag{A–6}$$

In like fashion, for polar moments of inertia and polar radii of gyration, we obtain the following analogous relations between any axis and a parallel centroidal axis:

$$\left.\begin{array}{l} J = \bar{J} + Ad^2 \\ k^2 = \bar{k}^2 + d^2 \end{array}\right\} \tag{A–7}$$

A–5   MOMENTS OF INERTIA BY INTEGRATION

In determining the moment of inertia by integration, it is desirable to choose the differential area so that either (1) all parts of the differential area are at the same distance from the reference axis* or (2) the moment of inertia of the differential area with respect to the reference axis is known (the moment of inertia of the area is then the summation of the moments of inertia of its elements).

As in the case of centroids, the moment of inertia of a composite figure may be found by combining the moments of inertia of its parts. When the evaluation of $\bar{I}$ for a particular part is known, the transfer formula (Art. A–4) is used to refer the moments of inertia of the various parts of the figure to a common reference axis.

ILLUSTRATIVE PROBLEMS

A1.   Determine the moment of inertia for a rectangle of base b and depth h with respect to (a) a centroidal axis parallel to the base and (b) an axis coinciding with the base.

Solution:

Centroidal Axis.   Select the differential element as shown in Fig. A–5. All parts of the element are at the same distance from the centroidal X_o axis. Applying Eq. (A–1), we find the centroidal moment

---

*When all parts of an element are at the same distance from an axis, this distance is really the radius of gyration for the element. See Fig. A–3.

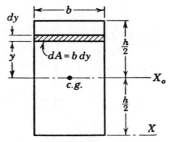

Figure A–5.

of inertia to be

$$\left[I_x = \int y^2 \, dA \right] \qquad \bar{I}_x = \int_{-h/2}^{h/2} y^2 b \, dy = b \left[\frac{y^3}{3} \right]_{-h/2}^{h/2} = \frac{bh^3}{12} \qquad Ans.$$

Axis Coinciding with the Base.   The preceding result can be transferred through the distance $h/2$ to the parallel base axis by applying the transfer formula as follows:

$$\left[I = \bar{I} + Ad^2 \right] \qquad I_x = \frac{bh^3}{12} + (bh)\left(\frac{h}{2} \right)^2 = \frac{bh^3}{3} \qquad Ans.$$

The moment of inertia of the parallelogram in Fig. A–6 has the same values as for a rectangle because the elemental strips composing the parallelogram have merely shifted their position laterally from the dashed rectangle of corresponding dimensions but have not altered their distances from the corresponding axes of inertia.

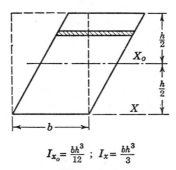

$$I_{x_o} = \frac{bh^3}{12} \; ; \; I_x = \frac{bh^3}{3}$$

Figure A–6.

A2.   Determine the moment of inertia for a triangle of base b and altitude h with respect to (a) an axis coinciding with its base and (b) a centroidal axis parallel to its base.

Solution:

Axis Coinciding with the Base. Select the differential element, as shown in Fig. A–7. From similar triangles, the length $x = (b/h)(h - y)$. The moment of inertia with respect to the X axis is obtained from

$$\left[I_x = \int y^2\, dA \right] \qquad I_x = \int_0^h y^2 \cdot x\, dy = \int_0^h y^2 \cdot \frac{b}{h}(h - y)\, dy$$

$$= \frac{b}{h}\left[\int_0^h hy^2\, dy - \int_0^h y^3\, dy \right]$$

$$= \frac{b}{h}\left[\frac{hy^3}{3} - \frac{y^4}{4} \right]_0^h$$

$$I_x = \frac{bh^3}{12} \qquad Ans.$$

Centroidal Axis. To obtain the centroidal moment of inertia $\bar{I}_x$, we transfer the known value of I_x from the base axis X to the parallel centroidal axis X_o. Since the transfer distance is $h/3$ as shown in Fig. A–8, we obtain

$$\left[I_x = \bar{I}_x + Ad^2 \right] \qquad \frac{bh^3}{12} = \bar{I}_x + \left(\frac{bh}{2} \right)\left(\frac{h}{3} \right)^2$$

$$\bar{I}_x = \frac{bh^3}{36} \qquad Ans.$$

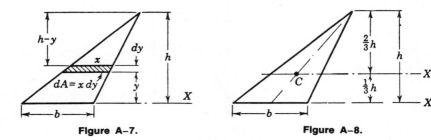

Figure A–7.   **Figure A–8.**

A3. Determine the moment of inertia of a circular area of radius r with respect to a diametral axis.

Solution: Using polar coordinates, select the differential element as shown in Fig. A–9. From the figure, $y = \rho \sin \theta$. The moment of inertia

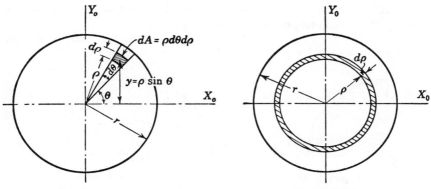

Figure A–9.   Figure A–10.

with respect to the diameter is

$$\left[I_x = \int y^2 \, dA \right] \qquad \bar{I}_x = \int_0^r \int_0^{2\pi} \rho^2 \sin^2 \theta \, \rho \, d\theta \, d\rho$$

$$= \int_0^r \int_0^{2\pi} \rho^3 \, d\rho \cdot \sin^2 \theta \, d\theta$$

$$= \frac{r^4}{4} \int_0^{2\pi} \sin^2 \theta \, d\theta = \frac{r^4}{4} \cdot \pi$$

$$\bar{I}_x = \frac{\pi r^4}{4} \qquad Ans.$$

An alternate and simpler solution is to use Fig. A–10 in which the differential element is taken as the shaded ring of area $dA = (2\pi\rho)(d\rho)$. The polar moment of inertia is

$$\left[J = \int \rho^2 \, dA \right] \qquad J = \int_0^r \rho^2 \cdot 2\pi\rho \, d\rho = \frac{\pi r^4}{2}$$

The rectangular moments of inertia $\bar{I}_x$ and $\bar{I}_y$ are obviously equal because of symmetry so that applying Eq. (A–3) yields

$$\left[\bar{J} = \bar{I}_x + \bar{I}_y \right] \qquad \frac{\pi r^4}{2} = \bar{I}_x + \bar{I}_x \quad \text{or} \quad \bar{I}_x = \frac{\pi r^4}{4} \qquad Check$$

PROBLEMS

A4. Determine the moment of inertia of a triangle of base b and altitude h with respect to an axis through the apex parallel to the base. Use the transfer formula and the results of Illustrative Problem A2.

$$Ans. \quad I = bh^3/4$$

A5.   Determine the moment of inertia of the quarter circle shown in Fig. P–A5 with respect to the given axes.

Ans. $I_x = I_y = \pi r^4 / 16$

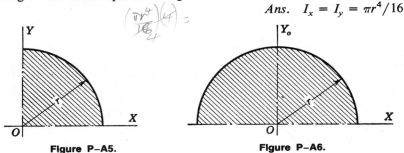

Figure P–A5.   **Figure P–A6.**

A6.   Determine the moment of inertia of the semicircle shown in Fig. P–A6 with respect to the given axes.   *Ans.* $I_x = \bar{I}_y = \pi r^4 / 8$

A7.   Show that the moment of inertia of a semicircle of radius r is $0.11 r^4$ with respect to a centroidal axis parallel to the diameter.

A8.   Determine the moment of inertia for the quarter circle shown in Fig. P–A5 with respect to a centroidal X axis.

Ans. $\bar{I}_x = 0.055 r^4$

A9.   Determine the moment of inertia with respect to the X axis for the area enclosed by the ellipse whose equation is $(x^2/a^2) + (y^2/b^2) = 1$. Also determine the radius of gyration.

Ans. $\bar{I}_x = \pi ab^3 / 4; \ \bar{k}_x = b/2$

A10.   Determine the moment of inertia and the radius of gyration, with respect to the Y axis, of the area cut from the first quadrant by the curve $y = 100 - 0.04 x^2$, where x and y are in millimeters.

Ans. $I_y = 1.667 \times 10^6 \text{ mm}^4; \ k_y = 22.4 \text{ mm}$

A11.   Determine the moment of inertia with respect to the X axis of the shaded parabolic area shown in Fig. P–A11.

Ans. $I_x = \frac{2}{15} ab^3$

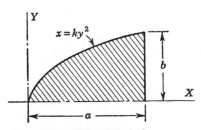

Figures P–A11 and P–A12.

A12.   Determine I_y for the shaded parabolic area in Fig. P–A11.

Ans. $I_y = \frac{2}{7} a^3 b$

A-6   MOMENTS OF INERTIA FOR COMPOSITE AREAS

When a composite area can be divided into geometric elements (rectangles, triangles, etc.) for which the moments of inertia are known, the moment of inertia for the composite area is the sum of the moments of inertia for the separate elements. Before the moments of inertia of the elements can be added, however, they must all be found with respect to the same axis.

In the illustrative problems which follow, the values of the moments of inertia for geometric elements can be taken from the results of the problems in Art. A-5; these are summarized in Table A-1. The properties of a geometric shape that is the cross section of a structural element such as a W shape beam or an angle or channel are given in the tables in Appendix B.

ILLUSTRATIVE PROBLEMS

A13.   Determine the moments of inertia with respect to the centroidal X and Y axes of the wide-flange beam section shown in Fig. A-11.

Solution: The moment of inertia of a composite area is the sum of the moments of inertia of the various parts of the area, all the moments of inertia being referred to the same axis of inertia before the addition is made.

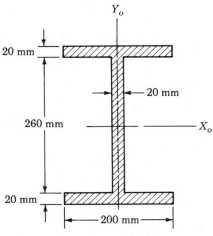

Figure A-11.

TABLE A–1.   Moments of Inertia for Geometric Shapes

SHAPE	MOMENT OF INERTIA	RADIUS OF GYRATION
Rectangle	$\bar{I}_x = \dfrac{bh^3}{12}$ $I_x = \dfrac{bh^3}{3}$	$\bar{k}_x = \dfrac{h}{\sqrt{12}}$ $k_x = \dfrac{h}{\sqrt{3}}$
Any triangle	$\bar{I}_x = \dfrac{bh^3}{36}$ $I_x = \dfrac{bh^3}{12}$	$\bar{k}_x = \dfrac{h}{\sqrt{18}}$ $k_x = \dfrac{h}{\sqrt{6}}$
Circle	$\bar{I}_x = \dfrac{\pi r^4}{4}$ $\bar{J} = \dfrac{\pi r^4}{2}$	$\bar{k}_x = \dfrac{r}{2}$ $\bar{k}_z = \dfrac{r}{\sqrt{2}}$
Semicircle	$I_x = \bar{I}_y = \dfrac{\pi r^4}{8}$ $\bar{I}_x = 0.11 r^4$	$k_x = \bar{k}_y = \dfrac{r}{2}$ $\bar{k}_x = 0.264 r$
Quarter circle	$I_x = I_y = \dfrac{\pi r^4}{16}$ $\bar{I}_x = \bar{I}_y = 0.055 r^4$	$k_x = k_y = \dfrac{r}{2}$ $\bar{k}_x = \bar{k}_y = 0.264 r$
Ellipse	$\bar{I}_x = \dfrac{\pi a b^3}{4}$ $\bar{I}_y = \dfrac{\pi b a^3}{4}$	$\bar{k}_x = \dfrac{b}{2}$ $\bar{k}_y = \dfrac{a}{2}$

With respect to the X_o axis, the simplest subdivision of the given area is to resolve it into a large rectangle 200 by 300 mm from which two smaller rectangles, each 90 by 260 mm, are subtracted. The centroidal axis for each of these parts coincides with the X_o axis of the figure; hence the transfer formula need not be used. Using the result listed in Table A–1, we therefore obtain

$$\left[\bar{I}_x = \frac{bh^3}{12} \right]$$

200- by 300-mm rectangle:
$$\bar{I}_x = \frac{200(300)^3}{12}$$
$$= 450.0 \times 10^6 \text{ mm}^4$$

Two 90- by 260-mm rectangles:
$$\bar{I}_x = 2\left[\frac{90(260)^3}{12} \right]$$
$$= 263.6 \times 10^6 \text{ mm}^4$$

Hence for shaded area:
$$\bar{I}_x = (450.0 - 263.6) \times 10^6$$
$$= 186.4 \times 10^6 \text{ mm}^4 \qquad Ans.$$

With respect to the Y_o axis, assume the figure to be composed of a 20- by 260-mm rectangle and two 20- by 200-mm rectangles. The Y_o axis is also the centroidal axis for each of these rectangles, so this subdivision of the area again eliminates the need for using the transfer formula. Referring again to Table A–1, we have

$$\left[\bar{I}_y = \frac{hb^3}{12} \right]$$

20- by 260-mm rectangle:
$$\bar{I}_y = \frac{260(20)^3}{12}$$
$$= 0.173 \times 10^6 \text{ mm}^4$$

Two 20- by 200-mm rectangles:
$$\bar{I}_y = 2\left[\frac{20(200)^3}{12} \right]$$
$$= 26.67 \times 10^6 \text{ mm}^4$$

Hence for shaded area:
$$\bar{I}_y = (0.173 + 26.67) \times 10^6$$
$$= 26.84 \times 10^6 \text{ mm}^4 \qquad Ans.$$

A14. Compute the moment of inertia for the composite area shown in Fig. A–12 with respect to the indicated X axis.

Solution: The area is composed of a semicircle (S) whose radius is 50 mm, a 100- by 240-mm rectangle (R), and a 75- by 240-mm triangle

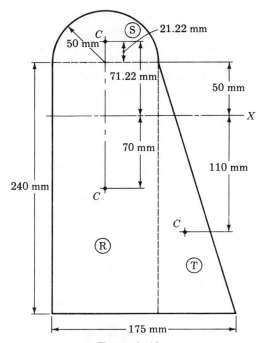

Figure A-12.

(T). With respect to the X axis, the moment of inertia for the area is the sum of the moments of inertia of these elements, each moment of inertia being referred to the X axis before addition:

$$I_x = I_R + I_S + I_T \qquad\qquad (a)$$

Expressing the moment of inertia of each element in terms of its centroidal moment of inertia plus a transfer term, we obtain

$$I_R = \bar{I}_R + (Ad^2)_R$$
$$I_S = \bar{I}_S + (Ad^2)_S$$
$$I_T = \bar{I}_T + (Ad^2)_T$$

Adding the left- and right-hand members of these equations results in

$$I_x = \sum \bar{I} + \sum Ad^2 \qquad\qquad (b)$$

Equation (b) indicates that the moment of inertia of a composite figure is the summation of the centroidal moments of inertia of the elements plus the summation of the transfer terms for these elements. This equation is readily adapted to tabular computation, as shown below. From Table A–1, the values of $\bar{I}$ for each of these elements are

$\bar{I}_R = bh^3/12$, $\bar{I}_S = 0.11r^4$, and $\bar{I}_T = bh^3/36$. The transfer distances are indicated on Fig. A–12.

ITEM	$\bar{I}$ (10^6 mm^4)	AREA (10^3 mm^2)	d (mm)	Ad^2 (10^6 mm^4)
Rectangle	115.20	24.00	70.00	117.6
Semicircle	0.69	3.93	71.22	19.9
Triangle	28.80	9.00	110.00	108.9
Totals	144.69			246.4

Taking the summations from the above table and substituting in Eq. (b), we obtain

$$\left[I_x = \sum \bar{I} + \sum Ad^2 \right] \qquad I_x = (144.69 + 246.4) \times 10^6$$

$$= 391.1 \times 10^6 \text{ mm}^4 \qquad Ans.$$

A15.   A girder is composed of four $150 \times 150 \times 13$ mm angles connected to a web plate 600 mm by 20 mm, plus two flange plates each 460 mm by 20 mm, as shown in Fig. A–13. The properties of the angle are $\bar{I}_x = \bar{I}_y = 8.05 \times 10^6 \text{ mm}^4$; the area = 3730 mm^2; and $\bar{x} = \bar{y} = 42.3$ mm. Compute the moment of inertia with respect to the centroidal X_o axis.

Solution: The tabular computation used in Illustrative Problem A14 is well suited for cases in which there are many elements. In the present problem, the elements are symmetrically placed, so a table is hardly justified.

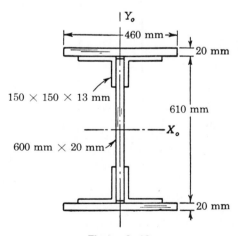

Figure A–13.

Direct application of the transfer formula $I = \bar{I} + Ad^2$ to each element gives

For web plate: $I = \dfrac{20(600)^3}{12} + (20 \times 600)(0)^2 = 360 \times 10^6 \text{ mm}^4$

For two flange plates: $I = 2\left[\dfrac{460(20)^3}{12} + (460 \times 20)(315)^2 \right]$

$= 1830 \times 10^6 \text{ mm}^4$

For four angles: $I = 4[(8.05 \times 10^6) + (3730)(305 - 42.3)^2]$

$= 1060 \times 10^6 \text{ mm}^4$

For entire figure: $\bar{I}_x = (360 + 1830 + 1060) \times 10^6$

$= 3250 \times 10^6 \text{ mm}^4 \qquad Ans.$

PROBLEMS

A16. Determine the moment of inertia of the T section shown in Fig. P–A16 with respect to its centroidal X_o axis.

$Ans. \quad \bar{y} = 87.5 \text{ mm}; \; \bar{I}_x = 113.5 \times 10^6 \text{ mm}^4$

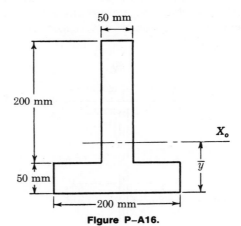

Figure P–A16.

A17. Determine the moment of inertia of the area shown in Fig. P–A17 with respect to its centroidal axes.

$Ans. \quad \bar{y} = 202 \text{ mm}; \; \bar{I}_x = 260 \times 10^6 \text{ mm}^4; \; \bar{I}_y = 50.8 \times 10^6 \text{ mm}^4$

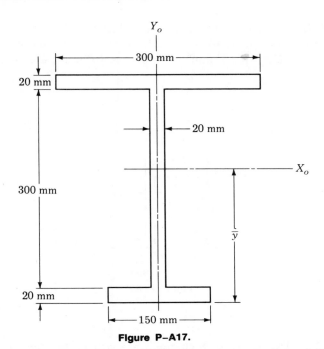

Figure P–A17.

A18. The base b of an equilateral triangle is horizontal. Show that the centroidal moments of inertia with respect to horizontal and vertical axes are equal.

A19. Compute the moment of inertia with respect to an axis passing through two opposite apexes of a regular hexagon of side a.

$$Ans. \quad I = (5\sqrt{3}/16)a^4$$

A20. Compute the moment of inertia of the 200- by 300-mm rectangle shown in Fig. P–A20 about the X axis to which it is inclined at an angle $\theta = \sin^{-1}(4/5)$. (*Hint*: Resolve the figure into parts A, B, and C.)

$$Ans. \quad I_x = 576 \times 10^6 \text{ mm}^4$$

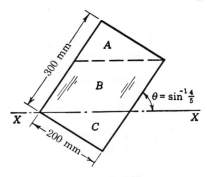

Figure P–A20.

A21. The cross section shown in Fig. P–A21 is that of a structural member known as a Z section. Determine the values of $\bar{I}_x$ and $\bar{I}_y$.
Ans. $\bar{I}_x = 17.55 \times 10^6$ mm^4; $\bar{I}_y = 6.91 \times 10^6$ mm^4; area = 5800 mm^2

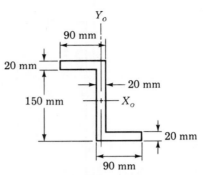

Figure P–A21.

A22. Two C200 × 28 channels are latticed together to form the section shown in Fig. P–A22. Determine how far apart the channels should be placed so that $\bar{I}_x = \bar{I}_y$ for the section. (Neglect the lattice bars which are indicated by the dashed lines.)   *Ans.* $d = 111$ mm

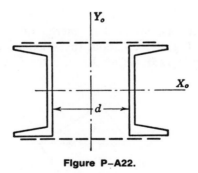

Figure P–A22.

A23. The area of the shaded section shown in Fig. P–A23 is 40×10^3 mm^2. If $I_{x_1} = 250 \times 10^6$ mm^4, determine I_{x_2}.
                              Ans. $I_{x_2} = 550 \times 10^6$ mm^4

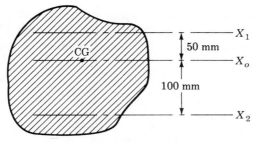

Figure P–A23.

A24.   The short legs of four $150 \times 100 \times 13$ mm angles are connected to a web plate 600 mm by 8 mm to form the plate and angle girder shown in Fig. P–A24. Compute the value of $\bar{I}_x$.

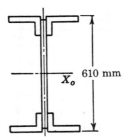

Figure P–A24.

A25.   A plate and angle column is composed of four $200 \times 100 \times 13$ mm angles with the short legs connected to a web plate 350 mm by 20 mm plus two flange plates, each 460 mm by 60 mm, as shown in Fig. P–A25. Determine the values of $\bar{I}_x$ and $\bar{I}_y$.

Ans. $\bar{I}_x = 2910 \times 10^6$ mm^4; $\bar{I}_y = 1140 \times 10^6$ mm^4

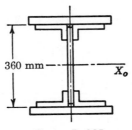

Figure P–A25.

A26. Determine the centroidal moments of inertia of the built-up column section shown in Fig. P–A26. It is composed of two 400- by 20-mm plates connected to two C310 × 31 channels.

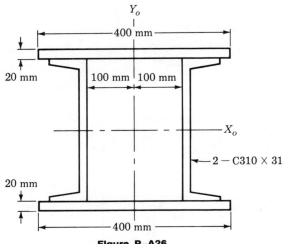

Figure P–A26.

A27. Four Z bars, each having the size and properties determined in Problem A21, are riveted to a 300- by 20-mm plate to form the section shown in Fig. P–A27. Determine the centroidal moments of inertia.　　　*Ans.*　$\bar{I}_x = 527 \times 10^6$ mm^4; $\bar{I}_y = 238 \times 10^6$ mm^4

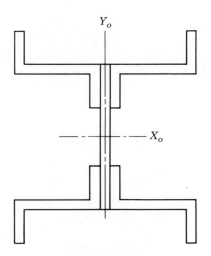

Figure P–A27.

A28.   A C250 × 23 channel is welded to the top of a W360 × 57 beam as shown in Fig. P–A28. Compute $\bar{y}$ and the moment of inertia about the centroidal X axis.

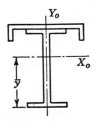

Figure P–A28.

A29.   Two C250 × 23 channels are welded together as shown in Fig. P–A29. Compute $\bar{y}$ and $\bar{I}_x$.

$Ans.$ $\bar{y} = 185.6$ mm; $\bar{I}_x = 48.5 \times 10^6$ mm^4

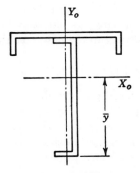

Figure P–A29.

A–7   PRODUCT OF INERTIA

The product of inertia is a mathematical expression of the form $\int xy \, dA$ and is denoted by the symbol P. The product of inertia is not used as often as the moment of inertia but is needed in such problems as determining maximum and minimum moments of inertia, unsymmetrical bending of beams, and structural analysis of indeterminate frames.

Units and Signs: The unit of the product of inertia is of the same form as that of the moment of inertia, namely, (length)^4. Unlike the moment of inertia, however, the sign for the product of inertia depends upon the location of the area relative to the axes, being positive if the area lies

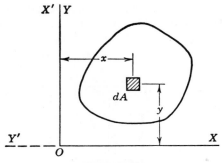

Figure A–14.

principally in the first or third quadrants and negative if the area lies principally in the second or fourth quadrants. For example, the area in Fig. A–14 lies in the first quadrant of the XY axes and $P_{xy} = \int xy \, dA$ is positive because all x and y coordinates of each differential area are positive. However, with respect to a new set of axes, marked X' and Y' and rotated 90° counterclockwise from the original set of axes, the area is in the fourth quadrant. The new coordinates of dA are $x' = y$ and $y' = -x$, so that with respect to the new axes the product of inertia is

$$P_{x'y'} = \int x'y' \, dA = \int y(-x) \, dA = -\int xy \, dA = -P_{xy}$$

Not only does this result confirm the rule of sign stated previously, but it indicates that during the rotation of the axes there will occur one critical position at which the product of inertia changes sign and will have a zero value. When in this position, the axes are known as the *principal axes* of the area. Their application is discussed in Art. A–12.

A–8   PRODUCT OF INERTIA IS ZERO WITH RESPECT TO AXES OF SYMMETRY

If an area has an axis of symmetry, this axis together with any axis perpendicular to it will form a set of axes for which the product of inertia is zero. Consideration of the symmetrical T section shown in Fig. A–15 will disclose that, for any differential area like that at A, there is a symmetrically placed equal differential area at B. With respect to the Y axis of symmetry, the x coordinates of A and B are equal but of opposite sign, whereas their y coordinates are equal and of the same sign regardless of the position of the X axis. Hence the sum of the products $xy \, dA$ for each such pair of symmetrically placed elements as A and B will be zero. It follows, therefore, that the value of $\int xy \, dA$ for

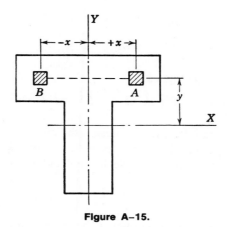

Figure A-15.

the entire area will be zero if either or both reference axes are axes of symmetry.

A-9 TRANSFER FORMULA FOR PRODUCT OF INERTIA

Consider any irregular area, such as that in Fig. A-16, whose cross-sectional area is A and whose product of inertia relative to the centroidal axes is denoted by $\overline{P}_{xy}$. Let a parallel set of axes X and Y be located so that the coordinates of the centroid of the given irregular area are $\overline{x}$ and $\overline{y}$ relative to these axes.

From the fundamental definition of product of inertia we have, with respect to centroidal axes,

$$\overline{P}_{xy} = \int x'y' \, dA \qquad (a)$$

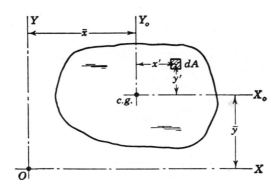

Figure A-16. Products of inertia between parallel sets of axes.

and with respect to any parallel set of X and Y axes,

$$P_{xy} = \int (x' + \bar{x})(y' + \bar{y})\, dA \qquad (b)$$

Expanding Eq. (b) gives

$$P_{xy} = \int x'y'\, dA + \bar{x} \int y'\, dA + \bar{y} \int x'\, dA + \bar{x}\bar{y} \int dA \qquad (c)$$

Note that the two middle terms represent the moment of area relative to the centroidal axes multiplied respectively by the constants $\bar{x}$ and $\bar{y}$. Since the moment of area relative to centroidal axes is zero, Eq. (c) finally reduces to

$$P_{xy} = \bar{P}_{xy} + A\bar{x}\bar{y} \qquad \textbf{(A–8)}$$

This equation, which is known as the transfer formula for products of inertia, forms the basis of the method of computing products of inertia for areas composed of simple geometric shapes. The signs of $\bar{x}$ and $\bar{y}$ in this equation may be taken either as the coordinates of the centroid relative to the $X–Y$ axes or as the coordinates of O with respect to the centroidal $X_o–Y_o$ axes. If the former, $\bar{x}$ and $\bar{y}$ in Fig. A–16 are both plus; if the latter, $\bar{x}$ and $\bar{y}$ are both minus: in either case their product is the same.

ILLUSTRATIVE PROBLEMS

A30. Determine the product of inertia of the right triangle shown in Fig. A–17 with respect to the X and Y axes.

Solution: In applying the definition of product of inertia, $P = \int xy\, dA$, observe that x and y represent the coordinates of the centroid of the differential area dA. For the right triangle illustrated, select the differen-

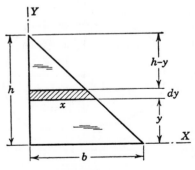

Figure A–17.

tial area as the shaded strip parallel to the base. The area of this strip is $dA = x\,dy$, and the coordinates of its centroid are $\frac{1}{2}x$ and y.

From consideration of similar triangles, it is evident that

$$x = \frac{b}{h}(h - y)$$

Hence

$$dA = x\,dy = \frac{b}{h}(h - y)\,dy$$

Applying the definition of product of inertia, we obtain

$$[P = \int xy \cdot dA] \qquad P_{xy} = \int_0^h \left[\frac{1}{2}\frac{b}{h}(h - y)\right] \cdot y \cdot \left[\frac{b}{h}(h - y)\,dy\right]$$

$$= \frac{b^2}{2h^2} \int_0^h (h^2 y - 2hy^2 + y^3)\,dy$$

$$= \frac{b^2}{2h^2}\left[\frac{h^2 y^2}{2} - \frac{2hy^3}{3} + \frac{y^4}{4}\right]_0^h$$

$$P_{xy} = +\frac{b^2 h^2}{24} \qquad Ans.$$

A31. Determine the product of inertia of the angle section shown in Fig. A–18 with respect to the indicated X and Y axes.

Solution: The angle section can be considered composed of a 100- by 20-mm rectangle plus a 160- by 20-mm rectangle. For the first rectangle, the centroidal axes parallel to the X and Y axes are axes of symmetry; hence, from Art. A–8, $\bar{P}_{xy}$ for this rectangle equals zero. The situation is similar for the other rectangle. Hence for the composite area, we

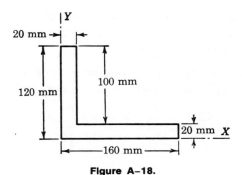

Figure A–18.

obtain

$$\left[P_{xy} = \overline{P}_{xy} + A\overline{x}\overline{y} \right]$$

100 × 20 mm rectangle: $P_{xy} = (100 \times 20) \times 10 \times 70$

$$= 1.40 \times 10^6 \text{ mm}^4$$

160 × 20 mm rectangle: $P_{xy} = (160 \times 20) \times 80 \times 10$

$$= 2.56 \times 10^6 \text{ mm}^4$$

For the composite area: $P_{xy} = (1.40 + 2.56) \times 10^6$

$$= 3.96 \times 10^6 \text{ mm}^4 \qquad Ans.$$

If the angle is rotated 90° counterclockwise to the position shown in Fig. A–19, the same value of P_{xy} but with opposite sign will be obtained; that is, $P_{xy} = -3.96 \times 10^6 \text{ mm}^4$. This result is equivalent to a 90° rotation of the axes in Fig. A–18, thereby showing that $P_{yx} = -P_{xy}$. The negative sign resulting from the interchange of subscripts is caused by choosing the Y axis as the first axis; then the second axis is the negative part of the X axis lying 90° counterclockwise from Y. In this respect, products of inertia are analogous to shearing stresses on perpendicular planes; i.e., $\tau_{xy} = -\tau_{yx}$, as discussed in the footnote on page 381.

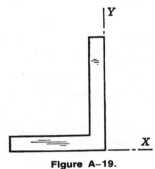

Figure A–19.

A32.   From the answer to Illustrative Problem A30, use the transfer formula to obtain the product of inertia of the right triangle shown in Fig. A-20 with respect to the indicated centroidal axes.

Solution: From the answer to Illustrative Problem A30, we have $P_{xy} = b^2h^2/24$. Applying the transfer formula, we obtain

$$[P_{xy} = \overline{P}_{xy} + A\overline{x}\overline{y}] \qquad \frac{b^2h^2}{24} = \overline{P}_{xy} + \frac{bh}{2} \times \frac{b}{3} \times \frac{h}{3}.$$

$$= \overline{P}_{xy} + \frac{b^2h^2}{18}$$

$$\overline{P}_{xy} = \frac{b^2h^2}{24} - \frac{b^2h^2}{18} = -\frac{b^2h^2}{72} \qquad Ans.$$

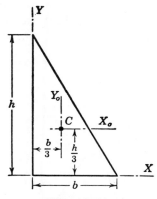

Figure A–20.

Note the minus sign carefully. It confirms the rule of sign stated on page 617 because here most of the area lies in the second and fourth quadrants of the centroidal axes. If the triangle is rotated through 90° from the position shown, the sign of $\bar{P}_{xy}$ will change to plus, but its magnitude will still be $b^2h^2/72$.

PROBLEMS

A33. For the angle section of Illustrative Problem A31 (Fig. A–18), determine the product of inertia with respect to centroidal axes parallel to the X and Y axes.   *Ans.* $\bar{P}_{xy} = -5.17 \times 10^6$ mm^4

A34. Determine the product of inertia of the Z section shown in Fig. P–A34 with respect to the centroidal X and Y axes.
                              Ans. $\bar{P}_{xy} = -8.19 \times 10^6$ mm^4

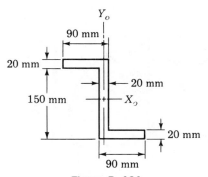

Figure P–A34.

A35. Compute the product of inertia of the triangular area shown in Fig. P–A35 with respect to the X and Y axes.

Ans. $P_{xy} = 242 \times 10^6 \text{ mm}^4$

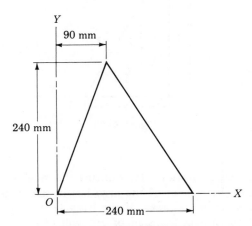

Figures P–A35 and P–A36.

A36. Compute the product of inertia of the triangular area of Prob. A35 with respect to centroidal axes parallel to the given X and Y axes.   *Ans.* $\overline{P}_{xy} = -11.4 \times 10^6 \text{ mm}^4$

A37. Determine the product of inertia of the quarter circular area shown in Fig. P–A37 with respect to the given X and Y axes.

Ans. $P_{xy} = r^4/8$

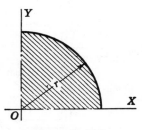

Figure P–A37.

A38.   Use the result of Problem A37 to determine the product of inertia of the shaded area described in Fig. P–A38 with respect to the given X and Y axes.   *Ans.* $P_{xy} = r^4/12$

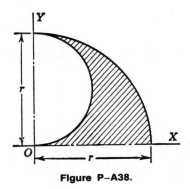

Figure   P–A38.

A–10   MOMENTS OF INERTIA WITH RESPECT TO INCLINED AXES

In some cases, it is necessary to determine the moment of inertia with respect to axes which are inclined to the usual axes. The moment of inertia in such cases can be obtained by formal integration, but a general formula is usually easier to use.

The problem may be stated as follows: Assuming the values of I_x, I_y, and P_{xy} with respect to the X and Y axes to be known, determine the values of I_u, I_v, and P_{uv} with respect to the U and V axes inclined at an angle α with the X and Y axes, as shown in Fig. A–21.

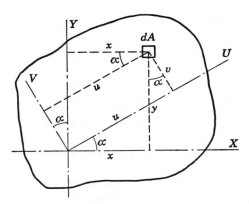

Figure A–21.   Moments of inertia with respect to inclined axes.

The coordinates for a typical differential area dA are given by x and y with respect to the X and Y axes, and by u and v relative to the U and V axes. The relations between these coordinates can be obtained by projecting x and y upon the U and V axes. This gives

$$\left. \begin{array}{l} v = y \cos \alpha - x \sin \alpha \\ u = y \sin \alpha + x \cos \alpha \end{array} \right\} \tag{a}$$

By definition $(I = \int \rho^2 \, dA)$, the values of I_u and I_v are

$$I_u = \int v^2 \, dA \tag{b}$$

$$I_v = \int u^2 \, dA \tag{c}$$

Replacing v in Eq. (b) by its value from Eq. (a), we obtain

$$I_u = \int (y^2 \cos^2 \alpha - 2xy \sin \alpha \cos \alpha + x^2 \sin^2 \alpha) \, dA$$

Since

$$I_x = \int y^2 \, dA, \quad I_y = \int x^2 \, dA, \quad \text{and} \quad P_{xy} = \int xy \, dA$$

this reduces to

$$I_u = I_x \cos^2 \alpha + I_y \sin^2 \alpha - P_{xy} \sin 2\alpha \tag{d}$$

If the relations

$$\cos^2 \alpha = \frac{1 + \cos 2\alpha}{2} \quad \text{and} \quad \sin^2 \alpha = \frac{1 - \cos 2\alpha}{2} \tag{e}$$

are substituted in Eq. (d), the result is

$$I_u = \frac{I_x + I_y}{2} + \frac{I_x - I_y}{2} \cos 2\alpha - P_{xy} \sin 2\alpha \tag{A-9}$$

Similarly, replacing u in Eq. (c) by its value from Eq. (a) gives

$$I_v = \int (y^2 \sin^2 \alpha + 2xy \sin \alpha \cos \alpha + x^2 \cos^2 \alpha) \, dA$$

This reduces to

$$I_v = I_x \sin^2 \alpha + I_y \cos^2 \alpha + P_{xy} \sin 2\alpha \tag{f}$$

The relations in Eq. (e) transform Eq. (f) into

$$I_v = \frac{I_x + I_y}{2} - \frac{I_x - I_y}{2} \cos 2\alpha + P_{xy} \sin 2\alpha \tag{A-10}$$

When values of I_x, I_y, and P_{xy} are known, Eqs. (A-9) and (A-10) permit the values of I_u and I_v, with respect to the U and V axes inclined at an angle α to the X and Y axes, to be determined without further integration. In a sense, these equations do for inclined axes what the transfer formula does for parallel axes.

Adding Eqs. (A–9) and (A–10) gives the relation

$$I_u + I_v = I_x + I_y$$

which shows that the sum of the moments of inertia with respect to any set of rectangular axes through the same point is a constant quantity. This conclusion could also have been obtained from Art. A–2, which shows that the polar moment of inertia J_z is the sum of the moments of inertia with respect to rectangular axes passing through the polar axis. Hence, since J_z is a constant, we obtain as before

$$J_z = I_x + I_y = I_u + I_v$$

To determine the product of inertia relative to the U and V axes, we note that P_{uv} is defined as

$$P_{uv} = \int uv \, dA \tag{g}$$

Substituting the values of u and v given in Eq. (a), we have

$$P_{uv} = \int (y^2 \sin \alpha \cos \alpha + xy \cos^2 \alpha$$
$$- xy \sin^2 \alpha - x^2 \sin \alpha \cos \alpha) \, dA$$
$$= \frac{I_x}{2} \sin 2\alpha + P_{xy} \cos^2 \alpha - P_{xy} \sin^2 \alpha - \frac{I_y}{2} \sin 2\alpha \tag{h}$$

whence by using the relation $\cos^2 \alpha - \sin^2 \alpha = \cos 2\alpha$, we obtain

$$P_{uv} = \frac{I_x - I_y}{2} \sin 2\alpha + P_{xy} \cos 2\alpha \tag{A–11}$$

The angles defining maximum and minimum moments of inertia may be found by differentiating Eq. (A–9) with respect to α and setting the derivative equal to zero. For these values of α, it will be found that the product of inertia is zero and the maximum and minimum moments of inertia are:

$$I_{\substack{\text{max.} \\ \text{min.}}} = \frac{I_x + I_y}{2} \pm \sqrt{\left(\frac{I_x - I_y}{2}\right)^2 + (P_{xy})^2} \tag{A–12}$$

A–11   MOHR'S CIRCLE FOR MOMENTS OF INERTIA

Except for a change of symbols, Eqs. (A–9) and (A–11) are identical with Eqs. (9–5) and (9–6), which express the variation in normal and shearing stress. It follows that the method described in Art. 9–7 for Mohr's circle of stress may be similarly applied to obtain a Mohr's circle of inertia. This construction will give a visual representation of all the possible values of I and P with respect to all axes passing through a specified point in an area. For convenience, the rules in Art. 9–7 are rephrased as follows to apply to moments of inertia:

1. On a set of rectangular coordinate axes choose one axis on which to plot numerical values of moments of inertia and the other on which to plot products of inertia; call these the I and P axes. Plot points having the coordinates (I_x, P_{xy}) and (I_y, $-P_{xy}$). The values for I_x, I_y, and P_{xy} are assumed to be already known. Note carefully that the value for P_{xy} with its real sign is associated with the value of I_x and that the value of P_{xy} with the opposite sign is associated with I_y. Actually P_{yx} should be paired with I_y, but $P_{yx} = -P_{xy}$, as we saw in Problem A31.

2. Join the points just plotted by a straight line. This line is the diameter of Mohr's circle having its center on the I coordinate axis. Draw the circle.

3. As different axes are passed through the selected point in the given area, the values of I and P relative to these axes are represented by the coordinates of points whose positions shift around the circumference of Mohr's circle.

4. The radius of the circle to any point on its circumference represents the axis of inertia corresponding to the I coordinate of that point.

5. The angle between any two radii on Mohr's circle is double the actual angle between the two axes of inertia represented by these two radii. The rotational sense of this angle corresponds to the rotational sense of the actual angle between the axes; that is, if the U axis of inertia is located at a counterclockwise angle α relative to the X axis of inertia, then on Mohr's circle the U radius is laid off at a counterclockwise angle 2α from the X radius.

ILLUSTRATIVE PROBLEM

A39. For the rectangle shown in Fig. A–22, compute the values of I_u, I_v, and P_{uv} with respect to the U and V axes. These axes are inclined 30° counterclockwise to the X and Y axes.

Solution: The moments of inertia and the product of inertia are first found with respect to the X and Y axes, as follows:

$$\left[I_x = \frac{bh^3}{12} \right] \qquad I_x = \frac{150(300)^3}{12} = 337.5 \times 10^6 \text{ mm}^4$$

$$\left[I_y = \frac{hb^3}{12} \right] \qquad I_y = \frac{300(150)^3}{12} = 84.4 \times 10^6 \text{ mm}^4$$

$$P_{xy} = 0 \qquad \text{because } X \text{ and } Y \text{ are axes of symmetry}$$

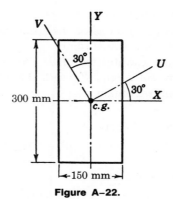

Figure A-22.

Following the rules given above, draw a set of rectangular coordinate axes and label them I and P, as shown in Fig. A-23. Using the values of I_x, I_y, and P_{xy}, plot points A and B whose coordinates are (337.5, 0) and (84.4, 0).

According to Rule 2, the diameter of Mohr's circle is AB. Its center C is midway between A and B. The I coordinate of C is 211.0. The radius of the circle is the distance $CA = 337.5 - 211.0 = 126.5$.

From Rule 4, the radius CA represents the axis of inertia corresponding to the I coordinate of A, in this case the X axis. Applying Rule 5, we find that the U axis of inertia is represented by the radius CD laid

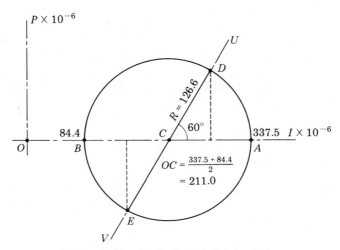

Figure A-23.   Application of Mohr's circle.

off 60° counterclockwise from the X axis (CA). Also, since V is actually 90° from U, the V axis (CE) is laid off 180° (i.e., double scale) from the U axis; D, C, and E form a straight line.

From Rule 3, the coordinates of D represent I_u and P_{uv}; the coordinates of E represent I_v and P_{uv} with the opposite sign. Accordingly, from the diagram we obtain

$$\left[\, I_u \times 10^{-6} = OC + CD \cos 60° \,\right]$$
$$I_u = (211.0 + 126.5 \cos 60°) \times 10^6 = 274.3 \times 10^6 \text{ mm}^4 \qquad Ans.$$

$$\left[\, I_v \times 10^{-6} = OC - CE \cos 60° \,\right]$$
$$I_v = (211.0 - 126.5 \cos 60°) \times 10^6 = 147.8 \times 10^6 \text{ mm}^4 \qquad Ans.$$

$$\left[\, P_{uv} \times 10^{-6} = CD \sin 60° \,\right]$$
$$P_{uv} = (126.5 \sin 60°) \times 10^6 = 109.6 \times 10^6 \text{ mm}^4 \qquad Ans.$$

A–12   MAXIMUM AND MINIMUM MOMENTS OF INERTIA: PRINCIPAL AXES

An inspection of Mohr's circle will show that the points whose coordinates indicate maximum and minimum moments of inertia are located on the I axis and have a zero product of inertia. Conversely, axes which have a zero product of inertia must be axes of maximum or minimum inertia. Such axes are called *principal axes*.

As we have already seen (Art. A–8), the products of inertia relative to axes of symmetry are zero. Hence we conclude that axes of symmetry must be principal axes because they always yield values of maximum and minimum moments of inertia. But many figures do not have axes of symmetry, although they do have principal axes with respect to which the product of inertia is zero. *Axes of symmetry are always principal axes, but the converse is not necessarily true.*

ILLUSTRATIVE PROBLEM

A40. A certain area is found to have the following values with respect to the X and Y axes: $I_x = 100 \times 10^6 \text{ mm}^4$, $I_y = 60 \times 10^6 \text{ mm}^4$, and $P_{xy} = 15 \times 10^6 \text{ mm}^4$. Determine the maximum and minimum moments of inertia and illustrate the position of the principal axes relative to the X and Y axes.

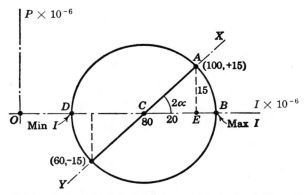

Figure A–24.   Maximum and minimum moments of inertia.

Solution: On a set of I and P axes, as shown in Fig. A–24, plot points having the following coordinates.

$$\begin{bmatrix} I_x = 100 \times 10^6 \\ P_{xy} = 15 \times 10^6 \end{bmatrix} \qquad \begin{bmatrix} I_y = -60 \times 10^6 \\ -P_{xy} = -15 \times 10^6 \end{bmatrix}$$

Note that the given value of P_{xy} is associated with I_x and that the value of P_{xy} with the opposite sign is associated with I_y. If P_{xy} had been negative originally, this value would have been associated with I_x; and the positive value of P_{xy} with I_y.

Plotting these points gives two points on Mohr's circle. Joining them gives the diameter of the circle shown in Fig. A–24. Obviously the radius of the circle is $CA = \sqrt{(20)^2 + (15)^2} = 25$. The maximum and minimum moments of inertia are located at B and D; hence

$$[\text{Max. } I \times 10^{-6} = OC + CB]$$

Max. $I = (80 + 25) \times 10^6 = 105 \times 10^6 \text{ mm}^4$ *Ans.*

$$[\text{Min. } I \times 10^{-6} = OC - CD]$$

Min. $I = (80 - 25) \times 10^6 = 55 \times 10^6 \text{ mm}^4$ *Ans.*

To go from the X axis to the axis of maximum inertia CB, we must rotate clockwise through an angle 2α. From the diagram

$$\left[\tan 2\alpha = \frac{AE}{CE} \right] \qquad \tan 2\alpha = \frac{15}{20} = 0.75$$

$$2\alpha = 36.9° \quad \text{and} \quad \alpha = 18.45° \qquad Ans.$$

Angle α, which locates the axis of the maximum moment of inertia (the U axis), is also rotated clockwise on the original reference axes; this gives the position shown in Fig. A–25. The axis of minimum moment of inertia (i.e., the V axis) is at $90°$ to the U axis.

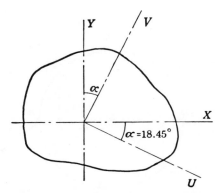

Figure A–25. Location of U and V axes of maximum and minimum moments of inertia.

PROBLEMS

A41. For a certain area it is known that $I_x = 80 \times 10^6$ mm^4, $I_y = 40 \times 10^6$ mm^4, and $P_{xy} = 0$. Find the moment of inertia of this area with respect to a U axis which is rotated 30° counterclockwise from the X axis. *Ans.* $I_u = 70 \times 10^6$ mm^4

A42. A certain area has the following properties: $I_x = 40 \times 10^6$ mm^4; $I_y = 100 \times 10^6$ mm^4; $P_{xy} = 40 \times 10^6$ mm^4. Determine the maximum and minimum moments of inertia, and also the angle that the axis of maximum moment of inertia makes with the X axis. Illustrate by a diagram. *Ans.* Max. $I = 120 \times 10^6$ mm^4; min. $I = 20 \times 10^6$ mm^4; $\theta = 63.4°$

A43. A right triangle has a base of 300 mm and an altitude of 600 mm. Determine the maximum and minimum moments of inertia with respect to principal axes passing through the centroid.
Ans. Max. $\bar{I} = 1936 \times 10^6$ mm^4; min. $\bar{I} = 314 \times 10^6$ mm^4

A44. The properties of a given area are $A = 8000$ mm^2, $\bar{I}_x = 16.0 \times 10^6$ mm^4, $\bar{I}_y = 340 \times 10^6$ mm^4, and $\bar{P}_{xy} = -14.0 \times 10^6$ mm^4. Compute the minimum radius of gyration for axes through the centroid.

A45. Determine the maximum and minimum moments of inertia of the Z section described in Problem A21 on page 614 with respect to the principal axes passing through the centroid. From Problem A34 it is known that $\bar{P}_{xy} = -8.19 \times 10^6$ mm^4.
Ans. Max. $\bar{I} = 22.0 \times 10^6$ mm^4; min. $\bar{I} = 2.46 \times 10^6$ mm^4

A46. Show that the moment of inertia for the area of any regular polygon is a constant with respect to all axes in the plane of the area which pass through its centroid.

A47. Show that the moment of inertia for the area of a quarter circle with respect to its axis of symmetry is $(\pi - 2)r^4/16$, where r is the radius.

A48. The figure for Problem A20 is redrawn as shown in Fig. P–A48. Check the answer to Problem A20 (i.e., find I_x) by first computing $\bar{I}_u$ and $\bar{I}_v$, then using Mohr's circle to find $\bar{I}_x$, and finally transferring this value to the X axis.

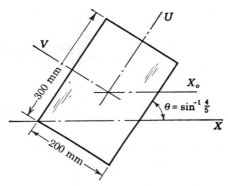

Figure P–A48.

A49. Use the method discussed in Problem A48 to compute the value of I_x for the area shown in Fig. P–A49.

Ans. $I_x = 1040 \times 10^6$ mm^4

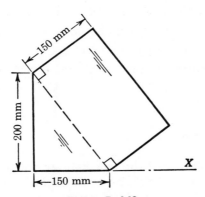

Figure P–A49.

Appendix B

Tables

Acknowledgment

Data for Tables B-2 through B-6 are taken from *Metric Structural Steel Design Data*, 1978, by permission of the Canadian Institute of Steel Construction. Wide-flange (W shapes) and angle sections are designated by the Canadian Standards Association (CSA) Standard CAN3-G312.3-M78.

TABLE B-1. Average Physical Properties of Common Metals

METAL	DENSITY (kg/m³)	TEMP. COEFF OF LINEAR EXPANSION [μm/(m·°C)]	PROPORTIONAL LIMIT (MPa)[a]		ULTIMATE STRENGTH (MPa)			MODULUS OF ELASTICITY (GPa)		PERCENTAGE OF ELONGATION (in 50 mm)
			Tension	Shear	Tension	Comp.	Shear	Tension, E	Shear, G	
Steel, 0.2% carbon, hot rolled	7 850	Varies from 11.0 to 13.2 Average is 11.7	240	150	410	b	310	200	80	35
Steel, 0.2% carbon, cold rolled	7 850		420	250	550	b	420	200	80	18
Steel, 0.6% carbon, hot rolled	7 850		420	250	690	b	550	200	80	15
Steel, 0.8% carbon, hot rolled	7 850	11.7	480	290	830	b	730	200	80	10
Gray cast iron	7 200	10.8	c	d	140	520	d	100	40	Slight
Malleable cast iron	7 200	11.9	250	160	370	b	330	170	90	18
Wrought iron	7 700	12.1	210	130	350	b	240	190	70	35
Aluminum, cast	2 650	23.1	60		90	b	70	70	30	20
Aluminum alloy 17ST	2 700	23.1	220	150	390	b	220	71	30	—
Brass, rolled (70% Cu, 30% Zn)	8 500	18.7	170	110	380	b	330	100	40	30
Bronze, cast	8 200	18.0	140		230	390	—	80	35	10
Copper, hard-drawn	8 800	16.8	260	160	380	b	—	120	40	4

[a]The proportional limit and modulus of elasticity for compression may be assumed equal to these values for tension except for cast iron where proportional limit is approximately 180 MPa.
[b]The ultimate compressive strength for ductile materials may be taken as the yield point, which is slightly greater than the proportional limit in tension.
[c]Not well defined; approximately 40 MPa.
[d]Cast iron fails by diagonal tension.

TABLE B-2. Properties of Wide-Flange Sections (W Shapes)

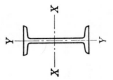

DESIGNATION	THEORETICAL MASS (kg/m)	AREA (mm²)	DEPTH (mm)	FLANGE WIDTH (mm)	FLANGE THICKNESS (mm)	WEB THICKNESS (mm)	AXIS X-X I (10⁶ mm⁴)	AXIS X-X $S=\frac{I}{c}$ (10³ mm³)	AXIS X-X $r=\sqrt{I/A}$ (mm)	AXIS Y-Y I (10⁶ mm⁴)	AXIS Y-Y $S=\frac{I}{c}$ (10³ mm³)	AXIS Y-Y $r=\sqrt{I/A}$ (mm)
W920 ×446	447.2	57 000	933	423	42.7	24.0	8 470	18 200	386	540	2 550	97.3
×417	418.1	53 300	928	422	39.9	22.5	7 880	17 000	385	501	2 370	97.0
×387	387.0	49 300	921	420	36.6	21.3	7 180	15 600	382	453	2 160	95.8
×365	364.6	46 400	916	419	34.3	20.3	6 710	14 600	380	421	2 010	95.2
×342	342.4	43 600	912	418	32.0	19.3	6 250	13 700	379	390	1 870	94.6
×313	312.7	39 800	932	309	34.5	21.1	5 480	11 800	371	170	1 100	65.4
×289	288.6	36 800	927	308	32.0	19.4	5 040	10 900	370	156	1 020	65.2
×271	271.7	34 600	923	307	30.0	18.4	4 720	10 200	369	145	946	64.8
×253	253.7	32 300	919	306	27.9	17.3	4 370	9 520	368	134	874	64.3
×238	238.3	30 400	915	305	25.9	16.5	4 060	8 880	366	123	806	63.6
×223	224.2	28 600	911	304	23.9	15.9	3 770	8 270	363	112	738	62.7
×201	201.3	25 600	903	304	20.1	15.2	3 250	7 200	356	94.4	621	60.7
W840 ×359	359.4	45 800	868	403	35.6	21.1	5 910	13 600	359	389	1 930	92.2
×329	329.4	42 000	862	401	32.4	19.7	5 350	12 400	357	349	1 740	91.2
×299	299.3	38 100	855	400	29.2	18.2	4 790	11 200	355	312	1 560	90.4
×226	226.6	28 900	851	294	26.8	16.1	3 400	7 990	343	114	774	62.8
×210	210.8	26 800	846	293	24.4	15.4	3 110	7 340	340	103	700	61.8
×193	193.5	24 700	840	292	21.7	14.7	2 780	6 630	336	90.3	618	60.5
×176	176.0	22 400	835	292	18.8	14.0	2 460	5 900	331	78.2	536	59.1

W760 ×314	314.4	40 100	786	384	33.4	19.7	4 270	10 900	327	316	1 640	88.8
×284	283.9	36 200	779	382	30.1	18.0	3 810	9 790	325	280	1 470	88.0
×257	257.6	32 800	773	381	27.1	16.6	3 420	8 840	323	250	1 310	87.3
×196	196.8	25 100	770	268	25.4	15.6	2 400	6 240	309	81.7	610	57.1
×185	184.8	23 500	766	267	23.6	14.9	2 230	5 820	308	75.1	563	56.5
×173	173.6	22 100	762	267	21.6	14.4	2 060	5 400	305	68.7	515	55.7
×161	160.4	20 400	758	266	19.3	13.8	1 860	4 900	302	60.7	457	54.5
×147	147.1	18 700	753	265	17.0	13.2	1 660	4 410	298	52.9	399	53.1
W690 ×265	264.5	33 700	706	358	30.2	18.4	2 900	8 220	294	231	1 290	82.8
×240	239.9	30 600	701	356	27.4	16.8	2 610	7 450	292	206	1 160	82.2
×217	217.8	27 700	695	355	24.8	15.4	2 340	6 740	291	185	1 040	81.7
×170	169.9	21 600	693	256	23.6	14.5	1 700	4 910	280	66.2	517	55.3
×152	152.1	19 400	688	254	21.1	13.1	1 510	4 380	279	57.8	455	54.6
×140	139.8	17 800	684	254	18.9	12.4	1 360	3 980	276	51.7	407	53.9
×125	125.6	16 000	678	253	16.3	11.7	1 190	3 500	272	44.1	349	52.5
W610 ×241	241.7	30 800	635	329	31.0	17.9	2 150	6 780	264	184	1 120	77.4
×217	217.9	27 800	628	328	27.7	16.5	1 910	6 070	262	163	995	76.7
×195	195.6	24 900	622	327	24.4	15.4	1 680	5 400	260	142	871	75.6
×174	174.3	22 200	616	325	21.6	14.0	1 470	4 780	257	124	761	74.7
×155	154.9	19 700	611	324	19.0	12.7	1 290	4 220	256	108	666	73.9
×140	140.1	17 900	617	230	22.2	13.1	1 120	3 630	250	45.1	392	50.3
×125	125.1	15 900	612	229	19.6	11.9	985	3 220	249	39.3	343	49.7
×113	113.4	14 400	608	228	17.3	11.2	875	2 880	246	34.3	300	48.7
×101	101.7	13 000	603	228	14.9	10.5	764	2 530	243	29.5	259	47.7
×92	92.3	11 800	603	179	15.0	10.9	646	2 140	234	14.4	161	35.0
×82	81.9	10 400	599	178	12.8	10.0	560	1 870	232	12.1	136	34.0

(Continued)

TABLE B–2. Properties of Wide-Flange Sections (W Shapes) (Continued)

DESIGNATION	THEORETICAL MASS (kg/m)	AREA (mm²)	DEPTH (mm)	FLANGE WIDTH (mm)	FLANGE THICKNESS (mm)	WEB THICKNESS (mm)	AXIS X-X I (10⁶ mm⁴)	AXIS X-X $S=\frac{I}{c}$ (10³ mm³)	AXIS X-X $r=\sqrt{I/A}$ (mm)	AXIS Y-Y I (10⁶ mm⁴)	AXIS Y-Y $S=\frac{I}{c}$ (10³ mm³)	AXIS Y-Y $r=\sqrt{I/A}$ (mm)
W530 ×219	218.9	27 900	560	318	29.2	18.3	1 510	5 390	233	157	986	75.0
×196	196.5	25 000	554	316	26.3	16.5	1 340	4 840	231	139	877	74.4
×182	181.7	23 100	551	315	24.4	15.2	1 240	4 480	231	127	808	74.2
×165	165.3	21 100	546	313	22.2	14.0	1 110	4 060	230	114	726	73.4
×150	150.6	19 200	543	312	20.3	12.7	1 010	3 710	229	103	659	73.2
×138	138.3	17 600	549	214	23.6	14.7	861	3 140	221	38.7	362	46.9
×123	123.2	15 700	544	212	21.2	13.1	761	2 800	220	33.8	319	46.4
×109	109.0	13 900	539	211	18.8	11.6	667	2 480	219	29.5	280	46.1
×101	101.4	12 900	537	210	17.4	10.9	617	2 300	219	26.9	256	45.6
×92	92.5	11 800	533	209	15.6	10.2	552	2 070	217	23.8	228	44.9
×82ᵃ	82.4	10 500	528	209	13.3	9.5	479	1 810	214	20.3	194	44.0
×85	84.7	10 800	535	166	16.5	10.3	485	1 810	212	12.6	152	34.2
×74	74.7	9 520	529	166	13.6	9.7	411	1 550	208	10.4	125	33.1
×66	65.7	8 370	525	165	11.4	8.9	351	1 340	205	8.57	104	32.0
W460 ×177	177.3	22 600	482	286	26.9	16.6	910	3 780	201	105	735	68.2
×158	157.7	20 100	476	284	23.9	15.0	796	3 350	199	91.4	643	67.4
×144	144.6	18 400	472	283	22.1	13.6	726	3 080	199	83.6	591	67.4
×128	128.4	16 400	467	282	19.6	12.2	637	2 730	197	73.3	520	66.9
×113	113.1	14 400	463	280	17.3	10.8	556	2 400	196	63.3	452	66.3
×106	105.8	13 500	469	194	20.6	12.6	488	2 080	190	25.1	259	43.2
×97	96.6	12 300	466	193	19.0	11.4	445	1 910	190	22.8	237	43.1
×89	89.3	11 400	463	192	17.7	10.5	410	1 770	190	20.9	218	42.9
×82	81.9	10 400	460	191	16.0	9.9	370	1 610	188	18.6	195	42.2
×74	74.2	9 450	457	190	14.5	9.0	333	1 460	188	16.6	175	41.9
×67ᵃ	68.1	8 680	454	190	12.7	8.5	300	1 320	186	14.6	153	40.9
×61ᵃ	60.9	7 760	450	189	10.8	8.1	259	1 150	183	12.2	129	39.6

× 68	68.5	8 730	459	154	15.4	9.1	297	1 290	184	9.41	122	32.8
× 60	59.6	7 590	455	153	13.3	8.0	255	1 120	183	7.96	104	32.4
× 52	52.0	6 630	450	152	10.8	7.6	212	943	179	6.34	83.4	30.9
W410 × 149	149.3	19 000	431	265	25.0	14.9	619	2 870	180	77.7	586	63.9
× 132	132.1	16 800	425	263	22.2	13.3	538	2 530	179	67.4	512	63.3
× 114	114.5	14 600	420	261	19.3	11.6	462	2 200	178	57.2	439	62.6
× 100	99.6	12 700	415	260	16.9	10.0	398	1 920	177	49.5	381	62.5
× 85	85.0	10 800	417	181	18.2	10.9	315	1 510	171	18.0	199	40.8
× 74	74.9	9 550	413	180	16.0	9.7	275	1 330	170	15.6	173	40.4
× 67	67.5	8 600	410	179	14.4	8.8	246	1 200	169	13.8	154	40.0
× 60	59.5	7 580	407	178	12.8	7.7	216	1 060	169	12.0	135	39.9
× 54	53.4	6 810	403	177	10.9	7.5	186	924	165	10.1	114	38.5
× 46	46.2	5 890	403	140	11.2	7.0	156	773	163	5.14	73.4	29.5
× 39	39.2	4 990	399	140	8.8	6.4	127	634	159	4.04	57.7	28.4
W360 × 1086	1087.9	139 000	569	454	125	78.0	5 960	20 900	207	1 960	8 650	119
× 990	991.0	126 000	550	448	115	71.9	5 190	18 900	203	1 730	7 740	117
× 900	902.2	115 000	531	442	106	65.9	4 500	17 000	198	1 530	6 940	116
× 818	819.0	104 000	514	437	97.0	60.5	3 920	15 300	194	1 360	6 200	114
× 744	744.3	94 800	498	432	88.9	55.6	3 420	13 700	190	1 200	5 550	112
× 677	677.8	86 300	483	428	81.5	51.2	2 990	12 400	186	1 070	4 990	111
× 634	634.3	80 800	474	424	77.1	47.6	2 740	11 600	184	983	4 630	110
× 592	592.6	75 500	465	421	72.3	45.0	2 500	10 800	182	902	4 280	109
× 551	550.6	70 100	455	418	67.6	42.0	2 260	9 940	180	825	3 950	108
× 509	509.5	64 900	446	416	62.7	39.1	2 050	9 170	178	754	3 630	108
× 463	462.8	59 000	435	412	57.4	35.8	1 800	8 280	175	670	3 250	107
× 421	421.7	53 700	425	409	52.6	32.8	1 600	7 510	172	601	2 940	106
× 382	382.4	48 700	416	406	48.0	29.8	1 410	6 790	170	536	2 640	105
× 347	347.0	44 200	407	404	43.7	27.2	1 250	6 140	168	481	2 380	104
× 314	313.4	39 900	399	401	39.6	24.9	1 100	5 530	166	426	2 120	103
× 287	287.6	36 600	393	399	36.6	22.6	997	5 070	165	388	1 940	103

(Continued)

639

TABLE B–2. Properties of Wide-Flange Sections (W Shapes) *(Continued)*

DESIGNATION	THEORETICAL MASS (kg/m)	AREA (mm²)	DEPTH (mm)	FLANGE WIDTH (mm)	FLANGE THICKNESS (mm)	WEB THICKNESS (mm)	AXIS X–X I (10^6 mm⁴)	$S = \frac{I}{c}$ (10^3 mm³)	$r = \sqrt{I/A}$ (mm)	AXIS Y–Y I (10^6 mm⁴)	$S = \frac{I}{c}$ (10^3 mm³)	$r = \sqrt{I/A}$ (mm)
W360 × 262	262.7	33 500	387	398	33.3	21.1	894	4 620	163	350	1 760	102
× 237	236.3	30 100	380	395	30.2	18.9	788	4 150	162	310	1 570	102
× 216	216.3	27 600	375	394	27.7	17.3	712	3 790	161	283	1 430	101
× 196	196.5	25 000	372	374	26.2	16.4	636	3 420	159	229	1 220	95.5
× 179	179.2	22 800	368	373	23.9	15.0	575	3 120	159	207	1 110	95.2
× 162	162.0	20 600	364	371	21.8	13.3	516	2 830	158	186	1 000	94.8
× 147	147.5	18 800	360	370	19.8	12.3	463	2 570	157	167	904	94.3
× 134	134.0	17 100	356	369	18.0	11.2	415	2 330	156	151	817	94.0
× 122	121.7	15 500	363	257	21.7	13.0	365	2 010	154	61.5	478	63.0
× 110	110.2	14 000	360	256	19.9	11.4	331	1 840	154	55.7	435	63.0
× 101	101.2	12 900	357	255	18.3	10.5	302	1 690	153	50.6	397	62.7
× 91	90.8	11 600	353	254	16.4	9.5	267	1 510	152	44.8	353	62.2
× 79	79.3	10 100	354	205	16.8	9.4	227	1 280	150	24.2	236	48.9
× 72	71.5	9 110	350	204	15.1	8.6	201	1 150	149	21.4	210	48.5
× 64	63.9	8 140	347	203	13.5	7.7	178	1 030	148	18.8	186	48.1
× 57	56.7	7 220	358	172	13.1	7.9	161	897	149	11.1	129	39.3
× 51	50.6	6 450	355	171	11.6	7.2	141	796	148	9.68	113	38.8
× 45	45.0	5 730	352	171	9.8	6.9	122	691	146	8.18	95.7	37.8
× 39	39.1	4 980	353	128	10.7	6.5	102	580	143	3.75	58.6	27.4
× 33	32.8	4 170	349	127	8.5	5.8	82.7	474	141	2.91	45.8	26.4

W310 ×500	500.4	63 700	427	340	75.1	45.1	1 690	7 910	163	494	2 910	88.0
×454	454.0	57 800	415	336	68.7	41.3	1 480	7 130	160	436	2 600	86.8
×415	415.1	52 900	403	334	62.7	38.9	1 300	6 450	157	391	2 340	86.0
×375	374.3	47 700	391	330	57.1	35.4	1 130	5 760	154	343	2 080	84.8
×342	343.3	43 700	382	328	52.6	32.6	1 010	5 260	152	310	1 890	84.2
×313	313.3	39 900	374	325	48.3	30.0	896	4 790	150	277	1 700	83.3
×283	283.0	36 000	365	322	44.1	26.9	787	4 310	148	246	1 530	82.6
×253	252.9	32 200	356	319	39.6	24.4	682	3 830	146	215	1 350	81.6
×226	226.8	28 900	348	317	35.6	22.1	596	3 420	144	189	1 190	81.0
×202	202.6	25 800	341	315	31.8	20.1	520	3 050	142	166	1 050	80.2
×179	178.8	22 800	333	313	28.1	18.0	445	2 680	140	144	919	79.5
×158	157.4	20 100	327	310	25.1	15.5	386	2 360	139	125	805	78.9
×143	143.1	18 200	323	309	22.9	14.0	348	2 150	138	113	729	78.6
×129	129.6	16 500	318	308	20.6	13.1	308	1 940	137	100	652	78.0
×118	117.5	15 000	314	307	18.7	11.9	275	1 750	136	90.2	588	77.6
×107	106.9	13 600	311	306	17.0	10.9	248	1 590	135	81.2	531	77.2
×97	96.8	12 300	308	305	15.4	9.9	222	1 440	134	72.9	478	76.9
×86	86.4	11 000	310	254	16.3	9.1	199	1 280	134	44.5	351	63.6
×79	78.9	10 100	306	254	14.6	8.8	177	1 160	133	39.9	314	63.0
×74	74.5	9 490	310	205	16.3	9.4	165	1 060	132	23.4	229	49.7
×67	66.8	8 510	306	204	14.6	8.5	145	949	131	20.7	203	49.3
×60	59.6	7 590	303	203	13.1	7.5	129	849	130	18.3	180	49.1
×52	52.3	6 670	317	167	13.2	7.6	118	747	133	10.3	123	39.2
×45	44.6	5 690	313	166	11.2	6.6	99.2	634	132	8.55	103	38.8
×39	38.7	4 940	310	165	9.7	5.8	85.1	549	131	7.27	88.1	38.4
×33	32.8	4 180	313	102	10.8	6.6	65.0	415	125	1.92	37.6	21.4
×28	28.4	3 610	309	102	8.9	6.0	54.3	351	123	1.58	31.0	20.9
×24	23.8	3 040	305	101	6.7	5.6	42.7	280	119	1.16	22.9	19.5
×21	21.1	2 690	303	101	5.7	5.1	37.0	244	117	0.983	19.5	19.1

(Continued)

TABLE B–2. Properties of Wide-Flange Sections (W Shapes) *(Continued)*

DESIGNATION	THEORETICAL MASS (kg/m)	AREA (mm²)	DEPTH (mm)	FLANGE WIDTH (mm)	FLANGE THICKNESS (mm)	WEB THICKNESS (mm)	AXIS X-X I (10⁶ mm⁴)	AXIS X-X $S=\frac{I}{c}$ (10³ mm³)	AXIS X-X $r=\sqrt{I/A}$ (mm)	AXIS Y-Y I (10⁶ mm⁴)	AXIS Y-Y $S=\frac{I}{c}$ (10³ mm³)	AXIS Y-Y $r=\sqrt{I/A}$ (mm)
W250 ×167	167.4	21 300	289	265	31.8	19.2	300	2 080	119	98.8	746	68.1
×149	148.9	19 000	282	263	28.4	17.3	259	1 840	117	86.2	656	67.4
×131	131.1	16 700	275	261	25.1	15.4	221	1 610	115	74.5	571	66.8
×115	114.8	14 600	269	259	22.1	13.5	189	1 410	114	64.1	495	66.2
×101	101.2	12 900	264	257	19.6	11.9	164	1 240	113	55.5	432	65.6
×89	89.6	11 400	260	256	17.3	10.7	143	1 100	112	48.4	378	65.1
×80	80.1	10 200	256	255	15.6	9.4	126	982	111	43.1	338	65.0
×73	72.9	9 280	253	254	14.2	8.6	113	891	110	38.8	306	64.6
×67	67.1	8 550	257	204	15.7	8.9	104	806	110	22.2	218	51.0
×58	58.2	7 420	252	203	13.5	8.0	87.3	693	108	18.8	186	50.4
×49	49.0	6 250	247	202	11.0	7.4	70.6	572	106	15.1	150	49.2
×45	44.9	5 720	266	148	13.0	7.6	71.1	534	111	7.03	95.1	35.1
×39	38.7	4 920	262	147	11.2	6.6	60.1	459	110	5.94	80.8	34.7
×33	32.7	4 170	258	146	9.1	6.1	48.9	379	108	4.73	64.7	33.7
×28	28.5	3 630	260	102	10.0	6.4	40.0	307	105	1.78	34.8	22.1
×25	25.3	3 230	257	102	8.4	6.1	34.2	266	103	1.49	29.2	21.5
×22	22.4	2 850	254	102	6.9	5.8	28.9	227	101	1.23	24.0	20.7
×18	17.9	2 270	251	101	5.3	4.8	22.4	179	99.3	0.913	18.1	20.0

Designation												
W200 × 100	99.5	12 700	229	210	23.7	14.5	113	989	94.5	36.6	349	53.8
× 86	86.7	11 100	222	209	20.6	13.0	94.7	853	92.6	31.4	300	53.3
× 71	71.5	9 110	216	206	17.4	10.2	76.6	709	91.7	25.4	246	52.8
× 59	59.4	7 560	210	205	14.2	9.1	61.1	582	89.9	20.4	199	51.9
× 52	52.3	6 660	206	204	12.6	7.9	52.7	512	89.0	17.8	175	51.8
× 46	46.0	5 860	203	203	11.0	7.2	45.5	448	88.1	15.3	151	51.2
× 42	41.7	5 310	205	166	11.8	7.2	40.9	399	87.7	9.00	108	41.2
× 36	35.9	4 580	201	165	10.2	6.2	34.4	342	86.7	7.64	92.6	40.9
× 31	31.4	4 000	210	134	10.2	6.4	31.4	299	88.6	4.10	61.1	32.0
× 27	26.6	3 390	207	133	8.4	5.8	25.8	249	87.3	3.30	49.6	31.2
× 22	22.4	2 860	206	102	8.0	6.2	20.0	194	83.6	1.42	27.8	22.3
× 19	19.4	2 480	203	102	6.5	5.8	16.6	163	81.7	1.15	22.6	21.6
× 15	15.0	1 900	200	100	5.2	4.3	12.7	127	81.8	0.869	17.4	21.4
W150 × 37	37.1	4 730	162	154	11.6	8.1	22.2	274	68.5	7.07	91.8	38.7
× 30	29.8	3 790	157	153	9.3	6.6	17.2	219	67.3	5.56	72.6	38.3
× 22	22.3	2 850	152	152	6.6	5.8	12.1	159	65.1	3.87	50.9	36.9
× 24	24.0	3 060	160	102	10.3	6.6	13.4	168	66.3	1.83	35.8	24.4
× 18	18.0	2 290	153	102	7.1	5.8	9.16	120	63.3	1.26	24.7	23.5
× 14	13.6	1 730	150	100	5.5	4.3	6.87	91.5	63.0	0.918	18.4	23.0
W130 × 28	28.2	3 590	131	128	10.9	6.9	11.0	168	55.3	3.81	59.6	32.6
× 24	23.7	3 020	127	127	9.1	6.1	8.84	139	54.1	3.11	49.0	32.1
W100 × 19	19.4	2 470	106	103	8.8	7.1	4.76	89.9	43.9	1.61	31.2	25.5

[a]Produced exclusively by Algoma Steel (Canada).

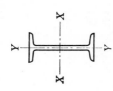

TABLE B-3. Properties of I-Beam Sections (S Shapes)

DESIGNATION	THEO-RETICAL MASS (kg/m)	AREA (mm²)	DEPTH (mm)	FLANGE WIDTH (mm)	FLANGE THICK-NESS (mm)	WEB THICK-NESS (mm)	AXIS X-X I (10⁶ mm⁴)	AXIS X-X $S=\frac{I}{c}$ (10³ mm³)	AXIS X-X $r=\sqrt{I/A}$ (mm)	AXIS Y-Y I (10⁶ mm⁴)	AXIS Y-Y $S=\frac{I}{c}$ (10³ mm³)	AXIS Y-Y $r=\sqrt{I/A}$ (mm)
S610 ×179	178.9	22 800	610	204	28.0	20.3	1 260	4 140	235	35.1	345	39.3
×157.6	158.0	20 100	610	200	28.0	15.9	1 180	3 870	242	32.8	328	40.4
×149	149.2	19 000	610	184	22.1	19.0	997	3 270	229	20.1	218	32.5
×134	133.9	17 100	610	181	22.1	15.8	938	3 070	234	18.9	209	33.3
×118.9	119.1	15 200	610	178	22.1	12.7	879	2 880	241	17.9	201	34.3
S510 ×141	141.8	18 100	508	183	23.3	20.3	674	2 660	193	21.0	230	34.1
×127	126.9	16 200	508	179	23.3	16.6	633	2 490	198	19.5	218	34.7
×112	111.9	14 300	508	162	20.1	16.3	532	2 100	193	12.5	154	29.6
×97.3	97.8	12 500	508	159	20.1	12.7	494	1 950	199	11.7	147	30.6
S460 ×104	104.7	13 300	457	159	17.6	18.1	387	1 690	170	10.3	129	27.7
×81.4	81.6	10 400	457	152	17.6	11.7	335	1 470	180	8.77	115	29.1
S380 ×74	74.6	9 500	381	143	15.8	14.0	203	1 060	146	6.60	92.3	26.4
×64	63.9	8 150	381	140	15.8	10.4	187	980	151	6.11	87.3	27.4

Designation												
S310 ×74	74.6	9 500	305	139	16.8	17.4	128	836	116	6.64	95.6	26.4
×60.7	60.8	7 750	305	133	16.8	11.7	114	747	121	5.71	85.9	27.1
×52	52.2	6 650	305	129	13.8	10.9	95.8	629	120	4.16	64.5	25.0
×47	47.4	6 040	305	127	13.8	8.9	91.1	597	123	3.94	62.1	25.5
S250 ×52	52.3	6 660	254	126	12.5	15.1	61.6	485	96.1	3.56	56.5	23.1
×38	37.8	4 820	254	118	12.5	7.9	51.4	405	103	2.84	48.2	24.3
S200 ×34	34.3	4 370	203	106	10.8	11.2	27.0	266	78.6	1.81	34.2	20.4
×27	27.5	3 500	203	102	10.8	6.9	24.0	237	82.9	1.59	31.1	21.3
S180 ×30	29.7	3 780	178	97	10.0	11.4	17.6	198	68.3	1.30	26.8	18.5
×22.8	22.7	2 890	178	92	10.0	6.4	15.3	172	72.7	1.08	23.6	19.4
S150 ×26	25.5	3 250	152	90	9.1	11.8	10.8	143	57.7	0.952	21.2	17.1
×19	18.4	2 340	152	84	9.1	5.8	9.08	119	62.3	0.750	17.9	17.9
S130 ×22	21.9	2 790	127	83	8.3	12.5	6.33	99.6	47.6	0.690	16.6	15.7
×15	14.8	1 880	127	76	8.3	5.3	5.11	80.4	52.1	0.507	13.4	16.4
S100 ×14.1	14.0	1 790	102	70	7.4	8.3	2.81	55.2	39.7	0.362	10.3	14.2
×11	11.3	1 440	102	67	7.4	4.8	2.52	49.4	41.9	0.311	9.28	14.7
S75 ×11	11.1	1 420	76	63	6.6	8.9	1.20	31.6	29.1	0.238	7.56	13.0
×8	8.4	1 070	76	59	6.6	4.3	1.04	27.4	31.2	0.190	6.43	13.3

TABLE B–4. Properties of Channel Sections

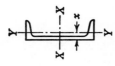

DESIGNATION	THEO-RETICAL MASS (kg/m)	AREA (mm²)	DEPTH (mm)	FLANGE WIDTH (mm)	FLANGE THICK-NESS (mm)	WEB THICK-NESS (mm)	AXIS X-X I (10⁶ mm⁴)	AXIS X-X $S=\frac{I}{c}$ (10³ mm³)	AXIS X-X $r=\sqrt{I/A}$ (mm)	AXIS Y-Y I (10⁶ mm⁴)	AXIS Y-Y $S=\frac{I}{c}$ (10³ mm³)	AXIS Y-Y $r=\sqrt{I/A}$ (mm)	x (mm)
C380 ×74	74.4	9 480	381	94	16.5	18.2	168	881	133	4.60	62.4	22.0	20.3
×60	59.4	7 570	381	89	16.5	13.2	145	760	138	3.84	55.5	22.5	19.7
×50	50.5	6 430	381	86	16.5	10.2	131	687	143	3.39	51.4	23.0	20.0
C310 ×45	44.7	5 690	305	80	12.7	13.0	67.3	442	109	2.12	33.6	19.3	17.0
×37	37.1	4 720	305	77	12.7	9.8	59.9	393	113	1.85	30.9	19.8	17.1
×31	30.8	3 920	305	74	12.7	7.2	53.5	351	117	1.59	28.2	20.1	17.5
C250 ×45	44.5	5 670	254	76	11.1	17.1	42.8	337	86.9	1.60	26.8	16.8	16.3
×37	37.3	4 750	254	73	11.1	13.4	37.9	299	89.4	1.40	24.3	17.1	15.7
×30	29.6	3 780	254	69	11.1	9.6	32.7	257	93.0	1.16	21.5	17.5	15.3
×23	22.6	2 880	254	65	11.1	6.1	27.8	219	98.2	0.922	18.8	17.9	15.9

Designation													
C230 ×30	29.8	3 800	229	67	10.5	11.4	25.5	222	81.9	1.01	19.3	16.3	14.8
×22	22.3	2 840	229	63	10.5	7.2	21.3	186	86.6	0.806	16.8	16.8	14.9
×20	19.8	2 530	229	61	10.5	5.9	19.8	173	88.6	0.716	15.6	16.8	15.1
C200 ×28	27.9	3 560	203	64	9.9	12.4	18.2	180	71.6	0.825	16.6	15.2	14.4
×21	20.4	2 600	203	59	9.9	7.7	14.9	147	75.8	0.627	13.9	15.5	14.0
×17	17.0	2 170	203	57	9.9	5.6	13.5	133	78.8	0.544	12.8	15.8	14.5
C180 ×22	21.9	2 780	178	58	9.3	10.6	11.3	127	63.7	0.568	12.8	14.3	13.5
×18	18.2	2 310	178	55	9.3	8.0	10.0	113	65.9	0.476	11.4	14.3	13.2
×15	14.5	1 850	178	53	9.3	5.3	8.86	99.6	69.3	0.405	10.3	14.8	13.8
C150 ×19	19.2	2 450	152	54	8.7	11.1	7.12	93.7	53.9	0.425	10.3	13.2	12.9
×16	15.5	1 980	152	51	8.7	8.0	6.22	81.9	56.1	0.351	9.13	13.3	12.6
×12	12.1	1 540	152	48	8.7	5.1	5.36	70.6	59.1	0.279	7.93	13.5	12.8
C130 ×13	13.3	1 700	127	47	8.1	8.3	3.66	57.6	46.5	0.252	7.20	12.2	11.9
×10	9.9	1 260	127	44	8.1	4.8	3.09	48.6	49.5	0.195	6.14	12.5	12.2
C100 ×11	10.8	1 370	102	43	7.5	8.2	1.91	37.4	37.3	0.174	5.52	11.3	11.5
×8	8.0	1 020	102	40	7.5	4.7	1.61	31.6	39.7	0.132	4.65	11.4	11.6
C75 ×9	8.8	1 120	76	40	6.9	9.0	0.85	22.3	27.4	0.123	4.31	10.5	11.4
×7	7.3	933	76	37	6.9	6.6	0.75	19.7	28.3	0.096	3.67	10.1	10.8
×6	6.0	763	76	35	6.9	4.3	0.67	17.6	29.6	0.077	3.21	10.1	10.9

TABLE B-5. Properties of Equal Angle Sections

SIZE AND THICKNESS (mm)	THEORETICAL MASS (kg/m)	AREA (mm²)	AXIS X–X AND AXIS Y–Y				AXIS Z–Z
			I (10^6 mm⁴)	$S = \dfrac{I}{c}$ (10^3 mm³)	$r = \sqrt{I/A}$ (mm)	x or y (mm)	$r = \sqrt{I/A}$ (mm)
200×200×30	87.1	11 100	40.3	290	60.3	60.9	39.0
×25	73.6	9 380	34.8	247	60.9	59.2	39.1
×20	59.7	7 600	28.8	202	61.6	57.4	39.3
×16	48.2	6 140	23.7	165	62.1	55.9	39.5
×13	39.5	5 030	19.7	136	62.6	54.8	39.7
×10	30.6	3 900	15.5	106	63.0	53.7	39.9
150×150×20	44.0	5 600	11.6	110	45.5	44.8	29.3
×16	35.7	4 540	9.63	90.3	46.0	43.4	29.4
×13	29.3	3 730	8.05	74.7	46.4	42.3	29.6
×10	22.8	2 900	6.37	58.6	46.9	41.2	29.8
125×125×16	29.4	3 740	5.41	61.5	38.0	37.1	24.4
×13	24.2	3 080	4.54	51.1	38.4	36.0	24.5
×10	18.8	2 400	3.62	40.2	38.8	34.9	24.7
×8	15.2	1 940	2.96	32.6	39.1	34.2	24.8
100×100×16	23.1	2 940	2.65	38.3	30.0	30.8	19.5
×13	19.1	2 430	2.24	31.9	30.4	29.8	19.5
×10	14.9	1 900	1.80	25.2	30.8	28.7	19.7
×8	12.1	1 540	1.48	20.6	31.1	28.0	19.8
×6	9.14	1 160	1.14	15.7	31.3	27.2	19.9
90×90×13	17.0	2 170	1.60	25.6	27.2	27.2	17.6
×10	13.3	1 700	1.29	20.2	27.6	26.2	17.6
×8	10.8	1 380	1.07	16.5	27.8	25.5	17.7
×6	8.20	1 040	0.826	12.7	28.1	24.7	17.9

Designation							
75×75×13	14.0	1 780	0.892	17.3	22.4	23.5	14.6
×10	11.0	1 400	0.725	13.8	22.8	22.4	14.6
×8	8.92	1 140	0.602	11.3	23.0	21.7	14.7
×6	6.78	864	0.469	8.68	23.3	21.0	14.8
×5	5.69	725	0.398	7.32	23.4	20.6	14.9
65×65×10	9.42	1 200	0.459	10.2	19.6	19.9	12.7
×8	7.66	976	0.383	8.36	19.8	19.2	12.7
×6	5.84	744	0.300	6.44	20.1	18.5	12.8
×5	4.91	625	0.255	5.45	20.2	18.1	12.9
55×55×10	7.85	1 000	0.268	7.11	16.4	17.4	10.7
×8	6.41	816	0.225	5.87	16.6	16.7	10.7
×6	4.90	624	0.177	4.54	16.9	16.0	10.8
×5	4.12	525	0.152	3.85	17.0	15.6	10.8
×4	3.33	424	0.125	3.13	17.1	15.2	10.9
×3	2.52	321	0.096	2.39	17.3	14.9	11.0
45×45×8	5.15	656	0.118	3.82	13.4	14.2	8.76
×6	3.96	504	0.094	2.98	13.7	13.4	8.79
×5	3.34	425	0.081	2.53	13.8	13.1	8.82
×4	2.70	344	0.067	2.07	13.9	12.7	8.87
×3	2.05	261	0.052	1.58	14.1	12.4	8.93
35×35×6	3.01	384	0.042	1.74	10.5	10.9	6.81
×5	2.55	325	0.036	1.49	10.6	10.6	6.83
×4	2.07	264	0.030	1.22	10.7	10.2	6.86
×3	1.58	201	0.024	0.940	10.8	9.86	6.91
25×25×5	1.77	225	0.012	0.724	7.39	8.06	4.87
×4	1.44	184	0.010	0.599	7.50	7.71	4.87
×3	1.11	141	0.008	0.465	7.63	7.35	4.89

TABLE B–6. Properties of Unequal Angle Sections

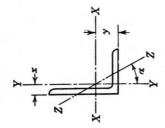

SIZE AND THICKNESS (mm)	THEORETICAL MASS (kg/m)	AREA (mm²)	AXIS X-X I (10⁶ mm⁴)	AXIS X-X $S=\frac{I}{c}$ (10³ mm³)	AXIS X-X $r=\sqrt{I/A}$ (mm)	AXIS X-X y (mm)	AXIS Y-Y I (10⁶ mm⁴)	AXIS Y-Y $S=\frac{I}{c}$ (10³ mm³)	AXIS Y-Y $r=\sqrt{I/A}$ (mm)	AXIS Y-Y x (mm)	AXIS Z-Z $r=\sqrt{I/A}$ (mm)	AXIS Z-Z $\tan\alpha$
200×150 ×25	63.8	8 120	31.6	236	62.3	66.3	15.1	139	43.2	41.3	32.0	0.543
×20	51.8	6 600	26.2	193	63.0	64.5	12.7	115	43.8	39.5	32.1	0.549
×16	42.0	5 340	21.6	158	63.5	63.1	10.5	93.8	44.3	38.1	32.3	0.554
×13	34.4	4 380	17.9	130	64.0	62.0	8.77	77.6	44.7	37.0	32.5	0.557
200×100 ×20	44.0	5 600	22.6	180	63.6	74.3	3.84	50.8	26.2	24.3	21.3	0.256
×16	35.7	4 540	18.7	147	64.2	72.8	3.22	41.8	26.6	22.8	21.4	0.262
×13	29.3	3 730	15.6	121	64.6	71.7	2.72	34.7	27.0	21.7	21.6	0.266
×10	22.8	2 900	12.3	94.8	65.1	70.5	2.18	27.4	27.4	26.5	21.8	0.271
150×100 ×16	29.4	3 740	8.40	84.8	47.4	50.9	3.00	40.4	28.3	25.9	21.6	0.434
×13	24.2	3 080	7.03	70.2	47.8	49.9	2.53	33.7	28.7	24.9	21.7	0.440
×10	18.8	2 400	5.58	55.1	48.2	48.8	2.03	26.6	29.1	23.8	21.9	0.445
×8	15.2	1 940	4.55	44.6	48.5	48.0	1.67	21.6	29.3	23.0	22.0	0.448

Designation												
125×90×16	25.0	3 180	4.84	58.5	39.0	42.2	2.09	32.0	25.6	24.7	19.2	0.499
×13	20.6	2 630	4.07	48.6	39.4	41.2	1.77	26.7	26.0	23.7	19.3	0.505
×10	16.1	2 050	3.25	38.2	39.8	40.1	1.42	21.1	26.4	22.6	19.5	0.511
×8	13.0	1 660	2.66	31.1	40.1	39.3	1.18	17.2	26.6	21.8	19.6	0.515
125×75×13	19.1	2 430	3.82	47.1	39.6	43.9	1.04	18.5	20.7	18.9	16.2	0.356
×10	14.9	1 900	3.05	37.1	40.0	42.8	0.841	14.7	21.0	17.8	16.3	0.363
×8	12.1	1 540	2.50	30.1	40.3	42.1	0.697	12.0	21.3	17.1	16.4	0.367
×6	9.14	1 160	1.92	23.0	40.6	41.3	0.542	9.23	21.6	16.3	16.6	0.372
100×90×13	18.1	2 300	2.17	31.4	30.7	31.1	1.66	25.9	26.8	26.1	18.4	0.796
×10	14.1	1 800	1.74	24.9	31.1	30.0	1.33	20.5	27.2	25.0	18.5	0.800
×8	11.4	1 460	1.43	20.3	31.4	29.3	1.10	16.8	27.5	24.3	18.6	0.802
×6	8.67	1 100	1.11	15.5	31.7	28.5	0.853	12.8	27.8	23.5	18.7	0.805
100×75×13	16.5	2 110	2.04	30.6	31.1	33.4	0.976	18.0	21.5	20.9	16.0	0.541
×10	13.0	1 650	1.64	24.2	31.5	32.3	0.791	14.3	21.9	19.8	16.1	0.549
×8	10.5	1 340	1.35	19.7	31.8	31.5	0.656	11.7	22.2	19.0	16.2	0.554
×6	7.96	1 010	1.04	15.1	32.1	30.8	0.511	9.01	22.4	18.3	16.3	0.559
90×75×13	15.5	1 980	1.51	24.8	27.6	29.3	0.946	17.8	21.9	21.8	15.6	0.672
×10	12.2	1 550	1.22	19.7	28.0	28.2	0.767	14.1	22.2	20.7	15.7	0.679
×8	9.86	1 260	1.01	16.1	28.3	27.5	0.636	11.6	22.5	20.0	15.8	0.683
×6	7.49	954	0.779	12.3	28.6	26.8	0.495	8.89	22.8	19.3	15.9	0.687
×5	6.28	800	0.660	10.4	28.7	26.4	0.421	7.50	22.9	18.9	16.0	0.689
90×65×10	11.4	1 450	1.16	19.2	28.3	29.8	0.507	10.6	18.7	17.3	13.9	0.506
×8	9.23	1 180	0.958	15.7	28.5	29.1	0.422	8.72	18.9	16.6	14.0	0.512
×6	7.02	894	0.743	12.1	28.8	28.4	0.330	6.72	19.2	15.9	14.2	0.518
×5	5.89	750	0.629	10.2	29.0	28.0	0.281	5.68	19.4	15.5	14.2	0.520

(Continued)

TABLE B-6. Properties of Unequal Angle Sections (Continued)

SIZE AND THICKNESS (mm)	THEORETICAL MASS (kg/m)	AREA (mm²)	AXIS X–X				AXIS Y–Y				AXIS Z–Z	
			I (10^6 mm⁴)	$S=\dfrac{I}{c}$ (10^3 mm³)	$r=\sqrt{I/A}$ (mm)	y (mm)	I (10^6 mm⁴)	$S=\dfrac{I}{c}$ (10^3 mm³)	$r=\sqrt{I/A}$ (mm)	x (mm)	$r=\sqrt{I/A}$ (mm)	$\tan\alpha$
80×60 ×10	10.2	1 300	0.808	15.1	24.9	26.5	0.388	8.92	17.3	16.5	12.8	0.543
×8	8.29	1 060	0.670	12.4	25.2	25.8	0.324	7.33	17.5	15.8	12.9	0.549
×6	6.31	804	0.522	9.50	25.5	25.1	0.254	5.66	17.8	15.1	13.0	0.555
×5	5.30	675	0.443	8.02	25.6	24.7	0.217	4.79	17.9	14.7	13.0	0.558
75×50 ×8	7.35	936	0.525	10.6	23.7	25.5	0.187	5.06	14.1	13.0	10.8	0.434
×6	5.60	714	0.410	8.15	24.0	24.7	0.148	3.92	14.4	12.2	10.9	0.441
×5	4.71	600	0.349	6.88	24.1	24.4	0.127	3.32	14.5	11.9	10.9	0.445
65×50 ×8	6.72	856	0.351	8.03	20.2	21.3	0.180	4.97	14.5	13.8	10.6	0.572
×6	5.13	654	0.275	6.19	20.5	20.6	0.142	3.85	14.7	13.1	10.7	0.580
×5	4.32	550	0.235	5.24	20.7	20.2	0.122	3.27	14.9	12.7	10.8	0.583
×4	3.49	444	0.192	4.25	20.8	19.9	0.100	2.66	15.0	12.4	10.8	0.587
55×35 ×6	3.96	504	0.152	4.23	17.4	19.0	0.048	1.85	9.77	9.04	7.55	0.396
×5	3.34	425	0.130	3.59	17.5	18.7	0.041	1.58	9.89	8.68	7.59	0.401
×4	2.70	344	0.107	2.92	17.7	18.3	0.034	1.29	10.0	8.31	7.65	0.406
×3	2.05	261	0.083	2.23	17.8	17.9	0.027	0.994	10.2	7.94	7.72	0.411
45×30 ×6	3.25	414	0.082	2.79	14.0	15.7	0.029	1.32	8.35	8.22	6.44	0.426
×5	2.75	350	0.070	2.37	14.2	15.4	0.025	1.13	8.46	7.86	6.47	0.433
×4	2.23	284	0.058	1.94	14.3	15.0	0.021	0.930	8.58	7.49	6.51	0.439
×3	1.70	216	0.045	1.49	14.5	14.6	0.016	0.717	8.72	7.12	6.57	0.445

Index

82 83 9 8 7 6 5

SI Units
(Système International d'Unités)

A. SELECTED SI UNITS

Quantity	Name	SI Symbol
Energy	joule	J (1 J = 1 N·m)
Force	newton	N (1 N = 1 kg·m/s^2)
Length	meter*	m
Mass	kilogram*	kg
Moment (torque)	newton meter	N·m
Plane angle	radian	rad
	degree	°
Rotational frequency	revolution per second	r/s
Stress (pressure)	pascal	Pa (1 Pa = 1 N/m^2)
Temperature	degree Celsius	°C
Time	second*	s
Power	watt	W (1 W = 1 J/s)

B. COMMONLY USED SI PREFIXES

Multiplying Factor	Prefix	SI Symbol
10^9	giga	G
10^6	mega	M
10^3	kilo	k
10^{-3}	milli	m
10^{-6}	micro	μ
10^{-9}	nano	n

*SI base unit